Wave Mot

Cambridge Texts in Applied Mathematics

Maximum and Minimum Principles
M.J. Sewell
Solitons: an Introduction
P.G. Drazin and R.S. Johnson
Kinematics of Mixing
J.M. Ottino
Introduction to Numerical Linear Algebra and Optimisation
P.G. Ciarlet
Integral Equations: from Spectral Theory to Applications
D. Porter and D.S.G. Stirling
Perturbation Methods
E.J. Hinch
Thermomechanics of Plasticity and Fracture
G.A. Maugin
Boundary Integral and Singularity Methods for Linearized Viscous Flow
C. Pozrikidis
Nonlinear Wave Processes in Acoustics
K. Naugolnykh and L. Ostrovsky
Nonlinear Systems
P.G. Drazin
Stability, Instability and Chaos
P. Glendinning
Analysis of the Navier–Stokes Equations
C. Doering and J.D. Gibbon
Viscous Flow
H. and J. Ockendon
Scaling, Self-Similarity and Intermediate Asymptotics
G.I. Barenblatt
First Course in Numerical Differential Equations
A. Iserles
Complex Variables: Introduction and Applications
M. Ablowitz and A. Fokas
Mathematical Modeling
A. Fowler
Thinking about Ordinary Differential Equations
R. O'Malley
A Modern Introduction to the Mathematical Theory of Water Waves
R.S. Johnson
Rarefied Gas Dynamics
C. Cercignani
Symmetry Methods for Differential Equations
P.E. Hydon
High Speed Flow
C.J. Chapman
Wave Motion
J. Billingham and A. King
An Introduction to Magnetohydrodynamics
P.A. Davidson
Linear Elastic Waves
J.G. Harris
Vorticity and Incompressible Flow
A. Majda and A. Bertozzi
Infinite Dimensional Dynamical Systems
J. Robinson

Wave Motion

J. BILLINGHAM
University of Birmingham

A. C. KING
University of Birmingham

PUBLISHED BY THE PRESS SYNDICATE OF THE UNIVERSITY OF CAMBRIDGE
The Pitt Building, Trumpington Street, Cambridge, United Kingdom

CAMBRIDGE UNIVERSITY PRESS
The Edinburgh Building, Cambridge, CB2 2RU, UK
40 West 20th Street, New York, NY 10011–4211, USA
10 Stamford Road, Oakleigh, VIC 3166, Australia
Ruiz de Alarcón 13, 28014 Madrid, Spain
Dock House, The Waterfront, Cape Town 8001, South Africa

http://www.cambridge.org

First published 2000

Printed in the United Kingdom at the University Press, Cambridge

Typeface Times 10/12pt *System* LaTeX [UPH]

A catalogue record of this book is available from the British Library

ISBN 0 521 63257 9 hardback
ISBN 0 521 63450 4 paperback

Contents

Introduction

Whenever we see or hear anything, we do so because of the existence of waves. Electromagnetic waves cover a spectrum from low frequency radio waves, through visible light to X- and gamma rays. Sound propagates as a wave through the air. When someone sings or plays a musical instrument, the standing waves in their vocal chords, guitar strings or drumskins produce a pressure change, or sound wave, which is audible. Although these examples alone would be sufficient to motivate their study, wave phenomena occur in many other physical systems. Waves can propagate both on the surface of solid bodies (for example, as earthquakes) and through the bulk of a solid (for example, in seismic oil prospecting). The surface of the sea is perhaps the most obvious example of a wave bearing medium. Water waves vary in size from the small ripples caused by raindrops, through shock waves such as the Severn bore, to enormous ocean waves that can capsize large ships. Waves in different media can interact, often with devastating effects, for example when an underwater earthquake causes a tsunami, a huge wall of water that can destroy coastal settlements, or when the waves generated by the wind blowing on a bridge produce a catastrophic resonance.

Wave phenomena emerge in unexpected contexts. The flow of traffic along a road can support a variety of wave-like disturbances as anybody who has sat in slowly moving traffic will know. The beat of your heart is regulated by spiral waves of chemical activity that swirl across its surface. You control the movement of your body through the action of electrochemical waves in your nervous system. Finally, quantum physics has revealed that, on a small enough scale, everything around us can only be described in terms of waves.

A question that we might reasonably ask at this point is 'what is a wave?' Well, the first place you might look is a dictionary, where waves are usually defined as 'disturbances propagating in water at a finite speed'. However, this is not satisfactory since standing waves do not propagate and wave propagation is possible in media other than water. Another feature that makes definitions difficult is the interaction of the wave with the medium through which it is passing. A wave on the surface of a pond passes by and leaves the medium unchanged. In contrast, a chemical wave usually leaves the reacting species involved in a different chemical state after it has passed by. For these reasons we believe that there is no single definition of waves, and choose to view them as a generic set of phenomena with many similarities.

To cover all of these topics would be an enormous task. We have written this textbook with the needs of advanced undergraduates in applied mathematics in mind, and have restricted the range of material covered accordingly. We hope that the book will also be of use to physics and engineering students. It is worth noting that this is not a book about numerical methods for wave propagation. This reflects both the need to keep the book to a manageable size and the interests of the authors. The emphasis is on analytical and asymptotic methods for the investigation of systems of equations with wave-like solutions. Much of the material is concerned with various aspects of fluid mechanics, but we also cover basic elastic, electromagnetic, traffic, chemical and electrochemical waves. We assume that the reader has taken basic courses in fluid mechanics, elasticity (for chapter 5), partial differential equations, vector calculus, phase plane methods (for chapter 9) and asymptotic methods. Throughout we have tried to emphasise the mathematical similarities between the disparate areas that we deal with, in terms of both the structure of the governing equations and the techniques available for their solution. At the end of each chapter there are exercises, designed to both reinforce and augment the material presented in the chapter. Bona fide teachers and instructors may obtain full worked solutions to many of these by emailing dtranah@cup.cam.ac.uk.

The book is divided into three parts. Linear waves are dealt with in part one. All of the analytical techniques of nineteenth century mathematics can be brought to bear on linear wave equations. The techniques of separation of variables, Fourier series and Fourier transforms are used to reveal the properties of linear waves on stretched strings (chapter 2), linear sound waves (chapter 3), linear water waves (chapter 4), linear waves in solids (chapter 5) and electromagnetic waves (chapter 6).

In part two we cross the threshold into the twentieth century and study nonlinear waves. We begin in chapter 7 by examining hyperbolic systems governed by the propagation of information on characteristics. As examples we use traffic flow and nonlinear gas dynamics. We then move on to study nonlinear water waves (chapter 8), an extension of the material presented in chapter 4, and then chemical and electrochemical waves (chapter 9).

The third and final part of the book covers more advanced topics. In chapter 10 we consider various physical systems that can be modelled using Burgers' equation. These include a more sophisticated model for the flow of traffic, and weakly nonlinear compressible gas dynamics. Chapter 11 is concerned with the analysis of scattering and diffraction of both scalar and vector waves, through apertures and past obstacles. In chapter 12 we describe the use of the inverse scattering transform to solve the Korteweg–de Vries equation (KdV) and the nonlinear Schrödinger equation (NLS). The KdV equation governs, amongst other things, the propagation of long waves on shallow water, whilst the NLS equation governs the propagation of dispersive wavepackets in a nonlinear medium, for example pulses of light in optical fibres. This allows us to illustrate some of the remarkable properties of solitons. These are localised waves that can retain their identity after nonlinear interactions.

Throughout the first two parts of the book, more advanced topics that do not fit naturally into part three, and which could be omitted at first reading, have been marked with an asterisk.

In writing this book we have benefited from the advice and encouragement of many colleagues. Particular gratitude is due to David Crighton for the invitation to write a book in this series. He also taught a stimulating course on waves to both authors as undergraduates. The various chapters of this book have been read and commented upon prior to publication by John Blake, Stephen Decent, Eammon Gaffney, Yulii Shikhmurzaev, Athanasios Yannacopoulos and Ray Jones (all at Birmingham), Tony Rawlins (Brunel), Peter Smith (Keele), Nigel Scott and Jean-Marc Vanden-Broeck (East Anglia), Howell Peregrine (Bristol), Sam Falle (Leeds), David Parker (Edinburgh) and Colin Pask and Rowland Sammut (New South Wales). Any remaining mistakes are, of course, our own. We also acknowledge and thank the copyright owners of the many photographs and drawings throughout the text: Lawrence Coates, Ken Elliott and Graham Westbrook (Birmingham), Neville Fletcher (ANU), Howell Peregrine (Bristol), Tim Rees (Transport Research Laboratory Ltd), Malcolm Bloor (Leeds), Geraint Thomas (Aberystwyth), Bernard Richardson (Cardiff), Martin Boeckmann (Magdeburg), Germany Aerospace Centre, Berlin, BAE Systems

Research, UK, the National Physical Laboratory, MIT Press, Hirzel-Verlag and Cambridge University Press. We would both like to thank the University of Birmingham for allowing us to take a semester of study leave during which most of the book was written. JB would like to thank the staff of University College, ADFA, University of New South Wales, and of the University of Adelaide for their hospitality during his study leave.

ACK and JB, Birmingham 2000

Part one

Linear Waves

1

Basic Ideas

In this chapter, we introduce some of the generic ideas and notation that underpin the analysis of waves in the physical, chemical and biological systems that we will study in subsequent chapters. We consider some typical, linear partial differential equations with wave-like solutions, and discuss some of their qualitative and quantitative features. We also present the method of stationary phase, which gives us a straightforward way of determining how the waves in a linear system behave a long time after they are generated.

A partial differential equation is **linear** if all of the dependent variables appear linearly. Consider, for example, the one-dimensional wave equation,

$$\frac{\partial^2 \phi}{\partial x^2} = \frac{1}{c^2}\frac{\partial^2 \phi}{\partial t^2}. \tag{1.1}$$

This governs, amongst other things, the propagation of small amplitude waves on a stretched string, as we shall see in chapter 2. Here ϕ is the amplitude of the displacement of the string, x position along the string and t time. This is a linear equation, since $\phi(x,t)$ only appears linearly (there are no **nonlinear** terms, such as ϕ^2 or $\phi\partial\phi/\partial x$).

The most important feature of solutions of linear equations is that there is a **principle of superposition**. If $\phi = \phi_1$ and $\phi = \phi_2$ are solutions of a linear equation, then $a_1\phi_1 + a_2\phi_2$ is also a solution for arbitrary constants a_1 and a_2. In addition, if $\phi(x,t;K)$ is a solution for any value of the constant K then

$$\int_{-\infty}^{\infty} \phi(x,t;K)dK$$

is also a solution. This feature allows us to build up the solution of a

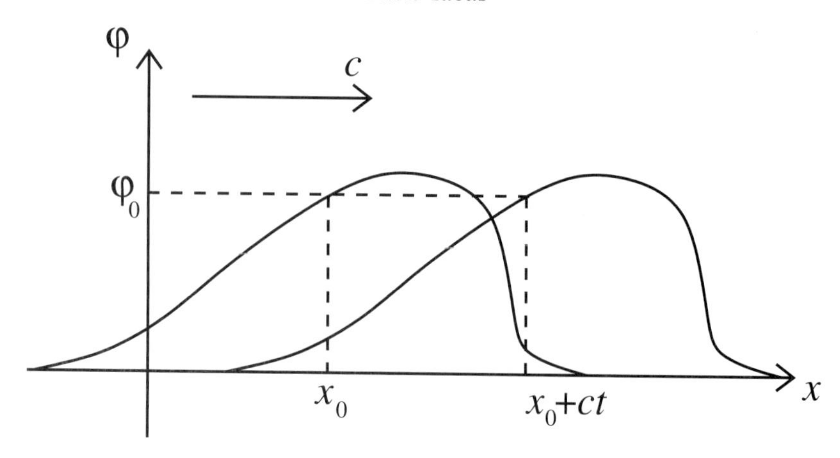

Fig. 1.1. A right-propagating travelling wave.

given boundary value problem in terms of sums or integrals of simple solutions, and is extremely important in the theory of linear waves.

We demonstrate in chapter 2 that the general solution of (1.1) is

$$\phi(x,t) = f(x-ct) + g(x+ct),$$

for arbitrary functions f and g. Let's consider a solution with $g = 0$ so that $\phi = f(x-ct)$. What does a solution like this represent? When $t = 0$, $\phi = f(x)$, so $f(x)$ is the initial value of ϕ. Consider the point $x = x_0$, at which $\phi = \phi_0 = f(x_0)$ when $t = 0$. Some time t later, for what values of x is ϕ equal to ϕ_0? This must be at the point $x = x_1$ at which $f(x_1 - ct) = \phi_0 = f(x_0)$, and hence $x_1 = x_0 + ct$. In other words, the point at which $\phi = \phi_0$ has moved a distance ct in the positive x-direction. Since, this argument holds for all points x, we conclude that the initial form of ϕ has simply been translated to the right through a distance ct, and hence that the initial form of ϕ propagates to the right without change of form at **wave speed** c, as shown in figure 1.1. The solution is a **travelling** or **progressive** wave. Similarly, the solution $\phi = g(x+ct)$ is a travelling wave that propagates in the negative x-direction at wave speed c. The general solution of the one-dimensional wave equation is simply the sum of a right- and a left-propagating wave, each with wave speed c. Note that, as we shall see in parts two and three, travelling wave solutions of *nonlinear* partial differential equations can also exist.

In order to study more general linear partial differential equations with wave-like solutions, we now define some terminology associated with **periodic**, or more specifically **harmonic**, waves. In general, we can look for solutions of any linear partial differential equation in the form

$$\phi = \text{Re}[A \exp\{i(kx - \omega t)\}] = \text{Re}(A)\cos(kx - \omega t) - \text{Im}(A)\sin(kx - \omega t).$$

This is often simply written as

$$\phi = A \exp\{i(kx - \omega t)\}, \tag{1.2}$$

with the understanding that we mean the real part of this expression, since this notation often makes calculations much easier. This is a periodic progressive or travelling wave that propagates in the positive x-direction for $k\omega > 0$. We call k the **wavenumber** and ω the **angular frequency**. For this expression to satisfy the one-dimensional wave equation we require that $k = \omega/c$. We can also define the **frequency**, $f = \omega/2\pi$, and the **wavelength**, $\lambda = 2\pi/k = 2\pi c/\omega = c/f$. The frequency, measured in s^{-1}, or equivalently hertz (Hz), is the number of complete oscillations that the wave makes during one second at a fixed position. The wavelength is the distance between successive maxima or minima of the waveform. The maxima are the **wave crests**, and the minima the **wave troughs**. The magnitude of ϕ at these points is the **amplitude**, $|A|$, as shown in figure 1.2.

Note that by adding two harmonic waves, identical in amplitude, but propagating in opposite directions, we can construct a **standing wave** solution. For example

$$\phi = A\cos(kx - \omega t) + A\cos(kx + \omega t) = 2A\cos kx \cos \omega t.$$

In such a standing wave, the solution has an **envelope** that is fixed in space, here $\cos kx$, which is modulated by a time dependent motion, here $\cos \omega t$. We will construct similar standing wave solutions when we consider the vibration of an elastic string with finite length in chapter 2, a column of air in chapter 3 and a body of water in chapter 4.

Many physical systems that exhibit wave motion can be modelled using linear equations other than the wave equation. These also have harmonic wave solutions of the form given by (1.2), but with the angular frequency a known function of the wavenumber, so that

$$\omega = \omega(k). \tag{1.3}$$

For the one-dimensional wave equation $\omega = ck$, and the wave crests move at speed c, independent of the wavenumber, k. However, in most systems, ω is not proportional to k, and the wave crests move with velocity $c_p(k) = \omega(k)/k$. This is known as the **phase velocity**, and is

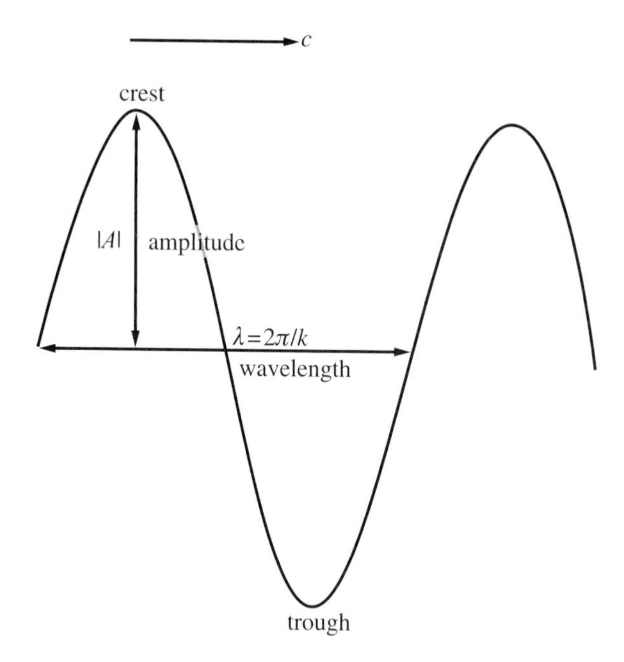

Fig. 1.2. The various quantities associated with a harmonic wave.

usually a function of the wavenumber. In other words, the wave crests move at different velocities for different wavenumbers, and hence also wavelengths. In solutions that are a combination of harmonic waves of different wavelengths, this eventually leads to a separation or **dispersion** of the various components. Such a system is said to be **dispersive** and (1.3) is the **dispersion relation**. As a particular example, we consider the **linearised Korteweg–de Vries**, or **KdV**, equation,

$$\frac{\partial u}{\partial t} + \alpha\frac{\partial u}{\partial x} = \beta\frac{\partial^3 u}{\partial x^3}. \tag{1.4}$$

This equation arises as a model for the propagation of long waves on shallow water, as we shall see in chapter 8. If we seek a harmonic wave solution of the form $u = A\exp\{i(kx - \omega t)\}$, we find that we require

$$\omega = \alpha k + \beta k^3, \tag{1.5}$$

so $\omega = \omega(k)$, and (1.5) is the dispersion relation for the linearised KdV equation. Note that the phase velocity is $c_{\mathrm{p}} = \alpha + \beta k^2$, so that short waves travel faster than long waves. Remember, the smaller the wavenumber, k, the longer the wavelength.

Well-known linear equations for which this type of analysis is relevant include the **linearised Klein–Gordon equation**, which arises in relativistic quantum mechanics, and also describes the motion of a stretched string attached to an elastic backing sheet. It is given by

$$\frac{\partial^2 u}{\partial t^2} - \alpha^2 \frac{\partial^2 u}{\partial x^2} + \beta^2 u = 0, \tag{1.6}$$

and the dispersion relation is

$$\omega = \pm\sqrt{\alpha^2 k^2 + \beta^2}. \tag{1.7}$$

In this case, waves can propagate in both the positive and negative x-directions with speeds dependent upon wavelength.

We can now generalise the idea of a harmonic wave solution by considering solutions that consist of two harmonic waves with the same amplitude but slightly different wavenumbers. Let's look for a solution of the linearised Klein–Gordon equation, (1.6), in the form

$$u = A\sin(kx - \omega(k)t) + A\sin((k + \delta k)x - \omega(k + \delta k)t),$$

where for each part of this superposition of waves to be a solution, ω must satisfy the dispersion relation, (1.7). After expanding the frequency as

$$\omega(k + \delta k) = \omega(k) + \delta k\omega'(k) + O((\delta k)^2),$$

for $\delta k \ll k$, we find that

$$u = 2A\cos\left\{\frac{1}{2}\delta k(x - \omega'(k)t)\right\}\sin\{kx - \omega(k)t\} + O(\delta k). \tag{1.8}$$

The form of this type of solution is interesting. The factor $\cos\left\{\frac{1}{2}\delta k(x - \omega'(k)t)\right\}$ modulates the amplitude of the basic wave, given by $\sin\{kx - \omega(k)t\}$, over a long period, of $O((\delta k)^{-1})$, with changes in the amplitude propagating at velocity $\omega'(k)$. This is called the **group velocity**. A typical solution of this form is shown in figure 1.3.

The group velocity has a fundamental significance in the theory of linear, dispersive waves. It is the velocity at which energy is transported by waves. To illustrate this we again use the linearised Klein–Gordon equation, (1.6), and form the energy equation associated with it. Multiplying through by $u_t = \partial u/\partial t$, and integrating between two arbitrary fixed points, x_1 and x_2, gives

$$\int_{x_1}^{x_2} \left(u_t u_{tt} + \beta^2 u_t u\right) dx = \int_{x_1}^{x_2} \alpha^2 u_t u_{xx} dx.$$

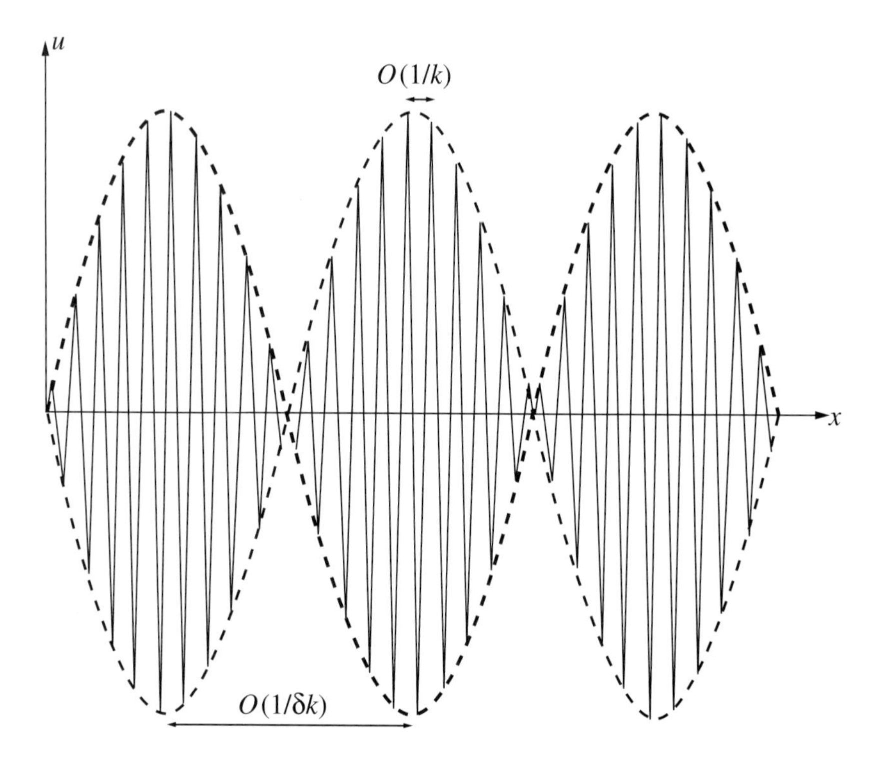

Fig. 1.3. The superposition of two time harmonic waves with wavenumbers differing by $\delta k \ll k$.

Integrating the right hand side by parts and rearranging leads to

$$\frac{\partial}{\partial t}\int_{x_1}^{x_2} \frac{1}{2}\left(u_t^2 + \alpha^2 u_x^2 + \beta^2 u^2\right) dx = \alpha^2 \left[u_t u_x\right]_{x_1}^{x_2}.$$

Now if $u_t = 0$ at $x = x_1$ and $x = x_2$, for example if the equation models the motion of a stretched string, with displacement u, attached to an elastic backing sheet fixed at these points, then the quantity

$$E = \int_{x_1}^{x_2} \frac{1}{2}\left(u_t^2 + \alpha^2 u_x^2 + \beta^2 u^2\right) dx, \tag{1.9}$$

is **conserved** during the motion, since $dE/dt = 0$. In fact, E is the energy, with $\frac{1}{2}u_t^2$ representing the total kinetic energy, $\frac{1}{2}u_x^2$ the elastic potential energy of the string, and $\frac{1}{2}u^2$ the strain energy. We will consider the case $\alpha = 1$, $\beta = 0$, when there is no backing sheet, in chapter 2.

Let's now return to the solution consisting of two harmonic waves of nearly equal wavenumbers and focus our attention on the single envelope between the points

$$x_1(t) = -\frac{\pi}{\delta k} + \omega'(k)t, \quad x_2(t) = \frac{\pi}{\delta k} + \omega'(k)t,$$

which move with the group velocity. By differentiating (1.9) we obtain

$$\frac{dE}{dt} = \int_{x_1(t)}^{x_2(t)} \left(u_t u_{tt} + \alpha^2 u_x u_{xt} + \beta^2 u u_t\right) dx$$
$$+\frac{1}{2}\omega'(k)\left[u_t^2 + \alpha^2 u_x^2 + \beta^2 u^2\right]_{x_1(t)}^{x_2(t)}.$$

Since $u(x,t)$ is a solution of (1.6), we can perform this integration and obtain

$$\frac{dE}{dt} = \left[\frac{1}{2}\omega'(k)\left(u_t^2 + \alpha^2 u_x^2 + \beta^2 u^2\right) + \alpha^2 u_x u_t\right]_{x_1(t)}^{x_2(t)}. \tag{1.10}$$

At leading order, we found earlier that

$$u \sim 2A\cos\left\{\frac{1}{2}\delta k(x - \omega'(k)t)\right\}\sin\{kx - \omega(k)t\}.$$

If we substitute this into (1.10), we find that $dE/dt = 0$ at leading order. The implication is that, if we travel at the group velocity, the amount of energy contained under this part of the envelope of the wave is unchanged, and we deduce that energy is transported at the group velocity. Another way of viewing this is that *wave motion is a mechanism for energy communication.* This can take place over large distances and, in many cases, without producing any permanent change in the state of the medium through which it passes. We will return to this idea in several places throughout the book and give a detailed physical justification.

As a further generalisation, we can look for solutions of linear partial differential equations in **wavepacket form**,

$$\phi(x,t) = \int_{-\infty}^{\infty} A(k)\exp[i\{kx - \omega(k)t\}]dk, \tag{1.11}$$

which is an integral over all possible wavenumbers. When $t = 0$

$$\phi(x,0) = \int_{-\infty}^{\infty} A(k)e^{ikx}dk, \tag{1.12}$$

and we can see that $A(k)$ is the Fourier transform of the initial conditions, with

$$A(k) = \frac{1}{2\pi}\int_{-\infty}^{\infty} \phi(x,0)e^{-ikx}dx. \tag{1.13}$$

For a general, localised initial disturbance, this means that all wavelengths contribute to the solution. The representation of the solution as an integral over all possible values of k, (1.11), is an inevitable consequence of this physical fact.

It is often useful to know how this solution develops after a long time. If we consider a point moving with constant velocity v, so that $x/t = v$, we can write

$$\phi = \int_{-\infty}^{\infty} A(k) \exp[it\{kv - \omega(k)\}]dk. \tag{1.14}$$

For $t \gg 1$, we can estimate this integral, whose integrand oscillates rapidly, using the **method of stationary phase**. The basic idea is that for an integral of the form

$$I(\lambda) = \int_{-\infty}^{\infty} f(s)e^{i\lambda g(s)}ds \tag{1.15}$$

with $\lambda \gg 1$, the integrand oscillates rapidly, and the positive and negative parts of the integrand cancel out, except in the neighbourhood of **points of stationary phase**, where $g'(s) = 0$. The whole integrand can be approximated by taking Taylor series expansions of f and g in the neighbourhood of each of these points. Assuming that there is only one such point, at $s = s_0$,

$$I(\lambda) \sim f(s_0)e^{i\lambda g(s_0)} \int_{-\infty}^{\infty} e^{i\frac{1}{2}\lambda g''(s_0)(s-s_0)^2} ds, \quad \text{for } \lambda \gg 1.$$

We also assume that this is not a degenerate case where $g''(s_0) = 0$. By making the substitution

$$z = (s - s_0)\sqrt{\frac{1}{2}\lambda\,|g''(s_0)|},$$

this becomes

$$I(\lambda) \sim f(s_0)e^{i\lambda g(s_0)} \left(\frac{2}{\lambda\,|g''(s_0)|}\right)^{1/2} \int_{-\infty}^{\infty} e^{iz^2 \operatorname{sgn}\{g''(s_0)\}} dz, \quad \text{for } \lambda \gg 1.$$

Using the change of variable $z = e^{\pm i\pi/4}Z$, this integral can be evaluated using standard complex variable methods as

$$\int_{-\infty}^{\infty} e^{\pm iz^2} dz = 2e^{\pm i\pi/4} \int_{0}^{\infty} e^{-Z^2} dZ = e^{\pm i\pi/4}\sqrt{\pi},$$

and hence

$$I(\lambda) \sim f(s_0)e^{i\lambda g(s_0)} \left(\frac{2\pi}{\lambda\,|g''(s_0)|}\right)^{1/2} e^{i\frac{\pi}{4}\operatorname{sgn}\{g''(s_0)\}}, \quad \text{for } \lambda \gg 1. \tag{1.16}$$

In (1.14), the points of stationary phase, $k = k_0$, satisfy $v = d\omega/dk$. If there is a single point of stationary phase,

$$\phi \sim A(k_0)e^{it\{k_0 v - \omega(k_0)\}} \left(\frac{2\pi}{t\,|\omega''(k_0)|}\right)^{1/2} e^{-i\frac{\pi}{4}\operatorname{sgn}\{\omega''(k_0)\}},$$

which, to an observer moving at the group velocity, represents a single wave with amplitude decreasing like $t^{-1/2}$ and a phase shift of $\pi/4$.

Again we can show that the energy associated with any given wavelength, or equivalently, wavenumber, travels at velocity $c_g = d\omega/dk$, the group velocity. For dispersive systems, the group velocity and the phase velocity are not equal. For example, in the linearised KdV equation, the phase velocity, $c_p = \omega/k = \alpha + \beta k^2$, is not equal to the group velocity, $c_g = d\omega/dk = \alpha + 3\beta k^2$.

The mathematical theory of wavelike solutions to linear partial differential equations can be developed further than we have done here. However this is a textbook primarily concerned with waves in real physical, chemical and biological systems to which we now turn our attention.

Exercises

1.1 Find the dispersion relation for the propagation of plane harmonic waves in the systems

(a) $i\dfrac{\partial u}{\partial t} + \dfrac{\partial^2 u}{\partial x^2} = 0$
(the free space Schrödinger equation, which arises in quantum mechanics),

(b) $\dfrac{\partial \phi}{\partial t} + \alpha\dfrac{\partial \phi}{\partial x} = \beta\dfrac{\partial^3 \phi}{\partial x^3} - \gamma\dfrac{\partial^5 \phi}{\partial x^5}$
(the linearised fifth order KdV equation).

In each case, determine both the phase and group velocities. For each equation, determine $\phi(x, t)$ if $\phi(x, 0) = \delta(x)$. Use the method of stationary phase to find an approximation to this solution for $t \gg 1$ with $v = x/t = O(1)$, v constant.

1.2 Consider the equation

$$\psi_{xxtt} = -g\alpha\psi_{xx},$$

where g and α are positive constants. This equation arises as a model for internal waves in a fluid. Calculate the dispersion relation, and hence determine the solution when

$$\psi(x,0) = \begin{cases} 0 & \text{for } |x| > 1, \\ 1 & \text{for } |x| \leq 1, \end{cases}$$

$$\psi_t(x,0) = 0.$$

1.3 Show that $E = \int_0^l \frac{1}{2}m\eta_t^2 + \frac{1}{2}B\eta_{xx}^2 dx$ is a conserved quantity if η satisfies the **linearised plate bending equation**,

$$m\frac{\partial^2 \eta}{\partial t^2} + B\frac{\partial^4 \eta}{\partial x^4} = 0,$$

subject to the boundary conditions $\eta_{xx} = \eta_t = 0$ at $x = 0, l$.

In fact, E is the bending energy of the plate. By considering two waves of nearly equal wavenumber, show that the bending energy propagates at the group velocity.

2

Waves on a Stretched String

Waves on stretched strings determine some of the fundamental properties of stringed musical instruments. They are also an excellent way of introducing the basic ideas and techniques of the theory of non-dispersive, linear waves. After deriving the governing, one-dimensional wave equation, we analyse waves on strings of finite and infinite length – typical standing and progressive waves. We also consider the reflection and transmission of waves at a step change in the density of the string. This is the most basic example of a class of phenomena that includes reflection, scattering and diffraction.

2.1 Derivation of the Governing Equation

Consider a thin, elastic string stretched between two fixed points at $x = 0$ and $x = L$, as shown in figure 2.1. In equilibrium, the string, which we treat as a line, lies at $y = 0$, has line density ρ in units of mass per length and tension T. We consider displacements in the y-direction small enough that changes in line density and tension are negligible, and hence that changes in length are negligible. Since a small element of length along the string is given by

$$ds = dx\sqrt{1 + \left(\frac{\partial y}{\partial x}\right)^2},$$

we require that $(\partial y/\partial x)^2 \ll 1$. In other words, we assume that the slope of the string, and hence ψ, is small. To find the partial differential equation that governs the motion of the string, we use Newton's second law in the vertical direction on a small element, as shown in figure 2.2. Since the

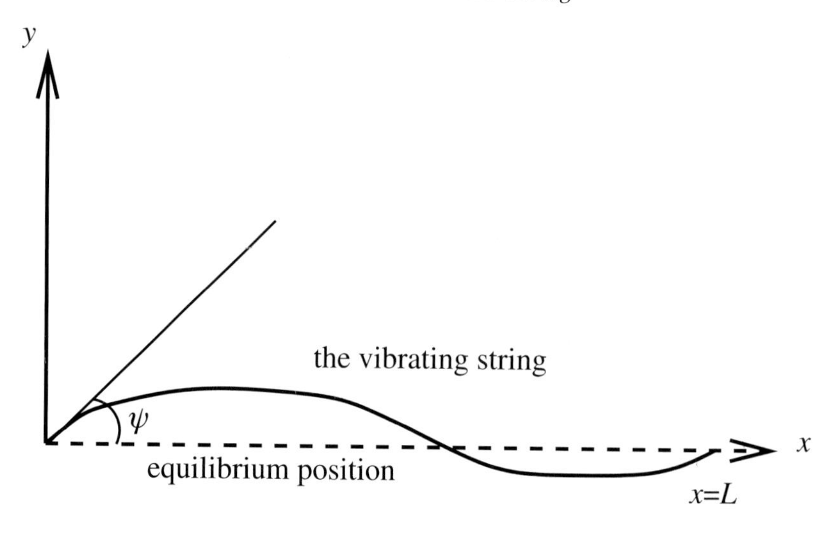

Fig. 2.1. A vibrating, stretched string.

element has mass $\rho\ \delta x$ and acceleration $\partial^2 y/\partial t^2$,

$$T\sin(\psi+\delta\psi)-T\sin\psi=\rho\ \delta x\frac{\partial^2 y}{\partial t^2}, \tag{2.1}$$

where $\psi=\psi(x)$ is the angle that the string makes with the horizontal, and $\psi+\delta\psi=\psi(x+\delta x)$. Since, for $\delta\psi\ll 1$,

$$\sin(\psi+\delta\psi)=\sin\psi\cos\delta\psi+\cos\psi\sin\delta\psi\approx\sin\psi+\cos\psi\,\delta\psi,$$

(2.1) shows that

$$T\cos\psi\,\delta\psi\approx\rho\delta x\frac{\partial^2 y}{\partial t^2}.$$

If we now take the limit $\delta\psi\to 0$, we arrive at

$$T\cos\psi\frac{\partial\psi}{\partial x}=\rho\frac{\partial^2 y}{\partial t^2}. \tag{2.2}$$

We also know that

$$\tan\psi=\frac{\partial y}{\partial x}, \tag{2.3}$$

by definition. Our main simplifying assumption, that $(\partial y/\partial x)^2\ll 1$, shows that $\tan\psi\ll 1$, and hence $\psi\ll 1$. Therefore

$$\psi\approx\frac{\partial y}{\partial x}\quad\text{and}\quad\frac{\partial\psi}{\partial x}\approx\frac{\partial^2 y}{\partial x^2}.$$

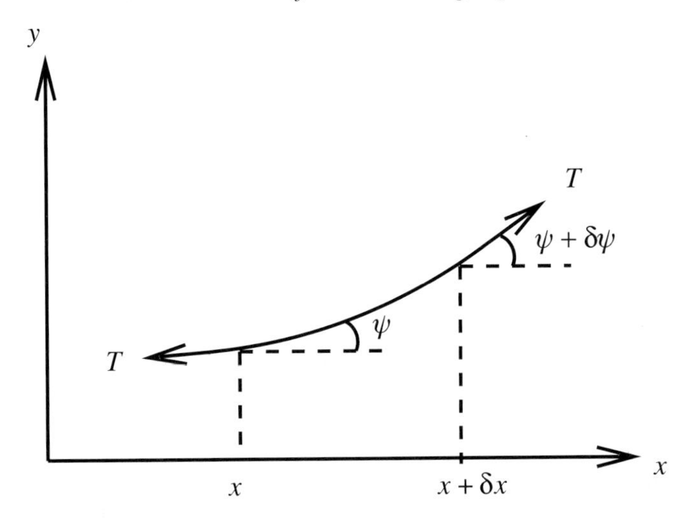

Fig. 2.2. The vertical force balance on a small element of a stretched string.

Using these results in (2.2) gives, at leading order,

$$T\frac{\partial^2 y}{\partial x^2} = \rho\frac{\partial^2 y}{\partial t^2}.$$

If we now define $c = \sqrt{T/\rho}$ we obtain

$$\frac{\partial^2 y}{\partial x^2} = \frac{1}{c^2}\frac{\partial^2 y}{\partial t^2}. \tag{2.4}$$

This is the **one-dimensional wave equation** with wave speed c. We can now investigate its properties in the context of vibrations on elastic strings, firstly for finite strings, and secondly for strings sufficiently long that we can make the idealisation that they are of infinite extent.

Before doing this, it is worth considering how energy is stored by the motion of the string. The storage of energy and its speed of propagation are themes that will recur frequently in the following chapters. The kinetic energy in the small amplitude vibrations of a stretched string is straightforward to calculate. Since the kinetic energy of a small element of length δx is, at leading order, $\frac{1}{2}\rho(\partial y/\partial t)^2\ \delta x$, the total energy is

$$K = \int \frac{1}{2}\rho\left(\frac{\partial y}{\partial t}\right)^2 dx. \tag{2.5}$$

The limits of the integral depend, of course, on the length of the string. In particular, if the string is infinitely long, the disturbance must either be confined to a finite length of the string or decay sufficiently rapidly as $x \to \pm\infty$ that the integral in (2.5) exists. If this is not the case, we can still consider *changes* in kinetic energy, or, for periodic waves, the kinetic energy in a single wavelength.

Energy is also stored as potential energy due to the stretching of the string. For a small element, this energy, δV, is equal to the work done in stretching the string from its undisturbed length, δx, to its stretched length, δs, so that

$$\delta V = T(\delta s - \delta x) = T\left\{\delta x\sqrt{1 + \left(\frac{\partial y}{\partial x}\right)^2} - \delta x\right\} \approx \frac{1}{2}T\left(\frac{\partial y}{\partial x}\right)^2 \delta x,$$

since $(\partial y/\partial x)^2 \ll 1$. Summing these contributions over the length of the string gives

$$V = \int \frac{1}{2}T\left(\frac{\partial y}{\partial x}\right)^2 dx. \tag{2.6}$$

The same considerations concerning the range of integration apply as for the kinetic energy. The total energy stored is simply $E = K + V$. Note that both potential and kinetic energies are quadratic in small quantities. As we shall see in the chapters to come, this is typical of linear waves, for which the governing equations are linear in small quantities.

As a typical example, for the D string of a violin, the tension is usually about 55 N in a steel string of density $7000\,\mathrm{kg\,m^{-3}}$, with a length of about 35 cm and diameter 0.5 mm. This gives a line density, $\rho \approx 1.4 \times 10^{-3}\,\mathrm{kg\,m^{-1}}$, and hence a wave speed $c \approx 200\,\mathrm{m\,s^{-1}}$.

2.2 Standing Waves on Strings of Finite Length

If the string is fixed at $x = 0$ and $x = L$, the solution, $y(x, t)$, of (2.4) must satisfy the boundary conditions

$$y(0, t) = y(L, t) = 0 \quad \text{for } t \geq 0. \tag{2.7}$$

We can proceed by looking for separable solutions of the form

$$y(x, t) = X(x)T(t).$$

Substituting into (2.4) we obtain

$$X''T = \frac{1}{c^2}XT'',$$

where a prime denotes the derivative with respect to the independent variable. Rearranging this gives

$$\frac{X''}{X} = \frac{1}{c^2}\frac{T''}{T} = -\kappa, \quad \text{the separation constant.}$$

Remember, if a function of x is equal to a function of t for all values of x and t, as is the case here, each function must be a constant, denoted here by $-\kappa$. The ordinary differential equation for X is

$$X'' + \kappa X = 0.$$

If $\kappa < 0$, this has solution

$$X = A\exp(\sqrt{-\kappa}x) + B\exp(-\sqrt{-\kappa}x).$$

The boundary conditions (2.7) show that $X(0) = X(L) = 0$, and hence that $A = B = 0$. We do not therefore obtain any useful solution if $\kappa < 0$. If $\kappa > 0$, the solution is

$$X = A\sin(\sqrt{\kappa}x) + B\cos(\sqrt{\kappa}x).$$

The boundary condition $X(0) = 0$ shows that $B = 0$, whilst $X(L) = 0$ means that $A\sin(\sqrt{\kappa}L) = 0$. In this case we can find a sequence of non-zero solutions by choosing κ so that $\sin(\sqrt{\kappa}L) = 0$ which gives

$$\kappa = \frac{n^2\pi^2}{L^2}, \quad \text{for } n = 1, 2, 3, \ldots.$$

The function $T(t)$ therefore satisfies the equation

$$T'' + \frac{n^2\pi^2c^2}{L^2}T = 0,$$

and hence

$$T = \overline{F}\sin\left(\frac{n\pi ct}{L}\right) + \overline{G}\cos\left(\frac{n\pi ct}{L}\right).$$

We have now constructed a sequence of solutions

$$y_n(x,t) = \sin\left(\frac{n\pi x}{L}\right)\left\{F\sin\left(\frac{n\pi ct}{L}\right) + G\cos\left(\frac{n\pi ct}{L}\right)\right\}, \tag{2.8}$$

for arbitrary constants $F = A\overline{F}$ and $G = A\overline{G}$ and n a positive integer. Each of these takes the form of a standing wave, with envelope $\sin(n\pi x/L)$ and angular frequency $n\pi c/L$. The envelope for $n = 1, 2, 3$ is shown in figure 2.3. The solution $y_1(x,t)$ is called the **fundamental mode** of the string, and is what produces most of the sound when you play a stringed instrument such as a violin or guitar. The solution $y_2(x,t)$ is called the

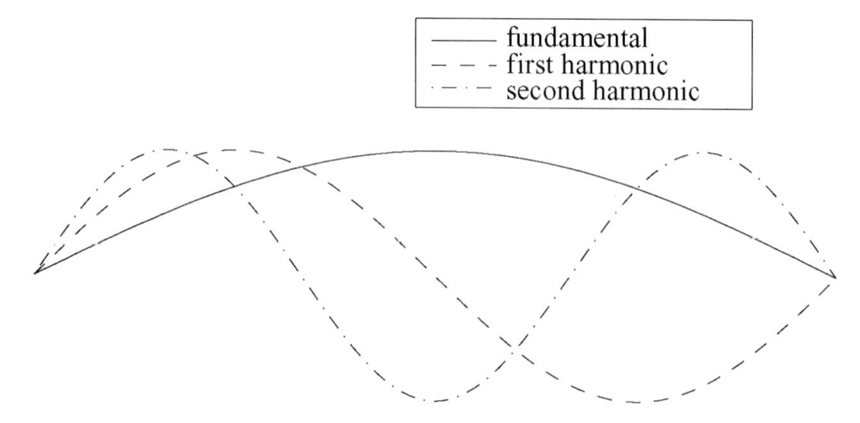

Fig. 2.3. The envelope of the fundamental, and first and second harmonics.

first harmonic, and has $y_2(\frac{1}{2}L, t) = 0$. This sounds an octave higher than the fundamental. A stringed instrument can be forced to vibrate in this way by lightly touching it halfway along the string to make it stationary there. Similar comments apply for the higher harmonics, $n = 3, 4, \ldots$, which generate all the notes of the musical scale. The procedure that we have just been through, of looking for separable solutions that turn out to exist as a sequence of discrete **eigensolutions** or **modes** ($y_n(x,t)$ for the finite string) with **eigenfrequencies** ω_n ($= n\pi c/L$), is applicable to most other linear, wave bearing systems, provided that they are finite in at least one direction (see sections 3.5, 5.4 and 6.8 which are concerned with wave propagation in waveguides).

Now that we know that a sequence of standing wave modes exists, can we determine which of these are excited by any particular initial disturbance? Since the one-dimensional wave equation is linear, the general solution is the linear combination

$$y(x,t) = \sum_{n=1}^{\infty} \sin\left(\frac{n\pi x}{L}\right) \left\{ F_n \sin\left(\frac{n\pi ct}{L}\right) + G_n \cos\left(\frac{n\pi ct}{L}\right) \right\}, \tag{2.9}$$

for constants F_n and G_n. If we now consider the initial value problem defined by (2.4) and (2.7) along with initial conditions

$$y(x,0) = y_0(x), \quad \frac{\partial y}{\partial t}(x,0) = v_0(x), \tag{2.10}$$

we can use (2.10) to determine F_n and G_n. The initial conditions (2.10) state that when $t = 0$ the shape of the string is given by $y_0(x)$ and the velocity by $v_0(x)$. Since (2.4) contains a second derivative with respect to t, we should not be surprised that we need two initial conditions to determine the solution. The general solution (2.9) shows that

$$y_0(x) = \sum_{n=1}^{\infty} G_n \sin\left(\frac{n\pi x}{L}\right), \quad v_0(x) = \sum_{n=1}^{\infty} \frac{n\pi c}{L} F_n \sin\left(\frac{n\pi x}{L}\right). \tag{2.11}$$

These are just the Fourier series representations of y_0 and v_0.

We can determine F_n and G_n by exploiting the orthogonality of the basis functions, given by

$$\int_0^L \sin\left(\frac{n\pi x}{L}\right) \sin\left(\frac{m\pi x}{L}\right) dx = \left\{ \begin{array}{ll} \frac{1}{2}L & \text{when } m = n, \\ 0 & \text{when } m \neq n. \end{array} \right\} \tag{2.12}$$

We simply multiply (2.11) by $\sin(m\pi x/L)$, integrate term by term from 0 to L and use (2.12) to obtain

$$F_m = \frac{2}{m\pi c} \int_0^L v_0(x) \sin\left(\frac{m\pi x}{L}\right) dx, \quad G_m = \frac{2}{L} \int_0^L y_0(x) \sin\left(\frac{m\pi x}{L}\right) dx. \tag{2.13}$$

As an example, consider the solution when

$$y_0(x) = \begin{cases} \dfrac{a}{h} x & \text{for } 0 \leq x \leq h, \\ \dfrac{a}{L-h}(L-x) & \text{for } h \leq x \leq L, \end{cases}$$

and $v_0(x) = 0$. This represents a string plucked at the point $x = h$. Note that the initial conditions must be consistent with the boundary conditions, so that $y_0(0) = y_0(L) = v_0(0) = v_0(L) = 0$. We can immediately see from (2.13) that $F_m = 0$ and after integrating by parts we find that

$$G_m = \frac{2a}{h(L-h)} \frac{L^2}{m^2\pi^2} \sin\left(\frac{m\pi h}{L}\right).$$

This leads to

$$y(x,t) = \frac{2aL^2}{h(L-h)\pi^2} \sum_{n=1}^{\infty} \frac{1}{n^2} \sin\left(\frac{n\pi h}{L}\right) \cos\left(\frac{n\pi ct}{L}\right) \sin\left(\frac{n\pi x}{L}\right).$$

Note that when $h = \frac{1}{2}L$ only the terms with n odd appear in the solution. In this case the initial condition is symmetric about $x = \frac{1}{2}L$, which means that only symmetric modes can be excited by the initial displacement. The amplitudes of the various modes in an experimental investigation of a plucked violin string are shown in figure 2.4, and are found to be in excellent agreement with our linear theory.

Finally, let's consider the energy stored in a string of finite length. For

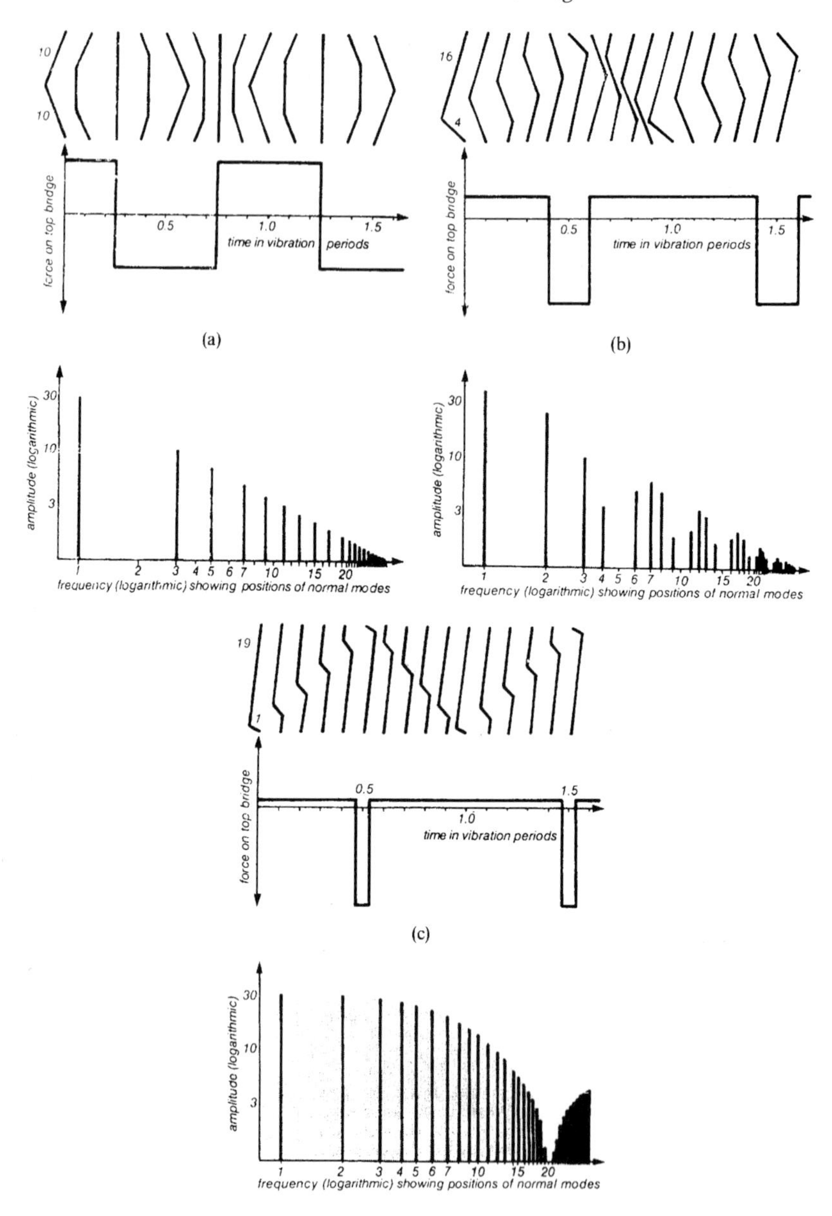

Fig. 2.4. Experimentally measured wave mode amplitudes for a plucked violin string, along with the theoretically predicted displacements (Fletcher, 1976). String plucked at its centre (a), at $\frac{1}{5}$ of its length (b), and at $\frac{1}{20}$ of its length (c) from the bridge.

each wave mode $y_n(x,t)$, given by (2.8), we simply substitute into (2.5) and (2.6) with range of integration $0 \leq x \leq L$, and find that

$$K_n = \frac{\rho n^2 \pi^2 c^2}{4L} \left\{ F_n \cos\left(\frac{n\pi ct}{L}\right) - G_n \sin\left(\frac{n\pi ct}{L}\right) \right\}^2, \tag{2.14}$$

$$V_n = \frac{\rho n^2 \pi^2 c^2}{4L} \left\{ F_n \sin\left(\frac{n\pi ct}{L}\right) + G_n \cos\left(\frac{n\pi ct}{L}\right) \right\}^2, \tag{2.15}$$

making use of $T = \rho c^2$. Using some simple trigonometry, we can also write this as

$$K_n = \frac{\rho \pi^2 c^2}{4L} R_n^2 \sin^2\left(\frac{n\pi ct}{L} + \kappa_n\right), \quad V_n = \frac{\rho \pi^2 c^2}{4L} R_n^2 \cos^2\left(\frac{n\pi ct}{L} + \kappa_n\right), \tag{2.16}$$

where

$$R_n = n\sqrt{F_n^2 + G_n^2}, \quad \kappa_n = \tan^{-1}\left(\frac{G_n}{F_n}\right). \tag{2.17}$$

We can now see that, as time passes, the energy stored in each wave mode is transferred back and forth between kinetic and potential energy, one form of energy reaching a maximum when the other is zero. Moreover, the time averages of each form of energy over a period of the motion are

$$\overline{K}_n = \overline{V}_n = \frac{\rho \pi^2 c^2}{8L} R_n^2, \tag{2.18}$$

since

$$\frac{nc}{2L} \int_{t_0}^{t_0 + \frac{2L}{nc}} \sin^2\left(\frac{n\pi ct}{L} + \kappa_n\right) dt$$

$$= \frac{nc}{2L} \int_{t_0}^{t_0 + \frac{2L}{nc}} \cos^2\left(\frac{n\pi ct}{L} + \kappa_n\right) dt = \frac{1}{2}.$$

This **equipartition of energy** occurs for all time harmonic, linear oscillators and wave systems, as we shall see over the next few chapters. From (2.18), the total energy stored in each wave mode is

$$E_n = \frac{\rho \pi^2 c^2}{4L} R_n^2. \tag{2.19}$$

When we come to consider the most general vibration of the string, we must substitute (2.9) into (2.5) and (2.6). At first sight, it looks like quite a tall order to calculate these integrals, since in each we need to square an infinite sum. However, the orthogonality of the standing wave modes, given by (2.12), means that most terms in the square of each infinite sum integrate to zero over the length of the string. The only terms that remain are those that give the energy stored in each of the individual standing wave modes, with

$$K = \sum_{n=1}^{\infty} K_n, \quad V = \sum_{n=1}^{\infty} V_n, \quad E = \sum_{n=1}^{\infty} E_n. \tag{2.20}$$

This shows that the wave modes each vibrate with their own individual amounts of energy, and there is no exchange of energy between them.

2.3 D'Alembert's Solution for Strings of Infinite Length

If a stretched string is very long, we can assume that it is effectively infinite and look for solutions of the initial value problem

$$\frac{\partial^2 y}{\partial x^2} = \frac{1}{c^2}\frac{\partial^2 y}{\partial t^2} \quad \text{for } -\infty < x < \infty,\ t \geq 0, \tag{2.21}$$

subject to

$$y(x,0) = y_0(x), \quad \frac{\partial y}{\partial t}(x,0) = v_0(x). \tag{2.22}$$

We are again specifying the initial displacement and velocity of the string, but now for $-\infty < x < \infty$. In order to solve this initial value problem, we introduce the variables

$$\xi = x + ct, \quad \eta = x - ct.$$

If we treat y as a function of these new variables, the chain rule tells us that

$$\frac{\partial^2 y}{\partial x^2} = \frac{\partial^2 y}{\partial \xi^2} + 2\frac{\partial^2 y}{\partial \xi \partial \eta} + \frac{\partial^2 y}{\partial \eta^2}, \quad \frac{\partial^2 y}{\partial t^2} = c^2\left(\frac{\partial^2 y}{\partial \xi^2} - 2\frac{\partial^2 y}{\partial \xi \partial \eta} + \frac{\partial^2 y}{\partial \eta^2}\right).$$

Equation (2.21) therefore becomes

$$\frac{\partial^2 y}{\partial \xi \partial \eta} = 0. \tag{2.23}$$

In this form, we can integrate the equation with respect to ξ and obtain

$$\frac{\partial y}{\partial \eta} = F(\eta),$$

where F is an arbitrary function of η. Now we can integrate again and find that

$$y = \int^{\eta} F(s)ds + g(\xi),$$

where g is an arbitrary function of ξ. If we now define

$$f(\eta) = \int^{\eta} F(s)ds,$$

we obtain $y = f(\eta) + g(\xi)$, and hence

$$y(x,t) = f(x - ct) + g(x + ct), \tag{2.24}$$

for arbitrary functions f and g. This is the solution that we discussed in chapter 1, where we showed that it represents the sum of left- and right-travelling waves with wave speed c. Note that the displacement of the string is perpendicular to the direction of propagation of the waves. Such waves are said to be **transverse**.

The energy stored in a simple right- or left-travelling wave is easy to calculate. If $y = f(x - ct)$, we can substitute this into (2.5) and (2.6), and find that

$$K = V = \frac{1}{2}\rho c^2 \int_{-\infty}^{\infty} \{f'(x-ct)\}^2 \, dx, \tag{2.25}$$

where we assume that f is localised or decays sufficiently rapidly as $x \to \pm\infty$ that the integral exists. As with standing waves on a finite string, we have equipartition of energy. This is also true for left-travelling waves, but is *not* usually true for the general solution, $y = f(x - ct) + g(x + ct)$.

We must now choose f and g to satisfy the initial conditions (2.22). Firstly,

$$y(x,0) = f(x) + g(x) = y_0(x). \tag{2.26}$$

Secondly, after noting that the chain rule gives us

$$\frac{\partial y}{\partial t} = -cf'(x - ct) + cg'(x + ct),$$

where the prime denotes differentiation with respect to the single independent variable, $x - ct$ or $x + ct$, we find that

$$\frac{\partial y}{\partial t}(x,0) = -cf'(x) + cg'(x) = v_0(x).$$

This can be integrated to give

$$-cf(x) + cg(x) = \int_a^x v(s)ds, \tag{2.27}$$

where a is an arbitrary constant. Equations (2.26) and (2.27) show that

$$f(x) = \frac{1}{2}y_0(x) - \frac{1}{2c}\int_a^x v(s)ds, \quad g(x) = \frac{1}{2}y_0(x) + \frac{1}{2c}\int_a^x v(s)ds,$$

and hence that

$$y(x,t) = \frac{1}{2}\{y_0(x-ct) + y_0(x+ct)\} + \frac{1}{2c}\int_{x-ct}^{x+ct} v_0(s)ds. \qquad (2.28)$$

This is known as **D'Alembert's solution of the one-dimensional wave equation**.

We consider two examples. Firstly,

$$v_0(x) = 0, \quad y_0(x) = H(x+a) - H(x-a), \qquad (2.29)$$

where $H(x)$ is the **Heaviside function**, defined by

$$H(x) = \left\{ \begin{array}{ll} 1 & \text{for } x > 0, \\ 0 & \text{for } x < 0. \end{array} \right\} \qquad (2.30)$$

The initial conditions (2.29) represent an initially stationary string with an initial displacement given by the **top hat function**. The top hat function is equal to 1 for $-a < x < a$ and zero otherwise. Clearly this type of initial condition is rather unsatisfactory from a physical point of view, since $y(x,0)$ is discontinuous. However, the solution is perfectly well-defined, and very instructive.

Substituting from (2.29) into (2.28), we find that

$$\begin{aligned} y(x,t) &= \frac{1}{2}\{H(x+ct+a) - H(x+ct-a)\} \\ &+ \frac{1}{2}\{H(x-ct+a) - H(x-ct-a)\}. \end{aligned} \qquad (2.31)$$

The first term is a left-travelling top hat, whilst the second term is a right-travelling top hat. These waves overlap for $0 \leq t \leq a/c$, and propagate separately for $t > a/c$, as illustrated in figure 2.5.

Secondly, consider the initial conditions

$$y_0(x) = a\sin kx, \quad v_0(x) = 0. \qquad (2.32)$$

The string is again initially stationary, but in this case is everywhere displaced sinusoidally. Substitution of (2.32) into (2.28) gives

$$y(x,t) = \frac{1}{2}a\{\sin k(x+ct) + \sin k(x-ct)\} = a\sin kx\cos kct. \qquad (2.33)$$

This is just a standing wave, and shows how such a wave can be generated from a pair of left- and right-travelling waves of equal amplitude, as we discussed in chapter 1.

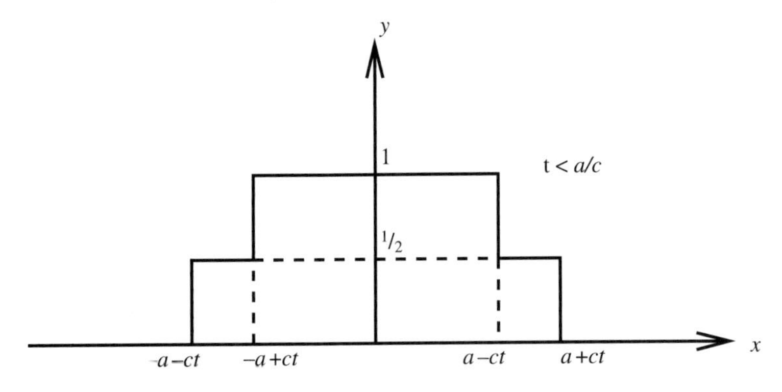

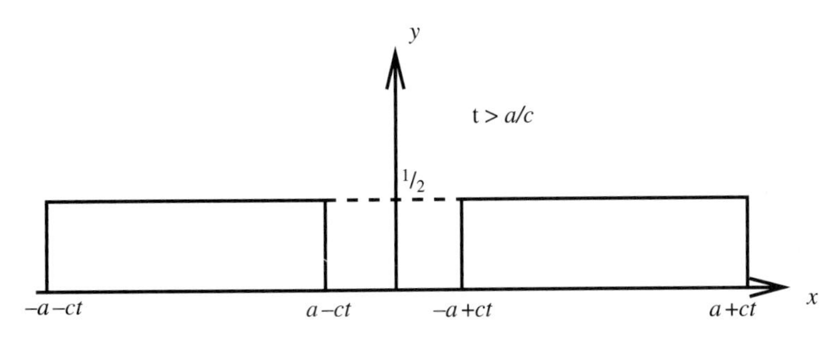

Fig. 2.5. The solution for an initial top hat displacement of a stationary, infinite string.

2.4 Reflection and Transmission of Waves by Discontinuities in Density

In this section we consider two situations that lead to the reflection of some of the energy of a wave incident on a point of discontinuity. The results that we derive here are applicable in a wide range of physical situations that can be modelled with linear wave equations.

2.4.1 A Single Discontinuity

Consider a string with density ρ_1, wave speed $c_1 = \sqrt{T/\rho_1}$ for $x < 0$ and density ρ_2, wave speed $c_2 = \sqrt{T/\rho_2}$ for $x > 0$. If a wave $f_I(x - c_1 t)$ is incident from the left, part of it will be reflected and part transmitted by the density discontinuity at $x = 0$. We seek a solution of the form

$$y(x,t) = \left\{ \begin{array}{cl} f_I(x - c_1 t) + f_R(x + c_1 t) & \text{for } x < 0, \\ f_T(x - c_2 t) & \text{for } x > 0. \end{array} \right\} \tag{2.34}$$

The reflected wave travels to the left, and is given by $f_R(x + c_1 t)$, whilst the transmitted wave travels to the right, and is given by $f_T(x - c_2 t)$. In order to determine f_R and f_T we need to consider two conditions at $x = 0$. The string does not break, so y must be continuous at $x = 0$, and hence

$$f_I(-c_1 t) + f_R(c_1 t) = f_T(-c_2 t). \tag{2.35}$$

In addition, since the tension in the string is constant, the slope, $\partial y/\partial x$, must also be continuous, and hence

$$f_I'(-c_1 t) + f_R'(c_1 t) = f_T'(-c_2 t).$$

This expression can be integrated once with respect to t to give

$$-\frac{1}{c_1} f_I(-c_1 t) + \frac{1}{c_1} f_R(c_1 t) = -\frac{1}{c_2} f_T(-c_2 t). \tag{2.36}$$

Note that we have assumed that each of the waves is sufficiently localised that they are simultaneously zero at some time, so that the constant of integration is zero. Solving (2.35) and (2.36) gives

$$f_T(-c_2 t) = \frac{2c_2}{c_1 + c_2} f_I(-c_1 t),$$

and hence

$$f_T(x - c_2 t) = \frac{2c_2}{c_1 + c_2} f_I\left(\frac{c_1}{c_2}(x - c_2 t)\right), \tag{2.37}$$

and similarly

$$f_R(x + c_1 t) = \frac{c_2 - c_1}{c_1 + c_2} f_I(-(x + c_1 t)). \tag{2.38}$$

The important point to note is that, relative to the incident wave, the transmitted wave has amplitude $2c_2/(c_1 + c_2)$ and the reflected wave has amplitude $(c_2 - c_1)/(c_1 + c_2)$. If $c_1 = c_2$ there is no discontinuity in density, and hence no reflection, as expected. When $c_1 \gg c_2$, so that $\rho_2 \gg \rho_1$, the transmitted wave has a small amplitude, whilst the reflected wave has the same amplitude as the incident wave at leading order, but interestingly, with the opposite sense. If y is positive in the incident wave it will be negative in the reflected wave. This case corresponds to an almost immobile string for $x > 0$, when we would expect almost complete reflection of the incident wave. When $c_2 \gg c_1$, so that $\rho_1 \gg \rho_2$, the reflected wave again has the same amplitude as the incident wave at leading order, and in this case, the same sense. The transmitted wave has an amplitude twice that of the incident wave. In this case the string for

$x > 0$ is much lighter than the string that carries the incident wave, so only a little energy is needed to excite a large transmitted wave.

The phenomenon of reflection and transmission of waves from points where the properties of the system change is a topic that we will return to for most of the linear systems that we study in part one of the book. We discuss more complicated variations on this theme in chapter 11. Note also that we can interpret the standing waves that we constructed in section 2.2 in terms of complete reflection of a wave at the points $x = 0$ and $x = L$, where the string is fixed. Each standing wave can be written as the sum of a left- and a right-travelling wave, and we can think of this as a single wave bouncing back and forth between $x = 0$ and $x = L$ (see also subsection 3.5.2).

2.4.2 Two Discontinuities: Impedance Matching

Let's now consider a situation where three strings with different densities are joined together, so that

$$c = \begin{cases} c_1 & \text{for } x < 0, \\ c_2 & \text{for } 0 < x < L, \\ c_3 & \text{for } x > L. \end{cases}$$

If a wave is incident from the left, there will, in general, be a reflected wave in $x < 0$, reflected and transmitted waves in $0 < x < L$, and a transmitted wave in $x > L$. We will restrict our attention to the case of an incident harmonic wave of angular frequency ω, and try to determine for what values of c_2 and L there is no wave reflected into $x < 0$. In this case all of the energy incident from the left is transmitted to the right, and we say that the two semi-infinite strings are **impedance matched**. The **impedance** of a material is simply a measure of how difficult it is to move it. For the stretched string, it is defined as

$$Z(x_0) = -\frac{\text{Vertical force exerted on the string in } x > x_0 \text{ by the string in } x < x_0}{\text{Vertical velocity of the string at } x = x_0},$$

and hence

$$Z = -\frac{T\partial y/\partial x}{\partial y/\partial t}.$$

For right-propagating solutions of the form $y = f(x - ct)$, this gives $Z = T/c = \sqrt{\rho T}$. Since T remains constant at leading order, $Z \propto \sqrt{\rho}$.

Equivalent definitions can be given for other linear systems, for example electrical circuits.

For our system of three stretched strings, we can write the solution in the form

$$y=\left\{\begin{array}{ll}\exp\left\{i\omega\left(t-\dfrac{x}{c_1}\right)\right\} & \text{for } x\le 0,\\ T_2\exp\left\{i\omega\left(t-\dfrac{x}{c_2}\right)\right\}+R_2\exp\left\{i\omega\left(t+\dfrac{x}{c_2}\right)\right\} & \text{for } 0\le x\le L,\\ T_3\exp\left\{i\omega\left(t-\dfrac{x}{c_3}\right)\right\} & \text{for } x\ge L.\end{array}\right\} \tag{2.39}$$

From the analysis of the previous subsection, both y and $\partial y/\partial x$ must be continuous at $x=0$ and $x=L$. Applying these continuity conditions leads to the four equations

$$\left.\begin{aligned} T_2+R_2&=1,\\ T_2-R_2&=\frac{c_2}{c_1},\\ T_2e^{-i\omega L/c_2}+R_2e^{i\omega L/c_2}&=T_3e^{-i\omega L/c_3},\\ T_2e^{-i\omega L/c_2}-R_2e^{i\omega L/c_2}&=\frac{c_2}{c_3}T_3e^{-i\omega L/c_3}.\end{aligned}\right\} \tag{2.40}$$

It is straightforward to eliminate T_2, R_2 and T_3, and, after taking the real part of the resulting expression, arrive at

$$\cos\left(\frac{2\omega L}{c_2}\right)=\left(\frac{c_3-c_2}{c_3+c_2}\right)\left(\frac{c_1+c_2}{c_1-c_2}\right). \tag{2.41}$$

The easiest way of choosing c_2 and L to satisfy this equation is to take $c_2=\sqrt{c_1c_3}$, and hence, since the tension in each string is the same, $\rho_2=\sqrt{\rho_1\rho_3}$. Equation (2.41) then becomes

$$\cos\left(\frac{2\omega L}{c_2}\right)=-1,$$

which can be satisfied by taking

$$L=\frac{\pi}{2}\frac{c_2}{\omega}=\frac{1}{4}\lambda_2,$$

where λ_2 is the wavelength of the disturbance in the middle string. In other words, we can impedance match the two semi-infinite strings by making the density of the middle string the geometric mean of the densities of the other two strings, and its length one quarter of a wavelength.

The joining of two dissimilar media by impedance matching, using a quarter wavelength of a medium where the linear wave speed is the geometrical mean of those in the dissimilar media, occurs in many different physical situations: anti-reflection coatings on military submarines to try to evade detection by radar, coatings on camera lenses to ensure that all of the incident light is transmitted, the reflectionless joining of transmission lines for electrical power, the transmission of a signal from an antenna to a television, the use of a transducing jelly to transmit ultrasound into the human body during medical imaging. For a related, but slightly different, example see exercise 3.2.

Exercises

2.1 Consider the finite string that we studied in section 2.2. When $t = 0$ the displacement and velocity of the string are given by

(a) $y(x,0) = x(L-x)$,
$\partial y/\partial t\,(x,0) = 0$,

(b) $y(x,0) = 0$,
$\partial y/\partial t\,(x,0) = 1$, for $|x-d| \le a$,
$\partial y/\partial t\,(x,0) = 0$, for $0 \le x \le d-a$ and $d+a \le x \le L$.

In each case determine the displacement of the string for $t > 0$ and the amount of energy stored in the string. Note that case (b) corresponds to a string struck by a hammer with width $2a$ and unit velocity at the point $x = d$.

2.2 The air resistance per unit length experienced by a vibrating string is equal to $\rho\kappa$ times the local velocity, $\partial y/\partial t$. By including this force in the derivation of the wave equation, show that y satisfies

$$\frac{\partial^2 y}{\partial t^2} + \kappa\frac{\partial y}{\partial t} = c^2\frac{\partial^2 y}{\partial x^2}.$$

For the string described in the previous exercise, write down the solution in cases (a) and (b) when the effect of air resistance is included. You can assume that $\kappa < 2\pi c/L$. What happens to the string and the energy stored in it as $t \to \infty$? Hint: Begin by looking for separable solutions, and then adapt the methods used when $\kappa = 0$.

2.3 Consider the infinite string that we studied in section 2.3. When $t = 0$ the displacement and velocity of the string are

(a) $y(x,0) = \sin x$, for $-\pi \le x \le \pi$,
$y(x,0) = 0$, for $x > \pi$, and $x < -\pi$,
$\partial y/\partial t\,(x,0) = 0$,

(b) $y(x,0) = 0,$
$\partial y/\partial t\,(x,0) = 0$, for $|x| > a$,
$\partial y/\partial t\,(x,0) = 1$, for $|x| \le a$.

In each case, use d'Alembert's solution to find the displacement for $t > 0$. Sketch the solutions at various times and describe what is happening to the string.

The solution that you constructed in exercise 2.1(b) should be identical to the solution you found in exercise 2.3(b), after a suitable change of coordinates and before any disturbance reaches the fixed ends of the string. Prove that this is true by showing that the solution that you have just constructed can be written as a Fourier series identical to the one that you found in exercise 2.1(b).

2.4 An infinitely long stretched string has density ρ for $x < 0$ and 4ρ for $x > 0$. When $t = 0$ the string is stationary and has displacement

$$y(x,0) = \begin{cases} 1 & \text{for } -2 < x < -1, \\ 0 & \text{otherwise.} \end{cases}$$

Determine and sketch the displacement of the string for $t > 0$.

2.5 An infinitely long stretched string with tension T, mass per unit length ρ and displacement $y(x,t)$ has a point mass, M, attached to it at $x = 0$. By considering Newton's second law applied to the mass, assuming that gravity is negligible, show that

$$M\frac{\partial^2 y}{\partial t^2}(0,t) = T\left\{\left(\frac{\partial y}{\partial x}\right)^+ - \left(\frac{\partial y}{\partial x}\right)^-\right\}, \tag{2.42}$$

where $(\partial y/\partial x)^+$ is the slope of the string to the right of the mass at $x = 0$ and $(\partial y/\partial x)^-$ is the slope to the left.

When $t = 0$ a wave with displacement $y(x,0)$ given by

$$y(x,t) = \left\{\begin{array}{cc} 0 & \text{for } x > -1, \\ \sin(\pi x), & \text{for } -2 \le x \le -1, \\ 0 & \text{for } x < -2 \end{array}\right\} \tag{2.43}$$

is travelling in the positive sense. The incident, reflected and transmitted waves are $f_{\mathrm{I}}(x-ct)$, $f_{\mathrm{R}}(x+ct)$ and $f_{\mathrm{T}}(x-ct)$ respectively, where $c = \sqrt{T/\rho}$. By using continuity of displacement at $x = 0$, and (2.42), show that

$$\frac{d}{dt}\left\{e^{\alpha t} f_{\mathrm{T}}(-ct)\right\} = \alpha e^{\alpha t} f_{\mathrm{I}}(-ct), \tag{2.44}$$

where $\alpha = 2T/Mc$. By solving (2.44), show that

$$f_{\mathrm{T}}(-ct) = \frac{\alpha}{c}e^{-\alpha t}\int_0^{ct} e^{\alpha u/c} f_{\mathrm{I}}(-u)\,du, \tag{2.45}$$

and write down a similar expression for $f_{\mathrm{R}}(ct)$. Using (2.43) for the incident wave, show that $f_{\mathrm{T}}(x - ct)$ is zero for $x \geq ct - 1$,

$$\frac{\alpha^2}{\alpha^2 + c^2\pi^2}\left[\frac{c\pi}{\alpha}e^{\alpha(1+x-ct)/c} + \frac{c\pi}{\alpha}\cos\{\pi(x - ct)\} + \sin\{\pi(x - ct)\}\right],$$

for $ct - 2 \leq x \leq ct - 1$, and

$$\frac{\alpha c\pi}{\alpha^2 + c^2\pi^2}\left(1 + e^{\alpha/c}\right)e^{\alpha(1+x-ct)/c},$$

for $x \leq ct - 2$, and find a similar expression for $f_{\mathrm{R}}(x + ct)$. Show that the displacement of the mass at $x = 0$ tends monotonically to zero as $t \to \infty$. Sketch the solution for $t > 2/c$. Show that the solution reduces to simple propagation of the incident wave when $M = 0$ ($\alpha = \infty$). What happens as $M \to \infty$?

3

Sound Waves

Sound and sources of sound are part of our everyday experience. Sound waves are essential for speech, generally bring pleasure when listening to music, but can also be a nuisance, for example if you live near an airport (see figure 3.1), or have noisy neighbours. Depending upon the circumstances, we may want to maximise or minimise the audibility of sound waves, but in either case we must firstly understand them and be able to model their propagation. In this chapter, after deriving the **three-dimensional wave equation**, which governs the propagation of sound, or **acoustic** waves, we will study the transmission of sound in tubes and horns. We also consider the **reflection** and **guiding** of acoustic waves, together with their **generation** by localised sources.

3.1 Derivation of the Governing Equation

Sound waves are small amplitude disturbances of a body of compressible gas. In order to study them, we must consider the equations that express the conservation of mass and momentum in an inviscid gas. These are

$$\frac{\partial \rho}{\partial t} + \nabla \cdot (\rho \mathbf{u}) = 0, \tag{3.1}$$

$$\frac{\partial \mathbf{u}}{\partial t} + \mathbf{u} \cdot \nabla \mathbf{u} = -\frac{1}{\rho} \nabla p, \tag{3.2}$$

where ρ is the gas density, p is the gas pressure, $\mathbf{u}$ is the gas velocity and t is time. We will neglect gravitational and viscous forces, and determine in section 3.2 when this is a valid approximation. For the moment we will assume that $\rho = \rho(p)$ instead of $\rho = \rho(p, T)$, where T is the absolute temperature, an assumption that we will justify in section 7.2.

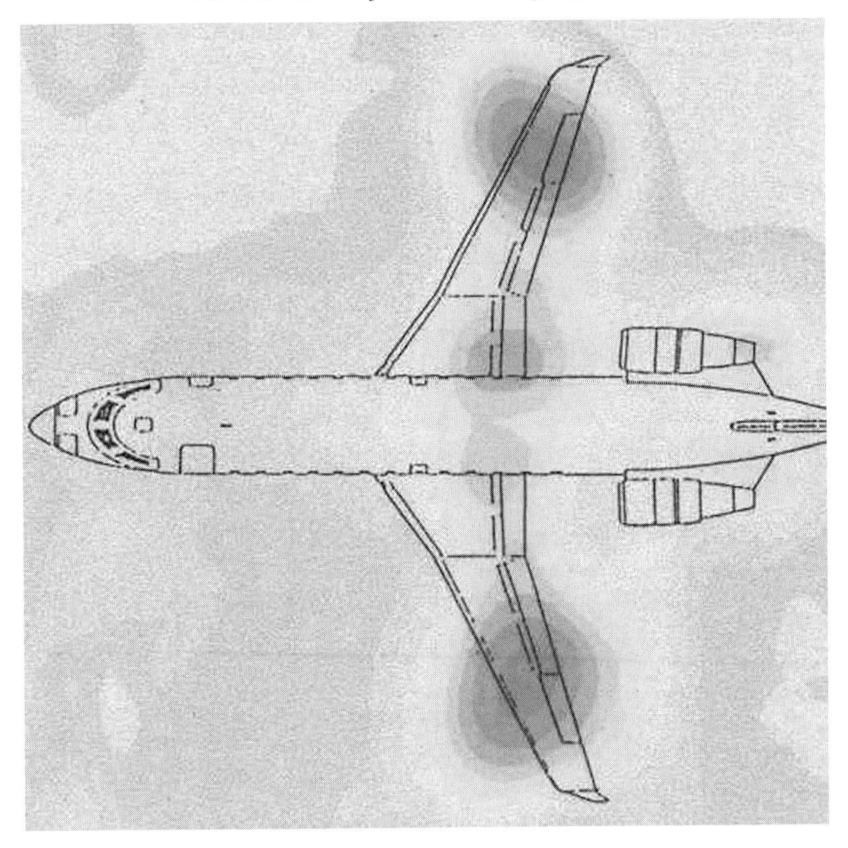

Fig. 3.1. A large aircraft and the sound field associated with it during landing.

Consider a small disturbance of a uniform, stationary body of gas, which we write as

$$p = p_0 + \tilde{p}, \quad \rho = \rho_0 + \tilde{\rho}, \quad \mathbf{u} = \tilde{\mathbf{u}}. \tag{3.3}$$

The uniform state has $p = p_0$, $\rho = \rho_0 = \rho(p_0)$, and $\mathbf{u} = 0$. The quantities $\tilde{p}$, $\tilde{\rho}$ and $\tilde{\mathbf{u}}$ are small amplitude disturbances of this state. Equations (3.1) and (3.2) become

$$\frac{\partial \tilde{\rho}}{\partial t} + \nabla \cdot \{(\rho_0 + \tilde{\rho})\tilde{\mathbf{u}}\} = 0, \tag{3.4}$$

$$\frac{\partial \tilde{\mathbf{u}}}{\partial t} + \tilde{\mathbf{u}} \cdot \nabla \tilde{\mathbf{u}} = -\frac{1}{\rho_0 + \tilde{\rho}} \nabla \tilde{p}. \tag{3.5}$$

Since we are considering small amplitude disturbances, we can neglect products of small quantities in these equations and obtain at leading order

$$\frac{\partial \tilde{\rho}}{\partial t} + \rho_0 \nabla \cdot \tilde{\mathbf{u}} = 0, \tag{3.6}$$

$$\frac{\partial \tilde{\mathbf{u}}}{\partial t} = -\frac{1}{\rho_0} \nabla \tilde{p}. \tag{3.7}$$

We know that $\rho_0 + \tilde{\rho} = \rho(p_0 + \tilde{p})$ and, since $|\tilde{p}| \ll p_0$, we can use a Taylor series expansion to obtain

$$\rho_0 + \tilde{\rho} \approx \rho(p_0) + \tilde{p}\frac{d\rho}{dp}(p_0).$$

Since $\rho_0 = \rho(p_0)$ we find that, at leading order,

$$\tilde{\rho} = \frac{d\rho}{dp}(p_0)\tilde{p}. \tag{3.8}$$

Equation (3.6) then becomes

$$\frac{1}{\rho_0}\frac{d\rho}{dp}(p_0)\frac{\partial \tilde{p}}{\partial t} = -\nabla \cdot \tilde{\mathbf{u}}.$$

If we differentiate this with respect to t and use (3.7) to eliminate $\partial \tilde{\mathbf{u}}/\partial t$, we obtain

$$\frac{1}{\rho_0}\frac{d\rho}{dp}(p_0)\frac{\partial^2 \tilde{p}}{\partial t^2} = -\frac{\partial}{\partial t}(\nabla \cdot \tilde{\mathbf{u}}) = -\nabla \cdot \left(\frac{\partial \tilde{\mathbf{u}}}{\partial t}\right) = \frac{1}{\rho_0}\nabla^2 \tilde{p}.$$

If we now define

$$c = \left\{\frac{d\rho}{dp}(p_0)\right\}^{-1/2}, \tag{3.9}$$

we find that

$$\nabla^2 \tilde{p} = \frac{1}{c^2}\frac{\partial^2 \tilde{p}}{\partial t^2}. \tag{3.10}$$

In other words, small pressure fluctuations in a gas satisfy the **three-dimensional wave equation** with wave speed c given by (3.9).

So what is the value of c in atmospheric air? For ideal gases

$$p = RT\rho/m, \tag{3.11}$$

where $R \approx 8.3\,\mathrm{J\,K^{-1}}$ is the universal gas constant, m is the molecular mass and T is the absolute temperature, measured in kelvins (K). If we assume that T is a constant (the flow is **isothermal**), then p/ρ is a constant, which must be equal to p_0/ρ_0. Hence $dp/d\rho = p_0/\rho_0$ and $c = \sqrt{p_0/\rho_0}$. In air at room temperature this gives $c \approx 290\,\mathrm{m\,s^{-1}}$. This is a calculation that Sir

Isaac Newton made in the seventeenth century, and it is flawed. In fact, measurements show that $c \approx 340\,\mathrm{m\,s^{-1}}$. As Laplace pointed out in the nineteenth century, pressure fluctuations in a sound wave are so rapid that the temperature of the gas does not remain constant. In fact, the compression and expansion of the gas are such that there is no heat loss (the flow is **adiabatic**). We shall return to this in section 7.2, where we show that for an adiabatic gas, p/ρ^{γ} is constant, where γ is the ratio of the specific heats of the gas. In air, $\gamma \approx 1.4$. Since $dp/d\rho = \gamma p_0/\rho_0$, this means that $c = \sqrt{\gamma p_0/\rho_0} \approx 340\,\mathrm{m\,s^{-1}}$, the observed value.

We also want to know the equation satisfied by the velocity fluctuation, $\tilde{\mathbf{u}}$. Since the background state of the gas is spatially uniform, we can write (3.7) as

$$\frac{\partial \tilde{\mathbf{u}}}{\partial t} = -\nabla\left(\frac{\tilde{p}}{\rho_0}\right). \tag{3.12}$$

If we now let

$$-\frac{\tilde{p}}{\rho_0} = \frac{\partial \tilde{\phi}}{\partial t},$$

for some function $\tilde{\phi}$,

$$\frac{\partial \tilde{\mathbf{u}}}{\partial t} - \nabla\left(\frac{\partial \tilde{\phi}}{\partial t}\right) = 0 \quad \Rightarrow \quad \frac{\partial}{\partial t}(\tilde{\mathbf{u}} - \nabla\tilde{\phi}) = 0,$$

and hence $\tilde{\mathbf{u}} - \nabla\tilde{\phi} = \mathbf{f}(\mathbf{x})$ for some function $\mathbf{f}$. If we assume that $\tilde{\mathbf{u}} = 0$ when $t = t_0$,

$$\tilde{\mathbf{u}} = \nabla(\tilde{\phi}(t) - \tilde{\phi}(t_0)).$$

If we now define $\phi = \tilde{\phi} - \tilde{\phi}(t_0)$,

$$\tilde{\mathbf{u}} = \nabla\phi. \tag{3.13}$$

The function ϕ is called the **acoustic velocity potential**. Note that

$$\nabla \times \tilde{\mathbf{u}} = \nabla \times \nabla\phi = 0,$$

and hence that acoustic disturbances of a stationary gas are irrotational. In addition, we have

$$\tilde{p} = -\rho_0 \frac{\partial \phi}{\partial t}. \tag{3.14}$$

Substituting these definitions of the velocity and pressure disturbances in terms of the velocity potential into the governing equations then gives

$$\nabla^2\phi = \frac{1}{c^2}\frac{\partial^2\phi}{\partial t^2}, \tag{3.15}$$

so the velocity potential also satisfies the three-dimensional wave equation with wave speed c. Once ϕ is known, $\tilde{\mathbf{u}}$ and $\tilde{p}$ can be calculated. In summary

$$\nabla^2\phi = \frac{1}{c^2}\frac{\partial^2\phi}{\partial t^2}, \quad \tilde{\mathbf{u}} = \nabla\phi, \quad \tilde{p} = -\rho_0\frac{\partial\phi}{\partial t}.$$

3.2 Plane Waves

Perhaps the simplest type of solution of the three-dimensional wave equation is one that varies in a single spatial direction, say the x-direction. The solution then only needs to satisfy the one-dimensional wave equation, and hence is of the form

$$\phi = f(x - ct). \tag{3.16}$$

This is a permanent form travelling wave solution that propagates in the positive x-direction. It is known as a **plane wave**, since the value of ϕ, and hence $\tilde{p}$ and $\tilde{\mathbf{u}}$, is constant in planes perpendicular to the direction of propagation. From the definition of ϕ, we can see that the velocity fluctuation is confined to the x-direction, and is given by

$$\tilde{u} = f'(x - ct), \tag{3.17}$$

whilst the pressure fluctuation is

$$\tilde{p} = \rho_0 c f'(x - ct), \tag{3.18}$$

and hence

$$\tilde{p} = \rho_0 c \tilde{u}. \tag{3.19}$$

This is our first example of a **longitudinal** wave, in which the disturbance of the background medium is in the direction in which the wave propagates. Waves on a stretched string are transverse, since the direction of the displacement of the string is perpendicular to the direction of propagation of the wave. Note that a more general form in which to write a three-dimensional plane wave solution is $f(\mathbf{k}\cdot\mathbf{x} - \omega t)$, where $\mathbf{k}$ is the **wavenumber vector**. The wave propagates in the direction of $\mathbf{k}$ with wave speed $c = \omega/|\mathbf{k}|$ and wavenumber $|\mathbf{k}|$.

We are now in a position to determine the range of validity of the assumptions that we made when we derived the three-dimensional wave equation, using a plane, harmonic wave as an example. We neglected (i) convective accelerations, (ii) viscosity and (iii) gravity. Consider the

x-momentum equation for a plane wave in the x-direction, assuming that gravity acts in the x-direction,

$$\rho\frac{\partial\tilde{u}}{\partial t} + \underset{(1)}{\rho\tilde{u}\frac{\partial\tilde{u}}{\partial x}} = -\frac{\partial\tilde{p}}{\partial x} + \underset{(2)}{\mu\frac{\partial^2\tilde{u}}{\partial x^2}} - \underset{(3)}{\rho g}.$$

(1) **Nonlinear convective accelerations**

In order to neglect term (1) relative to the pressure gradient we need $\rho_0\tilde{u}^2 \ll |\tilde{p}| = |\rho_0 c\tilde{u}|$, and hence $|\tilde{u}| \ll c$. In other words, the velocity fluctuations must have a magnitude much less than the speed of sound. This is often expressed in terms of $M = u_0/c$, where u_0 is the typical amplitude of the fluid velocity fluctuations. M is called the **Mach number**, and we require it to be small for linear acoustics to be a good approximation.

(2) **Viscosity**

In order to neglect term (2) relative to the pressure gradient we need $|\mu\partial^2\tilde{u}/\partial\tilde{x}^2| \ll |\partial\tilde{p}/\partial x|$. If the wavelength of the plane wave is λ, we need $\mu|\tilde{u}|/\lambda^2 \ll \rho_0 c|\tilde{u}|/\lambda$, and hence $\lambda \gg \mu/\rho_0 c$. In air, $\mu/\rho_0 c \approx 3\times 10^{-7}$ m, and hence we require that the wavelength is not too short, $\lambda \gg 3\times 10^{-7}$ m. This is not very restrictive in most situations. Since the frequency, f, of the wave is given by $f = c/\lambda$, we need $f \ll 10^9$ Hz. Audible sounds lie roughly in the range $20\,\text{Hz} < f < 2\times 10^4$ Hz.

(3) **Gravity**

An analysis of when the effect of gravity can be neglected is slightly more difficult, since the background pressure and density are not spatially uniform. At leading order

$$\frac{dp_0}{dx} = -\rho_0 g,$$

where gravity acts in the negative x-direction. By considering a small perturbation of the static state, $p = p_0(x) + \tilde{p}$, $\rho = \rho_0(x) + \tilde{\rho}$, and $\mathbf{u} = \tilde{\mathbf{u}}$, we can show that the pressure perturbation, $\tilde{p}$, satisfies the **modified wave equation**

$$\frac{1}{c^2}\frac{\partial^2\tilde{p}}{\partial t^2} = \nabla^2\tilde{p} + g\frac{d}{dx}\left(\frac{\tilde{p}}{c^2}\right),$$

where

$$c(x) = \left\{\frac{\partial\rho}{\partial p}(p_0(x))\right\}^{-1/2}.$$

By comparing the term due to gravity with the term $\nabla^2\tilde{p}$ we can see that the gravitational term is negligible if the wavelength, λ, satisfies $\lambda \ll \lambda_g$, where $\lambda_g = c^2/g \approx 11$ km in the atmosphere at ground level. The associated frequency is $f_g = g/c \approx 0.03$ Hz, well below the audible range.

3.3 Acoustic Energy Transmission

Sound waves, along with most other types of wave, transmit energy from one place to another. Whenever you speak to someone, the energy that you convert into sound waves is transmitted to their ear, where it moves bones (the **hammer**, **anvil** and **stirrup**) and the sound is heard. It is, therefore, of interest to know at what velocity the energy is transmitted, and also how much is stored in a given wave. For any element of the gas we will show that

$$\frac{d}{dt}(\text{kinetic energy} + \text{potential energy of compression})$$

$$= \text{rate at which pressure forces do work.}$$

We begin by taking the scalar product of the velocity, $\mathbf{u}$, with the equation for conservation of momentum (3.2), as

$$\mathbf{u}\cdot\left(\rho\frac{\partial\mathbf{u}}{\partial t} + \rho\mathbf{u}\cdot\nabla\mathbf{u}\right) = -\mathbf{u}\cdot\nabla p,$$

and hence

$$\rho\mathbf{u}\cdot\frac{\partial\mathbf{u}}{\partial t} + \rho\mathbf{u}\cdot\left\{\frac{1}{2}\nabla(\mathbf{u}\cdot\mathbf{u}) - \mathbf{u}\times(\nabla\times\mathbf{u})\right\} = -\mathbf{u}\cdot\nabla p,$$

using the identity $\mathbf{u}\times(\nabla\times\mathbf{u}) \equiv \frac{1}{2}\nabla(\mathbf{u}\cdot\mathbf{u}) - \mathbf{u}\cdot\nabla\mathbf{u}$. Since $\mathbf{u}\cdot\mathbf{u}\times(\nabla\times\mathbf{u}) = 0$, we have

$$\rho\frac{\partial}{\partial t}\left(\frac{1}{2}|\mathbf{u}|^2\right) + \rho\mathbf{u}\cdot\nabla\left(\frac{1}{2}|\mathbf{u}|^2\right) = -\mathbf{u}\cdot\nabla p,$$

and hence

$$\frac{\partial}{\partial t}\left(\frac{1}{2}\rho|\mathbf{u}|^2\right) - \frac{1}{2}|\mathbf{u}|^2\frac{\partial\rho}{\partial t} + \nabla\cdot\left(\frac{1}{2}\rho|\mathbf{u}|^2\mathbf{u}\right) - \frac{1}{2}|\mathbf{u}|^2\nabla\cdot(\rho\mathbf{u}) = -\nabla\cdot(\mathbf{u}p) + p\nabla\cdot\mathbf{u}.$$

But conservation of mass, given by (3.1), shows that $\partial\rho/\partial t + \nabla.(\rho\mathbf{u}) = 0$, and hence that

$$\nabla\cdot\mathbf{u} = -\frac{1}{\rho}\left(\frac{\partial\rho}{\partial t} + \mathbf{u}\cdot\nabla\rho\right) = -\frac{1}{\rho}\frac{D\rho}{Dt}.$$

Here $D/Dt \equiv \partial/\partial t + \mathbf{u} \cdot \nabla$ is the convective derivative. It measures the rate of change of a property of a given fluid element as it is convected by the flow, rather than the rate of change at a fixed point. This shows that

$$\frac{\partial}{\partial t}\left(\frac{1}{2}\rho|\mathbf{u}|^2\right) + \frac{p}{\rho}\frac{D\rho}{Dt} = -\nabla \cdot \left\{\left(p + \frac{1}{2}\rho|\mathbf{u}|^2\right)\mathbf{u}\right\}.$$

If we now integrate this equation over a fixed volume, V, with surface S, and use the divergence theorem on the right hand side, we obtain

$$\frac{d}{dt}\int_V \frac{1}{2}\rho|\mathbf{u}|^2 dV + \int_V \frac{p}{\rho}\frac{D\rho}{Dt}\,dV = -\int_S \left(p + \frac{1}{2}\rho|\mathbf{u}|^2\right)\mathbf{u}\cdot\mathbf{n}\,dS, \qquad (3.20)$$

where $\mathbf{n}$ is the outward unit normal to the surface S. Equation (3.20) can be expressed as

rate of change of (kinetic energy + potential energy of compression)

= work done by pressure on S + flux of kinetic energy into V.

The only term that requires further explanation is the one labelled 'potential energy of compression'. Consider a given mass, M, of gas, which increases its volume by a small amount ΔV. In doing so, its density changes by a small amount $\Delta\rho$. But the mass of the gas is constant, and hence

$$M = \rho V = (\rho + \Delta\rho)(V + \Delta V) \approx \rho V + V\Delta\rho + \rho\Delta V.$$

This shows that, at leading order, $\Delta V = -V\Delta\rho/\rho$. The work done in changing the volume of the gas is $-p\Delta V = pV\Delta\rho/\rho$. Hence the rate at which the potential energy stored in a given volume of gas changes is given by

$$\int_V \frac{p}{\rho}\frac{D\rho}{Dt}\,dV.$$

We can now consider what (3.20) tells us when we make the acoustic approximation that $\rho = \rho_0 + \tilde{\rho}$, $p = p_0 + \tilde{p}$, $\mathbf{u} = \tilde{\mathbf{u}}$ with $\tilde{\rho}$, $\tilde{p}$ and $\tilde{\mathbf{u}}$ small. At leading order, we find that

$$\int_V \frac{p_0}{\rho_0}\frac{D\tilde{\rho}}{Dt}\,dV = -\int_S p_0\tilde{\mathbf{u}}\cdot\mathbf{n}\,dS.$$

This is just an expression of conservation of mass. We only obtain information concerning the energy associated with the pressure fluctuations at next order. We find that, retaining terms quadratic in small quantities,

$$\frac{d}{dt}\int_V \frac{1}{2}\rho_0|\tilde{\mathbf{u}}|^2\,dV + \int_V \frac{\tilde{p}}{\rho_0}\frac{\partial\tilde{\rho}}{\partial t}\,dV = -\int_S \tilde{p}\tilde{\mathbf{u}}\cdot\mathbf{n}\,dS. \qquad (3.21)$$

Note firstly that the term that represents the flux of kinetic energy on the right hand side of (3.20) does not appear, since it is cubic in small quantities, and secondly that the term $\tilde{p}\tilde{\mathbf{u}} \cdot \mathbf{n}$ in (3.21) indicates that we are interested only in the work done by the pressure fluctuation, $\tilde{p}$, not in the work done by the uniform background pressure, p_0. Since $\tilde{p} = c^2\tilde{\rho}$, by definition, we can write

$$\frac{\tilde{p}}{\rho_0}\frac{\partial \tilde{\rho}}{\partial t} = \frac{c^2\tilde{\rho}}{\rho_0}\frac{\partial \tilde{\rho}}{\partial t} = \frac{\partial}{\partial t}\left(\frac{1}{2}\frac{c^2}{\rho_0}\tilde{\rho}^2\right).$$

This means that we can write (3.21) as

$$\frac{d}{dt}\int_V (K + U)\, dV = -\int_S \mathbf{I} \cdot \mathbf{n}\, dS, \tag{3.22}$$

where

$$K = \frac{1}{2}\rho_0|\tilde{\mathbf{u}}|^2 \tag{3.23}$$

is the kinetic energy per unit volume,

$$U = \frac{c^2\tilde{\rho}^2}{2\rho_0} = \frac{\tilde{p}^2}{2c^2\rho_0} \tag{3.24}$$

is the compressive potential energy per unit volume due to the pressure fluctuation, and

$$\mathbf{I} = \tilde{p}\tilde{\mathbf{u}} \tag{3.25}$$

is the rate of working of the pressure fluctuation at S. This is also equal to the flux of energy per unit area, and is known as the **acoustic intensity**. The acoustic intensity has units of $\mathrm{W\,m^{-2}}$ and is equal to the rate at which the pressure fluctuations work on a unit area of the fluid.

The greater the acoustic intensity in a sound wave, the louder it sounds to the human ear. However, the ear and brain sense equal differences in loudness for equal differences in $\log|\mathbf{I}|$. The loudness of a sound wave is defined as $120 + 10\log_{10}|\mathbf{I}|$ and is measured in **decibels**. The minimum audible level in the 500 to 8000 Hz range is 0 dB, which is equivalent to $10^{-12}\,\mathrm{W\,m^{-2}}$. The typical **acoustic power output** of the human voice is about 10^{-5} W in conversation. The loudness of any sound depends on the distance of the listener from the source of the sound, and decreases logarithmically with distance. Note the difference in units between acoustic intensity, loudness and acoustic power output.

Finally, consider the energy content and transmission rate in a plane wave. From (3.16) to (3.18) we have equipartition of energy, with

$$K = U = \frac{1}{2}\rho_0 f'^2.$$

Hence the total energy at any point in the wave is

$$K + U = \rho_0 f'^2 = \rho_0 \tilde{u}^2,$$

whilst the acoustic intensity in the x-direction is

$$I = \tilde{p}\tilde{u} = \rho_0 c \tilde{u}^2.$$

Hence

$$\frac{\text{Rate of transport of energy per unit area}}{\text{Total energy per unit volume}} = \frac{\rho_0 c \tilde{u}^2}{\rho_0 \tilde{u}^2} = c,$$

so the energy propagates at the wave speed, c. We shall see that this is not true for all types of linear wave when we study water waves.

3.4 Plane Waves In Tubes

Consider a straight tube with a constant cross-sectional area, with the x-axis parallel to its wall. A tightly fitting, flat piston at $x = x_0(t)$ will generate a motion of the gas in the tube. If the velocity of the piston is small relative to the speed of sound, it will generate small amplitude sound waves, and the motion of the gas can be described using the three-dimensional wave equation. Continuity of normal velocity at the surface of the piston gives the boundary condition

$$\tilde{u}_x = \frac{\partial \phi}{\partial x} = \frac{dx_0}{dt}, \quad \text{at } x = x_0(t).$$

If we assume that the face of the piston remains close to $x = 0$ and undergoes motions with a small amplitude, we can linearise this boundary condition as

$$\frac{\partial \phi}{\partial x} = \frac{dx_0}{dt}, \quad \text{at } x = 0. \tag{3.26}$$

The remaining boundary condition is given by continuity of normal velocity as

$$\frac{\partial \phi}{\partial n} = 0 \quad \text{at the tube walls.} \tag{3.27}$$

We also need a **radiation condition**, which arises because we are considering a situation where waves are generated by the motion of the piston, and do not expect any incoming waves from infinity. We can therefore look for a plane wave solution of the form $\phi = f(x - ct)$, since this has no spatial variation perpendicular to the walls of the tube, and thus satisfies (3.27), and represents a right-travelling wave, thereby satisfying the radiation condition. The solution that satisfies (3.26) is

$$\phi = -cx_0\left(t - \frac{x}{c}\right). \tag{3.28}$$

The piston therefore generates a plane acoustic wave that propagates down the tube without change of form.

We can now consider some properties of such a plane wave. The volume flux of gas at any point in the wave is $Q = A\tilde{u}$, where A is the constant cross-sectional area of the tube and, since $\tilde{p} = \rho_0 c\tilde{u}$,

$$Q = Y\tilde{p}, \tag{3.29}$$

where

$$Y = \frac{A}{\rho_0 c}$$

is called the **characteristic admittance** of the tube. The impedance of the tube is $1/Y$. The energy flux down the tube is

$$IA = \tilde{p}\tilde{u}A = \tilde{p}Q = Y\tilde{p}^2.$$

The characteristic admittance is a property of the tube and the gas, but not of the particular wave, and is therefore a useful way of describing the system.

What happens when a plane wave is transmitted down a tube and meets a junction with a different tube, possibly containing a different gas? Let's assume that the junction is at $x = 0$, and that a wave propagates in the positive x-direction and is incident upon the junction. We expect that there will be a plane wave transmitted into the second tube, and a plane wave reflected into the first tube, propagating in the negative x-direction, as shown in figure 3.2. We can express this mathematically by looking for a solution as a sum of the three plane waves, incident, reflected and transmitted, in the form

$$\tilde{p} = \left\{ \begin{array}{ll} f_{\mathrm{I}}(x - c_1 t) + f_{\mathrm{R}}(x + c_1 t), & \text{for } x < 0, \\ f_{\mathrm{T}}(x - c_2 t), & \text{for } x > 0. \end{array} \right\} \tag{3.30}$$

This analysis is completely equivalent to that for the reflection of a wave on a string by a change in density, which we studied in section 2.4. In order to determine the amplitude of the reflected and transmitted waves

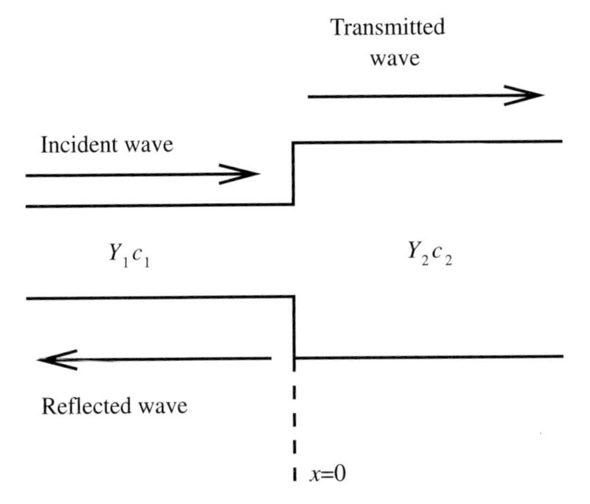

Fig. 3.2. Incident, reflected and transmitted plane waves at a junction between two tubes.

relative to that of the incident wave we note that the pressure must be continuous at the junction, $x = 0$, so that

$$f_{\mathrm{I}}(-c_1 t) + f_{\mathrm{R}}(c_1 t) = f_{\mathrm{T}}(-c_2 t),$$

and that the flux of gas must also be continuous at $x = 0$, so that

$$Y_1 f_{\mathrm{I}}(-c_1 t) - Y_1 f_{\mathrm{R}}(c_1 t) = Y_2 f_{\mathrm{T}}(-c_2 t).$$

Note that the reflected wave transports gas in the negative x-direction, hence the minus sign. Solving these two equations leads to

$$f_{\mathrm{R}}(c_1 t) = \frac{Y_1 - Y_2}{Y_1 + Y_2} f_{\mathrm{I}}(-c_1 t), \quad f_{\mathrm{T}}(-c_2 t) = \frac{2Y_1}{Y_1 + Y_2} f_{\mathrm{I}}(-c_1 t).$$

In terms of the actual reflected and transmitted waves,

$$f_{\mathrm{R}}(x + c_1 t) = \frac{Y_1 - Y_2}{Y_1 + Y_2} f_{\mathrm{I}}(-x - c_1 t),$$

$$f_{\mathrm{T}}(x - c_2 t) = \frac{2Y_1}{Y_1 + Y_2} f_{\mathrm{I}}\left(\frac{c_1}{c_2}(x - c_2 t)\right).$$

The reflected and transmitted waves therefore have the same shape as the incident wave, but with modified amplitudes. We also note that we expect that the flux of energy due to the incident wave is split between the reflected and transmitted waves. A little algebra does indeed show that $Y_1 f_{\mathrm{I}}^2 = Y_1 f_{\mathrm{R}}^2 + Y_2 f_{\mathrm{T}}^2$.

Note that, as for the interaction of an incident wave with a density discontinuity in a string,

(i) when $Y_1 = Y_2$ there is no reflected wave, as you should expect,
(ii) when $Y_2 \gg Y_1$ (a small tube carries the wave into a large tube) the transmitted wave is small,
(iii) when $Y_1 \gg Y_2$ (a large tube carries the wave into a small tube) the transmitted wave has twice the amplitude of the reflected wave. This is perhaps the most surprising result, but of course the transmitted wave only carries a small amount of energy in this case.

We can also consider plane, standing waves in tubes of finite length. These can be generated in organ pipes and flutes. Consider a straight tube of length L with one end open and one end closed. At the closed end the axial velocity of the gas must be zero, whilst at the other end, which is open to the atmosphere, the fluctuation in the pressure must be zero. One way of seeing this is to consider the above analysis, which shows that at $x = 0$, $\tilde{p} \to 0$ as $Y_2/Y_1 \to \infty$. We seek a solution of the one-dimensional wave equation with boundary conditions

$$\tilde{u} = \frac{\partial \phi}{\partial x} = 0 \quad \text{at } x = 0, \tag{3.31}$$

and

$$\tilde{p} = 0 \Rightarrow \frac{\partial \phi}{\partial t} = 0 \quad \text{at } x = L. \tag{3.32}$$

If we look for separable solutions we obtain

$$\phi = \phi_n = A \cos\left(\frac{\omega x}{c}\right) \cos \omega_n t, \tag{3.33}$$

with

$$\omega_n = \frac{c}{L}\left(n - \frac{1}{2}\right)\pi, \quad n = 1, 2, \ldots. \tag{3.34}$$

We can therefore express the solution as a sum of Fourier components, which represent the fundamental and the various harmonics, just as we did for a finite vibrating string in chapter 2. We conclude that plane waves in straight walled tubes behave very much like stretched strings, with similar musical properties.

What about musical instruments that are not straight uniform tubes? Consider a tube with cross-sectional area $A(x)$. We restrict our attention to the case where the wavelength of the disturbance in the tube is much longer than a typical radius of the tube, and where $A(x)$ varies on a length scale much longer than this wavelength. In a narrow, slowly varying tube like this, for example a clarinet or oboe, we can assume that locally the disturbance is a plane wave.

If we consider a small section of the tube with length Δx, the rate of change of mass is the flow in minus the flow out,

$$\frac{\partial}{\partial t}(\rho A \Delta x) = (\rho A u)_x - (\rho A u)_{x+\Delta x},$$

and hence in the limit $\Delta x \to 0$,

$$\frac{\partial}{\partial t}(A\rho) = -\frac{\partial}{\partial x}(A\rho u).$$

If we now linearise this we obtain

$$A\frac{\partial \tilde{p}}{\partial t} = -\rho_0 \frac{\partial}{\partial x}(A\tilde{u}). \tag{3.35}$$

This is the analogue of (3.6) when variation only occurs in the x-direction, but with a changing cross-sectional area. Similarly, the linearised equation for conservation of momentum is

$$\rho_0 \frac{\partial \tilde{u}}{\partial t} = -\frac{\partial \tilde{p}}{\partial x}, \tag{3.36}$$

which is the analogue of (3.7). Combining these two equations gives us the modified wave equation

$$\frac{1}{c^2}\frac{\partial^2 \tilde{p}}{\partial t^2} = \frac{1}{A}\frac{\partial}{\partial x}\left(A\frac{\partial \tilde{p}}{\partial x}\right). \tag{3.37}$$

As an example, consider the **exponential horn**, $A(x) = A_0 \exp(\alpha x)$. In this case, (3.37) becomes

$$\frac{1}{c^2}\frac{\partial^2 \tilde{p}}{\partial t^2} = \frac{\partial^2 \tilde{p}}{\partial x^2} + \alpha\frac{\partial \tilde{p}}{\partial x}. \tag{3.38}$$

We can look for plane, harmonic wave solutions of the form

$$\tilde{p} = a(x)\exp(i\omega t),$$

remembering that this notation implies that we take the real part of the solution. Substituting this form into (3.38) gives

$$\frac{d^2 a}{dx^2} + \alpha\frac{da}{dx} + \frac{\omega^2}{c^2}a = 0,$$

and hence

$$a \propto \exp\left\{-i\left(\frac{\omega^2}{c^2} - \frac{1}{4}\alpha^2\right)^{1/2} x - \frac{1}{2}\alpha x\right\},$$

for waves propagating in the positive x-direction. If $\omega > \frac{1}{2}\alpha c$ the waves propagate with speed

$$c_0(\omega) = \omega\left(\frac{\omega^2}{c^2} - \frac{1}{4}\alpha^2\right)^{-1/2},$$

and amplitude proportional to $\exp\left(-\frac{1}{2}\alpha x\right) = A^{-1/2}$. In other words, the amplitude decays (we say the wave is **attenuated**) as x increases. If $\omega < \frac{1}{2}\alpha c$ there is no propagation, just attenuation, since

$$a \propto \exp\left[-\left\{\frac{1}{2}\alpha + \left(\frac{1}{4}\alpha^2 - \frac{\omega^2}{c^2}\right)^{1/2}\right\}x\right].$$

These are known as **evanescent waves**. The **cut-off frequency**, $\omega = \frac{1}{2}\alpha c$, is the critical frequency below which no plane wave propagation is possible. We will meet a similar phenomenon in a different context in the next section.

3.5 Acoustic Waveguides

We have just seen how plane acoustic waves can propagate along tubes. Is there any other type of acoustic wave that that can propagate along a tube? In order to answer this question, we begin by considering the effect of a flat, rigid wall on an incident plane wave.

3.5.1 *Reflection of a Plane Acoustic Wave by a Rigid Wall*

Consider a plane wave, with amplitude A_I and frequency ω_I, incident on a rigid, plane wall at $x = 0$. If we take the y-axis to lie perpendicular to the direction of propagation of the wave, the wavenumber vector, which points in the direction of propagation of the wave, is

$$\mathbf{k}_\mathrm{I} = \left(-\frac{\omega_\mathrm{I}}{c}\cos\theta_\mathrm{I}, 0, \frac{\omega_\mathrm{I}}{c}\sin\theta_\mathrm{I}\right).$$

The potential of the incident wave is

$$\phi_\mathrm{I} = A_\mathrm{I}\exp\{i(\omega_\mathrm{I}t - \mathbf{k}_\mathrm{I}\cdot\mathbf{x})\} = A_\mathrm{I}\exp\left\{i\omega_\mathrm{I}\left(t - \sin\theta_\mathrm{I}\frac{z}{c} + \cos\theta_\mathrm{I}\frac{x}{c}\right)\right\},$$

where θ_I is the **angle of incidence**, as shown in figure 3.3.

We expect there to be another plane acoustic wave reflected from the rigid wall, and can satisfy the boundary condition that

$$\frac{\partial\phi}{\partial x} = 0 \text{ at } x = 0, \tag{3.39}$$

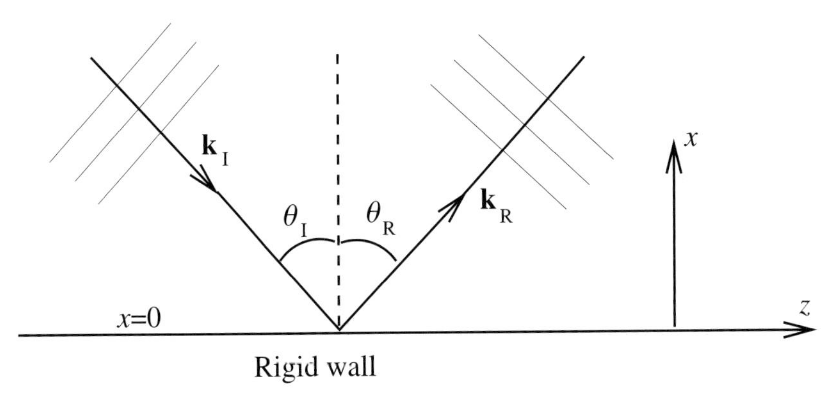

Fig. 3.3. Incident and reflected plane, acoustic waves at a rigid wall.

by seeking a solution of the form

$$\phi = \phi_{\mathrm{I}} + \phi_{\mathrm{R}},$$

where

$$\phi_{\mathrm{R}} = A_{\mathrm{R}} \exp\left\{i\omega_{\mathrm{R}}\left(t - \sin\theta_{\mathrm{R}}\frac{z}{c} - \cos\theta_{\mathrm{R}}\frac{x}{c}\right)\right\}$$

is the potential of the reflected wave and θ_{R} is the **angle of reflection**. Now (3.39) becomes

$$A_{\mathrm{I}}\cos\theta_{\mathrm{I}}\exp\left\{i\omega_{\mathrm{I}}\left(t - \sin\theta_{\mathrm{I}}\frac{z}{c}\right)\right\} - A_{\mathrm{R}}\cos\theta_{\mathrm{R}}\exp\left\{i\omega_{\mathrm{R}}\left(t - \sin\theta_{\mathrm{R}}\frac{z}{c}\right)\right\} = 0.$$

This can only hold for all z and t if the functional forms of the incident and reflected waves are identical, and hence $\omega_{\mathrm{I}} = \omega_{\mathrm{R}}$, $A_{\mathrm{I}} = A_{\mathrm{R}}$ and $\theta_{\mathrm{I}} = \theta_{\mathrm{R}}$. In other words, the wave is reflected without change of frequency or amplitude, and the angle of incidence equals the angle of reflection.

3.5.2 A Planar Waveguide

Consider a region bounded by two rigid, parallel, plane walls at $x = 0$ and $x = d$. What types of acoustic wave can propagate in this **planar waveguide**, assuming that there is no variation in the y-direction? We can answer the question using what we have just found out about the reflection of plane waves. We already know that a plane wave can propagate along the waveguide with wavenumber vector in the z-direction. We can also imagine a plane wave propagating in a different direction, being continually reflected from the walls, bouncing back and

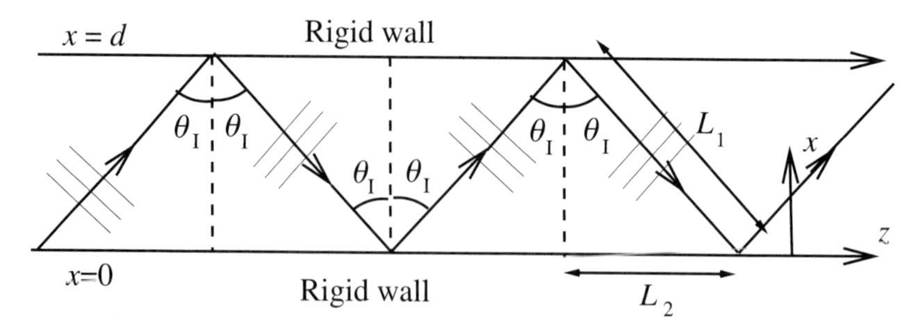

Fig. 3.4. Plane waves zig-zagging along a planar waveguide.

forth along the guide, with angle of incidence and reflection, $\theta_{\rm I}$, as shown in figure 3.4. This can still be represented as the sum of two plane waves,

$$\phi = A \exp\left\{i\omega\left(t - \sin\theta_{\rm I}\frac{z}{c} + \cos\theta_{\rm I}\frac{x}{c}\right)\right\}$$
$$+ A \exp\left\{i\omega\left(t - \sin\theta_{\rm I}\frac{z}{c} - \cos\theta_{\rm I}\frac{x}{c}\right)\right\}.$$

However, we also need to satisfy

$$\frac{\partial\phi}{\partial x} = 0 \text{ at } x = d,$$

and hence

$$\exp\left(i\cos\theta_{\rm I}\frac{\omega d}{c}\right) = \exp\left(-i\cos\theta_{\rm I}\frac{\omega d}{c}\right).$$

This implies that

$$\cos\theta_{\rm I}\frac{\omega d}{c} = -\cos\theta_{\rm I}\frac{\omega d}{c} + 2n\pi,$$

for $n = 0, 1, 2, \ldots$, and hence

$$\cos\theta_{\rm I} = \frac{n\pi c}{\omega d}.$$

Since $\cos\theta_{\rm I} \leq 1$, there is a discrete set of $n_{\max}$ angles $\theta_{\rm I}$ at which a zig-zagging plane wave can propagate down the waveguide, where $n_{\max}$ is the largest positive integer such that $n_{\max} \leq \omega d/\pi c$. Each of these is called one of the **modes**, or **wave modes**, of the waveguide.

We can write the acoustic potential for the nth mode as

$$\phi_n = 2A\cos\frac{n\pi x}{d}e^{i(\omega t - k_n z)}, \text{ for } n = 0, 1, 2, \ldots, \quad (3.40)$$

where

$$k_n = \sqrt{\frac{\omega^2}{c^2} - \frac{n^2\pi^2}{d^2}}, \tag{3.41}$$

is the wavenumber in the z-direction. Note that the solution with $n = 0$ represents a plane wave propagating in the z-direction with speed c. For $0 < n < n_{\text{max}}$ the wave has the character of a standing wave in the x-direction whilst propagating in the z-direction with wavenumber $k_n < \omega/c$. For $n > n_{\text{max}}$, even though this solution cannot be described physically in terms of the sum of two plane waves, (3.40) remains valid, with k_n purely imaginary. The solution decays exponentially as $z \to \infty$, so these are evanescent wave modes. Alternatively, for fixed wavenumber, k, a wave can only propagate along the waveguide at frequencies greater than the cut-off frequency,

$$\omega_{\text{c}} = \frac{\pi c}{d},$$

and there is a sequence of **eigenfrequencies**, or **resonant frequencies**, at which the gas can be excited, given by

$$\omega_n = c\sqrt{k^2 + \frac{n^2\pi^2}{d^2}}.$$

In the direction of propagation, the wave crests of the nth mode move at the phase speed, $c_{\text{p}} = \omega_n/k > c$. Since the mode is composed of two plane waves propagating with angle of incidence θ_{I}, the distance between the wavecrests is increased from the free space wavelength, λ, to $\lambda/\sin\theta_{\text{I}}$. The frequency is unchanged, and hence the phase speed increases to $c/\sin\theta_{\text{I}}$. The energy in each mode is propagating by bouncing back and forth between the rigid walls. The distance between 'bounces' is $L_1 = d/\cos\theta_{\text{I}}$, as shown in figure 3.4, which the energy traverses in a time $T = L_1/c = d/c\cos\theta_I$. However, in this time, the distance the energy has propagated in the z-direction is $L_2 = d\tan\theta_I$. The velocity at which energy propagates along the waveguide, the group velocity, is given by $c_{\text{g}} = L_2/T = c\sin\theta_I$ and is less than the free space wave speed, c. Moreover, $c_{\text{g}}c_{\text{p}} = c^2$, which we could also have deduced from the formula, $c_{\text{g}} = d\omega_n/dk$.

3.5.3 A Circular Waveguide

A stethoscope consists of a pair of circular tubes through each of which the acoustic waves generated by the patient are propagated to an ear of the doctor. We can crudely model this by considering a rigid, circular

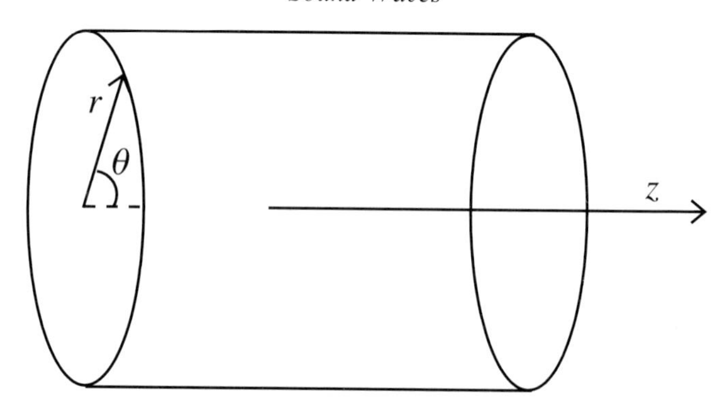

Fig. 3.5. The cylindrical polar coordinate system for the circular waveguide.

cylinder whose interior is given, in cylindrical polar coordinates, by $0 \le r \le a$, $-\infty < z < \infty$, $0 \le \theta < 2\pi$, as shown in figure 3.5. In this case, we are unable immediately to write down the solution based on our knowledge of the physics of the problem. However, as we shall see, the solution has all of the qualitative features of the simpler planar solution, (3.40).

In terms of cylindrical polar coordinates, the three-dimensional wave equation is

$$\nabla^2\phi = \frac{\partial^2\phi}{\partial r^2} + \frac{1}{r}\frac{\partial\phi}{\partial r} + \frac{\partial^2\phi}{\partial z^2} + \frac{1}{r^2}\frac{\partial^2\phi}{\partial\theta^2} = \frac{1}{c^2}\frac{\partial^2\phi}{\partial t^2}. \tag{3.42}$$

Since there can be no normal velocity at the solid walls of the tube, we have

$$\frac{\partial\phi}{\partial r} = 0 \quad \text{at } r = a. \tag{3.43}$$

Although non-axisymmetric solutions exist, we will restrict our attention to axisymmetric solutions and look for separable solutions of the form $\phi(r, z, t) = R(r)Z(z)T(t)$. If we substitute this into (3.42) and rearrange, we obtain the first separation,

$$\frac{R''}{R} + \frac{1}{r}\frac{R'}{R} + \frac{Z''}{Z} = \frac{1}{c^2}\frac{T''}{T} = -\frac{\omega^2}{c^2}. \tag{3.44}$$

This choice of separation constant, $-\omega^2/c^2$, makes sense when we note that $T'' + \omega^2 T = 0$, and hence

$$T = \exp(\pm i\omega t).$$

We can now rearrange (3.44) to obtain the second separation

$$\frac{R''}{R} + \frac{1}{r}\frac{R'}{R} = -\frac{\omega^2}{c^2} - \frac{Z''}{Z} = -\frac{\omega^2}{c^2} + k^2. \tag{3.45}$$

Our choice of separation constant gives $Z'' + k^2 Z = 0$, and hence

$$Z = \exp(\pm ikz).$$

This shows that k is the wavenumber for waves propagating in the z-direction. Finally, (3.45) shows that

$$R'' + \frac{1}{r}R' + \left(\frac{\omega^2}{c^2} - k^2\right) R = 0. \tag{3.46}$$

When $\omega^2 < k^2c^2$ it can be shown that there is no bounded solution that satisfies (3.43). If we now assume that $\omega^2 > k^2c^2$ and let

$$\bar{r} = \left(\frac{\omega^2}{c^2} - k^2\right)^{1/2} r,$$

equation (3.46) becomes

$$\frac{d^2R}{d\bar{r}^2} + \frac{1}{\bar{r}}\frac{dR}{d\bar{r}} + R = 0, \tag{3.47}$$

which is Bessel's equation of order zero. The general solution is therefore

$$R = AJ_0(\bar{r}) + BY_0(\bar{r}),$$

for arbitrary constants A and B. However, as $\bar{r} \to 0$, $Y_0 \to -\infty$, so the only solution that is bounded at $\bar{r} = 0$ is

$$R = AJ_0\left(\left(\frac{\omega^2}{c^2} - k^2\right)^{1/2} r\right).$$

The Bessel function of order zero, $J_0(r)$, is illustrated in figure 3.6. We also need to satisfy the boundary condition that $\partial\phi/\partial r = 0$ at $r = a$. Therefore we must choose k so that

$$J_0'\left(\left(\frac{\omega^2}{c^2} - k^2\right)^{1/2} a\right) = 0,$$

and hence

$$\left(\frac{\omega^2}{c^2} - k^2\right)^{1/2} a = m_n, \quad n = 0, 1, 2, \ldots,$$

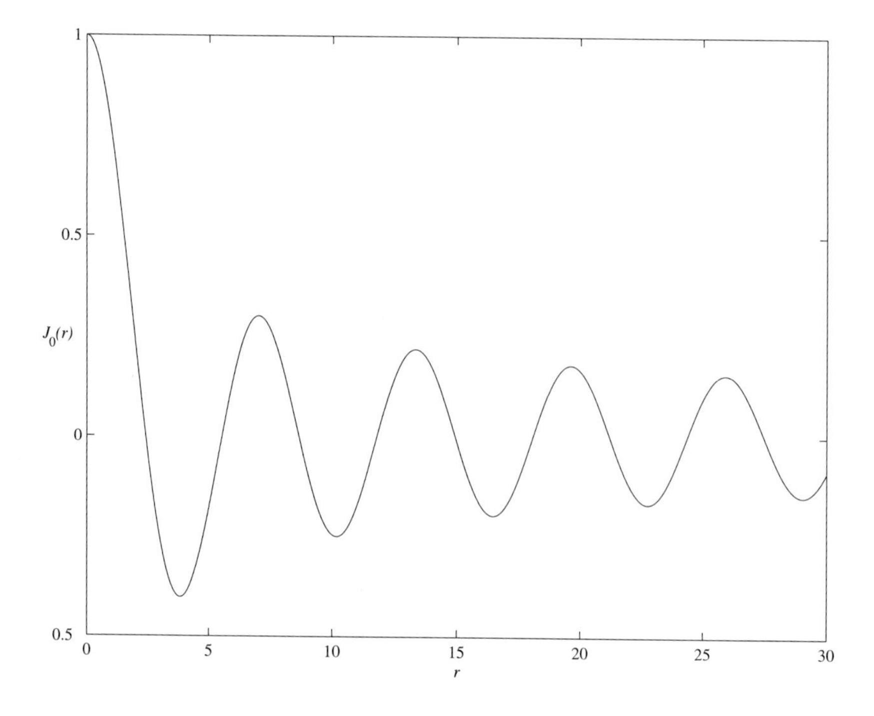

Fig. 3.6. The Bessel function of order zero, $J_0(\bar{r})$.

where m_n are the turning points of J_0. Note that $m_0 = 0$. For a given angular frequency, ω,

$$k_n = \sqrt{\frac{\omega^2}{c^2} - \frac{m_n^2}{a^2}}, \quad n = 0, 1, 2, \ldots. \tag{3.48}$$

Each value of n corresponds to a different wave mode, represented by a solution of the form

$$\phi_n(r, z, t) = \exp\{-i(\omega t - k_n z)\} J_0(m_n r/a), \tag{3.49}$$

for waves propagating in the positive x-direction. Apart from the quantitative differences in the functional form of the spatial variation across the waveguide, and the value of the eigenfrequencies, the qualitative form of the solution is as expected from our analysis of the planar waveguide. In particular, the lowest order mode, with $n = 0$, is a plane wave propagating in the z-direction, and is the mode that carries most of the acoustic energy in a stethoscope. The next order mode, with $n = 1$, has $m_1 \approx 3.831$ and varies radially, as shown in figure 3.7. The energy in the nth order mode propagates at the group velocity, $c_g = d\omega_n/dk < c$, whilst the wave

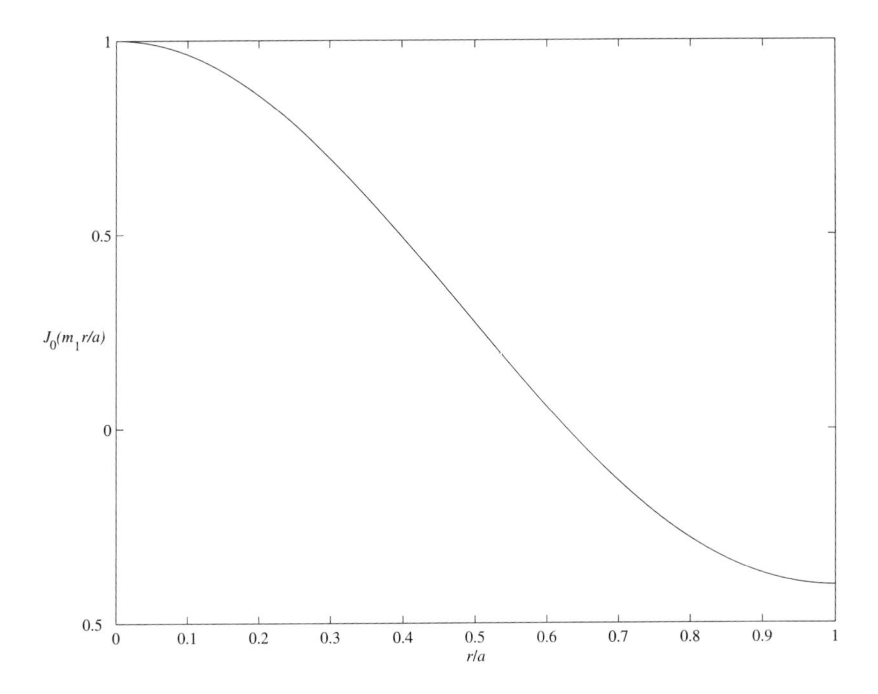

Fig. 3.7. The radial variation of the first order mode, with $n = 1$.

crests propagate at the phase velocity, $c_{\mathrm{p}} = \omega_n/k$. A simple differentiation shows that, as with the planar waveguide, $c_{\mathrm{g}}c_{\mathrm{p}} = c^2$.

All of the above analysis is also relevant to electromagnetic waves, in particular microwaves, which in practice are transmitted in waveguides as we shall see in section 6.8. The microwaves must have a frequency above the cut-off frequency if they are to propagate as anything other than plane waves.

3.6 Acoustic Sources

Most sources of sound do not transmit the same amount of energy in each direction. We say that there is a **directivity** to the sound generated, and this directivity usually depends upon its frequency. As an example, figure 3.8 show the directivity of the sound generated by a violinist.

Although it is not possible to write down a general solution of the three-dimensional wave equation, in many situations the solution, and its associated directivity, can be represented in terms of simple solutions

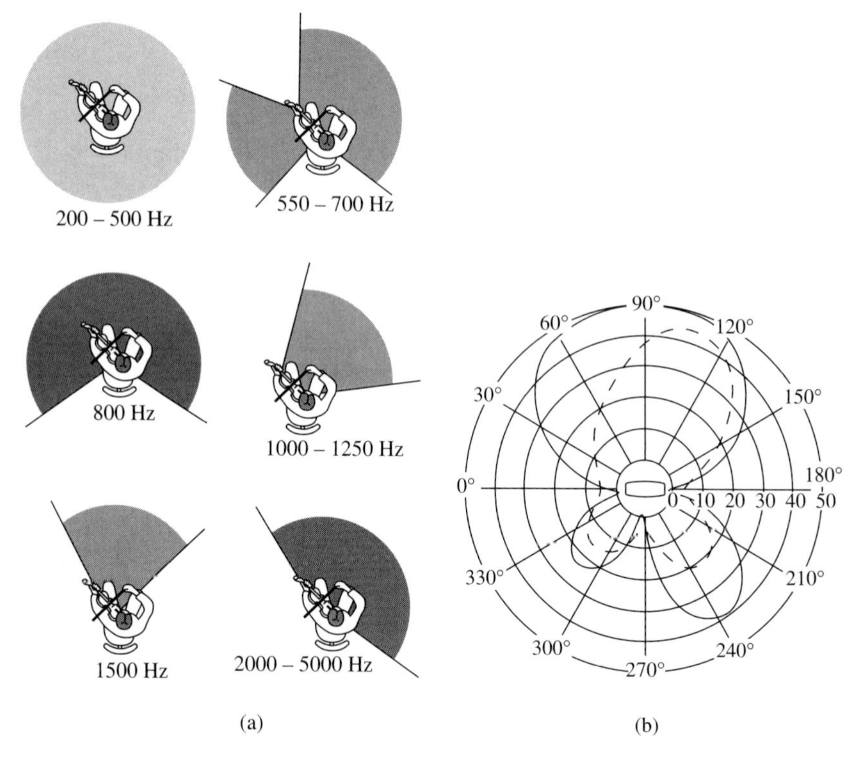

Fig. 3.8. The directivity of the sound generated by a violinist (a) in a horizontal plane at various frequencies, (b) in a longitudinal plane.

called sources. We will now construct these solutions and examine their properties.

3.6.1 The Acoustic Source

A three-dimensional, spherically symmetric acoustic potential satisfies the equation

$$\frac{\partial^2 \phi}{\partial r^2} + \frac{2}{r}\frac{\partial \phi}{\partial r} = \frac{1}{r}\frac{\partial^2}{\partial r^2}(r\phi) = \frac{1}{c^2}\frac{\partial^2 \phi}{\partial t^2},$$

and hence $r\phi$ satisfies the one-dimensional wave equation. Spherically symmetric waves incident from infinity are physically impossible, so the only meaningful solution is of the form

$$\phi = -\frac{1}{r} f\left(t - \frac{r}{c}\right). \tag{3.50}$$

This leads to a purely radial velocity field, with magnitude

$$u = \frac{\partial \phi}{\partial r} = \frac{1}{r^2} f\left(t - \frac{r}{c}\right) + \frac{1}{rc} \dot{f}\left(t - \frac{r}{c}\right), \tag{3.51}$$

where a dot denotes differentiation with respect to the argument of the function. This velocity field is singular as $r \to 0$. We can also determine the mass flux across a sphere of radius r as

$$4\pi r^2 \rho_0 u = 4\pi \rho_0 f\left(t - \frac{r}{c}\right) + 4\pi \frac{r}{c} \rho_0 \dot{f}\left(t - \frac{r}{c}\right).$$

As $r \to 0$ this mass flux tends to $q(t) = 4\pi\rho_0 f(t)$. We therefore write

$$\phi = -\frac{1}{4\pi\rho_0 r} q\left(t - \frac{r}{c}\right), \tag{3.52}$$

as the standard form for the velocity potential of an **acoustic point source**. This is also referred to as a **retarded potential**, since the behaviour of ϕ at any point is determined by the behaviour of the mass flux, q, a time r/c earlier. This reflects the finite time required for information to propagate at the wave speed, c. The corresponding pressure fluctuation is

$$\tilde{p} = -\rho_0 \frac{\partial \phi}{\partial t} = \frac{1}{4\pi r} \dot{q}\left(t - \frac{r}{c}\right). \tag{3.53}$$

The rate of change of the mass flux, $\dot{q}(t)$, is called the **strength** of the source, since it is this that affects the pressure fluctuations.

In a time harmonic source, with $q = q_0 \exp(i\omega t)$,

$$\tilde{p} = \frac{q_0}{4\pi r} i\omega \exp\left\{i\omega\left(t - \frac{r}{c}\right)\right\},$$

$$u = \frac{q_0}{4\pi\rho_0}\left(\frac{1}{r^2} + \frac{i\omega}{rc}\right) \exp\left\{i\omega\left(t - \frac{r}{c}\right)\right\}.$$

For sufficiently small values of r, specifically $r \ll c/\omega = \lambda/2\pi$, in other words at distances much less than a wavelength from the source,

$$u \approx \frac{q(t)}{4\pi\rho_0 r^2}.$$

This is just the expression for a simple point source in an incompressible fluid. Note that this is completely out of phase with the pressure fluctuation (if $u \propto \cos\omega t$, $\tilde{p} \propto \sin\omega t$), since

$$\tilde{p} \approx \frac{i\omega q(t)}{4\pi r}.$$

This is known as the **near field** of the acoustic source. The **far field**, where $r \gg c/\omega$, is very different in both phase and amplitude, since

$$u \approx \frac{i\omega q_0}{4\pi\rho_0 cr} \exp\left\{i\omega\left(t - \frac{r}{c}\right)\right\}.$$

This is in phase with the pressure fluctuation.

As an example, consider a spherical loudspeaker, which we can model as a spherically pulsating, solid sphere with radius $a(t) = a_0\{1 + \epsilon\exp(i\omega t)\}$. We assume that $\epsilon \ll 1$, and hence that at leading order we can apply the boundary condition at the undisturbed position of the sphere as

$$\frac{\partial\phi}{\partial r} = a_0\epsilon i\omega\exp(i\omega t) \quad \text{at } r = a_0. \tag{3.54}$$

The appropriate, outgoing, spherically symmetric solution of the three-dimensional wave equation is

$$\phi = -\frac{1}{r}f\left(t - \frac{r}{c}\right),$$

and to satisfy the boundary condition, we must have

$$\frac{1}{a_0^2}f\left(t - \frac{a_0}{c}\right) + \frac{1}{a_0 c}\dot{f}\left(t - \frac{a_0}{c}\right) = a_0\epsilon i\omega\exp(i\omega t).$$

If we now let $\tau = t - a_0/c$, we have

$$\frac{df}{d\tau} + \frac{c}{a_0}f = \epsilon a_0^2 i\omega c\exp\left\{i\omega\left(\tau + \frac{a_0}{c}\right)\right\},$$

and hence that

$$f = \frac{\epsilon a_0^2 i\omega c}{i\omega + \frac{c}{a_0}}\exp\left\{i\omega\left(\tau + \frac{a_0}{c}\right)\right\} + C\exp(-c\tau/a_0),$$

where C is a constant of integration. Since we are interested in solutions that are harmonic at all times, we want ϕ to be bounded as $\tau \to -\infty$, and hence set $C = 0$. This finally gives us the acoustic potential for the disturbance created by the loudspeaker as

$$\phi = -\frac{\epsilon a_0^2 i\omega c}{r\left(i\omega + \frac{c}{a_0}\right)}\exp\left\{i\omega\left(t - \frac{r - a_0}{c}\right)\right\}. \tag{3.55}$$

Now consider the case where $a_0 \ll c/\omega$. The spherical loudspeaker then has a radius much smaller than one wavelength of the sound generated, and is said to be **acoustically compact**. Far from the loudspeaker, where $r \gg a_0$,

$$\phi \approx -\frac{1}{r}\epsilon a_0^3 i\omega \exp\left\{i\omega\left(t-\frac{r}{c}\right)\right\}.$$

However, note that the volume of the sphere is

$$V(t) = \frac{4}{3}\pi a_0^3\{1+\epsilon \exp(i\omega t)\}^3,$$

and hence that the mass flux generated by the pulsation of the sphere is

$$q(t) = \rho_0 \dot{V} = 4\pi\rho_0 a_0^3 i\omega\epsilon\{1+\epsilon\exp(i\omega t)\}^2 \exp(i\omega t) \approx 4\pi\rho_0 a_0^3 i\omega\epsilon \exp(i\omega t).$$

This means that in the far field of an acoustically compact spherical loudspeaker,

$$\phi \approx -\frac{1}{4\pi\rho_0 r} q\left(t-\frac{r}{c}\right).$$

In other words, the far field is approximately that of a simple acoustic source with strength given by the rate of mass flux due to the volume change of the sphere. Although we will not pursue this further here, it can be shown that this is true for the far field of an arbitrary, acoustically compact source.

For a one-dimensional source, which generates plane waves, the excess pressure is proportional to the volume flux, $Q(t)$, as given by (3.29), and hence is also proportional to the mass flux $q(t) = \rho_0 Q(t)$. For a three-dimensional source we have shown that the excess pressure is proportional to the rate of change of mass flux, $\dot{q}(t)$, equation (3.53). Consider a one-dimensional acoustic source with mass flux per unit area of the form

$$q(t) = \frac{1}{\sqrt{1+t^2}}. \tag{3.56}$$

We can compare this with a three-dimensional acoustic source with mass flux $q(t)$. The qualitative form of the excess pressure at some fixed distance from the source is shown in figure 3.9. In one dimension this positive pulse of mass flux generates a positive pulse of pressure. In three dimensions the medium recoils, and a symmetrical negative fluctuation in the excess pressure is generated. This reflects the extra freedom that the background medium has in which to move in three dimensions.

What happens in two dimensions? The easiest way of generating a two-dimensional acoustic source is to consider a three-dimensional source of strength $\dot{q}(t)dz$ at the point $(0, 0, z)$ on the z-axis. The excess pressure at a point in the (x, y)-plane a distance $r = \sqrt{x^2 + y^2}$ from the origin is

$$d\tilde{p} = \frac{1}{4\pi\sqrt{r^2 + z^2}}\dot{q}\left(t - \frac{\sqrt{r^2 + z^2}}{c}\right) dz.$$

If we now consider a uniform line of three-dimensional sources along the z-axis, we can write its effect in terms of an integral of elementary sources, so that

$$\tilde{p} = \frac{1}{4\pi}\int_{-\infty}^{\infty} \frac{1}{\sqrt{r^2 + z^2}}\dot{q}\left(t - \frac{\sqrt{r^2 + z^2}}{c}\right) dz. \tag{3.57}$$

This line distribution of sources represents an axisymmetric solution, independent of z, which represents a two-dimensional acoustic source with mass flux per unit length $q(t)$, and generates cylindrical waves. This expression is clearly far more complicated than those for the equivalent one- and three-dimensional cases, (3.29) and (3.53). The excess pressure due to a two-dimensional source with mass flux given by (3.56) is also illustrated in figure 3.9. The qualitative form lies somewhere between that for a one-dimensional source and that for a three-dimensional source. There is some recoil, with an associated negative excess pressure, but not as much as in three dimensions.

3.6.2 Energy Radiated by Acoustic Sources and Plane Waves

The acoustic intensity vector, $\mathbf{I} = \tilde{p}\tilde{\mathbf{u}}$, for an acoustic source is purely radial, with

$$I_{\mathrm{r}} = \frac{q(t - r/c)\dot{q}(t - r/c)}{16\pi^2\rho_0 r^3} + \frac{\dot{q}^2(t - r/c)}{16\pi^2\rho_0 c r^2}. \tag{3.58}$$

The first term is dominant in the near field, whilst the second term determines the leading order behaviour in the far field. Consider a harmonic wave, with $q = q_0 \exp(i\omega t)$. We must remember to take the real part of this *before* forming any products, since $\mathrm{Re}(z_1 z_2) \neq \mathrm{Re}(z_1)\mathrm{Re}(z_2)$. In this case

$$I_{\mathrm{r}} = \frac{-q_0^2\omega^2 \sin\{\omega(t - r/c)\}\cos\{\omega(t - r/c)\}}{16\pi^2\rho_0 r^3} + \frac{q_0^2\omega^2 \sin^2\{\omega(t - r/c)\}}{16\pi^2\rho_0 c r^2}.$$

We can now calculate the average acoustic intensity over one period of the motion, from $t = t_0$ to $t = t_0 + 2\pi/\omega$, which we write as $\langle I_{\mathrm{r}}\rangle$. Since

$$\frac{\omega}{2\pi}\int_{t_0}^{t_0+\frac{2\pi}{\omega}} \sin\left\{\omega\left(t - \frac{r}{c}\right)\right\}\cos\left\{\omega\left(t - \frac{r}{c}\right)\right\} dt = 0$$

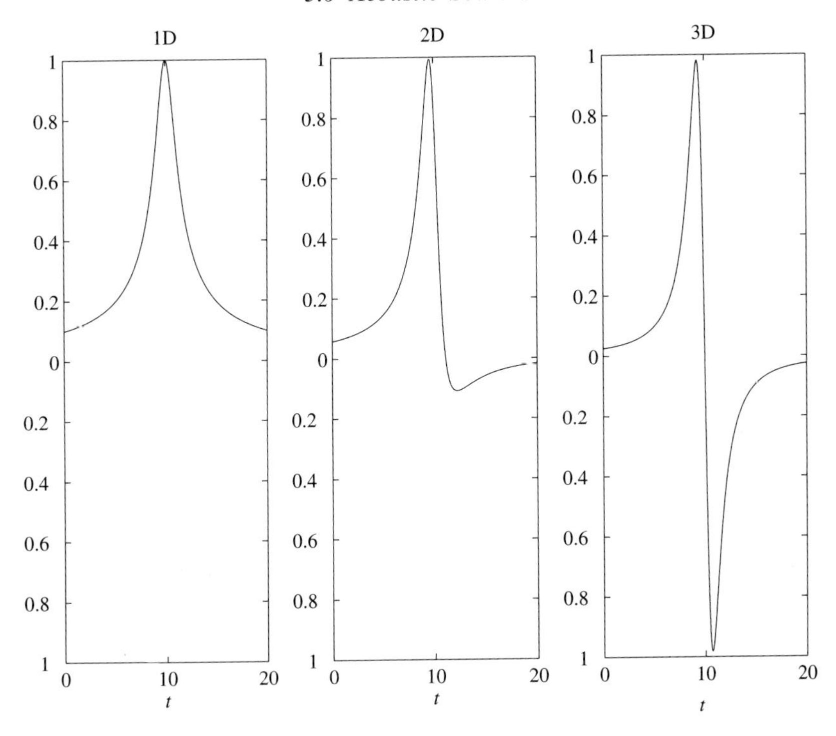

Fig. 3.9. A comparison of qualitative form of the excess pressure due to an acoustic source in one, two and three dimensions.

and

$$\frac{\omega}{2\pi}\int_{t_0}^{t_0+\frac{2\pi}{\omega}} \sin^2\left\{\omega\left(t-\frac{r}{c}\right)\right\} dt = \frac{1}{2},$$

$$\langle I_{\mathrm{r}}\rangle = \frac{\omega^2 q_0^2}{32\rho_0 c\pi^2 r^2}. \tag{3.59}$$

This means that the total, time averaged acoustic energy flux across a sphere of radius r is $\langle I_{\mathrm{r}}\rangle_{\mathrm{tot}} = 4\pi r^2\langle I_{\mathrm{r}}\rangle$, and

$$\langle I_{\mathrm{r}}\rangle_{\mathrm{tot}} = \frac{\omega^2 q_0^2}{8\pi\rho_0 c}. \tag{3.60}$$

We can compare this with the energy transmitted by a plane wave with the same mass flux, propagating in a tube of cross-sectional area, A. In such a wave, $\tilde{p} = \rho_0 c\tilde{u}$, and hence the magnitude of the acoustic intensity vector, which points along the tube, is $I = \rho_0 c\tilde{u}^2$. Since $\tilde{u} = q_0 \cos\omega t/\rho_0 A$, the time averaged flux of energy through any given cross-section of the tube is

$$\langle I \rangle_{\mathrm{tot}} = \frac{c q_0^2}{2\rho_0 A}.$$

Comparing this with the acoustic source gives

$$\frac{\langle I_{\mathrm{r}} \rangle_{\mathrm{tot}}}{\langle I \rangle_{\mathrm{tot}}} = \frac{A\omega^2}{4\pi c^2} = \frac{\pi A}{\lambda^2}. \tag{3.61}$$

For a given mass flux, the energy radiated by an acoustic source is much less than that radiated by a plane wave in a tube, provided that the typical radius of the tube is much less than the wavelength.

3.7 Radiation from Sources in a Plane Wall

As an example of radiation from non-compact sources we consider the acoustic waves generated in a half space by small amplitude motions of a bounding solid wall. We can think of this as a simple model for a typical loudspeaker. Let the bounding wall lie at $x = 0$, so that the boundary conditions are

$$\frac{\partial \phi}{\partial x} = u_0(y, z, t) \quad \text{at } x = 0, \tag{3.62}$$

along with the radiation condition. Now consider the case where a small, rectangular element of the wall at the point $(0, y, z)$, with sides of lengths dy and dz, moves uniformly at speed $u_0(y, z, t)$, whilst the rest of the wall is stationary. If we consider the acoustic disturbance due to the image of this small element in the bounding wall, as well as the element itself, then by symmetry we obtain a field with no normal velocity at $x = 0$, except at the small moving element. The mass outflow due to the motion of the small element and its image is $2\rho_0 u_0(y, z, t) dy dz$. If we consider a collection of infinitesimal elements of this type, we can superpose all of these linear, acoustic source solutions to obtain the pressure field in $x > 0$ as

$$\tilde{p}(x, y, z, t) = \int_{-\infty}^{\infty} \int_{-\infty}^{\infty} \frac{2\rho_0 \dot{u}_0(t - R/c, y', z')}{4\pi R} dy' dz', \tag{3.63}$$

where $R = \sqrt{x^2 + (y - y')^2 + (z - z')^2}$.

We will consider the case where $u_0 = 0$ for $|y'| > L$ or $|z'| > L$, for some length L. The region that generates the disturbance is bounded, but not necessarily acoustically compact. In the far field, $r =$

$\sqrt{x^2+y^2+z^2} \gg 1$ and

$$R = r\left\{1 - \frac{2yy'}{r^2} - \frac{2zz'}{r^2} + O\left(\frac{1}{r}\right)\right\}^{1/2} = r - \frac{yy'}{r} - \frac{zz'}{r} + O(r^{-1}).$$

For harmonic forcing, $u_0(y,z,t) = f(y,z)\exp(i\omega t)$, which gives

$$\tilde{p} \approx \frac{\rho_0}{2\pi r} i\omega \exp\{i\omega(t - r/c)\} \times \int_{-L}^{L}\int_{-L}^{L} f(y',z')\exp\left\{\frac{i\omega}{c}\left(\frac{yy'}{r} + \frac{zz'}{r}\right)\right\} dy'dz',$$

and hence

$$\tilde{p}(x,y,z,t) \approx \frac{\rho_0}{2\pi r} i\omega \exp\{i\omega(t - r/c)\}\hat{f}(m,n), \tag{3.64}$$

where $m = \omega y/rc$, $n = \omega z/rc$ and

$$\hat{f}(m,n) = \int_{-L}^{L}\int_{-L}^{L} f(y',z')\exp\{i(my' + nz')\}dy'dz' \tag{3.65}$$

is the two-dimensional Fourier transform of the amplitude of the forcing disturbance at the wall. For a compact source, $\omega L/c \ll 1$, $mL \ll 1$ and $nL \ll 1$, and hence

$$\tilde{p}(x,y,z,t) \approx \frac{\rho_0}{2\pi r} i\omega \exp\{i\omega(t - r/c)\}\hat{f}(0,0). \tag{3.66}$$

Note that

$$\hat{f}(0,0) = \int_{-L}^{L}\int_{-L}^{L} f(y',z')dy'dz'$$

is the integral of the amplitude of the motion over the moving part of the wall. Equation (3.66) therefore expresses the fact that in the far field this acoustically compact motion produces a sound field equivalent to that of an acoustic source with strength equal to twice the rate of change of mass flux due to the motion of the wall.

For non-compact sources, (3.64) contains directional information about the forcing at the wall. As an example, we consider the far field due to the uniform motion of a rectangular piston, with $u_0(y,z,t) = u_1\exp(i\omega t)$ for $|y| < a$ and $|z| < b$, $u_0(y,z,t) = 0$ elsewhere. In this case

$$\hat{f}(m,n) = u_1\int_{-b}^{b}\int_{-a}^{a} \exp\{i(my' + nz')\}dy'dz' = 4u_1\frac{\sin ma}{m}\frac{\sin nb}{n}.$$

Consider the case where $\omega b/c \ll 1$, but $\omega a/c \not\ll 1$, so that the piston is long and thin. The far field is given by

$$\tilde{p} \approx \frac{2\rho_0 i\omega b u_1}{\pi r}\frac{\sin ma}{m}\exp\{i\omega(t - r/c)\}, \tag{3.67}$$

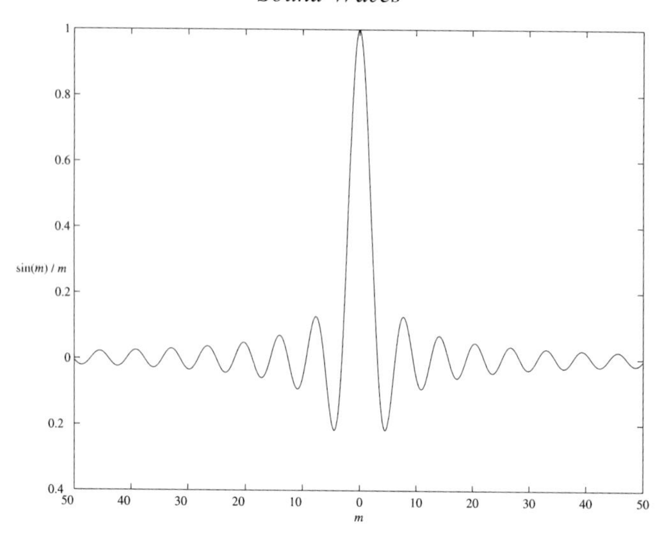

Fig. 3.10. The far field due to a long, thin piston.

which is shown in figure 3.10. The radiation is only substantial for $|ma| < \pi$, or $|y\omega a/rc| < \pi$, apart from minor side lobes. If $\omega a/c \gg 1$, the radiation will only exist in a small band with $y/r \ll 1$. Furthermore if in addition $\omega b/c \gg 1$, so that the piston is a large rectangle, the radiation would only be substantial in a small neighbourhood of the x-axis. This explains why loudspeakers are usually tall and narrow, so that the sound produced is largest in a horizontal band. Note that, since (3.65) can be inverted using the Fourier inversion formula, it is possible, at least in principle, to design a loudspeaker to give any desired far field directional dependence, $\hat{f}(m, n)$, by choosing the appropriate form for $f(y', z')$.

We could consider the equivalent two-dimensional case in the same manner, building up the solution in terms of elementary two-dimensional sources, but it is more instructive to construct the solution using Fourier integral transforms. We seek to solve the two-dimensional wave equation,

$$\frac{\partial^2 \phi}{\partial x^2} + \frac{\partial^2 \phi}{\partial y^2} = \frac{1}{c^2} \frac{\partial^2 \phi}{\partial t^2}, \tag{3.68}$$

in the half plane $y \geq 0$, subject to the boundary condition

$$\frac{\partial \phi}{\partial y} = u_0(x, t) \quad \text{at } y = 0, \tag{3.69}$$

along with the radiation condition. In this case, we will restrict our attention to time harmonic forcing, so that

$$u_0(x,t) = v_0(x)e^{-i\omega t},$$

and look for a solution

$$\phi(x,y,t) = \Phi(x,y)e^{-i\omega t}.$$

If we define the Fourier transform of Φ as

$$\tilde{\Phi}(w,y) = \int_{-\infty}^{\infty} \Phi(x,y)e^{iwx}dx,$$

we find that

$$\frac{\partial^2\tilde{\Phi}}{\partial y^2} + \left(k^2 - w^2\right)\tilde{\Phi} = 0,$$

and hence

$$\tilde{\Phi}(w,y) = \begin{cases} A(w)\exp(i\kappa y) + B(w)\exp(-i\kappa y) & \text{for } |w| < k, \\ A(w)\exp(-\gamma y) + B(w)\exp(\gamma y) & \text{for } |w| \geq k, \end{cases}$$

where

$$\kappa = \sqrt{k^2 - w^2} \quad \text{and} \quad \gamma = \sqrt{w^2 - k^2}, \quad k = \frac{\omega}{c}.$$

However, solutions proportional to $\exp(-i\kappa y - i\omega t)$ represent waves incoming from infinity, so we must take $B(w) = 0$ for $|w| < k$ in order to satisfy the radiation condition. Since $\exp(\gamma y) \to \infty$ as $y \to \infty$, we need $B(w) = 0$ for $|w| \geq k$, and hence for all w. We can then apply the transformed version of the boundary condition (3.69) and find that

$$\tilde{\Phi}(w,y) = \left\{ \begin{array}{ll} \tilde{v}_0(w)\exp(i\kappa y)/i\kappa & \text{for } |w| < k, \\ -\tilde{v}_0(w)\exp(-\gamma y)/\gamma & \text{for } |w| \geq k, \end{array} \right\} \tag{3.70}$$

where

$$\tilde{v}_0(w) = \int_{-\infty}^{\infty} v_0(x)e^{iwx}dx$$

is the Fourier transform of the velocity of the boundary. Finally, we can invert the Fourier transform, and write the solution as

$$\Phi(x,y) = -\frac{1}{2\pi}\int_{-\infty}^{-\omega/c} \frac{\tilde{v}_0(w)}{\gamma}e^{-iwx-\gamma y}\,dw$$
$$+\frac{1}{2\pi i}\int_{-\omega/c}^{\omega/c} \frac{\tilde{v}_0(w)}{\kappa}e^{-iwx+i\kappa y}\,dw - \frac{1}{2\pi}\int_{\omega/c}^{\infty} \frac{\tilde{v}_0(w)}{\gamma}e^{-iwx-\gamma y}\,dw. \tag{3.71}$$

We can now analyse this solution for $r = \sqrt{x^2 + y^2} \gg 1$. We can think of (3.71) as a single integral over the real w-axis, whose integrand oscillates rapidly for $r \gg 1$. In order to approximate such integrals, we can use the method of stationary phase, which we introduced in chapter 1. In the integral (3.71) there are clearly no stationary phase points for $|w| > k$, since the oscillating part of the integrand there is e^{iwx}. We need only consider the asymptotically dominant second term in (3.71). The limits of the integral can be extended to $\pm\infty$ since we will show below that is dominated by the contribution from a single point of stationary phase. In terms of polar coordinates, the integral that we wish to approximate is

$$\Phi(r,\theta) \sim \frac{1}{2\pi i}\int_{-\infty}^{\infty} \frac{\tilde{v}_0(w)}{\sqrt{\frac{\omega^2}{c^2} - w^2}} \times \exp\left\{ir\left(-w\cos\theta + \sin\theta\sqrt{k^2 - w^2}\right)\right\} dw, \tag{3.72}$$

for $r \gg 1$. In this case, using the definitions given by (1.15),

$$f(w) = \frac{1}{2\pi i}\frac{\tilde{v}_0(w)}{\sqrt{k^2 - w^2}}, \quad g(w) = -w\cos\theta + \sin\theta\sqrt{k^2 - w^2}.$$

Since

$$g'(w) = -\cos\theta - \frac{w\sin\theta}{\sqrt{k^2 - w^2}},$$

there is a unique point of stationary phase at $w = w_0$, where

$$w_0 = -k\cos\theta.$$

Since

$$g''(w_0) = -\frac{1}{k\sin^2\theta},$$

our estimate of the acoustic potential is

$$\phi(r,\theta,t) \sim -i\sqrt{\frac{1}{2\pi kr}}\,\tilde{v}_0(-k\cos\theta)\exp\left(ikr - i\omega t - i\frac{\pi}{4}\right) \quad \text{for } r \gg 1. \tag{3.73}$$

The first factor in this product shows that the amplitude of the disturbance decays like $r^{-1/2}$ as $r \to \infty$. The second term shows that the directional dependence is entirely determined by the forcing on the boundary, and the final factor that the solution is an outgoing cylindrical wave.

As a simple example, consider a piston of width $2a$ moving with a uniform, harmonic velocity $V_0 e^{-i\omega t}$, so that

$$v_0(x) = \begin{cases} V_0 & \text{for } -a < x < a, \\ 0 & \text{for } |x| \geq a. \end{cases}$$

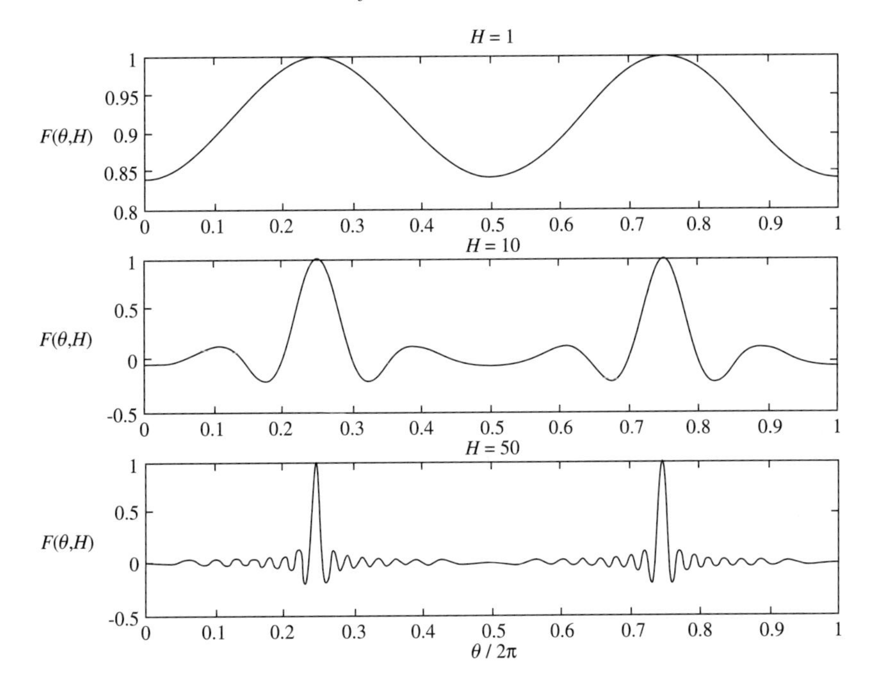

Fig. 3.11. The directivity function, $F(\theta, H)$, for a two-dimensional piston, given by (3.74), for $H = 1$, 10, 50.

The Fourier transform of this is

$$\tilde{v}_0(w) = \frac{2V_0}{w} \sin aw,$$

and hence

$$\phi \sim -i\,2V_0 a \sqrt{\frac{1}{2\pi kr}} \exp\left(ikr - i\omega t - i\frac{\pi}{4}\right) F(\theta, H) \quad \text{for } r \gg 1, \quad (3.74)$$

where $H = ak = 2\pi a/\lambda$ is called the **Helmholtz number**, and

$$F(\theta, H) = \frac{\sin(H \cos\theta)}{H \cos\theta},$$

is called the **directivity function**. When the piston is acoustically compact, $H \ll 1$ and $F \sim 1$. In this case, the disturbance is uniformly distributed in the θ-direction. When $H > \pi$, the θ-dependence has a multi-lobed pattern, since $F = 0$ when $\cos\theta = n\pi/H$ for $n = 1, 2, \ldots$, as shown in figure 3.11. This directional dependence is the same as that for the long, thin piston, given by (3.67), with $ma = H\cos\theta$. It is not possible, however, to obtain the r-dependence directly from (3.67), since $\tilde{p} \propto b/r$, and we would need to take the ill-defined limit $b \to \infty$ to obtain the two-dimensional piston.

It is worth noting here that all of the above analysis will prove to be relevant when we consider diffraction by apertures in plane screens in section 11.2. The reader who wishes to learn more about the science of acoustics should consult Dowling and Ffowcs Williams (1983). An introduction to the physics of musical instruments is provided by Fletcher and Rossing (1991).

Exercises

3.1 A semi-infinite tube with constant cross-sectional area A has characteristic admittance Y_1 for $x < 0$ and Y_2 for $x > 0$. At $x = 0$ there is a thin disc with mass M that fits exactly in the tube. The plane of the disc is perpendicular to the axis of the tube and can move freely in the axial direction. Write down the equation of motion for this disc in terms of the gas pressure on each of its two sides. A plane harmonic wave with amplitude A and angular frequency ω is incident on the disc from $x < 0$. Using linear acoustics, determine the amplitude of the reflected and transmitted waves. Show that the lower the frequency of the incident wave, the greater the amplitude of the transmitted wave. This explains why your neighbours are more likely to complain about your music if it involves lots of bass notes.

3.2 Two semi-infinite cylinders with characteristic admittance Y_1 are connected axially to a finite cylinder with characteristic admittance Y_2 and length L. The acoustic wave speed in each of the cylinders is c. The coordinate x measures distance along the tube, with the junctions between the tubes lying at $x = 0$ and $x = L$, as shown below. A plane wave enters the system from $x < 0$,

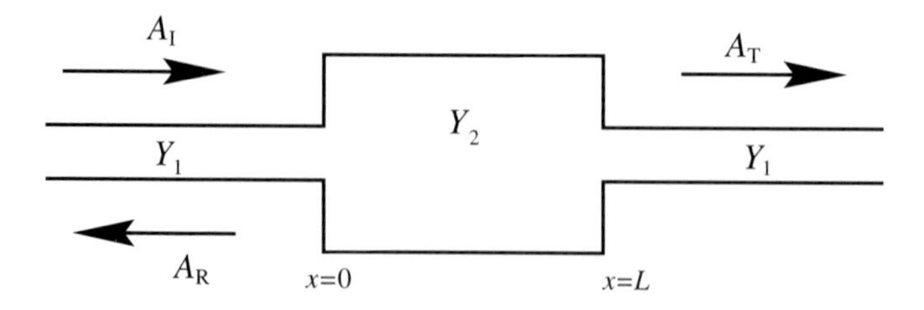

where there is also a reflected wave. A transmitted wave leaves the finite tube and propagates in $x > L$. The solution can be written as

$$\tilde{p} = A_{\mathrm{I}} e^{i\omega\left(t-x/c\right)} + A_{\mathrm{R}} e^{i\omega\left(t+x/c\right)}, \quad \text{for } x < 0,$$

$$\tilde{p} = A_{\mathrm{T}} e^{i\omega\left(t-x/c\right)}, \quad \text{for } x > L.$$

Write down a suitable form for the solution in $0 < x < L$. By applying continuity of pressure and continuity of volume flux at each of the junctions, show that

$$A_{\mathrm{T}} = \frac{2Y_1 Y_2 \left\{2Y_1 Y_2 \cos\left(\omega L/c\right) - i\left(Y_1^2 + Y_2^2\right)\sin\left(\omega L/c\right)\right\} e^{i\omega L/c}}{4Y_1^2 Y_2^2 + \left(Y_1^2 - Y_2^2\right)^2 \sin^2\left(\omega L/c\right)} A_{\mathrm{I}},$$

and find a similar expression for A_{R}. Show that $|A_{\mathrm{T}}|^2 + |A_{\mathrm{R}}|^2 = |A_{\mathrm{I}}|^2$, and explain why this should be so. Show that $|A_{\mathrm{T}}|^2$ is smallest when

$$\frac{\omega L}{c} = \left(n - \frac{1}{2}\right)\pi, \quad \text{for } n = 1, 2, \ldots.$$

When $\omega L/c = \pi/2$, determine the real part of the pressure disturbance in $0 < x < L$ in the form of a sum of two standing waves, and sketch each of them. Can you think of an everyday use for an arrangement of tubes like this?

3.3 A rectangular acoustic waveguide has rigid walls at $x = 0, a$ and $y = 0, b$, and has infinite extent in the z-direction. By looking for separable solutions of the wave equation with velocity potentials of the form $\phi = X(x)Y(y)\exp\{i(kz - \omega t)\}$, show that the angular frequency, ω, and wavenumber, k, must satisfy

$$\omega = c\sqrt{k^2 + \pi^2\left(\frac{n^2}{a^2} + \frac{m^2}{b^2}\right)},$$

for any $n = 1, 2, \ldots$, $m = 1, 2, \ldots$. Show that the solution with $m = n = 0$ is a plane wave.

For a mode with $n > 0$ and $m > 0$, calculate the potential and kinetic energy in the volume $0 \leq x \leq a$, $0 \leq y \leq b$, $z_0 \leq z \leq z_0 + 2\pi/k$, in other words, a single wavelength of the disturbance, averaged over one period of the motion. Calculate the integral of the z-component of the acoustic intensity over the cross-section of the waveguide, averaged over one period of the motion. Hence show that energy propagates along the waveguide at speed, c_g, where

$$c_g = \frac{kc^2}{\omega} = \frac{d\omega}{dk} < c.$$

3.4 By treating a woodwind instrument with all the fingerholes closed as a conical tube with a small angle and length L, at one end of which the cross-sectional area is effectively zero, and looking for solutions of the form $r\phi = F(r)e^{i\omega t}$, show that there is a set of eigenfrequencies

$$\omega = \frac{n\pi c}{L}, \quad \text{for } n = 1, 2, \ldots.$$

If instead the larger end is closed, so that the gas velocity is zero there, show that the eigenfrequencies are solutions of

$$\frac{\omega L}{c} = \tan\left(\frac{\omega L}{c}\right).$$

Note that one set of frequencies forms a musical scale, whilst the other does not.

3.5 The time varying pressure at the surface, $r = r_0$, of a sphere is given by the real part of

$$p(t) = p_0 + \tilde{p}_0(t)H(t),$$

where p_0 a constant and $H(t)$ the Heaviside function. A compressible gas is at rest outside the sphere when $t = 0$. Determine the acoustic velocity potential outside the sphere. If $\tilde{p}(t)$ is time harmonic, with constant angular frequency, ω, determine the mean acoustic intensity over one period of the disturbance.

3.6 A sphere with undisturbed radius a_0 is immersed in a gas that is at rest. The radius of the sphere changes in such a way that the change of volume per unit volume of the sphere is equal to $-K$ times the excess pressure at the surface. When $t = 0$ the sphere is at rest with radius $a = a_0 + a_1$, where $a_1 \ll a_0$. Determine $a(t)$.

3.7 A fluid of undisturbed density ρ and sound speed c occupies the half space $x > 0$. A diaphragm in the (y, z)-plane executes small amplitude motion in the x-direction with velocity

$$u = \begin{cases} \epsilon \sin \omega t & \text{for } |y| < a \text{ and } |z| < a, \\ 0 & \text{for } |y| \geq a \text{ or } |z| \geq a. \end{cases}$$

Obtain the leading order approximation to the pressure fluctuation in the far field. Determine W_1, the time averaged energy flux in the far field, assuming that the moving part of the diaphragm is acoustically compact. Determine W_2, the energy flux that would be produced if the diaphragm in the region $|y| < a$, $|z| < a$ formed part of the wall of an otherwise rigid, acoustically compact box, and the fluid occupied the whole of the space outside. What can you deduce by calculating the ratio W_1/W_2?

3.8 A fluid of undisturbed density ρ and sound speed c occupies the half space $z > 0$. A diaphragm in the plane $z = 0$ executes small amplitude motion in the z-direction with velocity

$$u = \begin{cases} \epsilon \cos \omega t & \text{for } r < a, \\ 0 & \text{for } r \geq a, \end{cases}$$

in terms of cylindrical polar coordinates, (r, θ, z). Obtain the leading order approximation to the pressure fluctuation, $\tilde{p}$, in the far field. Determine the leading order approximation to $\tilde{p}$ on the cone $z = Kr$ with K an $O(1)$ constant. Describe the far field solution when (i) $a\omega \ll c$, (ii) $a\omega \gg c$.

You will need to make use of

$$J_0(x) = \frac{1}{2\pi} \int_0^{2\pi} e^{ix\cos\theta} d\theta, \quad \int_0^1 xJ_0(kx)dx = \frac{1}{k} J_1(k).$$

3.9 Solve the two-dimensional wave equation,

$$\frac{\partial^2 \phi}{\partial x^2} + \frac{\partial^2 \phi}{\partial y^2} = \frac{1}{c^2} \frac{\partial^2 \phi}{\partial t^2},$$

in the half plane $x \geq 0$ subject to the boundary condition

$$\frac{\partial \phi}{\partial x} = ae^{-b|y|},$$

where a and b are positive constants, and a radiation condition that excludes incoming waves. What is the leading order asymptotic solution for $x^2 + y^2 \gg 1$? What is the directivity function for this type of source?

4

Linear Water Waves

Water waves manifest themselves as disturbances to the free surface of an incompressible fluid with mean depth h and constant, uniform density ρ. Whether the disturbance is generated by the wind, the passage of a ship or a sub-sea earthquake, gravity and/or surface tension will act as **restoring forces** that tend to drive the fluid towards its equilibrium state. It is the balance between fluid inertia and restoring forces that gives rise to free surface waves. In British coastal waters, where typical sea depths range from a few metres to a hundred metres, 40% of observed waves have amplitudes of 2 m or less, and much longer wavelengths, up to nearly a kilometre in some cases. It would therefore seem worthwhile to develop a linear theory for such waves, based on the assumption that their amplitude is much smaller than their wavelength. We will assume that the flow is laminar, so that there are no breaking waves, no white water and no turbulence. The system is illustrated in figure 4.1. After studying basic linear gravity waves, both progressive and standing, we will move on to consider the generation and propagation of water waves in a variety of situations. These include the harnessing of wave power using a simple mechanical device, the generation of waves by a moving ship and the refraction of waves by changes in bed topography. We will also briefly consider the effects of surface tension and viscosity.

4.1 Derivation of the Governing Equations

The equations for conservation of mass and momentum for an inviscid, incompressible fluid are

$$\nabla \cdot \mathbf{u} = 0, \tag{4.1}$$

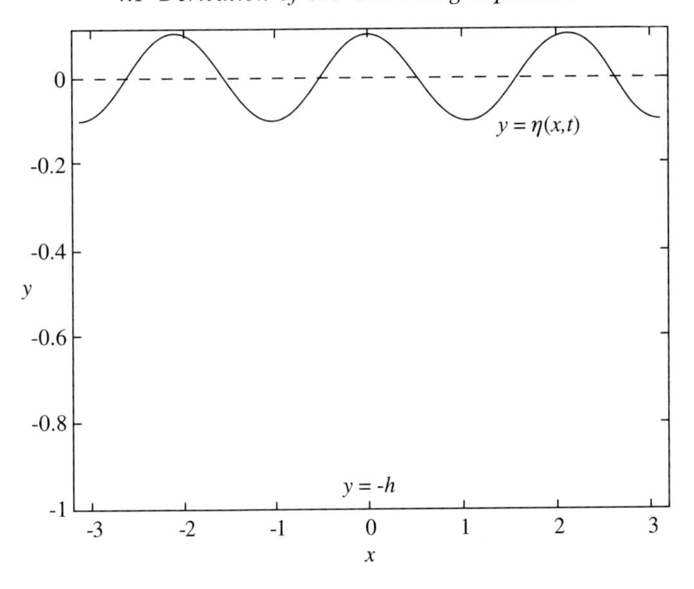

Fig. 4.1. The position of the free surface and rigid bed.

$$\frac{\partial \mathbf{u}}{\partial t} + \mathbf{u} \cdot \nabla \mathbf{u} = -\frac{1}{\rho}\nabla p + \mathbf{g}, \tag{4.2}$$

where $\mathbf{g}$ is the acceleration due to gravity. We also make the simplifying assumption that the flow is irrotational, and hence that the velocity field, $\mathbf{u}$, can be written as the gradient of a potential function, $\mathbf{u} = \nabla\phi$. The equation for conservation of mass then becomes Laplace's equation

$$\nabla^2\phi = 0. \tag{4.3}$$

Using the identity

$$\mathbf{u} \times (\nabla \times \mathbf{u}) = \nabla\left(\frac{1}{2}\mathbf{u} \cdot \mathbf{u}\right) - \mathbf{u} \cdot \nabla \mathbf{u}$$

we can write the convective acceleration as

$$\mathbf{u} \cdot \nabla \mathbf{u} = \nabla\left(\frac{1}{2}\nabla\phi \cdot \nabla\phi\right),$$

since the flow is irrotational ($\nabla \times \mathbf{u} = 0$). If gravity acts in the negative y-direction, we can write

$$\mathbf{g} = \nabla(-gy).$$

The momentum equation therefore becomes

$$\frac{\partial}{\partial t}(\nabla\phi)+\nabla\left(\frac{1}{2}\nabla\phi\cdot\nabla\phi\right)=-\frac{1}{\rho}\nabla p-\nabla(gy),$$

so that, with the usual assumptions of smoothness in $\phi=\phi(x,y,z,t)$,

$$\nabla\left\{\frac{\partial\phi}{\partial t}+\frac{1}{2}\nabla\phi\cdot\nabla\phi+\frac{p}{\rho}+gy\right\}=0,$$

and hence

$$\frac{\partial\phi}{\partial t}+\frac{1}{2}\nabla\phi\cdot\nabla\phi+\frac{p}{\rho}+gy=C(t),$$

for some function $C(t)$. However, we can take $C(t)=0$ by using the simple transformation

$$\phi\mapsto\phi+\int_0^t C(s)ds,$$

which does not affect the velocity field. For this type of flow we obtain **Bernoulli's equation**,

$$\frac{\partial\phi}{\partial t}+\frac{1}{2}\nabla\phi\cdot\nabla\phi+\frac{p}{\rho}+gy=0. \tag{4.4}$$

Once we have determined ϕ, Bernoulli's equation tells us the pressure field, p.

We now turn our attention to the surface of the fluid. We assume that the flow is two-dimensional, so that the surface lies at $y=\eta(x,t)$, where the coordinate x measures distance along the interface. For the moment we will neglect surface tension effects so that the pressure at this **free surface** is atmospheric, $p=p_{\text{atm}}$. We can take $p=0$ at the free surface using the simple transformation $p\mapsto p-p_{\text{atm}}$, which does not change the basic Euler equations which depend upon ∇p. Bernoulli's equation (4.4) at the free surface then gives us the boundary condition

$$\frac{\partial\phi}{\partial t}+\frac{1}{2}\nabla\phi\cdot\nabla\phi+g\eta=0\quad\text{at } y=\eta(x,t). \tag{4.5}$$

This is also known as the **dynamic boundary condition at the free surface**. To obtain the other boundary condition that we need there, we note that a fluid particle on the surface remains there. Mathematically, this means that

$$\frac{D}{Dt}\{y-\eta(x,t)\}=\frac{\partial}{\partial t}\{y-\eta(x,t)\}+\mathbf{u}\cdot\nabla\{y-\eta(x,t)\}=0,$$

which gives

$$\frac{\partial \phi}{\partial y} = \frac{\partial \eta}{\partial t} + \frac{\partial \phi}{\partial x}\frac{\partial \eta}{\partial x} \quad \text{at } y = \eta(x,t). \tag{4.6}$$

This is known as the **kinematic boundary condition at the free surface**. Finally, the fluid is bounded by a rigid surface at $y = -h(x)$. There is no normal velocity through this surface, and hence we have the **bed condition**,

$$\frac{\partial \phi}{\partial n} = 0 \quad \text{at } y = -h(x). \tag{4.7}$$

In order to obtain the boundary value problem for linear water waves, we assume that the amplitude of the disturbance of the free surface is much less than its wavelength, and hence that we can discard products of terms in the governing equations. We will return to the question of what restrictions are necessary for this approximation to be accurate later in this chapter. This leads to the boundary value problem for linear, two-dimensional water waves over a flat, rigid bed (h constant) as

$$\nabla^2 \phi = 0 \quad \text{for } -\infty < x < \infty,\ -h \le y \le 0, \tag{4.8}$$

subject to the linearised Bernoulli condition,

$$\frac{\partial \phi}{\partial t} + g\eta = 0 \quad \text{at } y = 0, \tag{4.9}$$

the linearised kinematic condition,

$$\frac{\partial \phi}{\partial y} = \frac{\partial \eta}{\partial t} \quad \text{at } y = 0, \tag{4.10}$$

and the bed condition

$$\frac{\partial \phi}{\partial y} = 0 \quad \text{at } y = -h. \tag{4.11}$$

In addition we need initial conditions on ϕ and η, and appropriate radiation conditions. More generally, for a given depth of fluid, h, and fluid density, ρ, we want to find a relationship between the wavelength and the speed of propagation for a harmonic wave. As we saw in chapter 1, this is known as the dispersion relation, and the phenomenon of waves with different wavelengths propagating at different speeds is called dispersion. This is in contrast to the non-dispersive systems, for example the vibrating string and linear sound waves, that we studied in chapters 2 and 3 (with the exception of waveguide modes), where waves of all wavelengths propagate at the same speed.

4.2 Linear Gravity Waves

4.2.1 Progressive Gravity Waves

We begin our analysis of linear water waves by looking for a solution of the form

$$\phi(x, y, t) = F(x - ct)Y(y).$$

This is a permanent form travelling wave that propagates in the positive x-direction with phase velocity c. The function $Y(y)$ determines how the velocity varies with depth. If we substitute this into Laplace's equation (4.3) we obtain

$$\frac{\partial^2 \phi}{\partial x^2} + \frac{\partial^2 \phi}{\partial y^2} = F''Y + FY'' = 0,$$

and hence

$$\frac{F''}{F} = -\frac{Y''}{Y} = -k^2.$$

This choice of separation constant gives us exponential behaviour in the y-direction and periodic behaviour in the x-direction, since

$$F'' + k^2 F = 0 \quad \text{and} \quad Y'' - k^2 Y = 0$$

leads to

$$F = A\cos k(x - ct) + B\sin k(x - ct) \quad \text{and} \quad Y = Ce^{ky} + De^{-ky}.$$

In order to satisfy $\partial\phi/\partial y = 0$ at $y = -h$ we need

$$k(Ce^{-kh} - De^{kh}) = 0,$$

and hence

$$Y = \overline{D}\cosh k(y + h),$$

where $\overline{D} = 2De^{kh}$. We can therefore write the solution in the form

$$\phi(x, y, t) = \cosh k(y + h)\left\{\overline{A}\cos k(x - ct) + \overline{B}\sin k(x - ct)\right\}, \qquad (4.12)$$

where $\overline{A} = 2A\overline{D}$ and $\overline{B} = 2B\overline{D}$.

We now note that we can combine the free surface conditions, (4.9) and (4.10), by differentiating (4.9) with respect to time, t, to give

$$\frac{\partial^2 \phi}{\partial t^2} + g\frac{\partial \eta}{\partial t} = 0 \quad \text{at } y = 0,$$

and using (4.10) to eliminate $\partial\eta/\partial t$, which leads to

$$\frac{\partial^2 \phi}{\partial t^2} + g\frac{\partial \phi}{\partial y} = 0 \quad \text{at } y = 0. \qquad (4.13)$$

If we now substitute the solution (4.12) into (4.13) we obtain

$$\left(-k^2c^2\cosh kh + gk\sinh kh\right)\left\{\overline{A}\cos k(x-ct) + \overline{B}\sin k(x-ct)\right\} = 0.$$

This must hold for all values of $x - ct$, so

$$-k^2c^2\cosh kh + gk\sinh kh = 0,$$

and hence

$$c^2 = \frac{g}{k}\tanh kh. \tag{4.14}$$

This is the dispersion relation, which shows that harmonic waves with different wavenumbers, and hence different wavelengths, propagate at different wave speeds. In terms of the wavelength, $\lambda = 2\pi/k$, the dispersion relation is

$$c = \pm\sqrt{\frac{g\lambda}{2\pi}\tanh\left(\frac{2\pi h}{\lambda}\right)}, \tag{4.15}$$

and in terms of the angular frequency,

$$\omega^2 = gk\tanh kh. \tag{4.16}$$

The two possible values of c indicate that waves can propagate in either horizontal direction, as we would expect.

In shallow water, $h \ll \lambda$, $\tanh(2\pi h/\lambda) \sim 2\pi h/\lambda$, and hence

$$c^2 \sim gh, \quad \text{for } h \ll \lambda. \tag{4.17}$$

This shows that **linear shallow water waves** are not dispersive, since the wave speed, c, is independent of the wavelength, λ. In deep water, $h \gg \lambda$, $\tanh(2\pi h/\lambda) \sim 1$, and hence

$$c^2 \sim \frac{g\lambda}{2\pi} \quad \text{for } h \gg \lambda. \tag{4.18}$$

This shows that **linear deep water waves** are dispersive, and the square of the wave speed increases linearly with the wavelength.

Finally, we need to determine the amplitude of the waves, which we do by calculating the position of the surface, $\eta(x,t)$. Without loss of generality, we will take $\overline{A} = 0$ in the solution (4.12), so that

$$\phi(x,y,t) = \overline{B}\cosh k(y+h)\sin k(x-ct). \tag{4.19}$$

This simply amounts to a shift in the definition of the point where $x = 0$

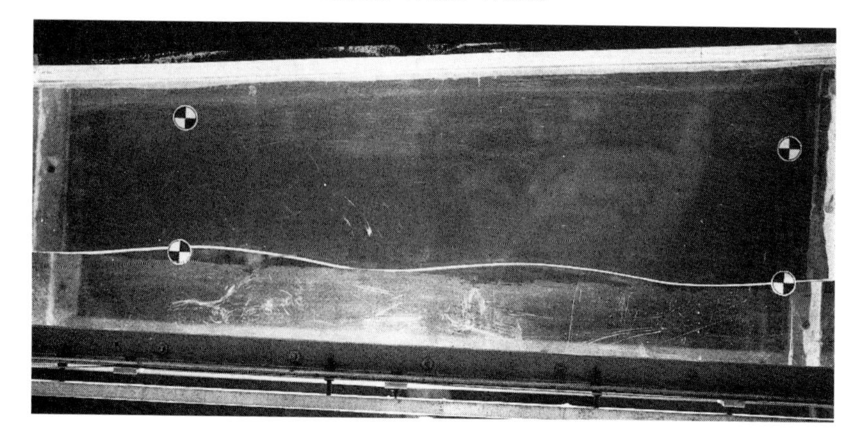

Fig. 4.2. A typical, linear, progressive gravity wave.

or the time when $t = 0$. If we substitute the solution (4.19) into the dynamic free surface condition, (4.9), we obtain

$$\eta(x,t) = \overline{B}\frac{kc}{g}\cosh kh \cos k(x - ct).$$

If we now choose $\overline{B}$ so that

$$\overline{B}\frac{kc}{g}\cosh kh = a,$$

we finally obtain the full solution as

$$\left.\begin{aligned} &\eta(x,t) = a\cos k(x - ct), \qquad c^2 = \frac{g}{k}\tanh kh. \\ &\phi(x,y,t) = \frac{ag}{kc\cosh kh}\cosh k(y+h)\sin k(x - ct). \end{aligned}\right\} \qquad (4.20)$$

A typical, linear, progressive gravity wave is shown in figure 4.2.

Having found this solution, let's return to the question of what restrictions on the flow are necessary for this to be an accurate approximation. In the kinematic boundary condition we retained the term $\eta_t = O(ack)$, but neglected the term $\phi_x\eta_x = O(a^2gk/c)$. For this to be valid we need $a^2gk/c \ll ack$, and hence $ak \ll \tanh kh$. In deep water, $\tanh kh \sim 1$, so we need $ak \ll 1$, or equivalently $a/\lambda \ll 1$, so that the amplitude of the wave must be much less than its wavelength. In shallow water, $\tanh kh \sim kh$ and we need $a/h \ll 1$, so that the amplitude must be much less than the depth. If we insist that $a \ll \lambda$ and $a \ll h$, the small amplitude approximation will be appropriate. These restrictions always need to be borne in mind when applying the small amplitude theory. We consider extensions to the theory when the amplitude is not small compared to the wavelength or depth in chapter 8.

We can now determine how a fluid particle moves under the influence of this **progressive gravity wave**. Consider the particle initially at $(x, y) = (X_0, Y_0)$. In a linear theory, such as we have developed above, its motion away from this initial point is small compared to its initial displacement, so we can describe its subsequent motion using $(x, y) = (X_0 + X_1(t), Y_0 + Y_1(t))$, with $|X_1| \ll |X_0|$ and $|Y_1| \ll |Y_0|$. This means that, at leading order

$$\frac{dX_1}{dt} = \frac{\partial \phi}{\partial x}(X_0, Y_0) = \frac{ag}{c \cosh kh} \cosh k(Y_0 + h) \cos k(X_0 - ct), \qquad (4.21)$$

$$\frac{dY_1}{dt} = \frac{\partial \phi}{\partial y}(X_0, Y_0) = \frac{ag}{c \cosh kh} \sinh k(Y_0 + h) \sin k(X_0 - ct), \qquad (4.22)$$

subject to $X_1(0) = Y_1(0) = 0$. On integrating once and eliminating t, we find that

$$\left(\frac{X_1}{\cosh k(Y_0 + h)} - \frac{ag}{kc^2 \cosh kh} \sin kX_0\right)^2$$
$$+ \left(\frac{Y_1}{\sinh k(Y_0 + h)} + \frac{ag}{kc^2 \cosh kh} \cos kX_0\right)^2 = \left(\frac{ag}{kc^2 \cosh kh}\right)^2. \qquad (4.23)$$

Fluid particles therefore move clockwise around a small ellipse with the same period as the progressive wave that drives the motion. The ellipse has its axes vertical and horizontal, with lengths proportional to $\sinh k(Y_0 + h)/\cosh kh$ and $\cosh k(Y_0 + h)/\cosh kh$ respectively. As we approach the bed of the fluid, and $Y_0 \to -h$, the ellipses become progressively thinner in the vertical direction, since the motion must adjust to satisfy the condition of no normal velocity there, as shown for a typical case in figure 4.3. At the bed, the fluid particles slip back and forth horizontally. Note that, since all of the particle paths are closed loops, there is no net mass transport by the wave, at least using the leading order solution for waves of small amplitude (see figure 4.4). There is actually a small net mass transport due to the correction to the leading order solution, which we discuss at the end of section 8.2.

How fast does the energy associated with a progressive gravity wave propagate? We have already seen in section 3.3 that the energy associated with a linear acoustic wave propagates at the wave speed $c = \omega/k$ and that the mean energy is divided equally between compressive potential energy and kinetic energy. For progressive gravity waves we can define the kinetic and gravitational potential energies of an elementary area

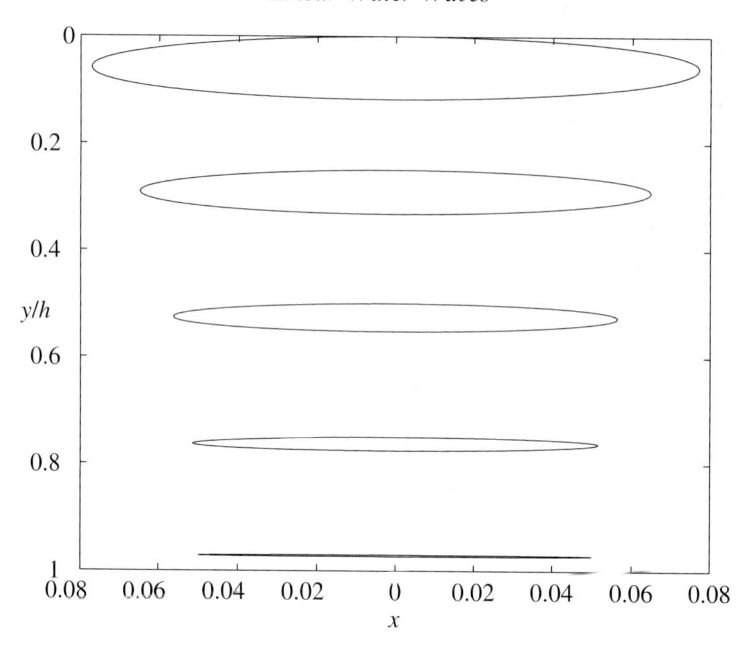

Fig. 4.3. Typical motions of fluid particles driven by a progressive gravity wave.

$dxdy$ by direct analogy with particle dynamics. These can be integrated vertically to give the energies in a strip of width dx, as shown in figure 4.5. Of particular interest to us are the kinetic and potential energy densities, or energies per unit horizontal length, which we define as

$$\text{Kinetic energy density} = \int_{-h}^{0} \frac{1}{2}\rho \left|\nabla\phi\right|^2 dy,$$

$$\text{Potential energy density} = \int_{0}^{\eta} \rho g y dy = \frac{1}{2}\rho g \eta^2.$$

(Note that the integral for the kinetic energy has its upper limit at the undisturbed position of the free surface, which is its leading order value for a wave of small amplitude, whilst the gravitational potential energy is given by displacements from the undisturbed level, $y = 0$.) These quantities can be averaged over a wavelength to give

$$\text{Mean kinetic energy density} = \frac{\rho}{2\lambda}\int_{0}^{\lambda}\int_{-h}^{0} \left|\nabla\phi\right|^2 dydx = \overline{K},$$

$$\text{Mean potential energy density} = \frac{\rho g}{2\lambda}\int_{0}^{\lambda} \eta^2 dx = \overline{V}.$$

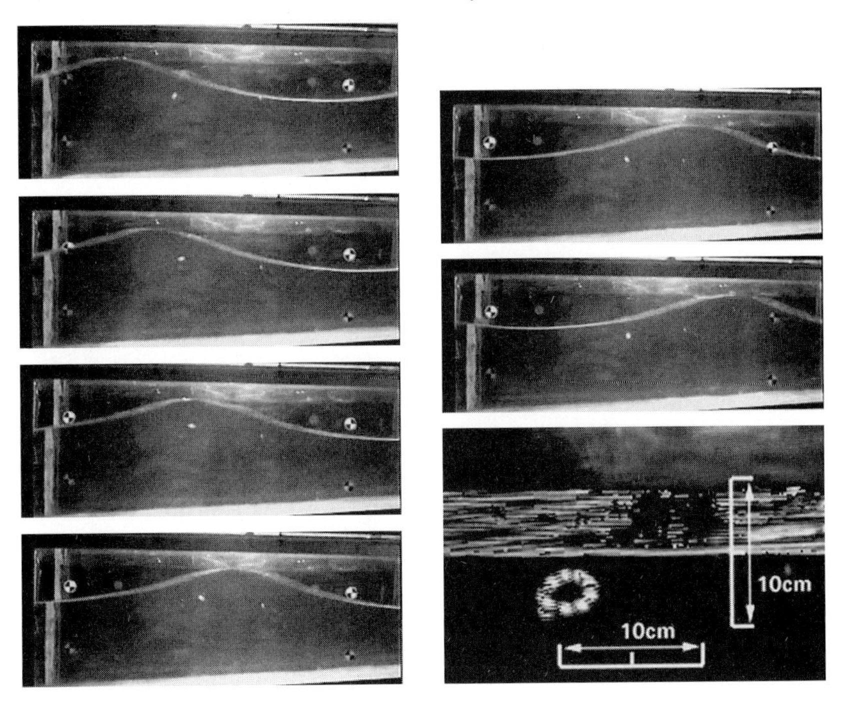

Fig. 4.4. The paths of neutrally buoyant particles under a progressive gravity wave.

For the progressive wave solutions derived above, these integrals are easily calculated, and we find that

$$\overline{K} = \overline{V} = \frac{1}{4}\rho g a^2,$$

so that the total mean energy density in a single wavelength is

$$H = \frac{1}{2}\rho g a^2.$$

The mean energy flux, E_F, through a vertical plane is equal to the time averaged rate of working of the pressure forces there during one period of the motion, since the flux of kinetic energy is negligibly small in this linear theory. Hence

$$\begin{aligned} E_F &= \frac{\omega}{2\pi}\int_{t_0}^{t_0+\frac{2\pi}{\omega}}\int_{-h}^{0} p\frac{\partial \phi}{\partial x}dydt \\ &= \frac{\omega}{2\pi}\int_{t_0}^{t_0+\frac{2\pi}{\omega}}\int_{-h}^{0}\left(p_{\text{atm}} - \rho g y - \rho\frac{\partial \phi}{\partial t}\right)\frac{\partial \phi}{\partial x}dydt. \end{aligned}$$

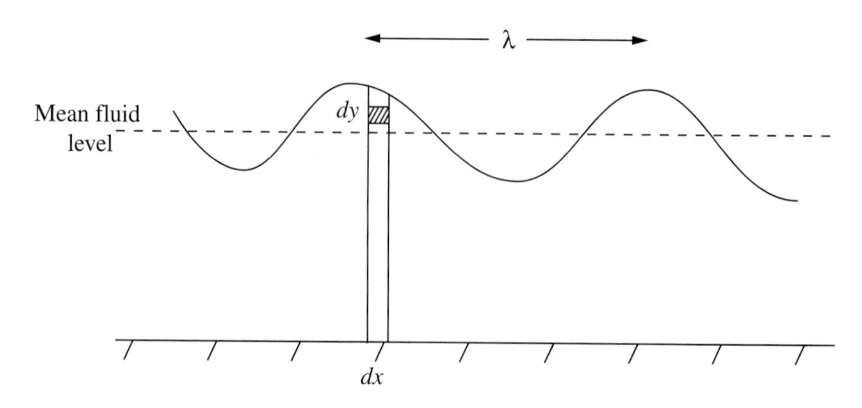

Fig. 4.5. The basic strip in the wave from which we calculate the kinetic and gravitational potential energy densities.

Since

$$\int_{t_0}^{t_0+\frac{2\pi}{\omega}} \frac{\partial \phi}{\partial x} dt \propto \int_{t_0}^{t_0+\frac{2\pi}{\omega}} \cos k\,(x - ct)\, dt = 0,$$

we have

$$E_F = -\rho \frac{\omega}{2\pi} \int_{t_0}^{t_0+\frac{2\pi}{\omega}} \int_{-h}^{0} \frac{\partial \phi}{\partial t} \frac{\partial \phi}{\partial x} dy dt.$$

We can evaluate this integral, using the fact that the average value of $\sin^2 X$ over a period is $\frac{1}{2}$, to give

$$E_F = \frac{\rho a^2 k^2 c^3}{4 \sinh^2 kh} \left(h + \frac{\sinh 2kh}{2k} \right).$$

The energy that passes through a vertical plane during a time τ is $E_F \tau$. If we define U to be the velocity at which energy flows horizontally, the energy contained in a horizontal length $U\tau$ is $U\tau H$, where H is the energy density. Since $E_F \tau = U\tau H$, $U = E_F/H$. After using the dispersion relation, $k^2c^2 = gk \tanh kh$, this gives

$$U = \frac{2E_F}{\rho g a^2} = \frac{1}{2} c \left(1 + \frac{2kh}{\sinh 2kh} \right) = c_g.$$

In other words, the energy in the wave propagates at the group velocity, $c_g = d\omega/dk = d(kc)/dk$. This has important implications, which we will return to later in this chapter.

Finally, let's consider a real water wave and work out some numbers for the physical quantities associated with it. For example, if you are out swimming in water of depth 4 m, you may see a wave of height 25 cm, and hence amplitude 12.5 cm, with a wavelength of 10 m and a lateral extent of 50 m. The elevated area of this wave is

$$\int_{-\pi/2k}^{\pi/2k} a \cos kx dx = \frac{2a}{k} = \frac{a\lambda}{\pi} \approx 0.40\text{m}^2.$$

The mass of the water in the elevated part of the wave is

$$\rho L \frac{a\lambda}{\pi} \approx 20 \times 10^3 \text{ kg}.$$

The mean energy per wavelength is

$$\frac{1}{2}\rho g a^2 L \approx 3.9 \times 10^3 \text{kg m s}^{-2},$$

and the wave speed is

$$c = \sqrt{\frac{g}{k} \tanh kh} \approx 3.96 \text{ m s}^{-1}.$$

To put this into some perspective, a Ford Fiesta of length 3 m and weight 1200 kg travelling at the same speed has a kinetic energy per unit length of 3.1×10^3 kg m s^{-2}, so the wave is heavier and carries more energy than the car!

4.2.2 Standing Gravity Waves

Another important type of water wave is a **standing gravity wave**, as illustrated in figure 4.6. These are solutions of the form $\phi = X(x)Y(y)T(t)$. Substituting this into Laplace's equation, (4.3), gives

$$X''YT + XY''T = 0,$$

and hence

$$\frac{X''}{X} = -\frac{Y''}{Y} = -k^2.$$

This gives

$$X'' + k^2 X = 0, \quad \text{and} \quad Y'' - k^2 Y = 0.$$

The solution that satisfies $\partial\phi/\partial y = 0$ at $y = -h$ is

$$Y = B \cosh k(y + h),$$

the general solution for X is

$$X = C \cos kx + D \sin kx,$$

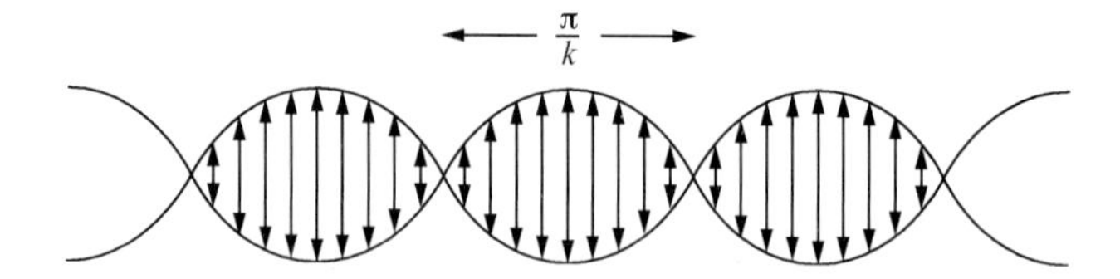

Fig. 4.6. A typical standing gravity wave. At any fixed position the free surface moves up and down like $\sin \omega t$. The envelope of the wave is $a \cos kx$.

and hence

$$\phi(x, y, t) = \cosh k(y + h) \left(\overline{C} \cos kx + \overline{D} \sin kx\right) T(t). \tag{4.24}$$

If we now substitute this solution into the combined free surface condition, (4.13), we obtain

$$(\cosh kh \; T'' + gk \sinh kh \; T) \left(\overline{C} \cos kx + \overline{D} \sin kx\right) = 0.$$

This must hold for all values of x, so that

$$T'' + \omega^2 T = 0,$$

where ω is given by the dispersion relation (4.16). The general solution for T is

$$T = F \cos \omega t + G \sin \omega t.$$

After suitable shifts of the position of the spatial and temporal origins this leads to

$$\phi(x, y, t) = L \cosh k(y + h) \cos kx \cos \omega t. \tag{4.25}$$

Finally, we can substitute this into the dynamic free surface condition, (4.9), to obtain

$$\eta(x, t) = \frac{L\omega}{g} \cosh kh \cos kx \sin \omega t.$$

If we define $a = L\omega \cosh(kh)/g$ we arrive at the full solution

$$\eta(x, t) = a \cos kx \sin \omega t, \quad \omega^2 = gk \tanh kh,$$

$$\phi(x, y, t) = \frac{ag}{\omega \cosh kh} \cosh k(y + h) \cos kx \sin \omega t. \tag{4.26}$$

The envelope of the standing wave is $\eta = a \cos kx$, its frequency is given by the dispersion relation, (4.16), and the particle paths are straight lines (see exercise 4.5).

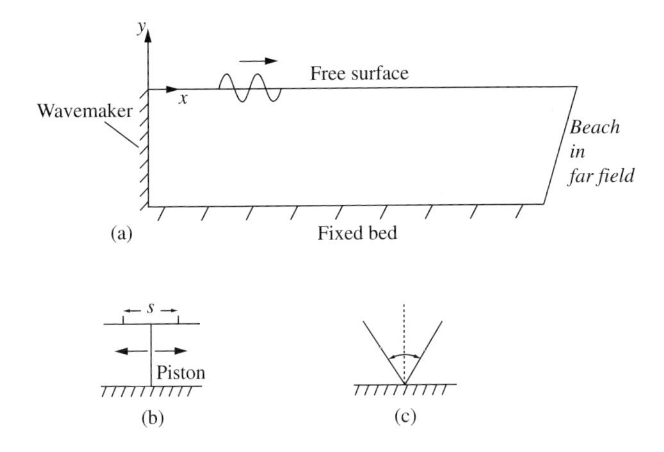

Fig. 4.7. (a) Schematic of the wavemaker geometry, (b) with a piston, (c) with a paddle (note that the small angle through which the paddle is displaced has been exaggerated).

4.2.3 The Wavemaker

It is easy to make water waves in a laboratory using a **wavemaker**. This consists of a long tank, filled with water, which is sealed at one end and has a piston or paddle at the other end. The walls of the tank are usually made of perspex, so that the waves generated can be seen. The piston or paddle is made to oscillate sinusoidally with a small amplitude, and produces wave trains that travel down the tank. A beach is usually put at the far end of the tank to absorb the wave energy and stop reflections and splashing, as shown in figure 4.7(a).

Water initially occupies the region $0 \leq x \leq \infty$, $0 \leq y \leq -h$ and is at rest. As usual, $y = 0$ is a free surface and we consider the waves generated when the wavemaker is switched on at $t = 0$. The surface of the piston or paddle is at

$$x = X(y, t) = \frac{1}{2}s(y)\cos\sigma t,$$

where s is a constant for the piston and $s = s_0(y + h)$ for the paddle, as shown in figure 4.7(b) and (c). We must solve Laplace's equation, (4.3), subject to the bed condition, (4.7), and the combined free surface condition, (4.13). At the surface of the paddle or piston there is continuity of normal velocity. We assume that the amplitude of the oscillation is small enough that the normal can be treated as horizontal, and that we can apply the condition at $x = 0$ to give

$$\frac{\partial \phi}{\partial x} = \frac{dX}{dt} = -\frac{1}{2}\sigma s(y) \sin \sigma t, \quad \text{at } x = 0. \tag{4.27}$$

We can solve this boundary value problem using separation of variables. We begin by looking for a progressive wave solution of the form given by (4.12) with $k > 0$. To obtain the time dependence given by the wavemaker condition, (4.27), we need

$$\phi = A \cosh k(y + h) \cos k(x - ct), \tag{4.28}$$

with $kc = \sigma$. The dispersion relation (4.14) shows that k must satisfy

$$\sigma^2 = gk \tanh kh. \tag{4.29}$$

Since the curves $f(k) = \sigma^2/gk$ and $g(k) = \tanh kh$ only intersect once for $k > 0$, (4.29) has a unique solution, $k = k_0$. However, (4.27) shows that we need

$$-kA \sin \sigma t \cosh k(y + h) = \frac{1}{2} s(y) \sigma \sin \sigma t, \tag{4.30}$$

which cannot be satisfied in general. Clearly, the solution has a more complicated structure than our first guess of a simple progressive wave.

Let's see if we can construct a solution that is confined to the neighbourhood of the piston or paddle. We proceed as we did in subsection 4.2.2, by looking for a solution of the form $\phi = X(x)Y(y)T(t)$. We choose $T(t) = \sin \sigma t$ to give ourselves a chance of satisfying (4.27) so that

$$Y'' = -k^2 Y, \quad X'' = k^2 X, \tag{4.31}$$

and hence

$$Y = A \cos ky + B \sin ky, \quad X = Ce^{kx} + De^{-kx}. \tag{4.32}$$

We have chosen the separation coefficient with the opposite sign to the standing wave solution of subsection 4.2.2, which leads to exponential decay in the x-direction, and periodic behaviour in the y-direction. From the bed condition, (4.7), we need $Y = F \cos k(y + h)$. For the solution to be finite as $x \to \infty$, we must take $C = 0$, so that the velocity potential is

$$\phi(x, y, t) = Ae^{-kx} \cos k(y + h) \sin \sigma t. \tag{4.33}$$

The combined surface condition, (4.13) then shows that

$$-Ae^{-kx}[\sigma^2 \sin \sigma t \cos kh + gk \sin kh \sin \sigma t] = 0, \tag{4.34}$$

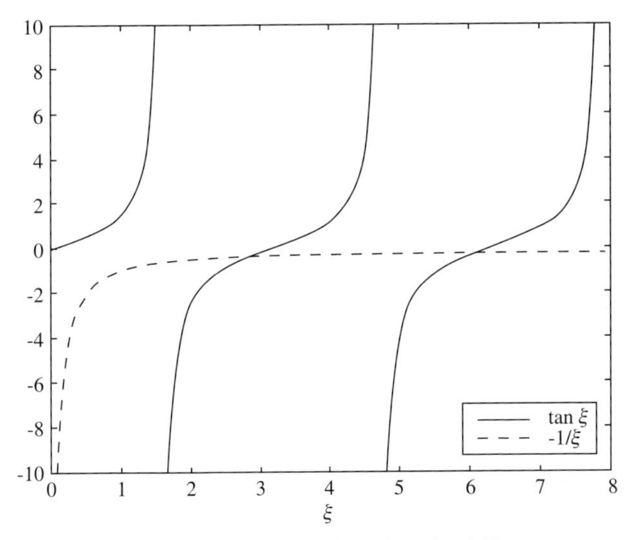

Fig. 4.8. Graphs of $\tan\xi$ and $-1/\xi$.

so that $\sigma^2 = -gk\tan kh$. Putting $\xi = kh$ gives

$$\tan\xi = -\frac{\sigma^2 h}{g\xi}. \tag{4.35}$$

Figure 4.8 shows the functions $-\sigma^2 h/g\xi$ and $\tan\xi$ for the typical case $\sigma^2 h/g = 1$. We can see that there are infinitely many values of ξ for which these curves intersect, and hence that there is a countably infinite sequence of suitable wavenumbers, $k = k_1, \ldots, k_n, \ldots$. The most general solution that is confined to the neighbourhood of $x = 0$ is therefore

$$\phi = \sum_{n=1}^{\infty} A_n e^{-k_n x}\cos k_n(y+h)\sin\sigma t. \tag{4.36}$$

However, this solution is not sufficient as it tends to zero for large x, whereas we expect a wave propagating away from $x = 0$.

The final step is to combine the unique progressive wave solution, (4.28), with the confined solution, (4.36), to give the most general solution,

$$\phi(x,y,t) = A_0\cosh k_0(y+h)\cos(k_0x-\sigma t) + \sum_{n=1}^{\infty} A_n e^{-k_n x}\cos k_n(y+h)\sin\sigma t \tag{4.37}$$

where k_n for $n \geq 1$ are the positive roots of $\sigma^2 = -gk\tan kh$ and k_0 is the unique positive root of $\sigma^2 = gk\tanh kh$. The remaining unsatisfied boundary condition, (4.27), requires that

$$-\frac{1}{2}\sigma s(y) = k_0 A_0 \cosh k_0(y+h) - \sum_{n=1}^{\infty} k_n A_n \cos k_n(y+h) \text{ for } -h \le y \le 0. \tag{4.38}$$

Although we will not prove it here, the functions $\cosh k_0(y+h)$, $\cos k_n(y+h)$ form a complete, orthogonal set, and any well-behaved function $s(y)$ can be represented in this form. In particular,

$$\left.\begin{aligned} &\int_{-h}^{0} \cos k_n(y+h) \cos k_m(y+h)\, dy = 0 \quad \text{for } n \neq m, \\ &\int_{-h}^{0} \cosh k_0(y+h) \cos k_n(y+h)\, dy = 0 \quad \text{for } n = 1, 2, \ldots. \end{aligned}\right\} \tag{4.39}$$

To find A_0, we therefore multiply (4.38) by $\cosh k_0(y+h)$ and integrate between $-h$ and 0 to obtain

$$\int_{-h}^{0} -\frac{1}{2}\sigma s(y) \cosh k_0(y+h)\, dy = A_0 k_0 \int_{-h}^{0} \cosh^2 k_0(y+h)\, dy. \tag{4.40}$$

Similarly, to find A_n, we multiply by $\cos k_n(y+h)$ and integrate between $-h$ and 0 to give

$$\int_{-h}^{0} \frac{1}{2}\sigma s(y) \cos k_n(y+h) = A_n k_n \int_{-h}^{0} \cos^2 k_n(y+h)\, dy. \tag{4.41}$$

For the piston wavemaker, with $s(y) = s_0$,

$$A_0 = \frac{-\sigma s_0 \sinh k_0 h}{k_0\left(k_0 h + \frac{1}{2}\sinh 2k_0 h\right)}, \quad A_n = \frac{\sigma s_0 \sin k_n h}{k_n\left(k_n h + \frac{1}{2}\sin 2k_n h\right)} \quad \text{for } n \ge 1. \tag{4.42}$$

The free surface elevation, η, can now be found from (4.9).

As x increases, the exponentially decaying part of the solution (4.38) decreases rapidly, with the dominant term of $O(e^{-k_1 x})$. Since $k_1 x > \pi x/2h$, we can neglect the exponentially decaying part of the solution when $\pi x/2h \gg 1$. For practical purposes $x = 2h$ will be sufficient for the standing waves to have become very small and we can calculate η from the travelling wave component only as

$$\eta \approx -\frac{\sigma A_0 \cosh k_0 h}{g} \sin(kx - \sigma t). \tag{4.43}$$

We can now compare the amplitude of this wave, $a = \sigma A_0 \cosh k_0 h/g$, with the amplitude of the piston's oscillation, s_0 and find that

$$\frac{a}{s_0} = \frac{2\sinh^2 k_0 h}{2k_0 h + \sinh 2k_0 h}. \tag{4.44}$$

In deep water $a/s_0 \sim 1$ whilst in shallow water $a/s_0 \sim \frac{1}{2}k_0 h \ll 1$.

It is also of some interest to calculate the **hydrodynamic loading** on the piston. As the linearised Bernoulli condition is $p = -\rho\partial\phi/\partial t$, ignoring the **hydrostatic pressure**, $-\rho g y$, the total hydrodynamic force on the piston is

$$F_h = -\rho \int_{-h}^{0} \left.\frac{\partial \phi}{\partial t}\right|_{x=0} dy.$$

Substituting for ϕ from (4.37) gives

$$F_h = \frac{-\rho\sigma^2 s_0 \sinh^2 k_0 h \sin\sigma t}{k_0^2\left(k_0 h + \frac{1}{2}\sinh 2k_0 h\right)} - \sum_{k=1}^{\infty} \frac{\rho\sigma^2 s_0 \sin^2 k_n h \cos\sigma t}{k_n^2\left(k_n h + \frac{1}{2}\sin 2k_n h\right)}. \tag{4.45}$$

This can be expressed in terms of the velocity, $\dot{x}_p$, and acceleration, $\ddot{x}_p$, of the piston as

$$\begin{aligned} F_h &= \frac{2\rho\sigma \sinh^2 k_0 h}{k_0^2\left(k_0 h + \frac{1}{2}\sinh 2k_0 h\right)}\dot{x}_p + \sum_{k=1}^{\infty} \frac{2\rho \sin^2 k_n h}{k_n^2\left(k_n h + \frac{1}{2}\sin 2k_n h\right)}\ddot{x}_p \\ &\equiv D_p \dot{x}_p + M_p \ddot{x}_p. \end{aligned} \tag{4.46}$$

We can now consider all of the forces and accelerations on the piston, which has mass m. If the driving force is F_p, by Newton's second law

$$m\ddot{x}_p = F_p - (D_p\dot{x}_p + M_p\ddot{x}_p), \tag{4.47}$$

since the hydrodynamic force opposes the motion of the piston. This can be written as

$$(m + M_p)\ddot{x}_p + D_p\dot{x}_p = F_p. \tag{4.48}$$

The quantity M_p is called the **added mass** of the piston, and D_p is the **added damping**. The added mass represents the work done by the piston in moving the fluid ahead of it. The added damping represents the work done in producing the waves that radiate into the fluid.

4.2.4 The Extraction of Energy from Water Waves

As we have seen earlier in this chapter, water waves carry energy with them as they propagate. The commercial extraction of this hydrodynamic energy by converting it into a more useful form, such as electrical energy, is an attractive proposition, since there are none of the problems associated with pollution that arise when fossil fuels are burnt or after nuclear fission. The construction of a device that can achieve the conversion of wave power into electrical power is a considerable engineering challenge. We will now consider a simple device that resembles the wavemaker

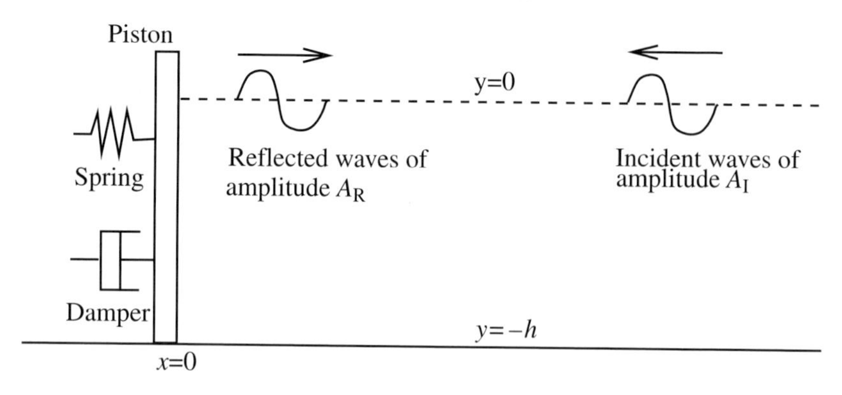

Fig. 4.9. A device for absorbing the energy in water waves.

of the previous section, and see how it can most efficiently be used to generate wave power.

Let's suppose that we have a tank of water with a piston at one end. Waves are *incident* upon the piston, which is connected to the device that extracts the energy by a linkage that is equivalent to a spring and a damper. For motion with a sufficiently small amplitude, any device can be represented in this way, with the spring representing the total stiffness and the damper the sum of all frictional effects. The piston moves in response to the incident wave, and some of the incident energy is reflected back down the tank, as illustrated in figure 4.9. It is convenient in our analysis of this system to use the complex exponential form of the solutions for linear water waves, as this readily allows us to deal with any phase shifts between the incident wave, the reflected wave, and the displacement of the piston. The appropriate form of the velocity potential is, as before,

$$\phi = A_{\mathrm{I}} \cosh k_0(y+h)e^{-i(k_0 x+\sigma t)} + A_{\mathrm{R}} \cosh k_0(y+h)e^{-i(k_0 x-\sigma t)} + \sum_{n=1}^{\infty} A_n \cos k_n(y+h)e^{-k_n x - i\sigma t},$$

where A_{I} is the amplitude of the incident wave, which propagates from right to left, A_{R} is the amplitude of the reflected wave, which propagates from left to right, and the coefficients A_n, which are now complex, determine the standing wave that will be needed to satisfy the boundary condition at the piston. If the displacement of the piston is

$$x_{\mathrm{p}} = -\frac{1}{2}Se^{i\sigma t}, \tag{4.49}$$

we can use the orthogonality of cosh and cos as before, and show that

$$\frac{1}{4}i(A_{\mathrm{R}} - A_{\mathrm{I}})(\sinh 2k_0 h + 2k_0 h) = \frac{i\sigma S}{2k_0}\sinh k_0 h, \tag{4.50}$$

$$-\frac{1}{4}A_n(\sin 2k_n h + 2k_n h)\frac{i\sigma S}{2k_n}\sin k_n h, \quad \text{for } n = 1, 2, \ldots. \tag{4.51}$$

If we know the properties of the incident wave, A_{I} and k_0, the dispersion relation then determines σ, and (4.50) and (4.51) determine A_{R} and A_n as functions of S. The value of S can then be found from the equation of motion of the piston. The hydrodynamic force on the piston is

$$F_h = -D_{\mathrm{p}}\dot{x}_{\mathrm{p}} - M_{\mathrm{p}}\ddot{x}_{\mathrm{p}} - \frac{2i\rho\sigma A_{\mathrm{I}}}{k_0}\sinh k_0 h e^{-i\sigma t},$$

which, compared to the equivalent equation for the wavemaker, (4.46), has an extra term due to the incident wave. The equation of motion of the piston is

$$m\ddot{x}_{\mathrm{p}} + D\dot{x}_{\mathrm{p}} + \kappa x_{\mathrm{p}} = F_h,$$

where m is the mass of the piston, D is the friction factor due to the damper and κ the stiffness of the spring, and hence

$$(m + M_{\mathrm{p}})\ddot{x}_{\mathrm{p}} + (D + D_{\mathrm{p}})\dot{x}_{\mathrm{p}} + \kappa x_{\mathrm{p}} = -\frac{2i\rho\sigma A_{\mathrm{I}}}{k_0}\sinh k_0 h e^{-i\sigma t}. \tag{4.52}$$

If we now substitute from (4.49) for x_{p}, we can determine S. If S is real, the motion of the piston is in phase with the incoming waves, whilst if S is imaginary it is completely out of phase with the incoming waves.

We can now determine how we should design the piston to make an efficient device for extracting wave power. Within the context of this simple theory, we want as much of the incident energy as possible to be absorbed by the piston, and hence we should try to minimise the reflected wave. If we put $A_{\mathrm{R}} = 0$, (4.50) shows that

$$A_{\mathrm{I}} = \frac{-2\sigma S\sinh k_0 h}{k_0(\sinh k_0 h + 2k_0 h)}, \tag{4.53}$$

Now substituting (4.49) into (4.52) gives

$$-\sigma^2(m + M_{\mathrm{p}}) - i\sigma(D + D_{\mathrm{p}}) + \kappa = -2i\sigma D_{\mathrm{p}},$$

and hence, equating real and imaginary parts,

$$\kappa = \sigma^2(m + M_{\mathrm{p}}), \quad D = D_{\mathrm{p}}.$$

This choice of spring stiffness and damping gives a perfectly tuned system.

In practice, the random and probably nonlinear nature of the state of the sea means that only a less than perfect tuning can be achieved.

4.3 The Effect of Surface Tension: Capillary–Gravity Waves

All molecules in a fluid experience a force due to the presence of the surrounding fluid molecules. In the bulk of the fluid these intermolecular forces balance and the total force is zero. However at the surface there is an imbalance in intermolecular forces, which leads to a non-zero force exerted on each molecule by the bulk fluid. This force can, for example, support a needle placed carefully onto the surface of a body of water, and allow an insect to walk across a pond. It is known as **surface tension**. When a fluid is in contact with the walls of a container there is a change in the level of the fluid at the walls caused by surface tension effects. If gravity acts on such a fluid so that the surface has an equilibrium position (called a capillary surface), we can use the physical quantities, the fluid density, ρ, gravity, g, and surface tension, σ, to form a quantity with dimensions of length, $(\sigma/\rho g)^{1/2}$, often called the **capillary length**. For water, σ is easily measured, and is about $0.07\,\mathrm{N\,m^{-1}}$, which gives a capillary length of approximately a couple of millimetres. The equilibrium surface varies on this length scale, as is readily observed at the edge of a stationary cup of water.

Consider a small element of the free surface, as shown in figure 4.10. A simple force balance gives

$$p_{\mathrm{f}}\delta s = p_{\mathrm{atm}}\delta s + 2\sigma\sin(\delta\theta/2),$$

where σ is the surface tension. Since $\delta s \sim R\delta\theta$ as $\delta\theta \to 0$, this shows that

$$p_{\mathrm{f}} - p_{\mathrm{atm}} = \frac{\sigma}{R},$$

where R is the radius of curvature. For two-dimensional deformations

$$\frac{1}{R} = -\frac{\partial^2\eta/\partial x^2}{\{1+(\partial\eta/\partial x)^2\}^{3/2}}.$$

For linear waves we can neglect the quadratic term in the denominator, and hence

$$p_{\mathrm{f}} - p_{\mathrm{atm}} = -\sigma\frac{\partial^2\eta}{\partial x^2}. \qquad (4.54)$$

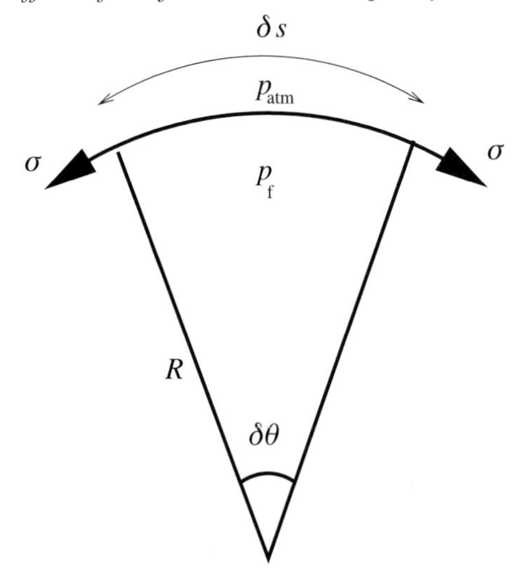

Fig. 4.10. Capillary forces on a small element of the free surface.

We can now incorporate the effect of surface tension into our linear theory of water waves. The linearised Bernoulli equation is

$$\frac{\partial \phi}{\partial t} + gy + \frac{p}{\rho} = 0.$$

Since $p = p_f - p_{atm}$ at $y = \eta$, this gives the dynamic boundary condition as

$$\frac{\partial \phi}{\partial t} + g\eta - \frac{\sigma}{\rho}\frac{\partial^2 \eta}{\partial x^2} = 0 \quad \text{at } y = \eta(x, t). \tag{4.55}$$

The kinematic condition, (4.10), is not affected by surface tension. If we combine the kinematic and modified dynamic boundary conditions, (4.10) and (4.55), using the same procedure that led to (4.13), we arrive at

$$\frac{\partial^2 \phi}{\partial t^2} + g\frac{\partial \phi}{\partial y} - \frac{\sigma}{\rho}\frac{\partial^3 \phi}{\partial y \partial x^2} = 0 \quad \text{at } y = \eta(x, t).$$

Since ϕ satisfies Laplace's equation, we know that $\partial^2\phi/\partial x^2 = -\partial^2\phi/\partial y^2$, and hence

$$\frac{\partial^2 \phi}{\partial t^2} + g\frac{\partial \phi}{\partial y} + \frac{\sigma}{\rho}\frac{\partial^3 \phi}{\partial y^3} = 0 \quad \text{at } y = \eta(x, t). \tag{4.56}$$

We could now go through the analysis of either progressive or standing **capillary–gravity waves**, as before, but we leave the details as exercise 4.1. Some typical capillary-gravity waves are shown in figure 4.11.

Fig. 4.11. Capillary–gravity waves around a rock in a stream.

If $\sigma \gg \rho g \lambda^2$, in other words the wavelength, λ, is much smaller than the capillary length, $(\sigma/\rho g)^{1/2}$, gravity is negligible compared to the effect of surface tension, and we can effectively set $g = 0$ and study **capillary waves**. At the other extreme, if $\lambda \gg (\sigma/\rho g)^{1/2}$, the effect of surface tension is negligible, and we have gravity waves.

Both progressive capillary–gravity and capillary waves are dispersive, and the dispersion relation is

$$\omega^2 = \left(g + \frac{\sigma}{\rho} k^2\right) k \tanh kh. \tag{4.57}$$

The phase speed, $c_{\mathrm{p}} = \omega/k$, for capillary–gravity waves is plotted in figure 4.12, along with the phase speed for capillary waves ($g = 0$) and gravity waves ($\sigma = 0$), on 10 cm of water as a function of the wavelength, λ. For capillary–gravity waves there is a minimum phase speed, c_{pmin}, at $\lambda = \lambda_{\mathrm{min}}$. The group velocity is not equal to the phase velocity for capillary–gravity waves, but varies with wavelength in a qualitatively similar manner. However, one important difference between capillary waves and gravity waves on deep water is that the group velocity for capillary waves is greater than the phase velocity, whilst the opposite

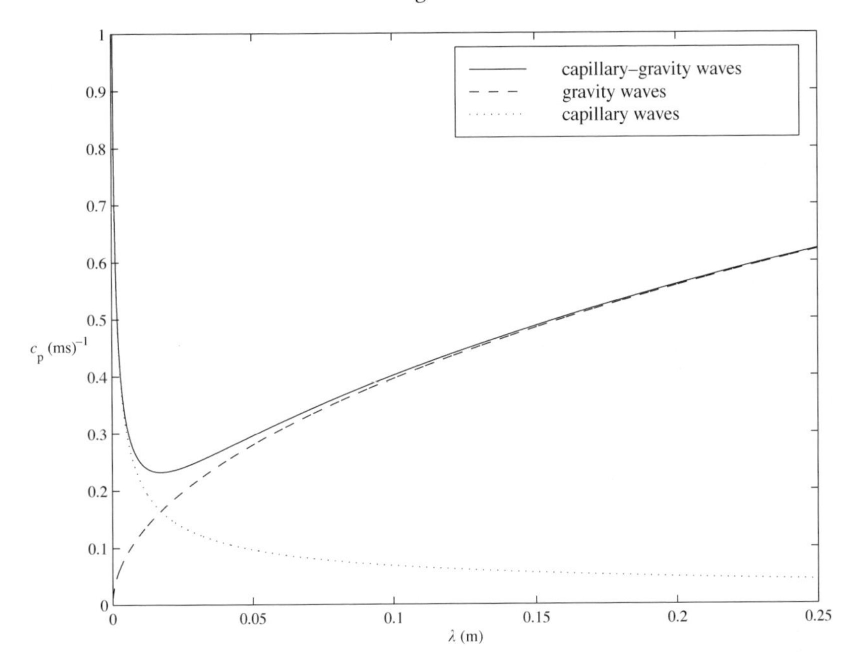

Fig. 4.12. The phase speed for capillary waves, gravity waves and capillary–gravity waves on 10 cm of water.

is true for gravity waves. One direct implication of this is discussed in section 4.5, where we consider waves generated by a moving ship.

4.4 Edge Waves

As an example of what happens with a different bed topography, we consider the propagation of **edge waves** in a simple model of what happens at a beach. If the fixed, rigid bed now makes an angle β with the horizontal, its equation is $y = -x \tan \beta$, where y and x are Cartesian coordinates, as before. In this case we use a three-dimensional set of Cartesian (x, y, z)-axes. The equilibrium position of the fluid surface is $y = 0$ for $x > 0$, as illustrated in figure 4.13. Since the unit normal to the sloping bed is $\mathbf{n} = (\sin \beta, \cos \beta, 0)$, the condition that there should be no normal velocity at the bed becomes

$$\sin \beta \frac{\partial \phi}{\partial x} + \cos \beta \frac{\partial \phi}{\partial y} = 0 \quad \text{at } y = -x \tan \beta. \tag{4.58}$$

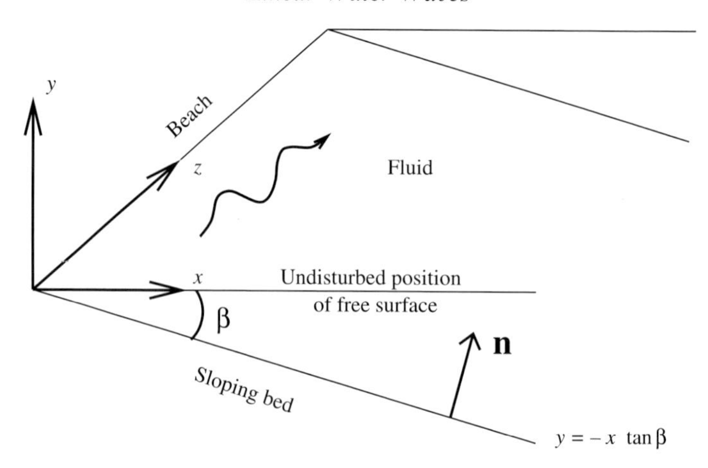

Fig. 4.13. The coordinate system and geometry for edge waves.

The kinematic and dynamic free surface boundary conditions, (4.9) and (4.10), remain the same, and we must now solve the three-dimensional version of Laplace's equation for ϕ. We seek a solution that propagates along the beach, in the z-direction, of the form $\phi = f(z - ct)F(x, y)$. Substituting this into Laplace's equation gives

$$fF_{xx} + fF_{yy} + f''F = 0,$$

and hence

$$\frac{F_{xx} + F_{yy}}{F} = -\frac{f''}{f} = k^2.$$

Since f satisfies $f'' + k^2 f = 0$, we can, without loss of generality, write f as

$$f = A\cos k(z - ct).$$

The function $F(x, y)$ satisfies the **modified Helmholtz equation**

$$F_{xx} + F_{yy} = k^2 F. \tag{4.59}$$

We seek a solution of the form $F = Be^{-bx+dy}$. On substituting this into equation (4.59) we find that

$$b^2 + d^2 = k^2,$$

Similarly, the bed condition (4.58) leads to

$$-b\sin\beta + d\cos\beta = 0.$$

Hence

$$b = k\cos\beta, \quad d = k\sin\beta,$$

and the solution takes the form

$$\phi(x, y, t) = C\exp(-kx\cos\beta + ky\sin\beta)\cos k(z - ct). \tag{4.60}$$

Since $\phi \to 0$ exponentially as $x \to \infty$, the edge wave is confined to the neighbourhood of the beach by the effect of the sloping bed. As β increases, the wave is confined in a broader band, until it becomes a deep water wave when $\beta = \pi/2$. If we substitute (4.60) into the combined free surface condition, (4.13), we obtain

$$C\exp(-kx\cos\beta)(-k^2c^2 + gk\sin\beta)\cos k(z - ct) = 0.$$

Since this must hold for all values of x and z we obtain the dispersion relation

$$c^2 = \frac{g\sin\beta}{k}. \tag{4.61}$$

Note that when $\beta = \pi/2$ we obtain the dispersion relation, (4.18), for deep water waves. These arguments can be generalised to allow for a wider variety of edge waves (see, for example, Evans and Kuznetsov (1997)).

4.5 Ship Waves

How does dispersion affect the wave pattern produced by a moving object? The most striking example of this is the wave pattern produced by an object, such as a duck or a ship, moving at a constant speed across deep water, as shown in figure 4.14. The wave patterns are confined to a wedge and are almost identical, in spite of their different sizes. These waves are usually referred to as **ship waves**, and were first studied by Lord Kelvin in 1887. Kelvin constructed a detailed solution by considering a point impulse moving across the surface of an undisturbed body of water, and then used the method of stationary phase to estimate where the amplitude of the disturbance would be greatest. We will not go into as much detail here (see exercise 4.10). Instead we use geometrical arguments, again due to Kelvin, that allow us to construct the shape of the lines of the wave crests. We treat the ship as a source of concentric waves of all wavelengths, and assume that it moves at constant velocity V in a straight line. (The amplitude of each of the wavelengths generated depends on the shape of the ship, and this is beyond the scope of this

Fig. 4.14. A boat moving through deep water at constant speed.

simple theory.) This approximation is similar to that made when applying **Huyghens' principle** in diffraction problems (see, for example, Kevorkian (1990) or Born and Wolf (1975)).

For deep water gravity waves we have seen that the dispersion relation is

$$\omega = \sqrt{gk},$$

and hence that the phase and group velocities are related by

$$c_{\mathrm{p}} = 2c_{\mathrm{g}} = \sqrt{\frac{g}{k}}.$$

The waves generated by the ship travel at different velocities depending on their wavelength, with the wave crests travelling twice as fast as the energy. If we assume that waves of all wavenumbers are generated by the ship, there is a component of the disturbance corresponding to any positive wave speed. We now examine the wave pattern that appears steady from the point of view of an observer on the ship. The component of the velocity of the ship in the direction of motion of the crests of waves must be equal to c_{p}. These waves are the ones that propagate at

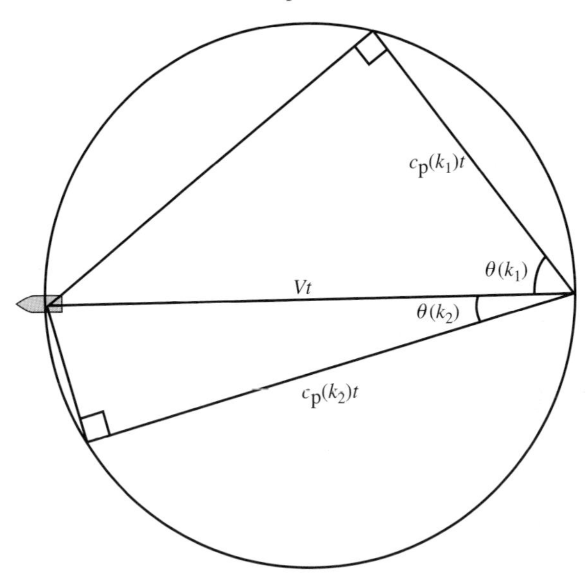

Fig. 4.15. The angle to the line of motion, $\theta(k)$, made by waves of wavenumber k whose crests appear stationary to an observer on the ship.

an angle $\theta(k)$ to the line of motion of the ship, where

$$\cos\theta(k) = \frac{c_p(k)}{V}.$$

This is illustrated in figure 4.15 for two different wavenumbers, k_1 and k_2. For each value of the wavenumber k, simple trigonometry shows that the triangles illustrated must be right-angled, and hence that the locus of all the points where the wave crests appear stationary is a circle of radius $\frac{1}{2}Vt$.

This argument would be fine, except for one flaw. The leading edge of the region disturbed by the motion of the ship at time $t = 0$ propagates at the group velocity $c_g(k) = \frac{1}{2}c_p(k)$, since this is the velocity at which energy is transmitted. This means that the wave crests travel faster than this leading edge, and disappear into it. We must modify our argument so that the wave crests that appear to be motionless lie on a circle of radius $\frac{1}{4}Vt$, as shown in figure 4.16. Circles that represent the position of the stationary wave crests emitted by the ship at later times are also shown in figure 4.16. These circles are bounded by a wedge with half angle α, known as the **Kelvin wedge**. A little trigonometry on the triangle BCD shown in figure 4.17 shows that $\alpha = \sin^{-1}(1/3) \approx 19.5°$, independent of V.

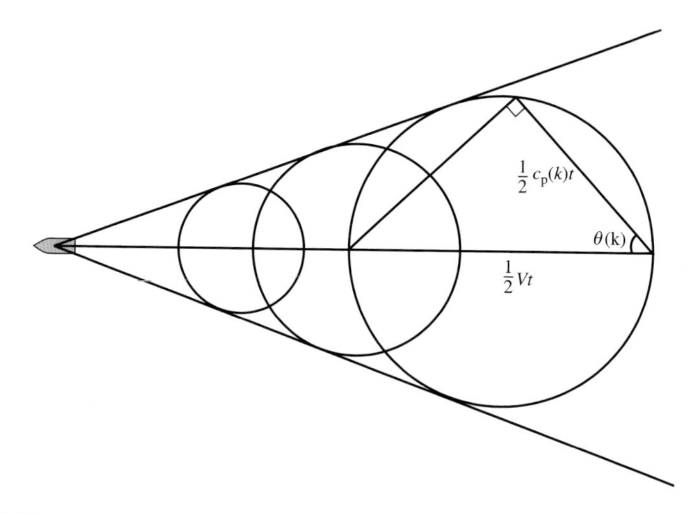

Fig. 4.16. The angle to the line of motion, $\theta(k)$, made by waves of wavenumber k whose crests appear stationary to an observer on the ship, taking into account the propagation of energy at the group velocity, $c_g(k) = \frac{1}{2}c_p(k)$.

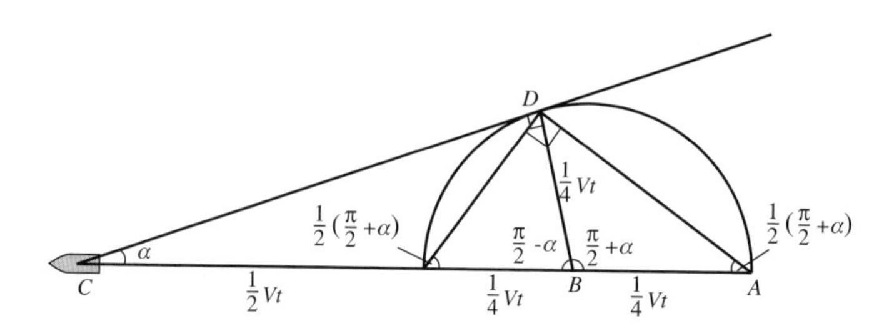

Fig. 4.17. The construction of the Kelvin wedge.

The wavelength of waves propagating at an angle θ to the direction of motion is

$$\lambda = 2\pi V^2 \cos^2\theta/g.$$

This means that the maximum wavelength, $\lambda_{\max} = 2\pi V^2/g$, is for waves that propagate in the same direction as the ship, $\theta = 0$. The wave crests at the edge of the Kelvin wedge make an angle $\frac{1}{2}\left(\frac{\pi}{2} + \alpha\right) \approx 55^\circ$ with the line of motion of the ship and have wavelength $\frac{2}{3}\lambda_{\max}$. These results are illustrated in figure 4.18.

We can now consider the pattern of the wave crests in more detail. Let the ship move at velocity $(-V, 0)$ along the x-axis of an (x, y)-coordinate system and lie instantaneously at the origin. Consider the waves generated

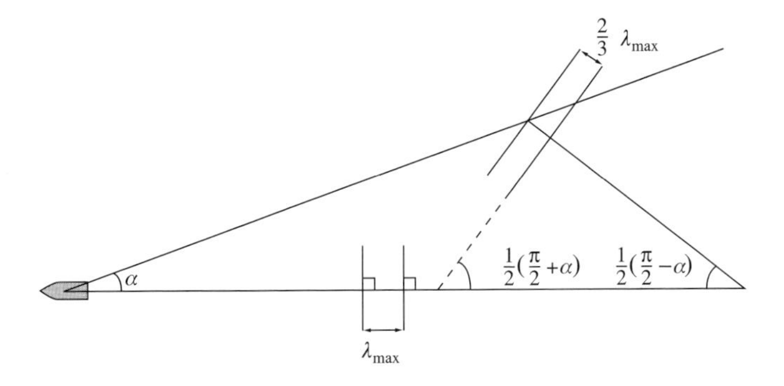

Fig. 4.18. The angles and wavelengths of the longest and bounding waves.

when the ship was at the point $(X, 0)$, propagating at an angle θ to the x-axis. Let the wave crest that appears stationary from the ship lie at the point (x_0, y_0). Some trigonometry on the triangle illustrated in figure 4.19 shows that

$$x_0 = X(\theta)\left(1 - \frac{1}{2}\cos^2\theta\right), \quad y_0 = \frac{1}{2}X(\theta)\cos\theta\sin\theta. \tag{4.62}$$

In order to consider the line of a given wave crest, which we parameterise using θ, we treat X as a function of θ since different parts of the wave crest will be due to waves emitted at different times, and hence different values of X. We can also see from figure 4.19 that the slope of the wave crest at (x_0, y_0) is

$$\frac{dy}{dx} = \cot\theta. \tag{4.63}$$

Using the chain rule, $dy/dx = (dy_0/d\theta)/(dx_0/d\theta)$ on the line of a wave crest. Calculating $dy_0/d\theta$ and $dx_0/d\theta$ from (4.62) and substituting into (4.63) leads, after some algebra, to

$$\frac{dX}{d\theta} = -X\tan\theta. \tag{4.64}$$

We can solve this separable ordinary differential equation to show that $X = A\cos\theta$, where A is a constant. Finally, substituting into (4.62) gives us the line of a wave crest as

$$x_0 = A\cos\theta\left(1 - \frac{1}{2}\cos^2\theta\right), \quad y_0 = \frac{1}{2}A\cos^2\theta\sin\theta. \tag{4.65}$$

The line is plotted in figure 4.20 for various values of A. Note that the pattern is bounded by the Kelvin wedge, where there is a cusp in the

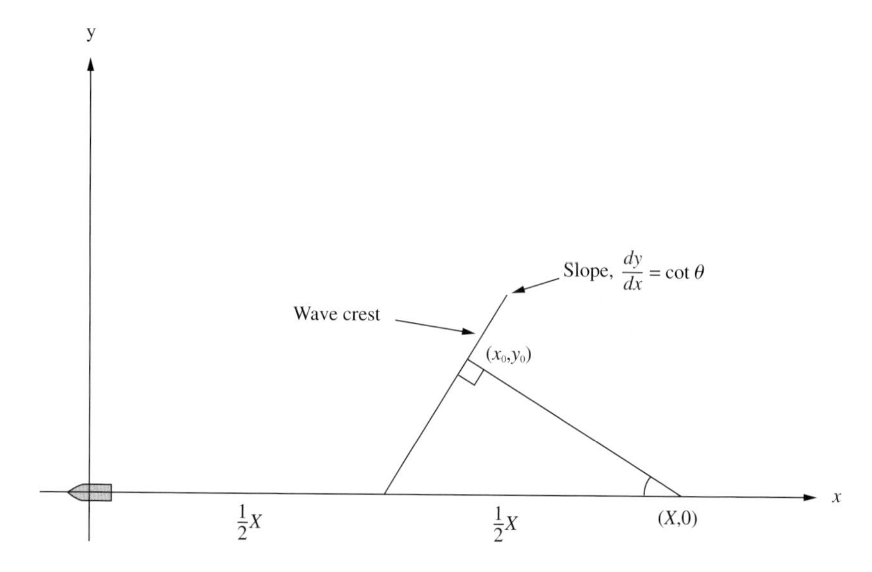

Fig. 4.19. The coordinate system and stationary wave crest for the calculation of the ship wave pattern.

line of the wave crests. The amplitude of the wave that lies along each of these lines is a function of the amplitudes generated by the ship, and cannot be determined from our simplified analysis. The photographs in figure 4.14 suggest that the amplitude is usually greatest close to the edge of the Kelvin wedge, and are in good agreement with the patterns shown in figure 4.20.

4.6 The Solution of Initial Value Problems

We now consider initial value problems where the fluid is initially at rest, but where the free surface is displaced from its flat, equilibrium state, so that

$$\phi = 0, \quad \frac{\partial \eta}{\partial t} = 0 \quad \text{and} \quad \eta = \eta_0(x) \quad \text{when } t = 0. \tag{4.66}$$

In addition, we only consider the limit of deep water, so that the bed condition (4.58) is replaced by

$$\phi \to 0 \quad \text{as } y \to -\infty. \tag{4.67}$$

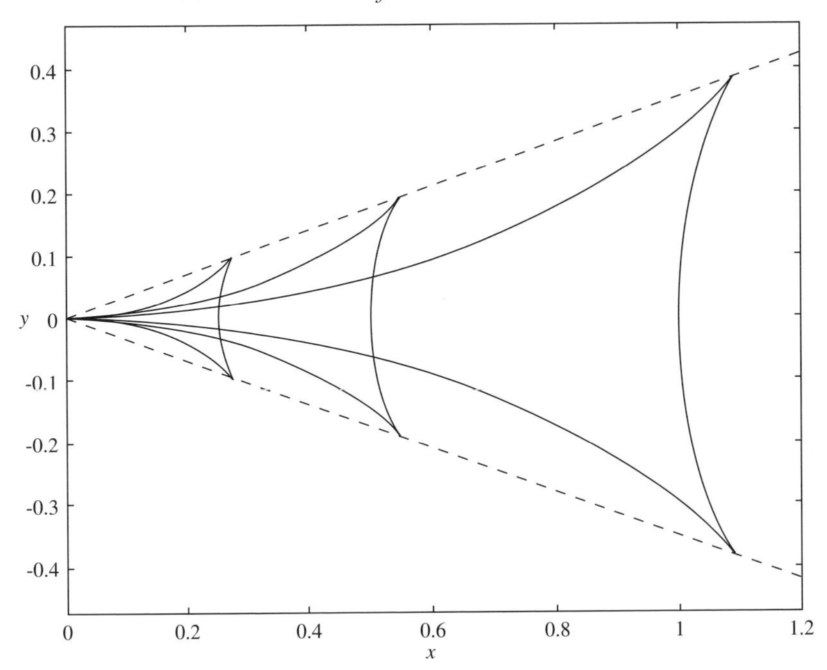

Fig. 4.20. The shape of the wave crests in the Kelvin ship wave pattern, obtained from (4.65) for various values of A. The pattern is bounded by the Kelvin wedge (dashed line).

To solve this problem we use Fourier transforms, defined by

$$\tilde{\phi}(k, y, t) = \int_{-\infty}^{\infty} e^{ikx}\phi(x, y, t)dx \quad \text{and} \quad \tilde{\eta}(k, t) = \int_{-\infty}^{\infty} e^{ikx}\eta(x, t)dx. \quad (4.68)$$

Note that we assume that the initial disturbance is sufficiently localised that $\eta \to 0$ and $\phi \to 0$ as $x \to \pm\infty$ rapidly enough that these Fourier transforms are well-defined. In effect, we are applying the radiation condition, which states that there are no waves incoming from infinity. Note that

$$\widetilde{\left(\frac{\partial \phi}{\partial x}\right)} = \int_{-\infty}^{\infty} e^{ikx}\frac{\partial \phi}{\partial x}dx = \left[e^{ikx}\phi\right]_{-\infty}^{\infty} - ik\int_{-\infty}^{\infty} e^{ikx}\phi dx = -ik\tilde{\phi},$$

and similarly

$$\widetilde{\left(\frac{\partial^2 \phi}{\partial x^2}\right)} = -k^2\tilde{\phi}, \quad \widetilde{\left(\frac{\partial^2 \phi}{\partial y^2}\right)} = \frac{\partial^2 \tilde{\phi}}{\partial y^2}.$$

The first step is to take a Fourier transform of Laplace's equation so that

$$\frac{\partial^2 \tilde{\phi}}{\partial y^2} - k^2 \tilde{\phi} = 0.$$

The solution of this equation that satisfies the boundary condition (4.67) is

$$\tilde{\phi} = A(k,t) e^{|k|y}. \tag{4.69}$$

The second step is to take a Fourier transform of the combined free surface condition, (4.13), which gives

$$\frac{\partial^2 \tilde{\phi}}{\partial t^2} + g \frac{\partial \tilde{\phi}}{\partial y} = 0 \quad \text{at } y = 0,$$

and hence from (4.69)

$$\frac{\partial^2 A}{\partial t^2} + g|k|A = 0.$$

This has the general solution

$$A = a(k)e^{i\omega t} + b(k)e^{-i\omega t},$$

where

$$\omega = \sqrt{g|k|}, \tag{4.70}$$

and hence

$$\tilde{\phi} = \left\{ a(k)e^{i\omega t} + b(k)e^{-i\omega t} \right\} e^{|k|y}. \tag{4.71}$$

The third step is to determine $a(k)$ and $b(k)$ from the initial conditions. Consider first the Fourier transform of the dynamic free surface boundary condition, (4.9),

$$\frac{\partial \tilde{\phi}}{\partial t} + g\tilde{\eta} = 0 \quad \text{at } y = 0. \tag{4.72}$$

Substituting the solution (4.71) into this equation gives

$$i\omega \left\{ a(k)e^{i\omega t} - b(k)e^{-i\omega t} \right\} + g\tilde{\eta} = 0,$$

and hence when $t = 0$

$$i\omega \left\{ a(k) - b(k) \right\} + g\tilde{\eta}_0 = 0, \tag{4.73}$$

where $\tilde{\eta}_0 = \tilde{\eta}(k, 0)$. Similarly, the Fourier transform of the kinematic free surface condition, (4.10), is

$$\frac{\partial \tilde{\phi}}{\partial y} = \frac{\partial \tilde{\eta}}{\partial t} \quad \text{at } y = 0,$$

and hence

$$a(k) + b(k) = 0. \tag{4.74}$$

The solution of (4.73) and (4.74) is

$$a(k) = -b(k) = \frac{ig\tilde{\eta}_0}{2\omega},$$

and hence

$$\tilde{\phi}(k, y, t) = \frac{g}{\omega}\frac{1}{2}i\left(e^{i\omega t} - e^{-i\omega t}\right)e^{|k|y}\tilde{\eta}_0(k). \tag{4.75}$$

The fourth and final step is to determine the free surface elevation from the solution, (4.75), for the velocity potential. Substituting (4.75) into (4.72) shows that

$$\tilde{\eta}(k, t) = \frac{1}{2}\left(e^{i\omega t} + e^{i\omega t}\right)\tilde{\eta}_0(k). \tag{4.76}$$

Finally, we can invert the transform and write

$$\eta(x, t) = \frac{1}{4\pi}\int_{-\infty}^{\infty}\tilde{\eta}_0(k)\left\{e^{-i(kx-\omega t)} + e^{-i(kx+\omega t)}\right\}dk. \tag{4.77}$$

This is a solution of **wavepacket form**. Each Fourier component of the original free surface elevation propagates in the positive and negative x-directions with speed $\omega/k = \sqrt{g/|k|}$.

In general, if we want to interpret what this integral representation of the solution tells us about the free surface, we must approximate it in some way. The most informative approach is to consider the asymptotic approximation to the integral (4.77) for $t \gg 1$ with x/t fixed. In other words, consider what happens a long time after the initial disturbance, at a point that moves with constant velocity, x/t. This is another integral that we can approximate using the method of stationary phase.

Firstly, we write (4.77) in the form

$$\eta(x, t) = \frac{1}{4\pi}\int_{-\infty}^{\infty}\tilde{\eta}_0(k) \times \left[\exp\left\{-it\left(k\frac{x}{t} - \omega\right)\right\} + \exp\left\{-it\left(k\frac{x}{t} + \omega\right)\right\}\right]dk. \tag{4.78}$$

Using the definitions given for equation (1.15), in our initial value problem, $f(k) = \tilde{\eta}_0(k)$, $g(k) = -\{k\frac{x}{t} - \omega(k)\}$, with x/t constant. Since

$$g'(k) = -\left\{\frac{x}{t} - \omega'(k)\right\},$$

the points of stationary phase occur when $x/t = d\omega/dk$. For deep water waves, $\omega^2 = g|k|$, and hence

$$\omega\frac{d\omega}{dk} = g\ \mathrm{sgn}(k) \quad \Rightarrow \quad \frac{d\omega}{dk} = \frac{g\ \mathrm{sgn}(k)}{2\sqrt{g|k|}}.$$

If we restrict out attention to points moving in the positive x-direction, so that $x/t > 0$ and $x > 0$, this means that $k > 0$ at points of stationary phase, and hence

$$\frac{x}{t} = \frac{1}{2}\sqrt{\frac{g}{k}}.$$

This gives a unique point of stationary phase at $k = k_0$, where

$$k_0 = \frac{gt^2}{4x^2}.$$

Note that for $x/t < 0$ and $x < 0$ there is a unique point of stationary phase at $k = -k_0$. At $k = k_0$,

$$g(k_0) = \frac{gt}{4x} \quad \text{and} \quad g''(k_0) = -\frac{2x^3}{gt^3},$$

and hence, from equation (1.16),

$$\int_{-\infty}^{\infty} \tilde{\eta}_0(k) \exp\left\{-it\left(k\frac{x}{t} - \omega\right)\right\}$$

$$\sim \tilde{\eta}_0\left(\frac{gt^2}{4x^2}\right)\left\{\frac{2\pi}{t(2x^3/gt^3)}\right\}^{1/2} \exp\left\{i\left(\frac{gt^2}{4x} - \frac{\pi}{4}\right)\right\}, \text{ for } t \gg 1.$$

Similarly

$$\int_{-\infty}^{\infty} \tilde{\eta}_0(k) \exp\left\{-it\left(k\frac{x}{t} + \omega\right)\right\} dk$$

$$\sim \tilde{\eta}_0\left(\frac{gt^2}{4x^2}\right)\left\{\frac{2\pi}{t(2x^3/gt^3)}\right\}^{1/2} \exp\left\{-i\left(\frac{gt^2}{4x} - \frac{\pi}{4}\right)\right\} \text{ for } t \gg 1.$$

Hence, as $t \to \infty$ with $x/t = O(1)$,

$$\eta(x,t) \sim \tilde{\eta}_0\left(\frac{gt^2}{4x^2}\right)\left\{\frac{1}{2\pi t(2x^3/gt^3)}\right\}^{1/2} \cos\left(\frac{gt^2}{4x} - \frac{\pi}{4}\right). \tag{4.79}$$

The amplitude of the free surface displacement decays like $t^{-1/2}$ as $t \to \infty$, whilst the wavelength is $8\pi x^2/gt^2$, a constant of $O(1)$.

Physically, this means that if you are sitting at a point moving with velocity x/t, the wavenumber of the dominant waves that you can observe there is k_0, where k_0 is such that $x/t = d\omega/dk$. This suggests that the energy associated with the Fourier component of the initial disturbance with wavenumber k propagates with the group velocity $c_g = d\omega/dk$. As we showed earlier, this is indeed the velocity at which energy propagates in progressive gravity waves. Note firstly that c_g is a function of the wavenumber k (the waves are dispersive), so that after a long time the different Fourier components spread out. This is why the amplitude of the disturbance tends to zero as $t \to \infty$. Secondly, note that the wave crests travel at the phase velocity, $c_p = \omega/k \neq c_g$. For deep water waves, $c_g = \frac{1}{2}c_p$. In other words, the energy travels at half the speed of the wave crests. This phenomenon is easily observed in a wave tank, where wave crests can be seen moving into the leading edge of a propagating disturbance and seemingly disappearing there. Similarly, if the piston in a wavemaker tank is suddenly turned off, there will be a trailing edge to the wavepacket that it has produced, at which wave crests appear and propagate forward. Note that for capillary waves the group velocity is greater than the phase velocity, and the opposite is true. In all wave bearing physical systems, it is the speed of propagation of the energy, the group velocity, that determines how fast the leading edge of any disturbance propagates, not the speed of the wave crests, the phase velocity. As we saw in chapter 1, these ideas are relevant to many other *linear* systems of partial differential equations that have wave-like solutions.

4.7 Shallow Water Waves: Linear Theory

We now reconsider the theory of two-dimensional water waves when the depth of the water is much less than the wavelength of the disturbance of the free surface. We will consider what happens when the depth of water below the undisturbed, flat free surface is a function $h_0(x)$ and the depth of water at any point is $h(x,t)$. The situation is illustrated in figure 4.21. We assume that (i) the vertical acceleration of the fluid is much smaller than the gravitational acceleration, g, and (ii) the horizontal component of the fluid velocity is approximately uniform along any vertical section through the fluid.

The equation for conservation of vertical momentum is

$$\frac{\partial v}{\partial t} + u\frac{\partial v}{\partial x} + v\frac{\partial v}{\partial y} = -\frac{1}{\rho}\frac{\partial p}{\partial y} - g,$$

where the horizontal and vertical components of the velocity field are u and v respectively. If we neglect the vertical accelerations of the fluid, we

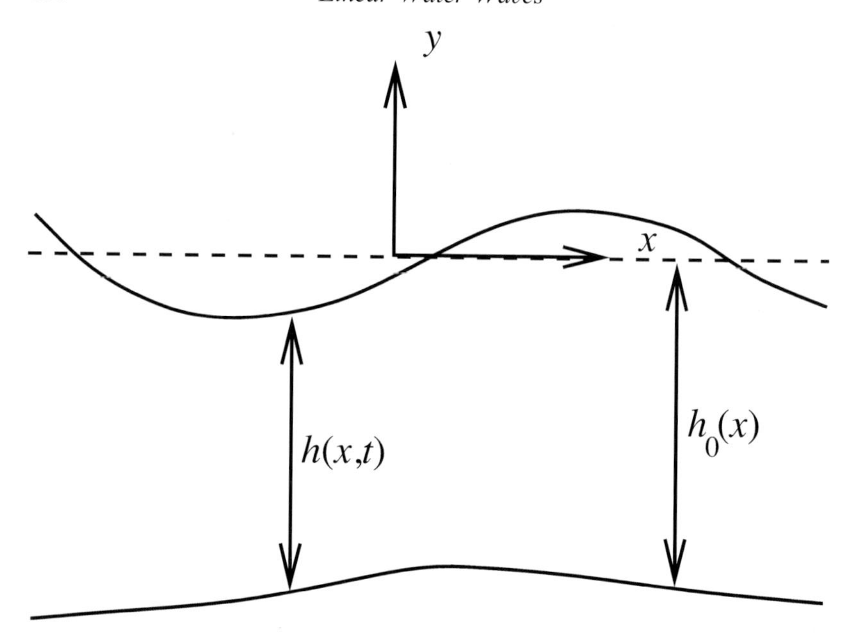

Fig. 4.21. The coordinate system and definitions of quantities involved in shallow water theory.

find that the pressure in the fluid is purely hydrostatic, with

$$\frac{\partial p}{\partial y} = -\rho g,$$

and hence

$$p = p_{\mathrm{atm}} + \rho g(h - h_0 - y).$$

The equation for conservation of horizontal momentum is

$$\frac{\partial u}{\partial t} + u\frac{\partial u}{\partial x} + v\frac{\partial u}{\partial y} = -\frac{1}{\rho}\frac{\partial p}{\partial x}.$$

If we write this in the form $Du/Dt = -(1/\rho)\,\partial p/\partial x$ then, as the right hand side is independent of y, Du/Dt must also be independent of y. If u has no y-dependence initially, it will remain independent of y throughout the fluid motion, which consequently reduces the horizontal momentum equation to

$$\frac{\partial u}{\partial t} + u\frac{\partial u}{\partial x} + g\frac{\partial}{\partial x}\{h - h_0(x)\} = 0. \tag{4.80}$$

The equation for conservation of mass is, by analogy with the argument that we gave in section 3.4 for conservation of mass in a non-uniform tube,

$$\frac{\partial h}{\partial t} + \frac{\partial}{\partial x}(uh) = 0. \tag{4.81}$$

Equations (4.80) and (4.81) are **nonlinear wave equations** and are known as the **shallow water equations**.

We will study solutions of these nonlinear equations in more detail in section 8.1. For the moment, we will consider the theory of linear shallow water waves, which is appropriate when $u \ll 1$ and $h - h_0(x) = \eta(x,t) \ll 1$. In this case equations (4.80) and (4.81) become

$$\frac{\partial u}{\partial t} + g\frac{\partial \eta}{\partial x} = 0, \tag{4.82}$$

$$\frac{\partial \eta}{\partial t} + \frac{\partial}{\partial x}(h_0 u) = 0. \tag{4.83}$$

We can eliminate u between these two equations by noting that

$$\frac{\partial^2}{\partial x \partial t}(h_0 u) = -g\frac{\partial}{\partial x}\left(h_0\frac{\partial \eta}{\partial x}\right) = -\frac{\partial^2 \eta}{\partial t^2},$$

which shows that η satisfies the modified wave equation

$$\frac{\partial^2 \eta}{\partial t^2} = g\frac{\partial}{\partial x}\left(h_0\frac{\partial \eta}{\partial x}\right), \tag{4.84}$$

which we have already met in the context of plane acoustic wave propagation along a non-uniform tube. In the present context, if the depth of the water varies exponentially with x, then the analysis of section 3.4 is relevant. When h_0 is constant, so that the fluid has a constant equilibrium depth,

$$\frac{\partial^2 \eta}{\partial t^2} = gh_0\frac{\partial^2 \eta}{\partial x^2}.$$

The elevation of the free surface satisfies the one-dimensional wave equation with wave speed $\sqrt{gh_0}$, and there is no dispersion. This is consistent with our earlier analysis of waves on shallow water of constant depth, where the wave speed is given by equation (4.17). The linear shallow water equations are frequently applied to the assessment of **tidal effects** on river estuaries and harbours, and we will concentrate on this application here.

Note that these equations cannot account for the generation of tides, as there is no forcing effect due to the moon and sun's gravitational fields, and no acknowledgement of the earth's rotation and the curvature of its surface. To get some idea of how accurate these equations may be in modelling small amplitude tidal waves in which the depth of the water is small compared with the wavelength, consider a plane, harmonic wave solution, $\eta = a\exp\{i(kx - \omega t)\}$, with $\omega = \sqrt{gh_0}k$. Using this dispersion relation, the period of the wave is $T = 2\pi/\omega = \lambda/\sqrt{gh_0}$, where λ is the wavelength. We can use this to write $h_0/\lambda = (1/T)\sqrt{h_0/g}$. For a tidal wave with a period of half a day, $T \approx 43\,500$ s, which, for $h_0 = 3$ m, gives $h_0/\lambda \approx 10^{-5}$, so we would expect the linear shallow water equations to be extremely accurate. Moreover, in this case $\lambda \approx 236$ km, which is small enough that we can plausibly neglect the curvature of the earth.

4.7.1 The Reflection of Sea Swell by a Step

Waves approaching the shoreline from the open sea both bring with them and remove sand and other debris. This can be a nuisance in a harbour, where a gradual silting up of deep water channels can occur. One simple way of trying to prevent this is to have a step down from the mouth of the harbour to the sea bed. This blocks the direct passage of sand, which is mainly transported in a layer close to the sea bed. Having decided to introduce this feature into the design of a harbour, it is prudent to consider what effect this may have on waves that are incident on the step. We now consider this problem in the context of linear shallow water theory, with orthogonal incidence of waves on the step, as illustrated in figure 4.22.

Let's suppose that the step is at $x = 0$, with the open sea lying in $x < 0$ and the harbour in $x > 0$. We will concentrate on the flow near the step and ignore all other boundaries. If the water depth in the open sea is h_1 and the depth in the harbour $h_2 < h_1$, we must solve the one-dimensional wave equation with wave speed $c_1 = \sqrt{gh_1}$ for $x < 0$ and $c_2 = \sqrt{gh_2} < c_1$ for $x > 0$. As the incident waves of amplitude a will be both reflected and transmitted, we look for a solution of the form

$$\eta = \left\{ \begin{array}{ll} a\exp\{i(k_1x - k_1c_1t)\} + Ra\exp\{i(-k_2x - k_2c_1t)\} & \text{for } x < 0, \\ Ta\exp\{i(k_3x - k_3c_2t)\} & \text{for } x > 0, \end{array} \right\} \tag{4.85}$$

where Ra and Ta are the amplitudes of the reflected and transmitted waves. We have also allowed for the possibility of a change in frequency by introducing the reflected and transmitted wavenumbers, k_2 and k_3. At $x = 0$ the fluid pressure, and hence the free surface elevation, must be

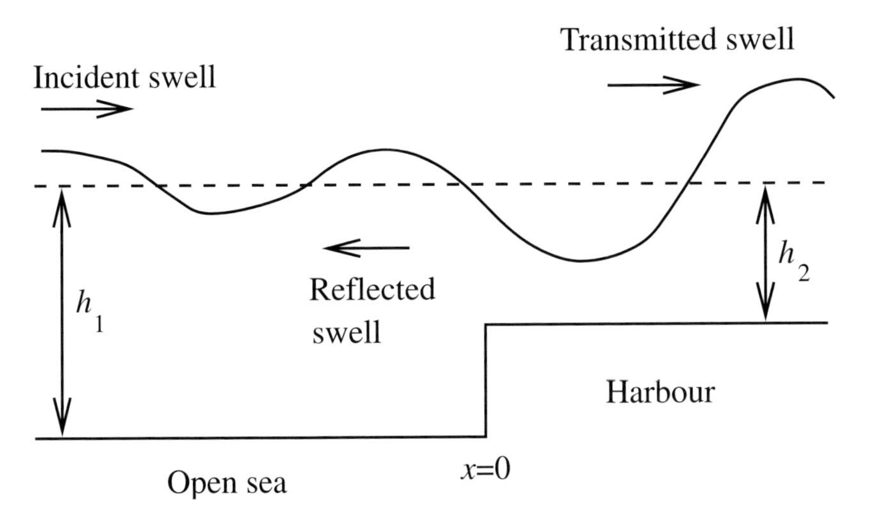

Fig. 4.22. The transmission and reflection of sea swell by a step at the mouth of a harbour.

continuous, so that

$$a \exp(-ik_1 c_1 t) + Ra \exp(-ik_2 c_1 t) = Ta \exp(-ik_3 c_2 t),$$

which can only be satisfied for all t if $k_1 c_1 = k_2 c_1 = k_3 c_2$, so all of the waves have the same frequency, and $T = R + 1$. In addition, we cannot lose any mass as the fluid crosses the step, so the flux of fluid must be continuous at $x = 0$. We can find the horizontal velocity by integrating $\partial u/\partial t = -g\partial\eta/\partial x$ to give

$$u = \begin{cases} \dfrac{ga}{c_1} \exp\{i(k_1 x - k_1 c_1 t)\} - \dfrac{gRa}{c_1} \exp\{i(-k_1 x - k_1 c_1 t)\} & \text{for } x < 0, \\ \dfrac{gTa}{c_2} \exp\{i(k_3 x - k_3 c_2 t)\} & \text{for } x > 0. \end{cases}$$

As the flux of fluid is uh, we obtain

$$\frac{1-R}{c_1} h_1 = \frac{T}{c_2} h_2,$$

and hence

$$R = \frac{1 - \sqrt{h_2/h_1}}{1 + \sqrt{h_2/h_1}} = \frac{c_1 - c_2}{c_1 + c_2}, \quad T = \frac{2}{1 + \sqrt{h_2/h_1}} = \frac{2c_1}{c_1 + c_2}. \tag{4.86}$$

We can now see that the transmitted waves, which move into the harbour, are of a larger amplitude than the incident waves, whilst the reflected

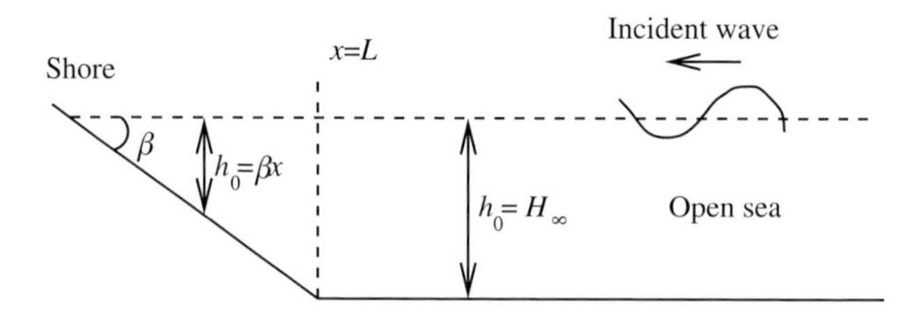

Fig. 4.23. Waves incident on a linearly sloping beach.

waves have a smaller amplitude. This is no great surprise, as the transmitted wave is moving into shallower water, and conservation of energy shows that its amplitude must increase. This analysis is, of course, identical to that given in sections 2.4 and 3.4 for reflection and transmission of waves on a string and plane sound waves in a tube.

4.7.2 Wave Amplification at a Gently Sloping Beach

We now consider the effect of a gently sloping beach on small amplitude waves approaching from the open sea. We take the equation of the sea bed to be

$$h_0(x) = \left\{ \begin{array}{cc} \dfrac{H_\infty}{L}x & \text{for } 0 \le x \le L, \\ H_\infty & \text{for } x \ge L, \end{array} \right\} \tag{4.87}$$

as shown in figure 4.23. We also assume that $H_\infty \ll L$, so that $H_\infty/L = \tan\beta \sim \beta$. In this case we can reasonably use linear shallow water theory, and hence look for solutions of the modified wave equation, (4.84). We find that

$$\frac{\partial^2\eta}{\partial t^2} = g\beta x\frac{\partial^2\eta}{\partial x^2} + g\beta\frac{\partial\eta}{\partial x}.$$

Now suppose that there is an incident wave, caused by tides or a distant storm, of the form

$$\eta(L,t) = a\cos(\omega t + \epsilon). \tag{4.88}$$

If we look for a periodic solution of the form $\eta = \cos(\omega t + \epsilon)H(x)$, we find that

$$x\frac{d^2H}{dx^2} + \frac{dH}{dx} + \frac{\omega^2}{g\beta}H = 0, \tag{4.89}$$

subject to $H(L) = a$. This is almost Bessel's equation. By making the change of variable $x = 2s^2$, so that

$$\frac{dH}{dx} = \frac{1}{4s}\frac{dH}{ds} \quad \text{and} \quad \frac{d^2H}{dx^2} = \frac{1}{16s^2}\frac{d^2H}{ds^2} - \frac{1}{16s^3}\frac{dH}{ds},$$

we arrive at

$$\frac{d^2H}{ds^2} + \frac{1}{s}\frac{dH}{ds} + \frac{8\omega^2}{g\beta}H = 0. \tag{4.90}$$

Recalling that Bessel's equation of order zero, $y'' + y'/x + y = 0$, has solutions of the form $y = AJ_0(x) + BY_0(x)$, and that $Y_0(x)$ is unbounded at $x = 0$, the appropriate solution of (4.90) is

$$H(s) = AJ_0\left(\sqrt{\frac{8\omega^2}{g\beta}}s\right).$$

The constant, A, is fixed by the condition $H = a$ at $x = L$, and the free surface elevation is therefore given by

$$\eta(x,t) = aJ_0\left(\frac{2\omega}{\sqrt{g\beta}}x^{1/2}\right) \Big/ J_0\left(\frac{2\omega}{\sqrt{g\beta}}L^{1/2}\right)\cos(\omega t + \epsilon). \tag{4.91}$$

This solution is illustrated in figure 4.24. The amplitude of the disturbance increases, and its wavelength decreases as the wave approaches the beach at $x = 0$. This can be observed in waves approaching a beach, although the assumptions that are made in shallow water theory may become invalid for x sufficiently small, where nonlinear effects cause the waves to break. The effect is dramatically illustrated by the behaviour of **tsunamis**. These are disturbances of the surface of the ocean generated by sub-sea earthquakes. In the open ocean they can be hundreds of kilometres long and only a few metres high, and so are almost impossible to detect with the naked eye. However, as they approach the coast, the effect of the sloping bed shortens the wavelength, and produces a destructive wall of water that can be tens of metres high (see González (1999)).

There is also the possibility of a resonance, which occurs when $2\omega\sqrt{L/g\beta}$ is a zero of J_0, and the solution (4.91) is unbounded. In this case, the incoming wave drives a standing wave close to the beach, but the analysis is too involved to present here.

Finally, in the open sea, the trigonometric form of the solution shows that the mean elevation of the surface is zero and that $H = O(a)$. The mean wave elevation above the beach may, however, be non-zero. If we

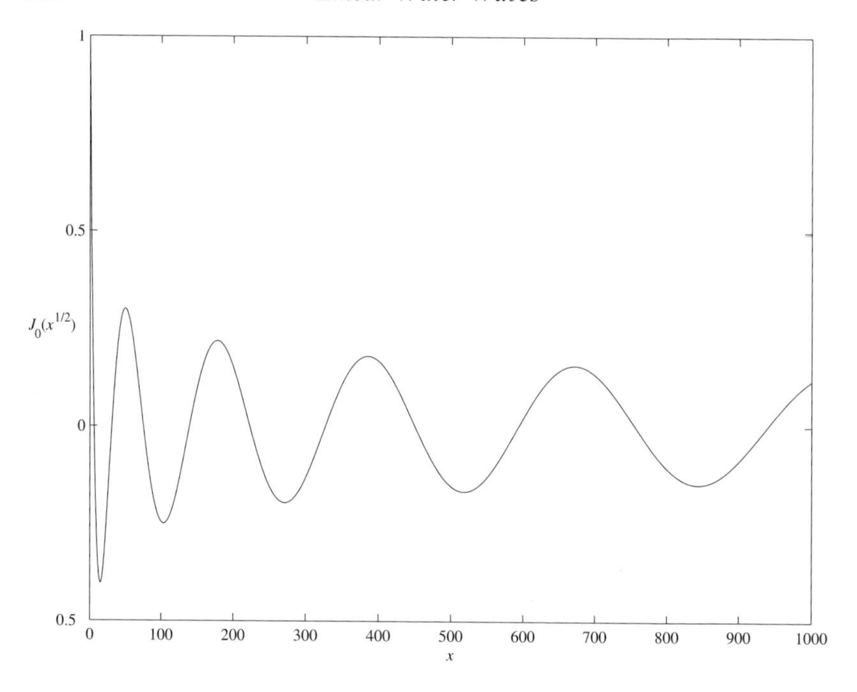

Fig. 4.24. A small amplitude disturbance of the free surface near a linearly sloping beach.

define this mean elevation to be

$$\overline{H} = \frac{1}{L}\int_0^L H(x)dx,$$

at first sight we are faced with a difficult integral involving a Bessel function. However, if we integrate (4.89) directly, we find that

$$\int_0^L xH''dx + \int_0^L H'dx + \frac{\omega^2}{g\beta}\int_0^L Hdx = 0.$$

Since

$$\int_0^L xH''dx = \left[xH'\right]_0^L - \int_0^L H'dx,$$

this gives

$$\int_0^L Hdx = -\frac{Lg\beta}{\omega^2}H'(L),$$

and hence

$$\overline{H} = -\frac{a}{L^{1/2}}\sqrt{\frac{g\beta}{\omega^2}}J_0'\left(2\omega\sqrt{\frac{L}{g\beta}}\right)\Big/J_0\left(2\omega\sqrt{\frac{L}{g\beta}}\right). \tag{4.92}$$

This shows that, provided that the frequency of the incident wave is not close to resonance,

$$\frac{\overline{H}}{L} = O\left(\frac{a}{L}\sqrt{\frac{gH_\infty}{\omega^2L^2}}\right). \tag{4.93}$$

Now linear shallow water theory is based upon $H/L = O(a/L) \ll 1$. Here, (4.93) shows that changes in mean surface elevation are an order of magnitude smaller than changes accounted for by the linear theory and can be neglected provided that $\omega L \gg \sqrt{gH_\infty}$.

4.8 Wave Refraction

When waves from the sea are incident upon the shoreline, there is a bending of the line of the wave crests. (The waves may also break, but that will not concern us here.) This bending is associated with a change in mean water depth, and hence in phase and group velocity, as the shoreline is approached. In shallower water, a wave crest travels more slowly than in deeper water, and there is an adjustment of the whole wave crest that follows the sea bed profile and leads to wave crests roughly parallel to the shoreline, as you can observe at any beach. This phenomenon is called **refraction** and can occur in most types of wave. In water waves it is associated with changes in mean depth. As we shall see in chapters 5 and 6, elastic and electromagnetic waves can be refracted if there is a change in the relevant properties of the material through which the wave is propagating. For example a ray of light will change its direction as it passes from air into glass. Two examples of the refraction of water waves are shown in figures 4.25 and 4.26.

In order to give a detailed account of the refraction of water waves, it is helpful to generalise the progressive gravity wave solution that we constructed in subsection 4.2.1. Let's consider uniform, three-dimensional, plane water waves propagating at an angle α to the horizontal, as shown in figure 4.27. It is convenient to take coordinates in the plane of the wave to be x, y with z as the vertical coordinate. By looking for a potential of the form $\phi = Z(z)\exp\{i(kx + ly - \omega t)\}$, we find that

$$\phi = \frac{a\omega}{|\mathbf{k}|}\frac{\cosh|\mathbf{k}|(z+h)}{\sinh|\mathbf{k}|h}\sin\chi, \quad \eta = a\cos\chi \tag{4.94}$$

Fig. 4.25. The refraction of water waves in an estuary.

where a is the amplitude, h is the mean depth and ω is the frequency, with $\omega^2 = g|\mathbf{k}|\tanh(|\mathbf{k}|h)$. The wavenumber vector is

$$\mathbf{k} = (k, l) = |\mathbf{k}|(-\cos\alpha, \sin\alpha),$$

which gives the direction of propagation of the wave crests. The **phase function** is

$$\chi = kx + ly - \omega t,$$

and is a constant on lines of constant free surface elevation.

4.8.1 The Kinematics of Slowly Varying Waves

For the solution (4.94) to be able to describe a refracted wave, we must modify it slightly. Firstly, the refracted wave field may have a non-zero mean surface elevation, $\bar{a}$. This can be dealt with easily by rewriting the solution in the form

$$\phi = \frac{a\omega}{|\mathbf{k}|}\frac{\cosh|\mathbf{k}|(z+h)}{\sinh|\mathbf{k}|d}\sin\chi, \quad \eta = \bar{a} + a\cos\chi, \tag{4.95}$$

where $d = \bar{a} + h$ is the total mean depth. However, as we saw in the previous section, this mean elevation will be very small for waves passing

Fig. 4.26. The refraction of water waves by a sand bar.

over gently changing topographies, and we will not consider it further. Secondly, we must recognise that while k, l, ω and χ vary very little over one wavelength or one period, they have a significant variation over longer length scales, such as the length of the shoreline. We now consider χ as the primary variable and look for solutions for the refracted waves in precisely the form of (4.95), in which

$$k = \frac{\partial \chi}{\partial x}, \quad l = \frac{\partial \chi}{\partial y}, \quad \omega = -\frac{\partial \chi}{\partial t},$$

a definition that is satisfied exactly for a uniform wave in water of constant depth. For the slowly varying wave assumption to apply we need to have a smooth variation of χ. Its mixed second derivatives must therefore be independent of the order of differentiation, which gives us the consistency conditions

$$\frac{\partial k}{\partial y} = \frac{\partial l}{\partial x}, \quad \frac{\partial k}{\partial t} + \frac{\partial \omega}{\partial x} = 0, \quad \frac{\partial l}{\partial t} + \frac{\partial \omega}{\partial y} = 0. \tag{4.96}$$

Now, if we regard ω as a function of the wavenumber vector in the refracted wave field, as well as spatial position and time, we write

$$\omega \equiv \Omega(k, l, x, y, t),$$

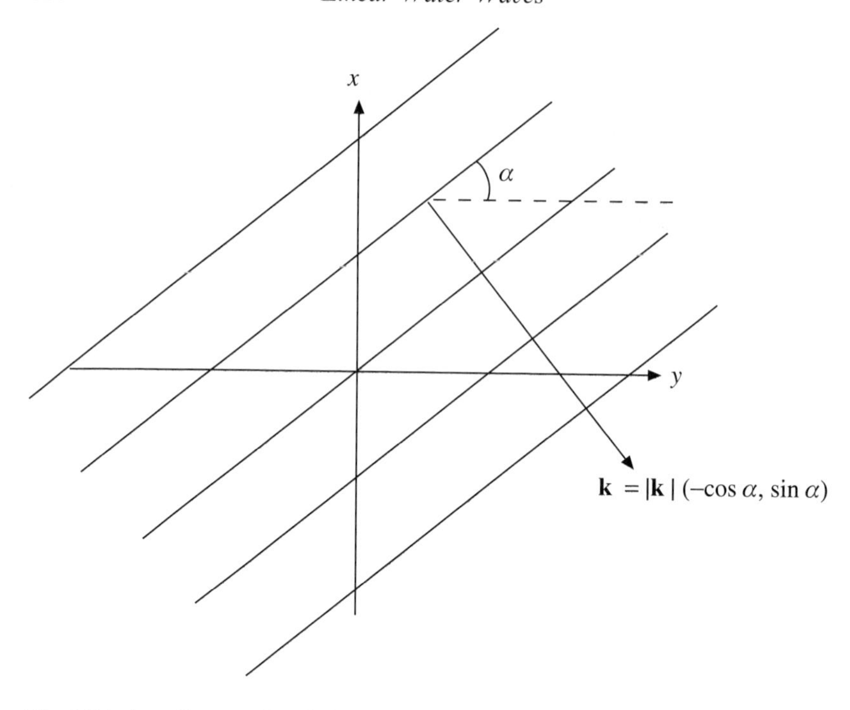

Fig. 4.27. A uniform train of plane, progressive gravity waves whose crests make an angle α with the y-axis.

and hence

$$\frac{\partial\omega}{\partial x} = \frac{\partial\Omega}{\partial x} + \frac{\partial\Omega}{\partial k}\frac{\partial k}{\partial x} + \frac{\partial\Omega}{\partial l}\frac{\partial l}{\partial x}. \tag{4.97}$$

Using (4.96), this can be written as

$$\frac{\partial k}{\partial t} + \frac{\partial\Omega}{\partial k}\frac{\partial k}{\partial x} + \frac{\partial\Omega}{\partial l}\frac{\partial k}{\partial y} = -\frac{\partial\Omega}{\partial x}, \tag{4.98}$$

and similarly,

$$\frac{\partial l}{\partial t} + \frac{\partial\Omega}{\partial k}\frac{\partial l}{\partial x} + \frac{\partial\Omega}{\partial l}\frac{\partial l}{\partial y} = -\frac{\partial\Omega}{\partial y}, \tag{4.99}$$

$$\frac{\partial\omega}{\partial t} + \frac{\partial\Omega}{\partial k}\frac{\partial\omega}{\partial x} + \frac{\partial\Omega}{\partial l}\frac{\partial\omega}{\partial y} = \frac{\partial\Omega}{\partial t}. \tag{4.100}$$

We can now define a curve $x = X(k,l)$, $y = Y(k,l)$, called a **ray**, as satisfying

$$\frac{dX}{dt} = \frac{\partial\Omega}{\partial k}, \quad \frac{dY}{dt} = \frac{\partial\Omega}{\partial l}. \tag{4.101}$$

On the rays, (4.98), (4.99) and (4.100) take the simple form

$$\frac{dk}{dt} = -\frac{\partial \Omega}{\partial x}, \quad \frac{dl}{dt} = -\frac{\partial \Omega}{\partial y}, \quad \frac{d\omega}{dt} = \frac{\partial \Omega}{\partial t}. \tag{4.102}$$

since their left hand sides are total derivatives. To understand the significance of (4.101) and (4.102), it is best to work with their vector form

$$\frac{d\mathbf{k}}{dt} = -\nabla_{\mathbf{x}}\Omega, \quad \frac{d\omega}{dt} = \frac{\partial \Omega}{\partial t} \quad \text{on the rays} \quad \frac{d\mathbf{X}}{dt} = \nabla_{\mathbf{k}}\Omega, \tag{4.103}$$

where $\mathbf{X} = (X, Y)$. From this we can see that changes of wavenumber and frequency are transmitted along the rays. The rays themselves will bend according to the value of

$$\nabla_{\mathbf{k}}\Omega = \left(\frac{\partial \Omega}{\partial k}, \frac{\partial \Omega}{\partial l}\right) = \mathbf{c}_g.$$

The **group velocity vector**, $\mathbf{c}_g$, determines the path of the ray, and is a direct generalisation of our earlier, one-dimensional definitions of group velocity. For uniform, plane waves at an angle α to the y-axis, as shown in figure 4.27, the group velocity vector is parallel to the wavenumber vector. For refracted waves, this is not the case.

4.8.2 Wave Refraction at a Gently Sloping Beach

We now apply the ray equations of the previous subsection to the refraction of a small amplitude wave approaching a gently sloping beach. We consider a beach with a straight, wave absorbing shoreline at $x = 0$. The sea occupies the region $x > 0$. For $x \geq L$ the sea has a uniform depth H_∞. For $0 \leq x \leq L$, the sea bed has depth $h = H_\infty x/L = \beta x$, as shown in figure 4.23. Out in the deep sea the wave has a period $2\pi/\omega$, and wavenumber vector $\mathbf{k}_\infty = |\mathbf{k}_\infty|(-\cos\alpha_\infty, \sin\alpha_\infty)$. We assume that $\beta \ll 1$, so that the beach slopes gently down to the sea bed, and also assume that we can use the dispersion relation, $\omega^2 = \Omega^2 = g|\mathbf{k}|\tanh|\mathbf{k}|h$, over the whole domain. The appropriate form of equations (4.103) in this case is

$$\frac{dl}{dt} = -\frac{\partial \Omega}{\partial y} = -\frac{\partial \Omega}{\partial h}\frac{\partial h}{\partial y} = 0, \tag{4.104}$$

$$\frac{dk}{dt} = -\frac{\partial \Omega}{\partial x} = -\frac{\partial \Omega}{\partial h}\frac{\partial h}{\partial x} = -\beta\frac{\partial \Omega}{\partial h}, \tag{4.105}$$

with the rays defined by

$$\frac{dy}{dt} = \frac{\partial \Omega}{\partial l} = \frac{\partial \Omega}{\partial |\mathbf{k}|}\frac{\partial |\mathbf{k}|}{\partial l} = \frac{\partial \Omega}{\partial |\mathbf{k}|}\frac{l}{|\mathbf{k}|}, \tag{4.106}$$

$$\frac{dx}{dt} = \frac{\partial \Omega}{\partial k} = \frac{\partial \Omega}{\partial |\mathbf{k}|}\frac{\partial |\mathbf{k}|}{\partial k} = \frac{\partial \Omega}{\partial |\mathbf{k}|}\frac{k}{|\mathbf{k}|}. \tag{4.107}$$

Equation (4.104) reveals that l is a constant, and hence that $|\mathbf{k}| \sin\alpha = |\mathbf{k}_\infty| \sin\alpha_\infty$. As ω is constant, we can use $c_\mathrm{p} = \omega/|\mathbf{k}|$ to write this in the form

$$\frac{c_\mathrm{p}}{\sin\alpha} = \frac{c_{\mathrm{p}\infty}}{\sin\alpha_\infty}. \tag{4.108}$$

This is our first encounter with **Snell's law**, which relates the phase speed of the wave to the angle made by the crests with the y-axis. We will find in chapters 5 and 6 that, in certain situations, it also applies to elastic and electromagnetic waves.

The Cartesian equation of the rays can be found by dividing equations (4.106) and (4.107) to give

$$\frac{dx}{dy} = \frac{k}{l} = \frac{k}{l_\infty}. \tag{4.109}$$

To find k we use $\omega^2 = g|\mathbf{k}| \tanh(|\mathbf{k}|\beta x)$. In general this equation has no analytical solution. However, near the shoreline $\tanh(|\mathbf{k}|\beta x) \approx |\mathbf{k}|\beta x$, so that $k^2 + l_\infty^2 \approx \omega^2/g\beta x$, and hence

$$\frac{dx}{dy} = \pm\sqrt{\frac{\omega^2}{g\beta l_\infty^2 x} - 1} \approx -\sqrt{\frac{\omega^2}{g\beta l_\infty^2 x}}. \tag{4.110}$$

We choose the minus sign so that the ray comes in to the beach. This separable equation has solution

$$y + A = -\frac{2}{3}\sqrt{\frac{g\beta l_\infty^2}{\omega^2}}\, x^{3/2}.$$

To find the equation for the wave crests we note that the tangent to the rays is in the direction (k, l), normal to the crests, so that the wave crests and rays are orthogonal. The lines of the wave crests therefore satisfy the equation

$$\frac{dx}{dy} = -\frac{l_\infty}{k} = \mp\left(\frac{\omega^2}{g\beta l_\infty^2 x} - 1\right)^{-1/2} \approx \sqrt{\frac{g\beta l_\infty^2 x}{\omega^2}}.$$

By integrating, and using the same constant of integration to ensure

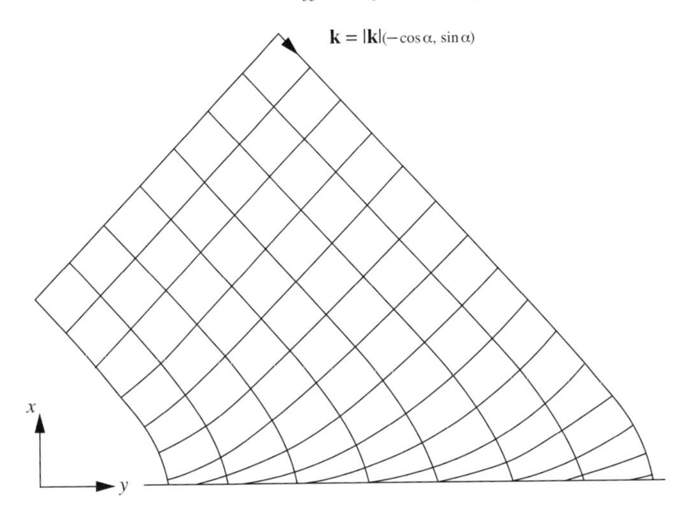

Fig. 4.28. A sketch of the wave crests and rays for waves approaching a gently sloping beach. The wave crests have positive slope, the rays negative slope.

that wave crest and ray pass through the same point, we find that the equation of the wave crests is

$$y + A = 2\sqrt{\frac{\omega^2 x}{g\beta l_\infty^2}},$$

and hence

$$x = \frac{g\beta l_\infty^2}{\omega^2}\frac{(y+A)^2}{4}.$$

The lines of the wave crests are therefore parabolas. Although we have neglected nonlinear effects, which will become important in the shallow water near the shore, this result agrees well with experiments. The lines of the wave crests and the rays are shown in figure 4.28.

4.9 The Effect of Viscosity

In choosing to ignore the effect of viscosity on the propagation of water waves, we are making a considerable mathematical simplification to the equations of motion. It is worthwhile considering over what time scale viscosity would actually damp out the waves. The linearised, two-dimensional Navier–Stokes equations, appropriate to small amplitude wave motions in deep water, are

$$u_t = -\frac{1}{\rho}p_x + \nu\left(u_{xx} + u_{yy}\right), \quad v_t = -\frac{1}{\rho}p_y - g + \nu\left(v_{xx} + v_{yy}\right). \quad (4.111)$$

The size of a typical term that we have retained in our inviscid theory is $u_t = O(U/T)$, whilst a typical viscous term, which we have neglected, is $\nu u_{xx} = O(\nu U/\lambda^2)$, where λ is a typical wavelength, U is a speed, and T is the time scale over which viscosity becomes important. Equating these two terms, we obtain $T = O(\lambda^2/\nu)$ as our viscous time scale.

In the case of capillary waves, we have $\sigma \gg \rho g\lambda^2$, so that $\lambda \ll (\sigma/\rho g)^{1/2} \approx 10^{-3}$ m for water. The viscosity of water is $\nu \approx 10^{-6}$ kg m^{-1} s^{-1}, so when $\lambda \approx 10^{-3}$ m, $T = O(1\text{ s})$. We conclude that pure capillary waves are rapidly damped by viscosity. We have also considered gravity waves, for which a typical wavelength is $\lambda \approx 1$ m. In this case, $T = O(10^6\text{s})$, and we can reasonably expect these waves to be unaffected by viscosity for several days and transmit the energy that generated them for hundreds of kilometres.

If there is another characteristic dimension in a wave propagation problem, for example a mean fluid depth, h, over a shallow sea or river bed, there will be a diffusion of viscous effects from the bed to the free surface. We can easily modify our analysis to give the viscous time scale as

$$T = O\left(\min\left(\frac{\lambda^2}{\nu}, \frac{h^2}{\nu}\right)\right).$$

For shallow water waves, $h \ll \lambda$, and we consequently expect sea or river bed effects to become important before the effects of the shearing motion in the bulk of the fluid.

Of course, this is just an order of magnitude analysis. To get a more accurate assessment of viscous effects it is necessary to solve (4.111) in any particular situation of interest. A more general analysis than is given here can be found in Lamb (1932). For solutions of initial value problems, see Bloor (1970).

Exercises

4.1 Taking into account the effect of both gravity and surface tension, deduce that

$$\phi(x, y, t) = (A\cos\omega t + B\sin\omega t)\cos kx\cosh k(y + h)$$

represents a possible velocity potential for the two-dimensional motion associated with standing waves on the surface of water having uniform undisturbed depth h. Find the relation between ω and k. Determine the equation of the particle paths.

4.2 Obtain the form of the velocity potential for a two-dimensional infinite train of progressive waves on the surface of water of uniform undisturbed depth h, including the effects of both gravity and surface tension. Calculate the velocities for both gravity and capillary waves on deep water.

4.3 A rectangular basin of infinite depth, bounded by sides $x = 0$, $x = a$, $z = 0$, $z = b$, contains water occupying the region $-\infty < y \leq 0$. Show that a standing wave solution exists in which the surface displacement is given by

$$\eta\,(x, z, t) = \sum_{n=0}^{\infty}\sum_{m=0}^{\infty} \cos\frac{n\pi x}{a}\cos\frac{m\pi z}{b} \times (A_{mn}\cos\omega_{mn}t + B_{mn}\sin\omega_{mn}t)\,,$$

where A_{mn} and B_{mn} are constants and ω_{mn} is an angular frequency. Neglecting surface tension, derive a dispersion relation for ω_{mn}. If initially $\eta = \eta_0\,(x)$ and $\partial\eta/\partial t = 0$, determine $\eta\,(x, z, t)$ for any subsequent time.

4.4 Find by separation of variables the velocity potential for axi-symmetric standing waves in a circular tank of radius a and undisturbed depth h. Determine the equation of the particle paths.

4.5 Show that there is equipartition of energy in the standing wave that we studied in subsection 4.2.1, and determine the particle paths.

4.6 A piston wavemaker is situated in a tank of water of depth h and length l. At the opposite end of the tank to the wavemaker there is a rigid, vertical wall at which all waves are reflected. Show that the solution of the linearised equations of motion can be written in the form

$$\phi = A_0\cosh k_0(y + h)\cos k_0(x - l)e^{-i\omega t} + \sum_{n=1}^{\infty} A_n\cos k_n(y + h)\cosh k_n(x - l)e^{-i\omega t}.$$

Show how to calculate the coefficients A_j, $j = 0, 1, \ldots$, in terms of the displacement of the piston, S.

How could you modify the above analysis if the wall at $x = l$ were to be flexible and dissipative, so as to absorb a fraction α of the incoming wave energy?

4.7 A water surface that extends to infinity in both positive and negative x-directions is given a small initial velocity so that, at time $t = 0$, $\eta = 0$ and $\partial\eta/\partial t = aH(x+b)H(b-x)$, where a and b are positive constants and H is the Heaviside step function. If surface tension can be neglected and the depth of the water is infinite, find the surface elevation η at time t as a Fourier integral.

Show that, at large distances $|x|$ and large times t such that $|x|/t$ is fixed, the surface elevation takes the form of a wave. Find the wavelength and frequency of the wave.

4.8 The wavelength and arrival time of waves generated by a distant storm are measured by a recorder on the surface of deep water. If waves of length λ_1 arrive at time $t = 0$ and waves of length λ_2 arrive at time $t = \tau$, calculate the approximate distance of the storm from the recorder.

4.9 A cylindrical vessel of radius a with vertical sides is sealed at one end and open at the other end. The water in the vessel has depth h, and is forced into motion by moving the vessel up and down with a displacement $A \sin \omega t$. The restoring force is gravity.

(a) By using h as a length scale and $\sqrt{h/g}$ as a time scale, show that the governing equations for this system can be written in the form

$$\nabla^2\phi = 0 \quad \text{in } 0 < r < \alpha,\ -1 + \epsilon \sin(\Omega t) < z < Z(r, t),$$

$$\phi_r = 0 \text{ on } r = \alpha,$$

$$\phi_z = \epsilon\Omega\cos(\Omega t) \text{ on } Z = -1 + \epsilon \sin(\Omega t),$$

$$\left.\begin{aligned} \phi_t + \frac{1}{2}|\nabla\phi|^2 + Z &= B(t), \\ \phi_z &= Z_t + \phi_r Z_r \end{aligned}\right\} \text{ on the free surface } z = Z(r, t).$$

Here $\Omega = \omega\sqrt{h/g}$ and $\epsilon = A/h$ are the dimensionless frequency and amplitude, $B(t)$ is the Bernoulli constant and $\alpha = a/h$ is the dimensionless radius of the vessel.

(b) By using the transformations $\bar{z} = z - \epsilon \sin(\Omega t)$, $\bar{\phi} = \phi -$

$\epsilon\Omega\cos(\Omega t)\,\bar{z}$, show that this problem transforms into the 'variable gravity' form

$$\nabla^2\bar{\phi} = 0 \quad \text{in } 0 < r < \alpha,\ -1 < \bar{z} < Z - \epsilon\sin(\Omega t),$$

$$\bar{\phi}_r = 0 \text{ on } r = \alpha,$$

$$\bar{\phi}_{\bar{z}} = 0 \text{ on } \bar{z} = -1,$$

$$\left.\begin{aligned} \bar{\phi}_t + \frac{1}{2}(\bar{\phi}_r^2 + \bar{\phi}_{\bar{z}}^2) + (1 - \epsilon\Omega^2\sin(\Omega t))Z &= 0, \\ \bar{\phi}_{\bar{z}} &= Z_t + \phi_r Z_r \end{aligned}\right\} \quad \text{on } \bar{z} = Z - \epsilon\sin(\Omega t).$$

What value has the Bernoulli constant been set to?

(c) Show that an appropriate expansion in small wave steepness leads to the linear system

$$\nabla^2\bar{\phi}_1 = 0 \quad \text{in } 0 < r < \alpha,\ -1 < \bar{z} < -\epsilon\sin(\Omega t),$$

$$\bar{\phi}_{1,r} = 0 \text{ on } r = \alpha,$$

$$\bar{\phi}_{1,\bar{z}} = 0 \text{ on } \bar{z} = -1,$$

$$\left.\begin{aligned} \bar{\phi}_{1,t} + (1 - \epsilon\Omega^2\sin(\Omega t))Z_1 &= 0, \\ \bar{\phi}_{1,\bar{z}} &= Z_{1,t} \end{aligned}\right\} \quad \text{on } \bar{z} = -\epsilon\sin(\Omega t).$$

(d) Derive the separable solution of this in the form

$$\bar{\phi}_1 = \sum_{n=1}^{\infty} A_n(t) J_0(k_n r)\cosh k_n(\bar{z} + 1),$$
$$Z_1 = \sum_{n=1}^{\infty} B_n(t) J_0(k_n r),$$

where $J_0'(k_n\alpha) = 0$, and state the equations satisfied by $A_n(t)$ and $B_n(t)$.

(e) The equations satisfied by $A_n(t)$ and $B_n(t)$ are difficult to solve. Some insight into the form of solution can be found by expanding $A_n(t)$ and $B_n(t)$ in the form $A_n(t) = A_{n,0}(t) + \epsilon A_{n,1}(t) + \cdots$. Show that for $\epsilon \ll 1$,

$$B_{n,0}' = k_n \sinh k_n A_{n,0}, \qquad A_{n,0}' \cosh k_n = -B_{n,0},$$

$$B_{n,1}' - k_n \sinh k_n A_{n,1} = -k_n^2 \cosh k_n \sin(\Omega t)\, A_{n,0},$$

$$\cosh k_n A'_{n,1} + B_{n,1} = k_n \sinh k_n \sin(\Omega t)\, A'_{n,0} + \Omega^2 \sin(\Omega t)\, B_{n,0}.$$

(f) If the natural frequencies of the system, ω_n, are given by $\omega_n^2 = k_n \tanh k_n$, show that a resonance will occur if $\Omega = 2\omega_n$. This is known as **Faraday resonance**. Note that there are further resonances not accounted for here (see Benjamin and Ursell (1954) for further details).

4.10 A ship moves on the surface of a deep sea with constant speed U. The wave pattern it makes is steady in a frame of reference moving with the ship.

(a) By introducing the transformation $\bar{x} = x + Ut$, show that the linearised unsteady equations transform to

$$\phi_{\bar{x}\bar{x}} + \phi_{yy} + \phi_{zz} = 0, \qquad \phi \to 0 \text{ as } z \to -\infty,$$

$$U\phi_{\bar{x}} + gZ = 0, \qquad \phi_z = UZ_{\bar{x}} \text{ on } z = 0.$$

(b) By introducing the far field variables $x^* = \epsilon\bar{x}$, $y^* = \epsilon y$, so that $x^*, y^* = O(1)$ when $\bar{x}, y = O\left(1/\epsilon\right) \gg 1$, show that the above boundary value problem can be put in the form

$$\phi_{x^*x^*} + \phi_{y^*y^*} + \frac{1}{\epsilon^2}\phi_{zz} = 0, \qquad \phi \to 0 \text{ as } z \to -\infty,$$

$$\frac{U^2}{g}\phi_{x^*x^*} + \frac{1}{\epsilon^2}\phi_2 = 0 \text{ on } z = 0.$$

(c) Solve this boundary value problem asymptotically by writing

$$\phi = A(x^*, y^*, z)\exp\left\{\frac{iB(x^*, y^*)}{\epsilon}\right\} + O(\epsilon)$$

and show that

$$B_{x^*} = \left(\frac{g}{U^2}\right)^{\frac{1}{2}} (B_{x^*}^2 + B_{y^*}^2)^{\frac{1}{4}}.$$

(d) Find a scaling transformation that removes g/U^2 from the above equation.

(e) Show that $B_{x^*} = k\cos\theta$ and $B_{y^*} = k\sin\theta$ are constant on the lines

$$\frac{dy^*}{dx^*} = \frac{\cos\theta\sin\theta}{\cos^2\theta - 2},$$

where $k^{\frac{1}{2}}\cos\theta = 1$.

(f) Deduce that B is given parametrically by

$$B = \frac{x\cos\theta + y\sin\theta}{\cos^2\theta}, \quad y(\cos^2\theta - 2) = x\cos\theta\sin\theta.$$

(g) Show that the waves are restricted to a wedge of semi-angle $\sin^{-1}(1/3)$ about the x^*-axis.

(h) Is the expansion uniformly valid within the wedge?

4.11 Show from first principles that the approximation for linear, shallow water, one-dimensional flow in a tidal estuary of constant depth h and variable breadth $b = b(x)$ leads to the equations

$$\frac{\partial u}{\partial t} + g\frac{\partial \eta}{\partial x} = 0, \quad \frac{\partial}{\partial t}(b\eta) + \frac{\partial}{\partial x}(bhu) = 0.$$

By eliminating the velocity, u, deduce that the free surface elevation, $\eta = \eta(x, t)$, satisfies the equation

$$\frac{\partial^2\eta}{\partial t^2} = \frac{gh}{b}\frac{\partial}{\partial x}\left(b\frac{\partial\eta}{\partial x}\right).$$

If the breadth of the estuary varies like $b = b_0 x$ for $0 \le x \le l$ and there is an open sea swell of the form $\eta = a\cos(\omega t + \epsilon)$ at $x = l$, show that the elevation in the estuary is

$$\eta(x, t) = a\frac{J_0(\alpha x)}{J_0(\alpha l)}\cos(\omega t + \epsilon),$$

where $\alpha^2 = \omega^2/gh$. Sketch the form of the waves and comment on the validity of the solution.

4.12 A deep layer of fluid of density ρ_2 flows with uniform horizontal speed U over another deep layer of fluid with density $\rho_1 > \rho_2$, which is at rest. By considering the fluids to be ideal and enforcing continuity of pressure and normal velocity at the interface between the fluids, show that small amplitude waves with gravity and surface tension acting as restoring forces can propagate along the interface with $\eta = a\exp\{i(kx - \omega t)\}$, provided that

$$(\rho_1 + \rho_2)\omega^2 - 2\rho_2 Uk\omega + \rho_2 U^2 k^2 - |k|\{k^2\sigma + (\rho_1 - \rho_2)g\} = 0.$$

Discuss the special cases $g = 0$ and $\sigma = 0$. This type of wave can lead to an instability of the interface, known as the **Kelvin–Helmholtz** instability.

5

Waves in Elastic Solids

The vibrations of panels in a car, the squeal of a train's brakes and the devastation left by an earthquake (see figure 5.1) are all examples of elastic wave propagation that are familiar to us. One of the features that these phenomena have in common is that the size of the elastic deformation is small compared to other length scales involved. The door panels of a car may only move by tenths of a millimetre to cause a noise, but the dimensions of the door itself may be of the order of a metre. A building need only be raised and tilted by a few tens of centimetres to cause considerable damage, but the length scale associated with the earth's crust is tens of kilometres. For this reason, most common elastic wave phenomena are approximately linear and we shall concentrate on these in this chapter.

5.1 Derivation of the Governing Equation

Consider the propagation of small amplitude waves in an **ideal elastic body**. By ideal, we mean that the stress–strain relationship is linear and isotropic and that the wave motion causes either no change in temperature or no heat flow within the body. Recall that the position vector, $\mathbf{x} = (x_1, x_2, x_3)$, of any point in the body after elastic deformation is related to its original position, $\mathbf{X} = (X_1, X_2, X_3)$, by a **displacement vector**, $\mathbf{u}$, through the relationship, $\mathbf{x} = \mathbf{X} + \mathbf{u}$. If we write $\mathbf{u} = (u_1, u_2, u_3)$ with $u_i = u_i(x_1, x_2, x_3, t)$, we are working in an Eulerian frame. Using the notation $u_{i,j} = \partial u_i / \partial x_j$, in this Eulerian frame the linearised, or small deformation, **strain tensor**, $e_{ij} = \frac{1}{2}(u_{i,j} + u_{j,i})$, is symmetric and the **rotation tensor**, $r_{ij} = \frac{1}{2}(u_{i,j} - u_{j,i})$, is anti-symmetric. The three independent components of this tensor can be written as a **rotation vector**, $\mathbf{r} = (r_1, r_2, r_3)$, with $r_k = \frac{1}{2}\epsilon_{ijk} r_{ij}$. Throughout this chapter, $\mathbf{r}$ is the rotation

Fig. 5.1. The deformation of the earth's surface after an earthquake.

vector, *not* a position vector. The strain tensor and rotation vector are the counterparts in solid mechanics of the rate of strain tensor and vorticity vector in fluid mechanics. The change in volume of an elastic element due to the deformation, known as the **dilatation**, is

$$\Delta = e_{kk} = \frac{\partial u_1}{\partial x_1} + \frac{\partial u_2}{\partial x_2} + \frac{\partial u_3}{\partial x_3}.$$

There are vector alternatives to the tensor definitions of dilatation and rotation, namely $\Delta = \nabla \cdot \mathbf{u}$, $\mathbf{r} = \frac{1}{2}\nabla \times \mathbf{u}$. We can write the stress–strain relationship in the form $\sigma_{ij} = \lambda e_{kk}\delta_{ij} + 2\mu e_{ij}$, where σ_{ij} is the **symmetric stress tensor** and λ and μ are the **Lamé constants**. The relationships

$$\lambda = \frac{E\nu}{(1+\nu)(1-2\nu)}, \qquad \mu = \frac{E}{2(1+\nu)}$$

between the **Young's modulus**, E, and **Poisson's ratio**, ν, of the material are derived by considering the elongation and lateral contraction of a rod of elastic material under uniaxial tension. For steel, typical measured values are $E \approx 200 \times 10^9\,\mathrm{N\,m^{-2}}$ and $\nu \approx 0.26$. This gives $\lambda \approx 8.6 \times 10^{10}\,\mathrm{N\,m^{-1}}$ and $\mu \approx 7.9 \times 10^{10}\,\mathrm{N\,m^{-2}}$.

The equations of motion for small displacements of an elastic body in the absence of body forces, which do not, in general, affect wave propagation, are

$$\rho \frac{\partial^2 u_i}{\partial t^2} = \frac{\partial \sigma_{ij}}{\partial x_j}, \tag{5.1}$$

where ρ is the undisturbed density of the elastic solid. This density does not change at leading order in a linearised theory. By substituting for σ_{ij} in this equation we obtain

$$\rho \frac{\partial^2 u_i}{\partial t^2} = \lambda \delta_{ij} \frac{\partial e_{kk}}{\partial x_j} + 2\mu \frac{\partial e_{ij}}{\partial x_j} = (\lambda + \mu) \frac{\partial}{\partial x_i} \left(\frac{\partial u_j}{\partial x_j} \right) + \mu \frac{\partial^2 u_i}{\partial x_j \partial x_j}. \tag{5.2}$$

This is conveniently written in vector form as

$$\rho \frac{\partial^2 \mathbf{u}}{\partial t^2} = (\lambda + \mu) \nabla (\nabla \cdot \mathbf{u}) + \mu \nabla^2 \mathbf{u}, \tag{5.3}$$

and is generally referred to as **Navier's equation**. This is the starting point for our study of elastic wave motions.

5.2 Waves in an Infinite Elastic Body

In an unbounded elastic body there are no finite surfaces on which boundary conditions must be applied. This greatly simplifies the analysis. We begin by using the identity $\nabla^2 \mathbf{u} = \nabla(\nabla \cdot \mathbf{u}) - \nabla \times (\nabla \times \mathbf{u})$ to write Navier's equation, (5.3), in invariant form as

$$\rho \frac{\partial^2 \mathbf{u}}{\partial t^2} = (\lambda + 2\mu) \nabla (\nabla \cdot \mathbf{u}) - \mu \nabla \times (\nabla \times \mathbf{u}). \tag{5.4}$$

Using the identity $\nabla \cdot \nabla \times \mathbf{u} \equiv 0$, and assuming that we can exchange the order of partial derivatives, the divergence of (5.4) is

$$\rho \frac{\partial^2}{\partial t^2} (\nabla \cdot \mathbf{u}) = (\lambda + 2\mu) \nabla^2 (\nabla \cdot \mathbf{u}).$$

Since $\nabla \cdot \mathbf{u} = \Delta$, the dilatation satisfies

$$\nabla^2 \Delta = \frac{1}{c_1^2} \frac{\partial^2 \Delta}{\partial t^2}, \qquad c_1^2 = \frac{(\lambda + 2\mu)}{\rho}, \tag{5.5}$$

which shows that volume changes within the elastic body propagate as solutions of the three-dimensional wave equation with speed $c_1 = \sqrt{(\lambda + 2\mu)/\rho}$. These are known as **dilatational** or **primary waves**.

Similarly, using the identity $\nabla \times \nabla(\nabla \cdot \mathbf{u}) \equiv 0$, the curl of (5.4) is

$$\rho \frac{\partial^2}{\partial t^2} (\nabla \times \mathbf{u}) = -\mu \nabla \times \{\nabla \times (\nabla \times \mathbf{u})\}.$$

If we now express this in terms of the rotation vector, $\mathbf{r} = \frac{1}{2}\nabla \times \mathbf{u}$, and use the fact that $\nabla \cdot \mathbf{r} = 0$, we find that

$$\nabla^2 \mathbf{r} = \frac{1}{c_2^2}\frac{\partial^2 \mathbf{r}}{\partial t^2}, \qquad c_2^2 = \frac{\mu}{\rho}. \tag{5.6}$$

The vector, $\mathbf{r}$, of infinitesimal rotations of the body therefore propagates as a solution of the three-dimensional vector wave equation with speed $\sqrt{\mu/\rho}$, giving rise to **rotational** or **secondary waves**. Since $c_1 > c_2$, rotational waves lag behind dilatational waves. If we cause dilatational and rotational waves to be initiated in some finite region, expansions and contractions will be felt before any shearing at a distant point in the body. To get some appreciation of the speed of these waves, a typical steel of density $7800\,\mathrm{kg\,m^{-3}}$ has $c_1 \approx 5600\,\mathrm{m\,s^{-1}}$ and $c_2 \approx 3180\,\mathrm{m\,s^{-1}}$. This should be compared with the speed of sound waves in air which is just $340\,\mathrm{m\,s^{-1}}$.

We can now analyse some simple solutions of (5.5) and (5.6) in an unbounded elastic body.

5.2.1 One-Dimensional Dilatation Waves

We begin by considering waves that vary in one direction only, firstly with displacements in the same direction, by taking $u_1 = u_1(x_1, t)$, $u_2 = u_3 = 0$. In this case we have $\Delta = \partial u_1/\partial x_1$ and $\mathbf{r} = 0$. The dilatation, Δ, obeys the one-dimensional wave equation

$$\frac{\partial^2 \Delta}{\partial x_1^2} = \frac{1}{c_1^2}\frac{\partial^2 \Delta}{\partial t^2}.$$

These waves are often called **longitudinal**, since there are displacements only in the direction of propagation. Note that there is no dispersion of longitudinal waves. Their properties are rather like those of a plane sound wave or a wave on a stretched string (although these are transverse), for example, being reflected and transmitted in a similar way by changes in composition of the wave bearing medium. Note that, although the displacement is one-dimensional, there are three non-zero components of stress,

$$\sigma_{11} = (\lambda + 2\mu)\frac{\partial u_1}{\partial x_1}, \quad \sigma_{22} = \sigma_{33} = \lambda\frac{\partial u_1}{\partial x_1}.$$

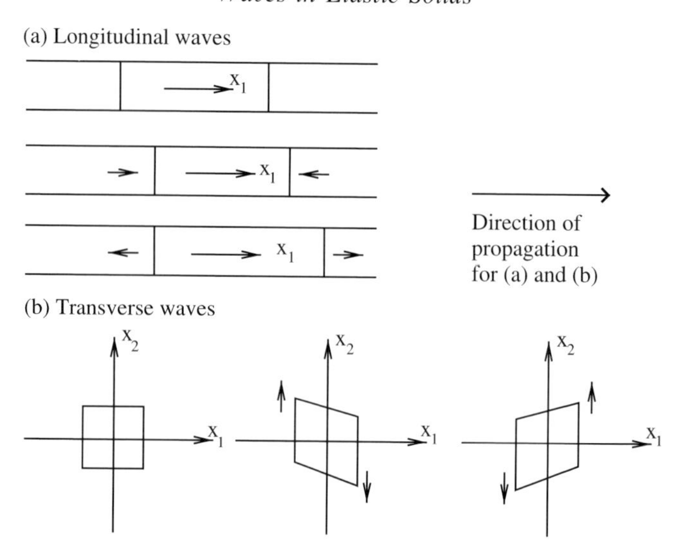

Fig. 5.2. The effect of the propagation of (a) longitudinal and (b) transverse waves on an element of elastic material.

5.2.2 One-Dimensional Rotational Waves

If the displacement is in the direction perpendicular to the direction of propagation, so that $u_2 = u_2(x_1, t)$, $u_1 = u_3 = 0$, we find that $\Delta = 0$ and $\mathbf{r} = \left(0, 0, \frac{1}{2}\partial u_2/\partial x_1\right)$. These waves are often called **transverse** or **shear waves**. The single non-zero component of the rotation vector satisfies the one-dimensional wave equation with speed c_2. These waves get their specific name from the fact that $e_{12} = e_{21} = \frac{1}{2}\partial u_2/\partial x_1$ and $\sigma_{12} = \sigma_{21} = 2\mu e_{12} = \mu \partial u_2/\partial x_1$ are the only non-zero components of the strain and stress tensors, which represents a purely shearing deformation propagating in the x_1-direction. An illustration of the deformation caused by one-dimensional longitudinal and transverse waves is shown in figure 5.2.

5.2.3 Plane Waves with General Orientation

We now look for a plane wave solution of Navier's equation, (5.3), of the form $\mathbf{u} = \mathbf{A}\exp\{i(\mathbf{k}\cdot\mathbf{x} - \omega t)\}$, where $\mathbf{k} = (k_1, k_2, k_3)$ and $\mathbf{x} = (x_1, x_2, x_3)$. This type of solution is constant on each plane $\mathbf{k}\cdot\mathbf{x} - \omega t = \text{constant}$, and represents a displacement wave moving in the direction $\mathbf{k}$ with amplitude

$\mathbf{A}$ and phase speed $\omega/|\mathbf{k}|$. Substitution into (5.3) gives

$$\rho\omega^2\mathbf{A} = (\lambda + \mu)(\mathbf{A} \cdot \mathbf{k})\mathbf{k} + \mu|\mathbf{k}|^2\mathbf{A}. \tag{5.7}$$

This is a vector equation which we can solve by taking the scalar product of both sides with $\mathbf{k}$ to give

$$(\mathbf{A} \cdot \mathbf{k})\{\rho\omega^2 - (\lambda + 2\mu)|\mathbf{k}|^2\} = 0. \tag{5.8}$$

There are two possible types of solution: those with $\mathbf{A}\cdot\mathbf{k} = 0$, in which the amplitude is any vector orthogonal to the wave vector, and a further class in which $\omega = \pm\sqrt{(\lambda + 2\mu)/\rho}\ |\mathbf{k}|$. Considering the first type, if $\mathbf{A} \cdot \mathbf{k} = 0$, (5.7) shows that $\omega = \pm\sqrt{\mu/\rho}\,|\mathbf{k}|$, so that these waves are rotational or transverse. For the second type, (5.7) shows that $\mathbf{A} = (\mathbf{A}\cdot\mathbf{k})\mathbf{k}/|\mathbf{k}|^2$, so that the amplitude is parallel to the wave vector. These waves are dilatational, or longitudinal. As the governing equation is linear, we can superimpose these two (or more) different solutions and make a general statement that three-dimensional elastic plane waves in an infinite medium will, in general, have both a dilatational and a rotational component. Indeed, for $\mathbf{k} = (k, 0, 0)$, a Fourier superposition of solutions of the first type gives plane transverse waves propagating in the x_1-direction, whilst a similar superposition using solutions of the second type gives shear waves propagating in the x_1-direction, which we described in the previous subsections.

5.3 Two-Dimensional Waves in Semi-infinite Elastic Bodies

We now consider a semi-infinite elastic body defined by $-\infty < x_1 < \infty$, $-\infty < x_2 \leq 0$, $-\infty < x_3 < \infty$. The existence of the surface $x_2 = 0$ gives us some freedom in choosing which boundary conditions to apply. We consider the rather different cases of a normally loaded surface and an unloaded surface. In each case we seek a two-dimensional solution with $u_1 = u_1(x_1, x_2, t)$, $u_2 = u_2(x_1, x_2, t)$ and $u_3 = 0$.

5.3.1 Normally Loaded Surface

Let us suppose that a uniform normal stress, $p(t)$, is applied to the surface $x_2 = 0$ beginning when $t = 0$. There is no shear stress on this surface, as shown in figure 5.3. The deformation produced by this type of stress is one-dimensional, with $u_2 = u_2(x_2, t)$, $u_1 = u_3 = 0$. As we have seen, in this case Navier's equation reduces to the one-dimensional wave equation

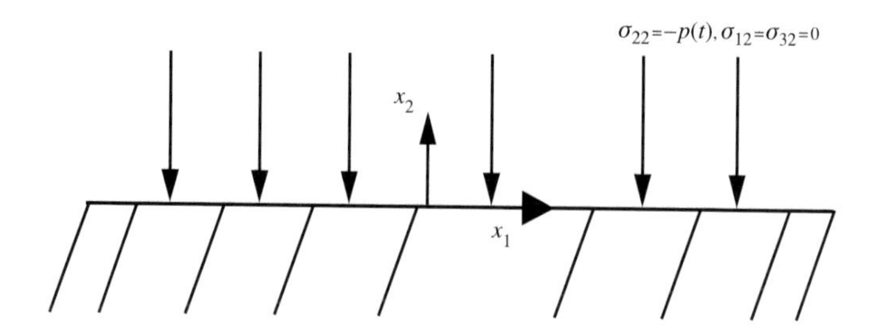

Fig. 5.3. A uniformly loaded elastic half-space.

with speed c_1, which we must solve subject to

$$\sigma_{22} = -p(t)H(t), \quad \sigma_{12} = \sigma_{13} = 0 \quad \text{on } x_2 = 0,$$

and a radiation condition that any waves produced by the loading at $x_2 = 0$ must travel in the direction of decreasing x_2 (there is no mechanism for reflecting a wave back towards $x_2 = 0$).

By taking the solution of the wave equation for u_2 that propagates in the negative x_2-direction in the form

$$u_2 = f(x_2 + c_1 t) = f\left\{c_1\left(t + \frac{x_2}{c_1}\right)\right\} = F\left(t + \frac{x_2}{c_1}\right), \tag{5.9}$$

we find that the stresses are

$$\sigma_{22} = \frac{(\lambda + 2\mu)}{c_1}F'\left(t + \frac{x_2}{c_1}\right), \qquad \sigma_{12} = \sigma_{13} = 0. \tag{5.10}$$

The normal stress boundary condition then leads to

$$\frac{(\lambda + 2\mu)}{c_1}F'(t) = -p(t)H(t). \tag{5.11}$$

Integration of this, setting the constant to zero to suppress rigid body translation, gives

$$F(t) = \frac{-c_1 H(t)}{(\lambda + 2\mu)}\int_0^t p(s)\, ds, \tag{5.12}$$

and we arrive at

$$u_2(x_2, t) = -c_1 H\left(t + x_2/c_1\right)\int_0^{t+(x_2/c_1)} p(s)\, ds \tag{5.13}$$

and

$$\sigma_{22}(x_2, t) = \frac{-c_1}{(\lambda + 2\mu)}H\left(t + \frac{x_2}{c_1}\right)p\left(t + \frac{x_2}{c_1}\right). \tag{5.14}$$

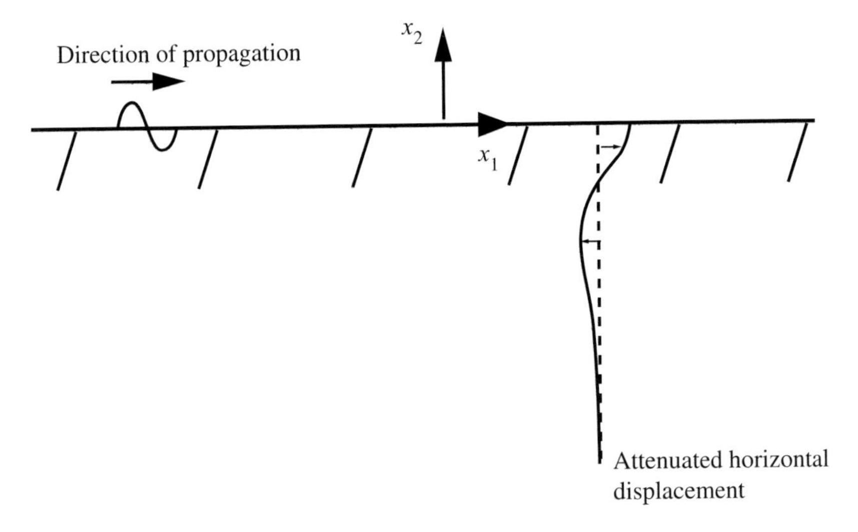

Fig. 5.4. Attenuated surface elastic waves.

Note that both the stress and displacement are zero for $x_2 < -c_1 t$ as we should expect.

5.3.2 *Stress-Free Surface*

When the surface $x_2 = 0$ is free from any normal or shear stress, we can distinguish two different types of wave propagation problem. Firstly, we can enquire as to the form of a wave that propagates in the x_1-direction only. Such a wave must be localised near the free surface to be physically realistic, as shown in figure 5.4. Secondly, we can consider what happens to a plane harmonic wave when it is incident on a stress-free surface, and determine how it is reflected.

Attenuated Waves Propagating along the Free Surface

Here we suppose that we initiate a deformation within an elastic half space when $t = 0$. After the decay of transient deformations, there is the possibility that, for large times, we could have a wave propagating along the free surface of the half space, $x_2 = 0$. We should expect that such a wave will be attenuated in the x_2-direction, and we now seek a solution of Navier's equation that will exhibit these properties. More specifically, we look for a solution in the form

$$\mathbf{u} = \mathbf{A} \exp\{i(kx_1 - \omega t) + \alpha x_2\} + \mathbf{B} \exp\{i(kx_1 - \omega t) + \beta x_2\}. \tag{5.15}$$

The choice of two independent amplitudes $\mathbf{A}$, $\mathbf{B}$ and decay rates α, β is the minimum possible with which we can satisfy the stress-free conditions on $x_2 = 0$.

Substitution into Navier's equation gives two vector equations of the form

$$-\rho\omega^2\mathbf{C} = (\lambda + \mu)(\mathbf{C} \cdot \mathbf{s})\mathbf{s} + \mu\mathbf{C}|\mathbf{s}|^2, \tag{5.16}$$

where $\mathbf{C}$ is equal to either $\mathbf{A}$ or $\mathbf{B}$ and $\mathbf{s}$ either (ik, α) or (ik, β). Recalling the solutions of this type of equation that we found in the previous section, it is convenient to choose, firstly, $\mathbf{A} \cdot \mathbf{s} = 0$, so that

$$A_1 = -\alpha A, \quad A_2 = ikA, \quad \omega^2 = c_2^2(k^2 - \alpha^2), \tag{5.17}$$

and secondly

$$\omega^2 = c_1^2(k^2 - \beta^2), \tag{5.18}$$

in order to get a solution that is a combination of dilatational and rotational components. By taking our displacement in the form

$$\mathbf{u} = (u_1, u_2, 0) = A(-\alpha, ik)\exp W(\alpha) + (B_1, B_2)\exp W(\beta), \tag{5.19}$$

where $W(\alpha) = i(kx_1 - \omega t) + \alpha x_2$, we can calculate

$$\sigma_{12} = \mu[-A(k^2 + \alpha^2)\exp W(\alpha) + (\beta B_1 + ikB_2)\exp W(\beta)], \tag{5.20}$$

$$\sigma_{22} = \lambda(ikB_1 + \beta B_2)\exp W(\beta) + 2\mu(\alpha Aik\exp W(\alpha) + \beta B_2\exp W(\beta)). \tag{5.21}$$

The stress-free boundary conditions on $x_2 = 0$ then lead to

$$\left.\begin{aligned} -A(k^2 + \alpha^2) + (\beta B_1 + ikB_2) &= 0, \\ \lambda ikB_1 + (\lambda + 2\mu)\beta B_2 + 2\mu\alpha ikA &= 0. \end{aligned}\right\} \tag{5.22}$$

We now have six equations (5.17), (5.18) and (5.22), for the nine unknowns w, k, α, β, A, A_1, A_2, B_1, B_2, which gives us rather a large family of waves. To be more specific, it is necessary to make a further assumption about the amplitude $\mathbf{B}$. If we take $\beta B_1 = ikB_2$ (can you see why this is indicated from the results of section 5.2.3?), equations (5.22) are reduced to

$$\left.\begin{aligned} -(\alpha^2 + k^2)A + 2ikB_2 &= 0, \\ 2\mu\alpha ikA + B_2\left\{\beta(\lambda + 2\mu) - \frac{\lambda k^2}{\beta}\right\} &= 0. \end{aligned}\right\} \tag{5.23}$$

For a non-trivial solution of these equations we require

$$\begin{vmatrix} -(\alpha^2 + k^2) & 2ik \\ 2\mu\alpha ik & \beta(\lambda + 2\mu) - \lambda k^2/\beta \end{vmatrix} = 0,$$

and hence

$$(\alpha^2 + k^2)\{\beta^2(\lambda + 2\mu) - \lambda k^2\} + 4k^2\mu\alpha\beta = 0, \tag{5.24}$$

or

$$(\alpha^2 + k^2)^2 \left\{\beta^2 \frac{c_1^2}{c_2^2} - \left(\frac{c_1^2}{c_2^2} - 2\right) k^2\right\}^2 = 16k^4\alpha^2\beta^2, \tag{5.25}$$

where we have written

$$\frac{c_1^2}{c_2^2} = \frac{\lambda + 2\mu}{\mu} > 2.$$

Substituting for α^2 and β^2 from (5.17) and (5.18), we can further simplify to

$$\left(2 - \frac{\omega^2}{c_2^2 k^2}\right)^4 = 16\left(1 - \frac{\omega^2}{c_2^2 k^2}\right)\left\{1 - \frac{c_2^2}{c_1^2}\left(\frac{\omega^2}{c_2^2 k^2}\right)\right\}. \tag{5.26}$$

Finally, if we write $X = \omega^2/c_2^2k^2$ and cancel a factor, we obtain the cubic equation

$$X^3 - 8X^2 = 16\left\{1 - \left(\frac{c_2}{c_1}\right)^2\right\} - \left\{24 - 16\left(\frac{c_2}{c_1}\right)^2\right\} X. \tag{5.27}$$

Both quantities in curly brackets are positive. By sketching the graphs of the simple cubic on the left hand side and the straight line on the right and looking for points of intersection, it is apparent that there is only one real root to this equation, $X = X_1$. From the sign of each side of (5.27) at $X = 0$ and $X = 1$, we find that $0 < X_1 < 1$. Hence we have $\omega = \pm X_1^{1/2} c_2 k$, and the waves are non-dispersive. The wave speed, $c_R = \omega/k = \pm X_1^{1/2} c_2$, is smaller than that of rotational waves, and is a function of the material parameters λ, μ and ρ alone. These waves are generally referred to as **Rayleigh waves** and their most important applications are to seismology and non-destructive testing. Figure 5.5 shows a typical seismogram measured during an earthquake. The arrival of the dilatational, or primary, waves, followed by the rotational, or secondary, waves and finally by the Rayleigh waves is clearly visible. Note that these Rayleigh waves have the largest amplitude of the three, and cause the most destruction. It is also of interest that points in an elastic body move on anticlockwise loops under the influence of a Rayleigh wave, in contrast to the clockwise motion of fluid particles due to a progressive gravity wave (see subsection 4.2.1).

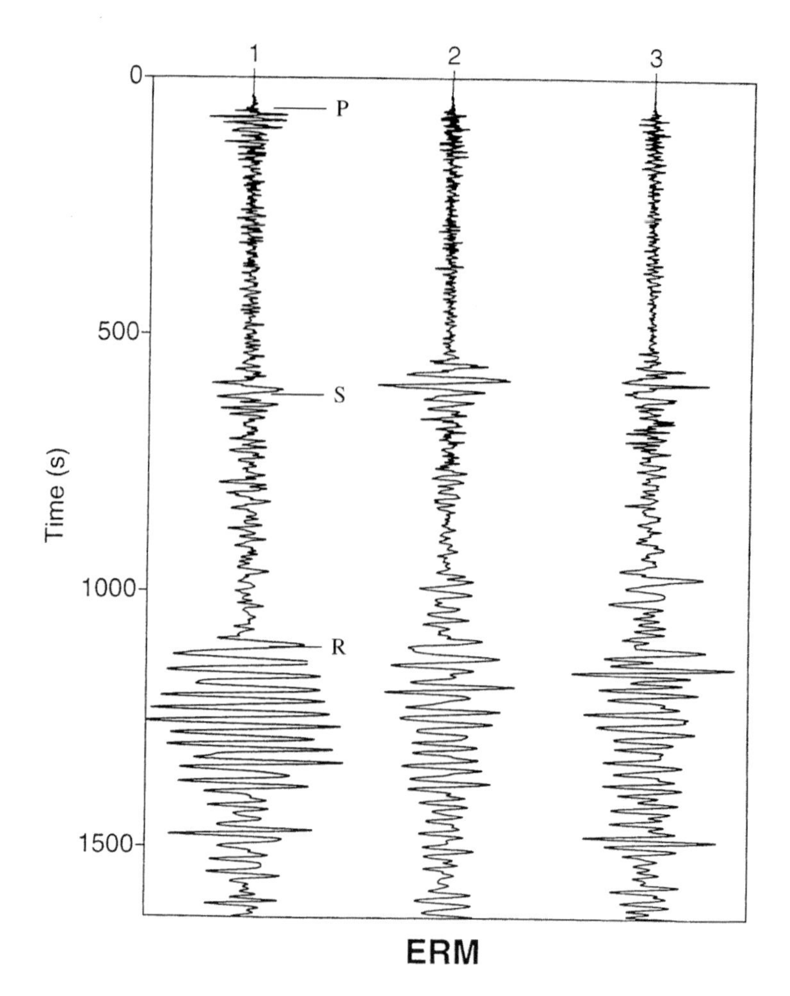

Fig. 5.5. A typical seismogram during an earthquake, showing the arrival of the P(rimary), S(econdary) and R(ayleigh) waves. Note that the Rayleigh waves have the largest amplitude.

The Reflection of Plane Waves at a Free Surface

We have seen in section 5.2 that plane elastic waves in an infinite body are, in general, composed of both dilatational and rotational components. If a pure dilatational or rotational wave is incident on a planar free surface, the reflected wave can be of the same type, the opposite type or a combination of the two types. This transition is governed, in any

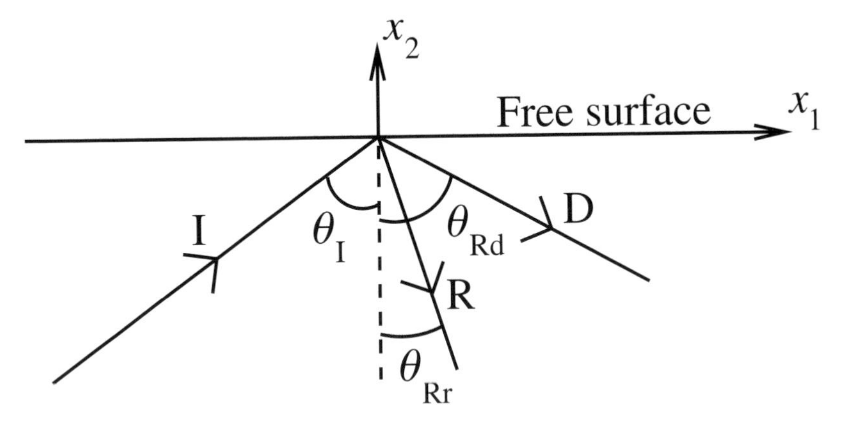

Fig. 5.6. The reflection of a dilatational plane wave at a free surface.

particular case, by the stress-free conditions on the surface $x_2 = 0$. To illustrate this further, we consider a dilatational wave of amplitude A_I incident at angle θ_I on the free surface. The reflected wave consists of a dilatational component, of amplitude A_D and angle of reflection θ_{Rd}, together with a rotational component of amplitude A_R and angle of reflection θ_{Rr} as shown in figure 5.6. (In the notation of seismologists, the incident dilatational wave is called a P wave, the reflected dilatational wave a PP wave and the reflected rotational wave a PS wave.)

To analyse this problem, we decompose our displacement into the incident and reflected components. Since the amplitude of dilatational waves of speed c_1 is parallel to the wavenumber vector and the amplitude of rotational waves of speed c_2 is orthogonal to the wavenumber vector, we write

$$\mathbf{u} = \mathbf{u}^{\mathrm{I}} + \mathbf{u}^{\mathrm{D}} + \mathbf{u}^{\mathrm{R}} \tag{5.28}$$

where

$$\begin{aligned}
\mathbf{u}^{\mathrm{I}} &= A_{\mathrm{I}}(\sin\theta_{\mathrm{I}}, \cos\theta_{\mathrm{I}}) \exp\{ik_{\mathrm{I}}(x_1 \sin\theta_{\mathrm{I}} + x_2 \cos\theta_{\mathrm{I}} - c_1 t)\}, \\
\mathbf{u}^{\mathrm{D}} &= A_{\mathrm{D}}(\sin\theta_{\mathrm{Rd}}, -\cos\theta_{\mathrm{Rd}}) \exp\{ik_D(x_1 \sin\theta_{\mathrm{Rd}} - x_2 \cos\theta_{\mathrm{Rd}} - c_1 t)\}, \\
\mathbf{u}^{\mathrm{R}} &= A_{\mathrm{R}}(\cos\theta_{\mathrm{Rr}}, \sin\theta_{\mathrm{Rr}}) \exp\{ik_R(x_1 \sin\theta_{\mathrm{Rr}} - x_2 \cos\theta_{\mathrm{Rr}} - c_2 t)\}.
\end{aligned}$$

When calculating the stress tensor prior to imposing the boundary conditions, it is clear that we will have three different exponential factors. If these are to combine together on $x_2 = 0$ to give stress-free conditions, each of the factors must be identical. This forces us to choose

$$k_{\mathrm{I}}c_1\left(\frac{x_1\sin\theta_{\mathrm{I}}}{c_1}-t\right)=k_D c_1\left(\frac{x_1\sin\theta_{\mathrm{Rd}}}{c_1}-t\right)=k_R c_2\left(\frac{x_1\sin\theta_{\mathrm{Rr}}}{c_2}-t\right), \tag{5.29}$$

which can only be satisfied for all x_1 and t by choosing $k_{\mathrm{I}}c_1 = k_{\mathrm{D}}c_1 = k_{\mathrm{R}}c_2 = \omega$, so that all waves have the same angular frequency, and

$$\frac{\sin\theta_{\mathrm{I}}}{c_1}=\frac{\sin\theta_{\mathrm{Rd}}}{c_1}=\frac{\sin\theta_{\mathrm{Rr}}}{c_2}. \tag{5.30}$$

This is another example of Snell's law (see subsections 4.8.2 and 6.7.3). Hence $\theta_{\mathrm{I}} = \theta_{\mathrm{Rd}}$ and the angle of incidence is equal to the angle of reflection for the dilatational waves. However, the rotational wave is inclined to the free surface at a larger angle than the dilatational waves (since $0 < c_2/c_1 < 1$ means that $0 \le \theta_{\mathrm{Rr}} \le \theta_{\mathrm{I}}$), with

$$\sin\theta_{\mathrm{Rr}}=\frac{c_2}{c_1}\sin\theta_{\mathrm{I}}.$$

We can now write the displacement in the form

$$\begin{aligned}\mathbf{u} = {} & A_{\mathrm{I}}(\sin\theta_{\mathrm{I}},\cos\theta_{\mathrm{I}})\exp\left\{i\omega\left(\frac{x_1\sin\theta_{\mathrm{I}}+x_2\cos\theta_{\mathrm{I}}}{c_1}-t\right)\right\}\\ &+A_{\mathrm{D}}(\sin\theta_{\mathrm{I}},-\cos\theta_{\mathrm{I}})\exp\left\{i\omega\left(\frac{x_1\sin\theta_{\mathrm{I}}-x_2\cos\theta_{\mathrm{I}}}{c_1}-t\right)\right\}\\ &+A_{\mathrm{R}}(\cos\theta_{\mathrm{Rr}},\sin\theta_{\mathrm{Rr}})\exp\left\{i\omega\left(\frac{x_1\sin\theta_{\mathrm{Rr}}+x_2\cos\theta_{\mathrm{Rr}}}{c_2}-t\right)\right\},\end{aligned} \tag{5.31}$$

from which we can calculate the stresses on $x_2 = 0$ as

$$\begin{aligned}\sigma_{12} &= \mu i\omega\left\{\frac{A_{\mathrm{I}}\sin 2\theta_{\mathrm{I}}}{c_1}-\frac{A_{\mathrm{D}}\sin 2\theta_{\mathrm{I}}}{c_1}-\frac{A_{\mathrm{R}}\cos 2\theta_{\mathrm{I}}}{c_2}\right\}\exp(\Omega_{\mathrm{I}}),\\ \sigma_{22} &= \lambda\frac{i\omega}{c_1}(A_{\mathrm{I}}+A_{\mathrm{D}})E+2\mu i\omega\left\{(A_{\mathrm{I}}+A_{\mathrm{D}})\frac{\cos 2\theta_{\mathrm{I}}}{c_1}-\frac{R\sin 2\theta_{\mathrm{Rr}}}{2c_2}\right\}\exp(\Omega_{\mathrm{I}}),\end{aligned} \tag{5.32}$$

where

$$\Omega_{\mathrm{I}} = i\omega\left(\frac{x_1\sin\theta_{\mathrm{I}}}{c_1}-t\right).$$

This means that, if the amplitude of the incident wave, A_{I}, is a known quantity, we can write the stress-free conditions in the form

$$\left.\begin{aligned}\left(\frac{A_{\mathrm{D}}}{A_{\mathrm{I}}}\right)\frac{\sin 2\theta_{\mathrm{I}}}{c_1}+\left(\frac{A_{\mathrm{R}}}{A_{\mathrm{I}}}\right)\frac{\cos 2\theta_{\mathrm{I}}}{c_2}=\frac{\sin 2\theta_{\mathrm{I}}}{c_1},\\ \left(\frac{A_{\mathrm{D}}}{A_{\mathrm{I}}}\right)\left(\frac{\lambda+2\mu\cos 2\theta_{\mathrm{I}}}{c_1}\right)-\left(\frac{A_{\mathrm{R}}}{A_{\mathrm{I}}}\right)\frac{\mu\sin 2\theta_{\mathrm{Rr}}}{c_2}=-\frac{\lambda+2\mu\cos 2\theta_{\mathrm{I}}}{c_1}.\end{aligned}\right\} \tag{5.33}$$

These equations determine A_D/A_I and A_R/A_I as real quantities, which shows that, on the free surface, both reflected waves are either in phase or exactly out of phase with the incident wave.

5.4 Waves in Finite Elastic Bodies

So far we have considered the propagation of waves in either infinite or semi-infinite bodies. All of the solutions we have found describe waves that are non-dispersive. This is no great surprise, and is easily predicted on dimensional grounds. Suppose we were to assume that $\omega = \omega(k, \lambda, \mu, \rho)$. It is then easy to show by dimensional analysis that $\omega = k\sqrt{\mu/\rho}\, f\left(\lambda/\mu\right)$, so that the phase speed, $c_p = \omega/k = \sqrt{\mu/\rho}\, f\left(\lambda/\mu\right)$, is independent of wavenumber, k. The situation is rather different in a finite elastic body in which there are one or more characteristic lengths. If we consider the simplest situation, in which the body has just one characteristic length, h, the assumption that $\omega = \omega(k, \lambda, \mu, \rho, h)$ gives, after the same dimensional analysis, $\omega = k\sqrt{\mu/\rho}\, g\left(\lambda/\mu, kh\right)$. In general, $c_p = c_p(kh)$, and the waves are dispersive. It is worth noting that for waves that have a small wavelength compared with the characteristic length, $kh \gg 1$ and $g\left(\lambda/\mu, kh\right) \sim f\left(\lambda/\mu\right)$, so that we recover the non-dispersive waves in this case.

There are very important technological reasons for the study of waves in finite elastic media. For example, surface elastic waves can excite **piezoelectric crystals** (elastic structures that produce electricity when they deform) so as to control the frequency of oscillators in watches and television sets. Rods of elastic material are used in projectors and hydrophones to generate and receive sound waves in water.

As we have seen for acoustic waves, a finite medium acts as a waveguide. In the context of elastic waves, this is a device or structure along which a wave can propagate without spreading in all directions. This lack of spreading means that the wave amplitude decreases less rapidly than in the unguided case. Waveguides are primarily used to transmit information from one place to another. More quantitatively, it can be shown that in three dimensions a dispersive wave has amplitude decaying like $r^{-\frac{3}{2}}$. This is reduced to r^{-1} in a two-dimensional waveguide, such as a plate, and further reduced to $r^{-\frac{1}{2}}$ for a one-dimensional waveguide, such as a rod (see Keller (1977)).

We now consider two types of waveguide, plates and rods, and some of the time harmonic dispersive waves that can propagate within them.

(For a detailed discussion on the initiation and decay of wavefronts in a waveguide, see the books by Miklowitz (1977) and Auld (1973).)

In order to simplify the analysis of this section it is timely to introduce the Helmholtz representation of a smooth vector field. Recall that we can write $\mathbf{u} = \nabla\phi + \nabla \times \mathbf{H}$, where ϕ is called the scalar potential and $\mathbf{H}$ the vector potential, along with the constraint $\nabla \cdot \mathbf{H} = 0$. If we substitute for $\mathbf{u}$ in (5.4), we obtain

$$\rho \frac{\partial^2}{\partial t^2}(\nabla\phi + \nabla \times \mathbf{H}) = (\lambda + 2\mu)\nabla(\nabla^2\phi) - \mu\nabla \times \{\nabla \times (\nabla \times \mathbf{H})\}, \quad (5.34)$$

where we have used the identity $\nabla\cdot(\nabla\times\mathbf{H}) = 0$. Commuting the operators and rearranging gives

$$\nabla\left\{\rho\frac{\partial^2\phi}{\partial t^2} - (\lambda + 2\mu)\nabla^2\phi\right\} + \nabla \times \left(\rho\frac{\partial^2\mathbf{H}}{\partial t^2} - \mu\nabla^2\mathbf{H}\right) = 0. \quad (5.35)$$

If we now take ϕ and $\mathbf{H}$ to be solutions of

$$\frac{\partial^2\phi}{\partial t^2} = c_1^2\nabla^2\phi, \quad \frac{\partial^2\mathbf{H}}{\partial t^2} = c_2^2\nabla^2\mathbf{H},$$

we can satisfy (5.35). We thereby reduce the problem of solving Navier's equations to that of finding solutions of two uncoupled wave equations, one scalar, one vector.

5.4.1 Flexural Waves in Plates

By a plate we will mean a solid body bounded by two parallel surfaces, $x_2 = \pm h$. We consider plane waves that propagate in the x_1-direction, in which there is displacement in both the x_1- and x_2-directions, but not in the x_3-direction – a situation of **plane strain**. Both surfaces of the plate are stress-free. If we take the vector potential in the form $\mathbf{H} = (0, 0, \psi(x_1, x_2, t))$ then $\nabla \cdot \mathbf{H} = 0$ and

$$\mathbf{u} = \left(\frac{\partial\phi}{\partial x_1} + \frac{\partial\psi}{\partial x_2}, \frac{\partial\phi}{\partial x_2} - \frac{\partial\psi}{\partial x_1}, 0\right).$$

Both ϕ and ψ satisfy wave equations,

$$\frac{\partial^2\phi}{\partial t^2} = c_1^2\left(\frac{\partial^2\phi}{\partial x_1^2} + \frac{\partial^2\phi}{\partial x_2^2}\right), \qquad \frac{\partial^2\psi}{\partial t^2} = c_2^2\left(\frac{\partial^2\psi}{\partial x_1^2} + \frac{\partial^2\psi}{\partial x_2^2}\right), \quad (5.36)$$

and the stresses relevant at $x_2 = \pm h$ are

$$\left.\begin{aligned} \sigma_{22} &= \lambda\left(\frac{\partial^2\phi}{\partial x_1^2} + \frac{\partial^2\phi}{\partial x_2^2}\right) + 2\mu\left(\frac{\partial^2\phi}{\partial x_2^2} - \frac{\partial^2\psi}{\partial x_1\partial x_2}\right), \\ \sigma_{12} &= \mu\left(2\frac{\partial^2\phi}{\partial x_1\partial x_2} + \frac{\partial^2\psi}{\partial x_2^2} - \frac{\partial^2\psi}{\partial x_1^2}\right). \end{aligned}\right\} \tag{5.37}$$

If we look for solutions of these wave equations in the form

$$\phi = Y_1(x_2)\exp\{i(kx_1 - \omega t)\}, \quad \psi = Y_2(x_2)\exp\{i(kx_1 - \omega t)\}, \tag{5.38}$$

we find that

$$\frac{d^2Y_1}{dx_2^2} + \left(\frac{\omega^2}{c_1^2} - k^2\right)Y_1 = 0, \quad \frac{d^2Y_2}{dx_2^2} + \left(\frac{\omega^2}{c_2^2} - k^2\right)Y_2 = 0. \tag{5.39}$$

In a **flexural** or **bending deformation**, when the top surface of the plate is in tension, the bottom surface is in compression, and vice versa, as shown in figure 5.7. To achieve this, we take u_1 to be an odd function of x_2 and u_2 to be even in x_2. Consideration of the expressions for u_1 and u_2 in terms of derivatives of the potentials shows that we must have ϕ odd in x_2 and ψ even in x_2. The appropriate solutions of (5.39) are therefore

$$Y_1 = A\sinh\alpha x_2, \qquad Y_2 = B\cosh\beta x_2, \tag{5.40}$$

$$\alpha^2 = k^2 - \frac{\omega^2}{c_1^2}, \quad \beta^2 = k^2 - \frac{\omega^2}{c_2^2},$$

with A and B constants to be determined. We can now calculate the stresses on the surface $x_2 = h$ as

$$\left.\begin{aligned} \sigma_{12} &= \mu\{2Aik\alpha\cosh\alpha h + B\cosh\beta h(k^2 + \beta^2)\}\exp\{i(kx_1 - \omega t)\}, \\ \sigma_{22} &= \{\lambda A(\alpha^2 - k^2)\sinh\alpha h \\ &\quad + 2\mu(A\alpha^2\sinh\alpha h - Bik\beta\sinh\beta h)\}\exp\{i(kx_1 - \omega t)\}. \end{aligned}\right\} \tag{5.41}$$

Setting these to zero gives two linear equations for A and B. For a non-trivial solution, we require

$$4\mu k^2\alpha\beta\sinh\beta h\cosh\alpha h = (\beta^2 + k^2)(\lambda(\alpha^2 - k^2) + 2\mu\alpha^2)\cosh\beta h\sinh\alpha h. \tag{5.42}$$

This can be rearranged into a more convenient form by noting that

$$\lambda(\alpha^2 - k^2) + 2\mu\alpha^2 = \mu(\beta^2 + k^2),$$

so that the dispersion relation, (5.42) can be written as

$$\frac{\tanh\beta h}{\tanh\alpha h} = \frac{(\beta^2 + k^2)^2}{4k^2\alpha\beta}. \tag{5.43}$$

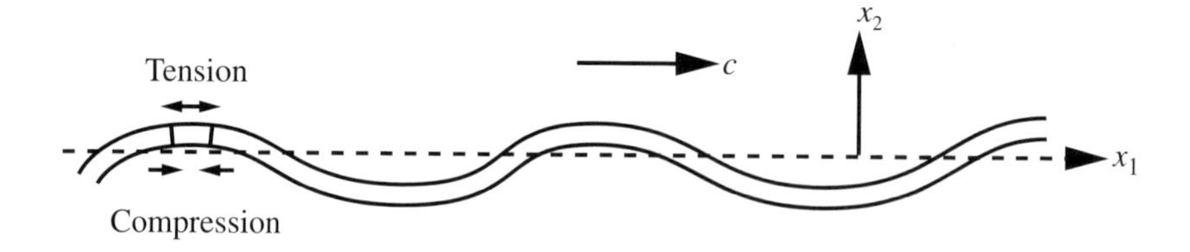

Fig. 5.7. The propagation of flexural waves in a plate.

In general, it is necessary to solve this relationship numerically. For a given value of k, the corresponding value of ω can be found by a combination of interval bisection and Newtonian iterative methods. The values of ω corresponding to short waves ($kh \gg 1$) and long waves ($kh \ll 1$) can, however, be found analytically.

Short Waves ($kh \gg 1$)

If we assume that $\omega = ak + o(k)$ as $k \to \infty$ then, at leading order,

$$\alpha = k\left(1 - \frac{a^2}{c_1^2}\right)^{\frac{1}{2}}, \quad \beta = k\left(1 - \frac{a^2}{c_2^2}\right)^{\frac{1}{2}},$$

and $\tanh \alpha h \sim 1$ and $\tanh \beta h \sim 1$ for $h = O(1)$. At leading order, the dispersion relation, (5.43), therefore gives

$$1 = \frac{\left(2 - \dfrac{a^2}{c_2^2}\right)^2}{4\left(1 - \dfrac{a^2}{c_1^2}\right)^{\frac{1}{2}}\left(1 - \dfrac{a^2}{c_2^2}\right)^{\frac{1}{2}}}. \tag{5.44}$$

This is equivalent to the dispersion relationship, (5.26), that we obtained for Rayleigh waves. This is because, if we take the wavelength of these short waves as our unit of length, the thickness of the plate appears to be very large. With the symmetry we have assumed for these flexural waves, we are actually asymptotically solving the problem in a half space with a stress-free boundary, and hence obtain the solution for Rayleigh waves.

Long Waves ($kh \ll 1$)

In this case things are more delicate! An inspection of the dispersion relation with $\alpha = \beta = k + O(k^2)$ as $k \to 0$ reveals an identity at leading order, so we must go to higher order terms. (This is usually a sign of some foul algebra to come.) Let's start by posing an expansion

$$\omega = dk^{\delta} + ek^{\delta+2} + o\left(k^{\delta+2}\right),$$

where d, e and δ are as yet unknown, and are to be found by asymptotic balance. That δ is even is clear, as the dispersion relation is invariant under the transformation $k \mapsto -k$. Using this expansion for ω, we find that

$$\beta = k\left\{1 - \frac{1}{2c_2^2}(d^2k^{2\delta-2} + 2dek^{2\delta}) - \frac{1}{8c_2^4}d^4k^{4\delta-4} + \cdots\right\},$$

and expanding each tanh in (5.43) to fifth order, we obtain

$$\begin{aligned}
&\left\{1 - \left(\frac{1}{2c_1^2} + \frac{1}{2c_2^2}\right)(d^2k^{2\delta-2} + 2dek^{2\delta}) + \left(\frac{1}{4c_1^2c_2^2} - \frac{1}{8c_2^4} - \frac{1}{8c_1^4}\right)d^4k^{4\delta-4}\right\} \\
&\times\left\{kh - \frac{hd^2}{2c_2^2}k^{2\delta-1} - \frac{deh}{c_2^2}k^{2\delta+1} - \frac{h^3k^3}{3} + \frac{h^3d^2}{2c_2^2}k^{2\delta+1} + h^5k^5 - \frac{hd^4}{8c_2^4}k^{4\delta-3}\right\} \\
&= \left\{1 - \frac{1}{c_2^2}(d^2k^{2\delta-2} + 2dek^{2\delta}) + \frac{1}{4c_2^4}d^4k^{4\delta-4}\right\} \\
&\times\left\{kh - \frac{hd^2}{2c_1^2}k^{2\delta-1} - \frac{deh}{c_1^2}k^{2\delta+1} - \frac{h^3k^3}{3} + \frac{h^3d^2}{2c_1^2}k^{2\delta+1} + h^5k^5 - \frac{hd^4}{8c_1^4}k^{4\delta-3}\right\}.
\end{aligned} \tag{5.45}$$

By collecting like terms, we find that at $O(k)$ we have an identity. The question now arises as to the size of δ. The richest balance occurs when $2\delta - 1 = 3$, so that $\delta = 2$. Balancing at $O(k^3)$ then gives

$$-\frac{h^3}{3} - hd^2\left(\frac{1}{2c_1^2} + \frac{1}{2c_2^2}\right) - \frac{hd^2}{2c_2^2} = -\frac{h^3}{3} - \frac{hd^2}{2c_1^2} - \frac{d^2h}{c_2^2}, \tag{5.46}$$

which is another identity. Proceeding to $O(k^5)$, we find that

$$\begin{aligned}
&h^5 + \frac{h^3d^2}{2c_2^2} - \frac{deh}{c_2^2} - \left(\frac{1}{2c_1^2} + \frac{1}{2c_2^2}\right)d^2\left(-\frac{h^3}{3} - \frac{hd^2}{2c_2^2}\right) \\
&- \left(\frac{1}{2c_1^2} + \frac{1}{2c_2^2}\right)2deh + d^4h\left(\frac{1}{4c_1^2c_2^2} - \frac{1}{4c_2^4} - \frac{1}{8c_1^4}\right) \\
&= h^5 + \frac{h^3d^2}{2c_1^2} - \frac{deh}{c_1^2} - \frac{d^2}{c_2^2}\left(-\frac{h^3}{3} - \frac{hd^2}{2c_1^2}\right) - \frac{2deh}{c_2^2} + d^4h\left(\frac{1}{4c_2^4} - \frac{1}{8c_1^4}\right).
\end{aligned} \tag{5.47}$$

After some algebra, we find that

$$d = \frac{2hc_2^2}{3^{1/2}}\sqrt{\frac{1}{c_2^2} - \frac{1}{c_1^2}},$$

and our low frequency limit is

$$\omega = \frac{2hc_2^2k^2}{3^{1/2}}\sqrt{\frac{1}{c_2^2} - \frac{1}{c_1^2}} + o\left(k^2\right).$$

The flexural waves are dispersive in this limit, with the group velocity approximately twice the phase velocity, so that the energy that caused the bending propagates twice as fast as the wave crests.

Finally, we note that, for any serious computation using the dispersion relationship, it is worth writing it in dimensionless form. If we define $\bar{\omega} = h\omega/c_2$, $\bar{k} = kh$ so that

$$\beta^2 = \frac{1}{h^2}(\bar{k}^2 - \bar{\omega}^2) = \frac{\bar{\beta}^2}{h^2}, \quad \alpha^2 = \frac{1}{h^2}\left(\bar{k}^2 - \frac{c_2^2}{c_1^2}\bar{\omega}^2\right) = \frac{\bar{\alpha}^2}{h^2},$$

we then need to solve the dimensionless dispersion relation

$$\frac{\tanh\bar{\beta}}{\tanh\bar{\alpha}} = \frac{(\bar{\beta}^2 + \bar{k}^2)^2}{4\bar{k}^2\bar{\alpha}\bar{\beta}},$$

which contains the single material parameter c_2^2/c_1^2. This can be written in terms of the Poisson ratio as

$$\frac{c_2^2}{c_1^2} = \frac{1 - 2\nu}{2(1 - \nu)}.$$

The dispersion relations we have derived for the short and long wave limits are shown in figure 5.8 for the typical steel we considered earlier, along with the dispersion relation determined numerically from (5.43). Figure 5.9 shows bending waves in the elastic plates that form the body of a guitar.

5.4.2 *Waves in Elastic Rods*

By an elastic rod, we mean a body of elastic material with a finite cross-sectional area. We restrict our attention to straight rods with circular cross-sections to avoid unnecessary complexity. It should be clear that with this type of three-dimensional body there are four modes of wave propagation. **Longitudinal** (down the axis of the rod), **torsional** (rotational only in the cross-section of the rod) and two **flexural** modes (bending of the rod about the two principal axes of the cross-section). This is in contrast to the study of waves in plates of the previous subsection, where only the two longitudinal and one flexural mode were present. It is convenient

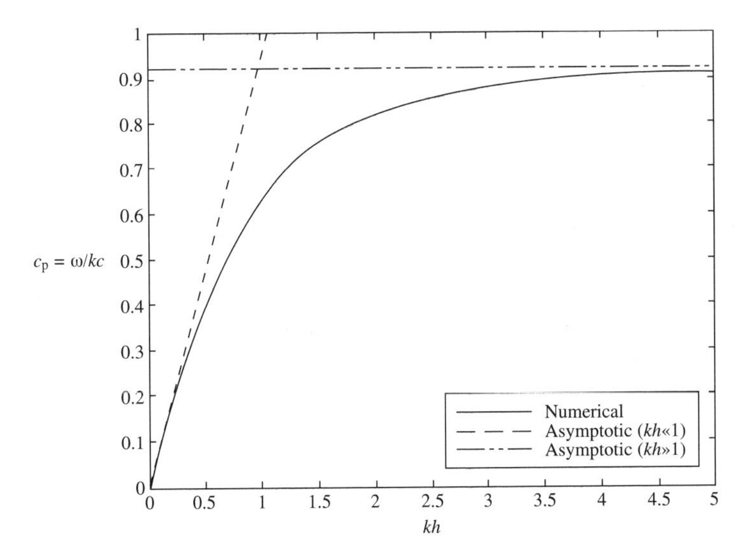

Fig. 5.8. The phase speed as a function of wavenumber for flexural waves with $\nu = 0.32$, as for a typical steel.

for subsequent analysis to state the components of Navier's equation in cylindrical polar coordinates (r, θ, z) with displacement (u, v, w) as

$$\left.\begin{aligned} \nabla^2 u - \frac{u}{r^2} - \frac{2}{r^2}\frac{\partial v}{\partial \theta} + \frac{1}{1-2\nu}\frac{\partial \Delta}{\partial r} &= \frac{1}{c_2^2}\frac{\partial^2 u}{\partial t^2}, \\ \nabla^2 v - \frac{v}{r^2} + \frac{2}{r^2}\frac{\partial u}{\partial \theta} + \frac{1}{1-2\nu}\frac{1}{r}\frac{\partial \Delta}{\partial \theta} &= \frac{1}{c_2^2}\frac{\partial^2 v}{\partial t^2}, \\ \nabla^2 w + \frac{1}{1-2\nu}\frac{\partial \Delta}{\partial z} &= \frac{1}{c_2^2}\frac{\partial^2 w}{\partial t^2}, \end{aligned}\right\} \tag{5.48}$$

where the Laplacian is

$$\nabla^2 = \frac{\partial^2}{\partial r^2} + \frac{1}{r}\frac{\partial}{\partial r} + \frac{1}{r^2}\frac{\partial^2}{\partial \theta^2} + \frac{\partial^2}{\partial z^2}$$

and the dilatation is

$$\Delta = \frac{\partial u}{\partial r} + \frac{u}{r} + \frac{1}{r}\frac{\partial v}{\partial \theta} + \frac{\partial w}{\partial z}.$$

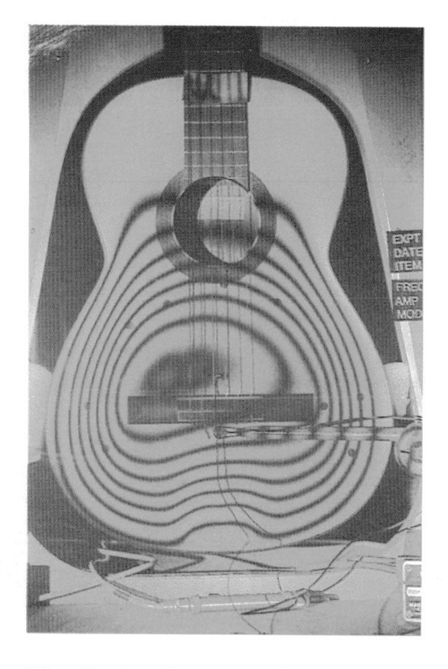

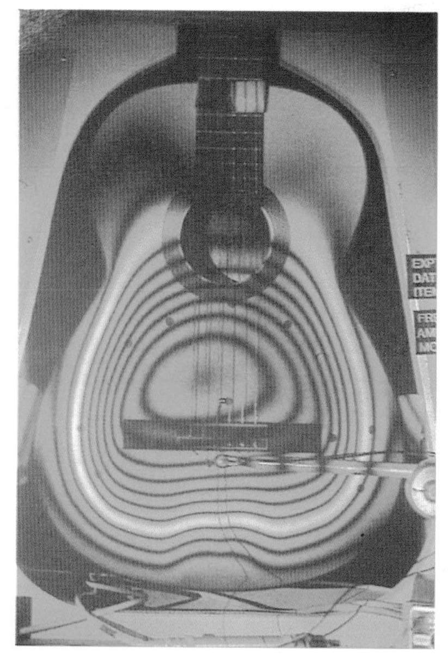

Fig. 5.9. Bending waves in the body of a guitar. A typical magnitude for the out of plane deformation is 1 μm.

The components of the stress tensor that we shall use most frequently are

$$\left.\begin{aligned} \sigma_{rr} &= \lambda\Delta + 2\mu\frac{\partial u}{\partial r}, & \sigma_{r\theta} &= \mu\left(\frac{1}{r}\frac{\partial u}{\partial \theta} - \frac{v}{r} + \frac{\partial v}{\partial r}\right), \\ \sigma_{\theta z} &= \mu\left(\frac{\partial v}{\partial z} + \frac{1}{r}\frac{\partial w}{\partial \theta}\right), & \sigma_{rz} &= \mu\left(\frac{\partial u}{\partial z} + \frac{\partial w}{\partial r}\right). \end{aligned}\right\} \tag{5.49}$$

5.4.3 *Torsional Waves*

Torsional waves involve a displacement in the circumferential direction only, as shown in figure 5.10. In this case the displacement vector is of the form $\mathbf{u} = (0, v(r,z,t), 0)$, and we can work with Navier's equations directly. Noting that $\Delta = 0$, the only equation we have to solve is

$$\nabla^2 v - \frac{v}{r^2} = \frac{1}{c_2^2}\frac{\partial^2 v}{\partial t^2}. \tag{5.50}$$

If we look for a solution of the form $v = F(r)\exp\{i(kz - \omega t)\}$ then

$$F'' + \frac{1}{r}F' + \left(\frac{\omega^2}{c_2^2} - k^2 - \frac{1}{r^2}\right)F = 0. \tag{5.51}$$

This is a scaled version of Bessel's equation, and the solution that is bounded as $r \to 0$ can be written in the form $F = AJ_1(sr)$ where $s^2 = (\omega^2/c_2^2) - k^2$. The only non-zero component of the surface stress is

$$\sigma_{r\theta} = \mu\left(\frac{\partial v}{\partial r} - \frac{v}{r}\right)$$

and setting this to zero on the surface of the rod $r = a$ gives

$$asJ_1'(sa) - J_1(sa) = 0. \tag{5.52}$$

Using the standard result $J_1'(x) = J_0(x) - (J_1(x))/x$, this becomes

$$asJ_0(as) = 2J_1(as).$$

This is a transcendental equation whose roots are tabulated (Abramowitz and Stegun, 1972). The first three roots are $as = 0, 5.136, 8.417$ ($= \sigma_1, \sigma_2, \sigma_3$). Although it is clear that an infinite set of frequencies is excited, the most important corresponds to $s = 0$. In this case the bounded solution of the differential equation is $F = \overline{A}r$ (where $\overline{A} = A/s$ in order to take the limit $s \to 0$ in the solution $F = AJ_1(sr)$) and our solution for the lowest torsional mode takes the simple, non-dispersive form

$$v = \overline{A}r\exp\{ik(z \pm c_2 t)\}. \tag{5.53}$$

This is the fundamental torsional mode of the rod, and is always able to propagate at the shear wave speed, c_2. The only non-zero stress in this case is $\sigma_{\theta z} = \mu\partial v/\partial z = \mu\overline{A}ikr\exp\{ik(z \pm c_2 t)\}$. If we take the total moment of these stresses about the centre of the rod,

$$\int_{\theta=0}^{2\pi}\int_{r=0}^{a} r\sigma_{\theta z} r\, d\theta\, dr = \frac{\pi}{2}\mu\overline{A}ika^4\exp\{ik(z \pm c_2 t)\}, \tag{5.54}$$

we can identify the quantity $\pi\mu a^4/2$ as the torsional rigidity of the rod. We interpret this as the couple per unit amplitude and wavelength that will balance the inertial twisting of the rod. As usual, for each of the other modes there is a cut-off frequency, below which the mode is evanescent, since for the jth mode

$$k_j = \pm\sqrt{\frac{\omega^2}{c_2^2} - \frac{\sigma_j^2}{a^2}}.$$

This gives real values for the wavenumber provided the frequency is larger than $\omega_j \equiv c_2\sqrt{\sigma_1^2/a^2}$. This is of some technological importance in designing elastic waveguides as is the issue of how to excite only the lowest mode (which propagates at a constant speed) when any such excitation

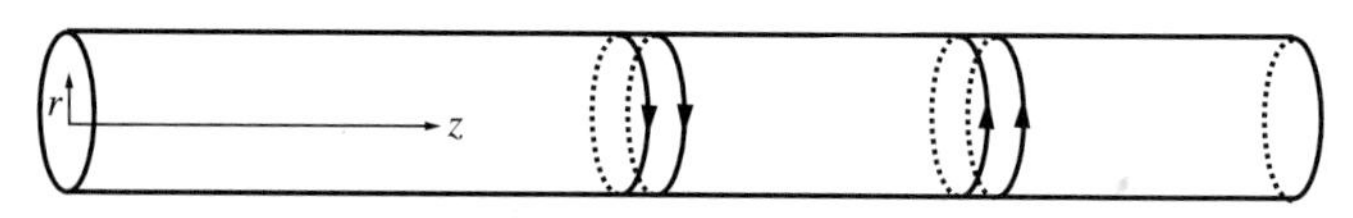

Fig. 5.10. Torsional oscillations of an elastic rod. Note that the sense of rotation changes along the axis of the rod as well as with time.

must produce a rotational displacement which is strictly proportional to the radius.

*The High Frequency Limit**

We can also study the propagation of elastic waves in the high frequency limit using an asymptotic method. These methods are extremely useful when investigating problems in which there is a more complicated geometry, such as a slight curvature of the rod or a non-circular cross-section, when it is difficult to find a separable solution of the equations. Details of solutions for curved rods can be found in Keller (1977). We now illustrate some of the ideas behind this work by re-examining the torsional wave problem above using a formal asymptotic method.

The appropriate equations are

$$\nabla^2 v - \frac{v}{r^2} = \frac{1}{c_2^2}\frac{\partial^2 v}{\partial t^2} \quad \text{in } 0 < r < a, \quad 0 < z < \infty, \tag{5.55}$$

with

$$\frac{\partial v}{\partial r} = \frac{v}{r} \quad \text{on } r = a.$$

If we look for a solution of the form $v = e^{-i\omega t}V(r,z)$, the function V satisfies

$$V_{rr} + \frac{1}{r}V_r + V_{zz} + \left(\frac{\omega^2}{c_2^2} - \frac{1}{r^2}\right)V = 0. \tag{5.56}$$

As ω is large, it is convenient to define a small parameter $\epsilon = c_2/\omega$ and seek a solution for $\epsilon \ll 1$ of the boundary value problem

$$V_{rr} + \frac{1}{r}V_r + V_{zz} + \left(\frac{1}{\epsilon^2} - \frac{1}{r^2}\right)V = 0, \qquad V_r = \frac{V}{r} \text{ on } r = a \tag{5.57}$$

in the domain $0 < r < a$, $0 < z < \infty$. Since (5.57) suggests that $V = 0$ at all algebraic orders of ϵ, the appropriate form of asymptotic expansion for V is

$$V = \exp\left[i\left\{\frac{V_0(r,z)}{\epsilon} + V_1(r,z) + \epsilon V_2(r,z) + \cdots\right\}\right]. \tag{5.58}$$

This is sometimes called the **WKB or WJKB expansion**, which was developed, from techniques pioneered in the nineteenth century by Green and Liouville, by Wentzel, Jeffreys, Kramers and Brillouin in the 1920s. Substituting (5.58) in (5.57) and collecting like terms, we arrive at

$$\left(\frac{\partial V_0}{\partial r}\right)^2 + \left(\frac{\partial V_0}{\partial z}\right)^2 = 1, \tag{5.59}$$

$$\frac{\partial V_0}{\partial r}(a, z) = 0, \tag{5.60}$$

at $O(1/\epsilon^2)$, and

$$-\frac{\partial V_0}{\partial r}\frac{\partial V_1}{\partial r} - \frac{\partial V_0}{\partial z}\frac{\partial V_1}{\partial z} + \frac{i}{2}\left(\frac{\partial^2 V_0}{\partial r^2} + \frac{\partial^2 V_0}{\partial z^2} + \frac{1}{r}\frac{\partial V_0}{\partial r}\right) = 0, \tag{5.61}$$

$$i\frac{\partial V_1}{\partial r}(a, z) = \frac{1}{a}, \tag{5.62}$$

at $O(1/\epsilon)$. Equation (5.59) is usually referred to as the **eikonal equation** and (5.61) as the **transport equation**, names that were first used in association with optics. The eikonal equation is a nonlinear first order partial differential equation and some care must be taken with its solution.

If we write $l = \partial V_0/\partial r$ and $m = \partial V_0/\partial z$, the equation may be represented in the form $F(l, m) = 0$, where $F = l^2 + m^2 - 1$. Differentiation of this with respect to z gives

$$\frac{\partial F}{\partial l}\frac{\partial l}{\partial z} + \frac{\partial F}{\partial m}\frac{\partial m}{\partial z} = 0 \quad \text{or} \quad \frac{\partial F}{\partial l}\frac{\partial m}{\partial r} + \frac{\partial F}{\partial m}\frac{\partial m}{\partial z} = 0, \tag{5.63}$$

using commutativity of partial derivatives. Now consider the changes in the values of l and m as we move by increments dr and dz along a path, usually called a **ray**, in the (r, z)-plane. If l and m are constant on this path, so that

$$dm = dr\frac{\partial m}{\partial r} + dz\frac{\partial m}{\partial z} = 0,$$

then

$$\frac{dr}{dz} = -\frac{\partial m/\partial z}{\partial m/\partial r} = \frac{\partial F/\partial l}{\partial F/\partial m}.$$

If we write $l = \sin\theta$, $m = \cos\theta$, then $F = 0$ is an identity and the paths of constant l and m are $dr/dz = l/m = \tan\theta$. We can now integrate $l = \partial V_0/\partial r = \sin\theta$ and $m = \partial V_0/\partial z = \cos\theta$ to get $V_0 = r\sin\theta + z\cos\theta + c$ and, by use of $dr/dz = r/z = \tan\theta$, could eliminate θ to get $V_0 = V_0(r, z)$. However, it is expedient to examine our boundary condition, $\partial V_0/\partial r(a,z) = 0$ or $l = 0$ on $r = a$. This clearly forces us to choose $\theta = 0$ or π and hence $V_0 \equiv \pm z + c$. With this choice of V_0, the transport equation, (5.61), reduces to $\partial V_1/\partial z = 0$ with solution $V_1 = V_1(r)$. To determine this (apparently) arbitrary function, we must go to higher order in our perturbation process.

At $O(1)$ we find

$$i\frac{d^2 V_1}{dr^2} - \left(\frac{dV_1}{dr}\right)^2 + i\frac{1}{r}\frac{dV_1}{dr} - 2\frac{\partial V_2}{\partial z} - \frac{1}{r^2} = 0, \tag{5.64}$$

which can be written as $\partial V_2/\partial z = f(r)$, and integrated to give $V_2 = zf(r) + g(r)$. However, the term $zf(r)$ grows as $z \to \infty$ and will destroy the validity of our asymptotic expansion unless we take $f(r) = 0$, so that

$$i\frac{d^2 V_1}{dr^2} - \left(\frac{dV_1}{dr}\right)^2 + i\frac{1}{r}\frac{dV_1}{dr} - \frac{1}{r^2} = 0. \tag{5.65}$$

Our system has now been closed by this secularity condition. To solve this equation, we write $G = -idV_1/dr$ to get

$$\frac{dG}{dr} = G^2 - \frac{G}{r} - \frac{1}{r^2}, \tag{5.66}$$

which is a form of Riccati's equation. In general, these are difficult to solve. The exception to this rule is when we can guess an exact solution, in which case the equation linearises. Since $G = 1/r$ is an exact solution, we now write $G = 1/r + 1/\hat{G}$, and $\hat{G}$ satisfies

$$\frac{d\hat{G}}{dr} + \frac{\hat{G}}{r} = -1, \tag{5.67}$$

with solution

$$\hat{G} = -\frac{1}{2}r + \frac{A}{r},$$

and hence

$$V_1 = i\log\left(\frac{r}{r^2 - 2A}\right) + B.$$

In order to satisfy the boundary condition

$$i\frac{dV_1}{dr}(a) = \frac{1}{a},$$

we must choose $A = 0$ and hence $V_1 = -i\log r$. Our complete high frequency solution takes the form

$$\begin{aligned} V &= \exp\left[i\left\{\frac{\pm z + c}{\epsilon} - i\log r + O(\epsilon)\right\}\right] \\ &= r\exp\{i\{(\pm z + c)/\epsilon\}\}\{1 + O(\epsilon)\}. \end{aligned} \tag{5.68}$$

This is clearly recognisable as the lowest mode, (5.53), of the solutions discussed earlier. This result is no surprise after we consider the wavenumber for the jth mode,

$$k_j = \left(\frac{\omega^2}{c_2^2} - \frac{\sigma_j^2}{a^2}\right)^{\frac{1}{2}},$$

which we determined earlier. Expanding for $\omega \gg 1$ gives $k = \pm\omega/c_2 + O(1)$, and the whole ensemble of different modes collapses onto the non-dispersive fundamental mode at leading order. The other dispersive modes are still there, but are at higher order in our WKB expansion.

5.4.4 Longitudinal Waves

Longitudinal waves propagate down the axis of the rod with no rotational displacement but displacement in both the radial and axial directions. As we discussed earlier, an alternative representation of the displacements is available to us via the Helmholtz representation, and it is convenient to make use of this to study longitudinal waves, for which $v = 0$. It is sufficient in this case to take $\mathbf{H} = (0, \psi(r,z,t), 0)$ and $\phi = \phi(r,z,t)$, so that

$$u = \frac{\partial\phi}{\partial r} - \frac{\partial\psi}{\partial z}, \quad w = \frac{\partial\phi}{\partial z} + \frac{1}{r}\frac{\partial}{\partial r}(r\psi),$$

and ϕ and ψ satisfy

$$\left.\begin{aligned} \frac{\partial^2\phi}{\partial r^2} + \frac{1}{r}\frac{\partial\phi}{\partial r} + \frac{\partial^2\phi}{\partial z^2} &= \frac{1}{c_1^2}\frac{\partial^2\phi}{\partial t^2}, \\ \nabla^2\psi - \frac{\psi}{r^2} &= \frac{1}{c_2^2}\frac{\partial^2\phi}{\partial t^2}. \end{aligned}\right\} \tag{5.69}$$

If we look for solutions of the wave equations satisfied by the scalar and vector potentials in the form

$$\phi = F(r)\exp\{i(kz - \omega t)\}, \qquad \psi = G(r)\exp\{i(kz - \omega t)\}, \tag{5.70}$$

we find that the bounded solutions take the form

$$F = AJ_0(pr), \quad G = BJ_1(sr),$$

with

$$p^2 = \frac{\omega^2}{c_1^2} - k^2, \quad s^2 = \frac{\omega^2}{c_2^2} - k^2.$$

The displacements are then given by

$$\left.\begin{aligned} u &= \{-pAJ_1(pr) - ikBJ_1(sr)\}\exp\{i(kz-\omega t)\} \\ v &= \{ikAJ_0(pr) + sBJ_0(sr)\}\exp\{i(kz-\omega t)\}. \end{aligned}\right\} \tag{5.71}$$

Stress-free conditions on the surface of the rod can be calculated by setting $\sigma_{rr} = \sigma_{rz} = 0$ on $r = a$, and lead to

$$\left.\begin{aligned} \left\{\frac{p}{a}J_1(pa) - \frac{1}{2}(s^2-k^2)J_0(pa)\right\}A + \left\{\frac{ik}{a}J_1(sa) - iksJ_0(sa)\right\}B = 0, \\ \{-2ikpJ_1(pa)\}A - \{(s^2-k^2)J_1(sa)\}\,B = 0. \end{aligned}\right\} \tag{5.72}$$

The condition for a non-trivial solution leads to the dispersion relation

$$\frac{2p}{a}(s^2+k^2)J_1(pa)J_1(sa) - (s^2-k^2)^2J_0(pa)J_1(sa) - 4k^2psJ_1(pa)J_0(sa) = 0, \tag{5.73}$$

a result due to Pochhammer (1876) and Chree (1889). As in the previous subsection, the dimensionless version of this reduces the relationship between the five variables ω, k, a, λ and μ to one between three dimensionless variables, which is the most convenient form for computation. An important limiting case is long waves, $ka \ll 1$, for which an expansion in the form $\omega = \alpha k + O(k^3)$ gives

$$\omega = \left\{\frac{\mu(3\lambda+2\mu)}{\rho(\lambda+\mu)}\right\}^{\frac{1}{2}} k + O(k^3). \tag{5.74}$$

Using

$$\frac{3\lambda+2\mu}{\lambda+\mu} = \frac{E}{\mu},$$

this gives a phase speed $c = \sqrt{E/\rho} + O(k^2)$, which we can relate to approximate theories of longitudinal waves. If the deformation in the rod is assumed to be one-dimensional, with a stress–strain relationship of the form $\sigma = Ee = E\partial u/\partial x$, the dynamic equations of motion take the form

$$\rho\frac{\partial^2 u}{\partial t^2} = \frac{\partial \sigma}{\partial x} = E\frac{\partial^2 u}{\partial x^2}.$$

This is a non-dispersive wave equation with a speed $\sqrt{E/\rho}$, as expected.

5.5 The Excitation and Propagation of Elastic Wavefronts*

The sudden application of a localised force or pressure to an elastic body results in the excitation and propagation of wavefronts, which initially travel away from the origin of the forcing. Earthquakes are a good example of this, and arise from a sudden motion in the tectonic plates, or slippage along a fault in the earth's crust. We now consider two initial value problems that have *some* of the features of an earthquake. These were first studied by Lamb in 1904. Firstly, we study the response of an infinite elastic body to a suddenly applied load, and secondly, we analyse how this response is modified by the presence of a free surface. Of course in an earthquake things are more complex than this due to the effects of the finite size, material anisotropy and internal structure of the earth and its surface. It should be clear that here we are just establishing some basic principles for the motion of elastic wavefronts.

5.5.1 *Wavefronts Caused by an Internal Line Force in an Unbounded Elastic Body*

Consider a line force, $\mathbf{F} = \mathbf{F}_0\delta(x_1)\delta(x_2)f(t)$, acting on an unbounded elastic body. The force is applied when $t = 0$, so that $f(t) = 0$ for $t < 0$. The coordinate system and direction of the force are shown in figure 5.11. Since this applied force is independent of the coordinate x_3 and there are no boundaries, the displacement in the x_3-direction vanishes as does any functional dependence of displacement, stress and strain on x_3. We therefore have a plane strain problem for which the governing equations are (5.36). As the material ahead of the force is in compression and that behind it is in tension, the response to this line force is odd in the variable x_2, so that the normal stress on the plane $x_2 = 0$ vanishes for $x_1 \neq 0$. For similar reasons the displacement in the x_1-direction is also odd in x_2. Hence we only need consider this problem in the half space $-\infty < x_1 < \infty$, $0 < x_2 < \infty$ subject to the boundary conditions

$$\sigma_{22}(x_1, 0, t) = -\frac{1}{2}F_0\delta(x_1)f(t), \quad u_1(x_1, 0, t) = 0.$$

(The factor of $\frac{1}{2}$ arises as we have split the whole space into two halves.) Since the body is at rest prior to the application of the force, we impose the initial conditions

$$u_i(x_1, x_2, 0) = \frac{\partial u_i}{\partial t}(x_1, x_2, 0) = 0 \quad \text{for } i = 1,\ 2.$$

Our task now is to solve the wave equations (5.36) for the vector and scalar potentials, subject to the boundary conditions appropriate to the problem. To do this we begin by taking a Laplace transform in time and a Fourier transform in the spatial variable x_1. Denoting by a bar a

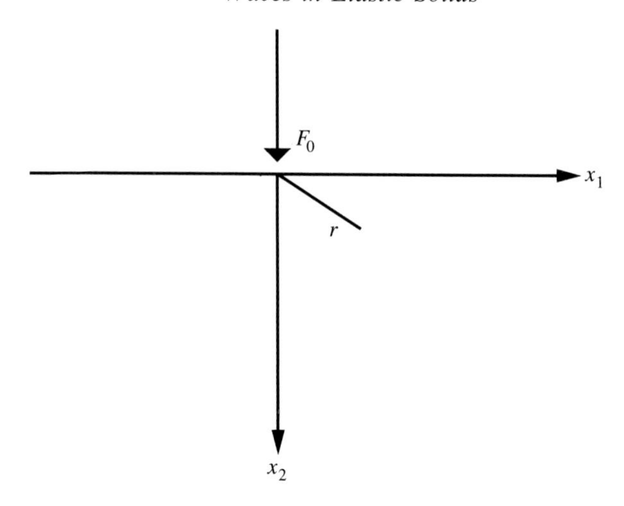

Fig. 5.11. A line load on an unbounded elastic body.

variable that has been Laplace transformed ($t \mapsto s$) and by a star one that is Fourier transformed ($x_1 \mapsto k$), the wave equations become

$$\frac{\partial^2 \bar{\phi}^*}{\partial x_2^2} - \left(k^2 + \frac{s^2}{c_1^2}\right)\bar{\phi}^* = 0, \quad \frac{\partial^2 \bar{\psi}^*}{\partial x_2^2} - \left(k^2 + \frac{s^2}{c_2^2}\right)\bar{\psi}^* = 0.$$

The solutions that decay as $x_2 \to \infty$ are

$$\bar{\phi}^*(k, x_2, s) = A(s,k)\exp\left\{-\left(k^2 + \frac{s^2}{c_1^2}\right)^{\frac{1}{2}} x_2\right\},$$

$$\bar{\psi}^*(k, x_2, s) = B(s,k)\exp\left\{-\left(k^2 + \frac{s^2}{c_2^2}\right)^{\frac{1}{2}} x_2\right\}.$$

Since

$$u_1 = \frac{\partial \phi}{\partial x_1} + \frac{\partial \psi}{\partial x_2},$$

and hence

$$\bar{u}_1^* = -ik\bar{\phi}^* + \frac{\partial \bar{\psi}^*}{\partial x_2},$$

the displacement boundary condition is

$$\frac{\partial \bar{\psi}^*}{\partial x_2} = ik\bar{\phi}^* \quad \text{on } x_2 = 0.$$

This means that

$$-\left(k^2+\frac{s^2}{c_2^2}\right)^{\frac{1}{2}} B(s,k) = ikA(s,k).$$

The stress boundary condition on $x_2 = 0$ can be written in the form

$$\lambda\left(\frac{\partial^2\phi}{\partial x_1^2}+\frac{\partial^2\phi}{\partial x_2^2}\right)+2\mu\left(\frac{\partial^2\phi}{\partial x_2^2}-\frac{\partial^2\psi}{\partial x_1\partial x_2}\right) = -\frac{1}{2}F_0\delta(x_1)f(t),$$

which, making use of the wave equation for ϕ, can be double transformed to

$$\frac{\lambda}{c_1^2}s^2\bar{\phi}^* + 2\mu\left(\frac{\partial^2\bar{\phi}^*}{\partial x_2^2}+ik\frac{\partial\bar{\psi}^*}{\partial x_2}\right) = -\frac{1}{2}F_0\bar{f}(s),$$

where $\bar{f}(s)$ is the Laplace transform of $f(t)$. Substitution for $\bar{\phi}^*$ and $\bar{\psi}^*$ gives

$$A(s,k) = \frac{-F_0}{2\rho s^2}\bar{f}(s), \quad B(s,k) = \frac{ikF_0\bar{f}(s)}{2\rho s^2\left(k^2+\frac{s^2}{c_2^2}\right)^{\frac{1}{2}}}.$$

If we now concentrate our attention on the displacement in the x_1-direction, we can calculate

$$\bar{u}_1^* = \frac{ikF_0\bar{f}(s)}{2\rho s^2}\left[\exp\left\{-\left(k^2+\frac{s^2}{c_1^2}\right)^{\frac{1}{2}}x_2\right\} - \exp\left\{-\left(k^2+\frac{s^2}{c_2^2}\right)^{\frac{1}{2}}x_2\right\}\right].$$

The inversion of double transforms is usually a rather difficult task. However in this case we can proceed in an operational manner by inverting first the Fourier transform and then the Laplace transform. Consider the Fourier part, for which we must evaluate two terms of the form

$$\begin{aligned}
&\frac{1}{2\pi s^2}\int_{-\infty}^{\infty} ike^{-ikx_1-(k^2+s^2/c_1^2)^{\frac{1}{2}}x_2}dk \\
&\quad = -\frac{\partial}{\partial x_1}\left\{\frac{1}{2\pi s^2}\int_{-\infty}^{\infty} e^{-ikx_1-(k^2+s^2/c_1^2)^{\frac{1}{2}}x_2}dk\right\} \\
&\quad = -\frac{\partial}{\partial x_1}\left\{\frac{1}{\pi s^2}\int_{0}^{\infty} \cos(kx_1)e^{-(k^2+s^2/c_1^2)^{\frac{1}{2}}x_2}dk\right\}.
\end{aligned}$$

By using a table of cosine transforms (for example, Erdélyi, (1954)) this can be written as

$$-\frac{\partial}{\partial x_1}\left\{\frac{c_1}{\pi s}\frac{x_2}{\sqrt{x_1^2+x_2^2}}K_1\left(\sqrt{x_1^2+x_2^2}\frac{s}{c_1}\right)\right\},$$

where K_1 is a modified Bessel function of the second kind. We now have

$$\bar{u}_1 = \frac{F_0\bar{f}(s)}{2\rho}\frac{\partial}{\partial x_1}(F_2 - F_1)$$

where

$$F_i = \frac{c_i}{\pi s}\frac{x_2}{\sqrt{x_1^2+x_2^2}}K_1\left(\sqrt{x_1^2+x_2^2}\frac{s}{c_i}\right).$$

The second part of our inversion can be done using a table of Laplace transforms. Denoting the Laplace transform by $L[\ ]$, the result

$$L\left[\frac{1}{\pi}\frac{x_2}{x_1^2+x_2^2}\left\{t^2-\left(\frac{x_1^2+x_2^2}{c_i^2}\right)\right\}^{\frac{1}{2}}H\left(t-\frac{\sqrt{x_1^2+x_2^2}}{c_i}\right)\right]$$
$$= \frac{c_i}{\pi s}\frac{x_2}{\sqrt{x_1^2+x_2^2}}K_1\left(\sqrt{x_1^2+x_2^2}\frac{s}{c_i}\right)$$

is a standard one (Erdélyi, 1954) and, using the Laplace convolution theorem, we can write

$$u_1 = \frac{F_0}{2\pi\rho}\frac{\partial}{\partial x_1}\left[f(t) * \left(\frac{x_2}{x_1^2+x_2^2}\left\{t^2-\left(\frac{x_1^2+x_2^2}{c_2^2}\right)\right\}^{\frac{1}{2}}H\left(t-\frac{\sqrt{x_1^2+x_2^2}}{c_2}\right)\right.\right.$$
$$\left.\left.-\frac{x_2}{x_1^2+x_2^2}\left\{t^2-\left(\frac{x_1^2+x_2^2}{c_1^2}\right)\right\}^{\frac{1}{2}}H\left(t-\frac{\sqrt{x_1^2+x_2^2}}{c_1}\right)\right)\right].$$

This form of solution is still rather complicated for a simple interpretation due to the presence of the convolution product, denoted by an asterisk. To simplify things let's choose f to be a Heaviside function, $H(t)$. Evaluating the convolution product and writing $r^2 = x_1^2 + x_2^2$ as the distance from a point to the origin of the forcing we find that

$$u_1 = \frac{F_0 x_1 x_2}{2\pi\mu\left(c_1/c_2\right)^2 r^4}$$
$$\times\begin{cases} 0 & \text{for } r \geq c_1 t, \\ c_1 t\left(c_1^2t^2-r^2\right)^{\frac{1}{2}} & \text{for } c_2 t < r < c_1 t, \\ c_1 t\left(c_1^2t^2-r^2\right)^{\frac{1}{2}} - \dfrac{c_1^2}{c_2}t\left(c_2^2t^2-r^2\right)^{\frac{1}{2}} & \text{for } 0 \leq r \leq c_2 t. \end{cases}$$

By following the method outlined above it is straightforward to obtain expressions for quantities such as the stress and vertical (x_2) displacement. All of these have a qualitatively similar structure. There is a cylindrical, dilatational wavefront at $r = c_1 t$ and a cylindrical, rotational wavefront at $r = c_2 t$. There is no disturbance ahead of the dilatational wavefront, only dilatational displacement between the two wavefronts and a mixture of both types of displacement behind the rotational wavefront. As $t \to \infty$ with x_1 and x_2 fixed, only the disturbance behind the rotational front is left within the body and

$$u_1 = \frac{F_0 x_1 x_2}{4\pi\mu r^2}\left\{1 - \left(\frac{c_2}{c_1}\right)^2\right\} + O\left(\frac{1}{t^2}\right),$$

which is the equilibrium displacement for a line force.

5.5.2 *Wavefronts Caused by a Point Force on the Free Surface of a Semi-infinite Elastic Body*

We now consider the displacement caused by a point force suddenly applied to a semi-infinite elastic body with a free surface. If the force is located within the body, one feature we would expect, in light of the above analysis, is that of dilatational and rotational wavefronts propagating outwards from the origin of the force. When these reach the free surface we would expect both reflection and the excitation of surface Rayleigh waves. The reflected primary waves are similar to those discussed earlier, although they are more complicated as they are not planar. An intriguing phenomenon occurs with the surface waves. Recall that for Rayleigh waves we need both dilatational and rotational components in order to satisfy the stress-free conditions on the free surface. When waves are generated by a line source in the interior of the body, we would expect to have two wavefronts moving on the free surface – one dilatational and one rotational. Between these there is only a dilatational displacement and, at first sight, we would appear to be unable to satisfy the stress-free boundary conditions at the free surface. A careful study of this initial value problem reveals the presence of a rather different type of wave, which connects the dilatational and rotational wavefronts. This is usually referred to as a **head wave**, and its structure is such that it will allow a type of transient Rayleigh waveform to move over the free surface of an elastic body in response to the forcing. To get further insight into this surface wave effect we now consider what happens when a point force is suddenly applied directly on the free surface of an elastic body. This avoids the reflection problems mentioned above for an internal force.

If a point force $\mathbf{F}_0 H(t)$ is suddenly applied to the free surface of a half space $z \geq 0$ at the origin, the wave motion is axisymmetric and we

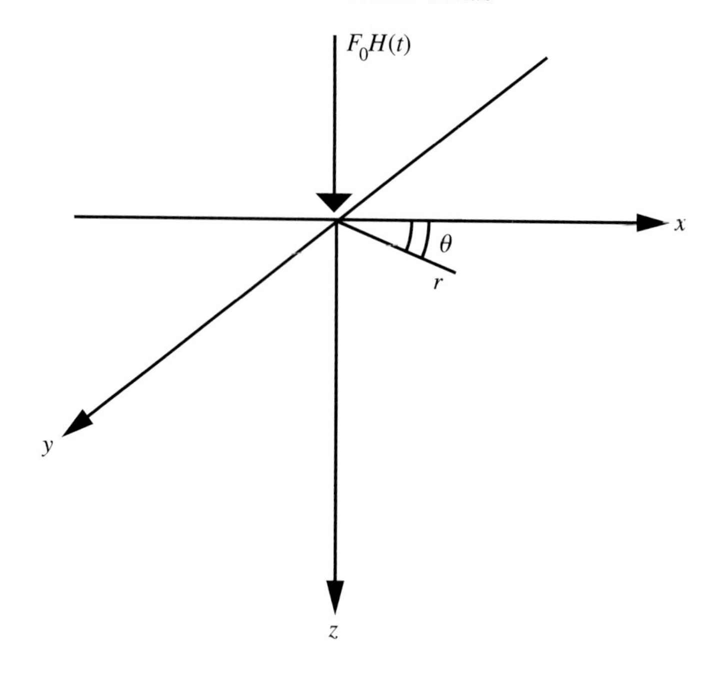

Fig. 5.12. A point force suddenly applied to the free surface of an elastic half space.

use the cylindrical coordinate system shown in figure 5.12 to describe the boundary value problem. The wave equations satisfied by the scalar and vector potentials, (5.69), are to be solved subject to the boundary conditions

$$\sigma_{rz} = 0, \quad \sigma_{zz} = -F_0 H(t)\frac{\delta(r)}{\pi r} \text{ on } z = 0,$$

and the initial conditions

$$\phi(r, z, 0) = \psi(r, z, 0) = \frac{\partial \psi}{\partial t}(r, z, 0) = \frac{\partial \phi}{\partial t}(r, z, 0) = 0.$$

To see where the normal stress boundary condition comes from, let's first write it in Cartesian form, as $\sigma_{zz} = -F_0 H(t)\delta(x)\delta(y)$. Now the two-dimensional delta function has the property that

$$\int_{-\infty}^{\infty}\int_{-\infty}^{\infty} \delta(x)\delta(y)dxdy = 1.$$

Transforming to polar coordinates, this becomes

$$\int_{\theta=0}^{2\pi}\int_{r=0}^{\infty} f(r)\delta(r)rdrd\theta = 1,$$

where we have written $\delta(x)\delta(y) = f(r)\delta(r)$ since the delta function acts isotropically. Evaluating the θ integral, we have

$$\int_0^\infty f(r)\delta(r)r\,dr = \frac{1}{2\pi},$$

which is satisfied if we choose $f(r) = 1/\pi r$, giving

$$\delta(x)\delta(y) = \frac{\delta(r)}{\pi r}.$$

If we take Laplace transforms with respect to time, the wave equations become

$$\frac{\partial^2\bar{\phi}}{\partial r^2} + \frac{1}{r}\frac{\partial\bar{\phi}}{\partial r} + \frac{\partial^2\bar{\phi}}{\partial z^2} = \frac{s^2}{c_1^2}\bar{\phi}, \quad \frac{\partial^2\bar{\psi}}{\partial r^2} + \frac{1}{r}\frac{\partial\bar{\psi}}{\partial r} + \frac{\partial^2\bar{\psi}}{\partial z^2} = \frac{s^2}{c_1^2}\bar{\psi}.$$

In order to proceed we need to use **Hankel transforms**. These arise naturally in the solution of axisymmetric problems in a half space, and to some extent can be anticipated by the forms of solution for waves in elastic rods, (5.70) and (5.71). The Hankel transform of order n is defined to be

$$f_n^*(k) = \int_0^\infty J_n(kr)rf(r)dr,$$

with inversion formula

$$f(r) = \int_0^\infty J_n(kr)kf_n^*(k)dk.$$

Taking transforms of order zero for the scalar potential and order one for the vector potential gives

$$\frac{\partial^2\bar{\phi}_0^*}{\partial z^2} - \left(k^2 + \frac{s^2}{c_1^2}\right)\bar{\phi}_0^* = 0, \quad \frac{\partial^2\bar{\psi}_1^*}{\partial z^2} - \left(k^2 + \frac{s^2}{c_2^2}\right)\bar{\psi}_1^* = 0.$$

Recall that a bar denotes a Laplace transformed variable. The bounded solutions of these equations are $\bar{\phi}_0^* = A(s,k)e^{-\alpha z}$, $\bar{\psi}_1^* = B(s,k)e^{-\beta z}$, where

$$\alpha = \left(k^2 + \frac{s^2}{c_1^2}\right)^{\frac{1}{2}}, \quad \beta = \left(k^2 + \frac{s^2}{c_2^2}\right)^{\frac{1}{2}},$$

and $A(s,k)$ and $B(s,k)$ are to be determined from the boundary conditions. The expressions

$$u = \frac{\partial\phi}{\partial r} - \frac{\partial\psi}{\partial z}, \quad w = \frac{\partial\phi}{\partial z} + \frac{1}{r}\frac{\partial}{\partial r}(r\psi)$$

transform to

$$\bar{u}^* = -k\bar{\phi}_0^* - \frac{\partial\bar{\psi}_1^*}{\partial z}, \quad \bar{w}^* = \frac{\partial\bar{\phi}_0^*}{\partial z} + k\bar{\psi}_1^*.$$

Performing these transformations is fairly straightforward, and we give details now only for u. Laplace transforming first gives

$$\bar{u} = \frac{\partial \bar{\phi}}{\partial r} - \frac{\partial \bar{\psi}}{\partial z}.$$

If we now multiply both sides by $rJ_1(kr)$ and integrate we have

$$\bar{u}^* = \int_0^\infty rJ_1(kr)\bar{u}dr = \int_0^\infty rJ_1(kr)\frac{\partial \bar{\phi}}{\partial r}dr - \frac{\partial}{\partial z}\int_0^\infty rJ_1(kr)\bar{\psi}dr.$$

The second of these integrals is $\bar{\psi}_1^*$, the Hankel transform of order one of $\bar{\psi}$. We deal with the first integral by integration by parts to give

$$\int_0^\infty rJ_1(kr)\frac{\partial \bar{\phi}}{\partial r}dr = \left[rJ_1(kr)\bar{\phi}\right]_0^\infty - \int_0^\infty \bar{\phi}\frac{\partial}{\partial r}\left(rJ_1(kr)\right)dr.$$

The contributions from zero and infinity vanish for a potential ϕ that is zero outside some finite region. Recalling that $\partial(rJ_1(r))/\partial r = rJ_0(r)$, we can recognise the remaining integral as a Hankel transform of zeroth order, and hence

$$\bar{u}^* = -k\bar{\phi}_0^* - \frac{\partial \bar{\psi}_1^*}{\partial z}.$$

The transforms of the normal and shear stresses are

$$\bar{\sigma}_{zz}^* = \mu\left\{\left(2k^2 + \frac{s^2}{c_2^2}\right)\bar{\phi}_0^* + 2k\frac{\partial \bar{\psi}_1^*}{\partial z}\right\},$$

$$\bar{\sigma}_{zr}^* = -\mu\left\{zk\frac{\partial \bar{\phi}_0^*}{\partial z} + \left(2k^2 + \frac{s^2}{c_2^2}\right)\bar{\psi}_1^*\right\}.$$

If we transform the boundary conditions on $z = 0$ we find that

$$\bar{\sigma}_{zz}^* = -\frac{F_0}{\pi s}, \quad \bar{\sigma}_{zr}^* = 0.$$

Substitution for $\bar{\phi}_0^*$ and $\bar{\psi}_1^*$ into the transformed expressions for stress gives rise to two simultaneous equations for $A(s,k)$ and $B(s,k)$ with solution

$$A(s,k) = -\frac{F_0\left(2k^2 + \frac{s^2}{c_2^2}\right)}{\pi\mu s D(s,k)}, \quad B(s,k) = -\frac{2\alpha k F_0}{\pi\mu s D(s,k)},$$

where

$$D(s,k) = \left(2k^2 + \frac{s^2}{c_2^2}\right)^2 - 4k^2\alpha\beta.$$

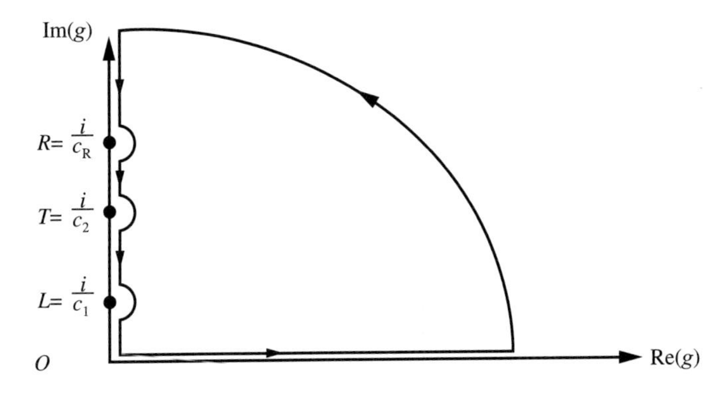

Fig. 5.13. The contour for inversion of the double transform (5.75).

We can now collect all the dependent variables together and find that the double transformed displacements take the form

$$\bar{u}^* = \frac{F_0}{\pi\mu}\left\{\left(2k^2 + \frac{s^2}{c_2^2}\right)e^{-\alpha z} - 2\alpha\beta e^{-\beta z}\right\}\frac{k}{sD(s,k)},$$

$$\bar{w}^* = \frac{F_0}{\pi\mu}\left\{\left(2k^2 + \frac{s^2}{c_2^2}\right)e^{-\alpha z} - 2k^2 e^{-\beta z}\right\}\frac{\alpha}{sD(s,k)}.$$

We now focus our attention on the vertical displacement of the free surface itself. By the Hankel inversion theorem,

$$\bar{w}(r,0,s) = \frac{F_0}{\pi\mu c_2^2}\int_0^\infty \frac{s\alpha}{D(s,k)} J_0(kr) k\,dk.$$

It is convenient to introduce a new variable $\eta = k/s$, which transforms this integral to

$$\bar{w}(r,0,s) = \frac{F_0}{\pi\mu c_2^2}\int_0^\infty \frac{\eta\left(\eta^2 + \frac{1}{c_1^2}\right)^{\frac{1}{2}} J_0(s\eta r)}{\left\{\left(2\eta^2 + \frac{1}{c_2^2}\right)^2 - 4\eta^2\left(\eta^2 + \frac{1}{c_1^2}\right)^{\frac{1}{2}}\left(\eta^2 + \frac{1}{c_2^2}\right)^{\frac{1}{2}}\right\}} d\eta.$$

The denominator of this expression,

$$\bar{D}(\eta) = \left(2\eta^2 + \frac{1}{c_2^2}\right)^2 - 4\eta^2\left(\eta^2 + \frac{1}{c_1^2}\right)^{\frac{1}{2}}\left(\eta^2 + \frac{1}{c_2^2}\right)^{\frac{1}{2}},$$

is very similar to the equation for the Rayleigh wave speed. Indeed if we replace η by i/η some simple manipulation leads to (5.26) for the zeros of $D(i/\eta)$. We deduce that the zeros of $D(\eta)$ are simply $\pm i/c_R$.

In order to proceed we note that the Bessel function of order zero can be represented as (Watson, 1922)

$$J_0(x) = \frac{2}{\pi}\mathrm{Im}\left\{\int_1^\infty \frac{e^{ixp}}{(p^2-1)^{\frac{1}{2}}}dp\right\},$$

so that we can write

$$\bar{w}(r,0,s) = \frac{2F_0}{\pi^2\mu c_2^2}\mathrm{Im}\left\{\int_0^\infty \frac{\eta\left(\eta^2+\frac{1}{c_1^2}\right)^{\frac{1}{2}}}{\bar{D}(\eta)}\int_1^\infty \frac{e^{is\eta rp}}{(p^2-1)^{\frac{1}{2}}}dp\,d\eta\right\}. \quad (5.75)$$

Our objective now is to express the second of the two integrals above as a standard Laplace transform so as to make the Laplace inversion of the double transform straightforward (this is known as the **Cagniard–de Hoop technique**). To do this we must use Cauchy's theorem on the contour indicated in figure 5.13. There are branch points on the imaginary axis at $\eta = i/c_1$ and i/c_2, and a pole at $\eta = i/c_R$. The contributions to the integral from the indentations around the branch points vanish as the singularity is weak. The contribution from the semi-circle around the Rayleigh pole at i/c_R does not vanish, but will not appear as it makes a real-valued contribution to the integral. We therefore have

$$\bar{w}(r,0,s) = -\frac{2F_0}{\pi^2\mu c_2^2}\mathrm{Im}\left\{\int_0^\infty \frac{\left(-q^2+\frac{1}{c_1^2}\right)^{\frac{1}{2}}q}{\bar{D}(iq)}\int_1^\infty \frac{e^{-sqrp}}{(p^2-1)^{\frac{1}{2}}}dp\,dq\right\}.$$

Now

$$L\left[\frac{H(t-qr)}{(t^2-q^2r^2)^{\frac{1}{2}}}\right] = \int_1^\infty \frac{e^{-sqrp}}{(p^2-1)^{\frac{1}{2}}}dp$$

is a standard result (Erdélyi, 1954), which means that we can now invert the Laplace part of the double transform to give, using $L^{-1}\left[\mathrm{Im}\int_0^\infty\right] = \mathrm{Im}\int_0^\infty L^{-1}[\ ]$,

$$w(r,0,t) = \frac{-2F_0}{\pi^2\mu c_2^2}\mathrm{Im}\left\{\int_0^{t/r} \frac{\left(-q^2+\frac{1}{c_1^2}\right)^{\frac{1}{2}}q}{\bar{D}(iq)(t^2-q^2r^2)^{\frac{1}{2}}}dq\right\}.$$

There is still some work to be done in interpreting the solution in this form. The quantities q and $(t^2-q^2r^2)^{\frac{1}{2}}$ are both real, so $w(r,0,t)$ is zero unless $E(q,r,t) \equiv \left(-q^2+1/c_1^2\right)^{\frac{1}{2}}/\bar{D}(iq)$ has a non-zero imaginary part.

On the line segment OL this quantity is real, so $w(r,0,t) \equiv 0$ for $r > c_1 t$. On the line segments LT and TR, it takes the form

$$E(q,r,t) = \begin{cases} \dfrac{i\left(q^2 - \frac{1}{c_1^2}\right)^{\frac{1}{2}}}{\left(\frac{1}{c_2^2} - 2q^2\right)^2 + 4iq^2\left(q^2 - \frac{1}{c_1^2}\right)^{\frac{1}{2}}\left(q^2 - \frac{1}{c_2^2}\right)^{\frac{1}{2}}} & \text{on } LT, \\[2ex] \dfrac{i\left(q^2 - \frac{1}{c_1^2}\right)^{\frac{1}{2}}}{\left(\frac{1}{c_2^2} - 2q^2\right)^2 - 4q^2\left(q^2 - \frac{1}{c_1^2}\right)^{\frac{1}{2}}\left(q^2 - \frac{1}{c_2^2}\right)^{\frac{1}{2}}} & \text{on } TR. \end{cases}$$

This means that the vertical displacement can be written in the form

$$w(r,0,t) = \begin{cases} 0 & \text{for } r > c_1 t, \\[1ex] -\dfrac{2F_0}{\pi^2 \mu c_2^2 r} F\left(\dfrac{t}{r}\right) & \text{for } c_2 t < r < c_1 t, \\[1ex] -\dfrac{2F_0}{\pi \mu c_2^2 r}\left\{F\left(\dfrac{1}{c_2}\right) + G\left(\dfrac{t}{r}\right)\right\} & \text{for } r < c_2 t, \end{cases}$$

where

$$F\left(\frac{t}{r}\right) = \int_{1/c_1}^{t/r} \frac{q\left(q^2 - \frac{1}{c_1^2}\right)^{\frac{1}{2}}\left(\frac{1}{c_2^2} - 2q^2\right)^2\left(\frac{t^2}{r^2} - q^2\right)^{-\frac{1}{2}} dq}{\left(\frac{1}{c_2^2} - 2q^2\right)^4 + 16q^4\left(q^2 - \frac{1}{c_1^2}\right)\left(q^2 - \frac{1}{c_2^2}\right)},$$

$$G\left(\frac{t}{r}\right) = \int_{1/c_2}^{t/r} \frac{q\left(q^2 - \frac{1}{c_1^2}\right)^{\frac{1}{2}}\left(\frac{t^2}{r^2} - q^2\right)^{-\frac{1}{2}} dq}{\left(\frac{1}{c_2^2} - 2q^2\right)^2 - 4q^2\left(q^2 - \frac{1}{c_1^2}\right)^{\frac{1}{2}}\left(q^2 - \frac{1}{c_2^2}\right)^{\frac{1}{2}}}.$$

In order to interpret this solution, consider a point on the free surface a distance r from the point of application of the force. There is no disturbance until $t = r/c_1$ when a dilatational wavefront arrives. This is followed when $t = r/c_2$ by a rotational wavefront. The form of displacement between these times is proportional to $F\left(t/r\right)$, which contains both dilatational and rotational components so as to satisfy the stress-free conditions. At a later time, $t = r/c_R$, the Rayleigh wave arrives, carrying with it a singularity in the displacement. This singularity occurs because of the delta function forcing of the free surface, and it arrives at the Rayleigh speed because the top limit of integration coincides with a zero in the denominator of the integrand. The Rayleigh wave is the largest part of the disturbance to the free surface. After it has passed, the

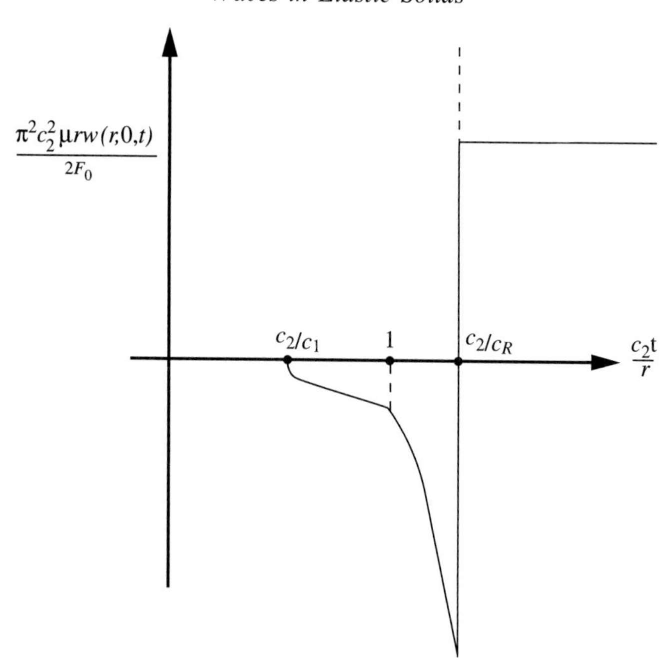

Fig. 5.14. The vertical displacement due to a point force F_0 at a free surface at a fixed position. The dilatational wavefront arrives at a dimensionless time c_2/c_1, the rotational wavefront at time 1, and the Rayleigh wavefront at time c_2/c_R.

disturbance falls to its equilibrium value, as sketched in figure 5.14. It is worth emphasising that these displacements are attenuated with increasing values of r and, at some distance from the applied force, it is only the Rayleigh component that will have any appreciable size, something that is confirmed by seismological data.

Exercises

5.1 A sphere of elastic material of radius a vibrates at a frequency ω. By taking a displacement field of the form $\mathbf{u} = (u(r,t), 0, 0)$, where r is the distance from the centre of the sphere, show that:

(a) The motion is longitudinal ($\nabla \times \mathbf{u} = 0$).

(b) Navier's equation reduces to the single scalar equation,

$$\frac{\partial^2 u}{\partial t^2} = c_1^2 \frac{\partial}{\partial r}\left(\frac{1}{r^2}\frac{\partial}{\partial r}(r^2 u)\right).$$

(c) The assumption that the solution takes the form $u = U(r)e^{-i\omega t} = (d\phi/dr)e^{-i\omega t}$ leads to

$$\frac{1}{r^2}\frac{d}{dr}\left(r^2\frac{d\phi}{dr}\right) = -k^2\phi, \quad \text{where } k = \omega/c_1.$$

(d) The bounded solution of this equation is

$$\phi = \frac{A\sin kr}{r}.$$

(e) If the normal stress,

$$\sigma_{rr} = (\lambda + 2\mu)\frac{\partial U}{\partial r} + \frac{2\lambda U}{r},$$

vanishes on $r = a$ then

$$\frac{\tan ka}{ka} = \frac{4c_2^2}{4c_2^2 - k^2a^2c_1^2}.$$

The roots of this equation are the characteristic frequencies of vibration of the sphere.

5.2 A two-dimensional plane rotational wave with amplitude I and angle of incidence θ_{I} is reflected from the stress-free boundary of an elastic half space. If the dilatational reflected component has amplitude D and angle of reflection θ_{Rd} and the rotational reflected component has amplitude R and angle of reflection θ_{Rr}, show that

$$\frac{\sin\theta_{\mathrm{I}}}{c_2} = \frac{\sin\theta_{\mathrm{Rr}}}{c_2} = \frac{\sin\theta_{\mathrm{Rd}}}{c_1}.$$

Show that the reflected dilatational wave can vanish if $\theta_{\mathrm{I}} = 0$ or $\pi/4$, provided certain conditions, which you should determine, are satisfied by the amplitudes of the other two waves.

If another elastic material in $x_2 \geq 0$ were to be bonded to the material in $x_2 \leq 0$, what effect would this have? What happens for $x_2 \geq 0$?

5.3 A strip of elastic material with depth h and material properties μ and ρ is bonded to a half space of a different elastic material, with properties μ' and ρ', along an interface $x_2 = 0$. Waves propagate along the interface in the x_1-direction, causing a deformation in the x_3-direction only. Show that:

(a) The assumption of a displacement field $\mathbf{u} = (0, 0, u_3(x_1, x_2))$ in each material reduces the equations of motion to

$$\rho \frac{\partial^2 u_3}{\partial t^2} = \mu \nabla^2 u_3, \qquad \rho' \frac{\partial^2 u_3'}{\partial t^2} = \mu' \nabla^2 u_3'.$$

(b) Solutions of the form

$$u_3 = (A \cos sx_2 + B \sin sx_2) \cos\{k(x_1 - ct)\} \text{ for } 0 < x_2 < h,$$
$$u_3 = Ce^{s' x_2} \cos\{k(x_1 - ct)\} \text{ for } -\infty < x_2 < 0$$

are possible, provided that

$$\frac{s^2}{k^2} = \frac{c^2}{c_2^2} - 1, \qquad \frac{(s')^2}{k^2} = 1 - \frac{c^2}{(c_2')^2},$$

where $c_2^2 = \mu/\rho$ and $(c_2')^2 = \mu'/\rho'$.

(c) By enforcing continuity of both displacement and the one non-zero component of the stress, σ_{32}, show that

$$A = B \cot(sh) = C, \qquad \tan^2(sh) = \left(\frac{\mu'}{\mu}\right)^2 \frac{\left(1 - \frac{c^2}{c_2'^2}\right)}{\left(\frac{c^2}{c_2^2} - 1\right)}.$$

(d) The wave velocity, c, lies between the two transverse wave velocities c_2 and c_2'. These are known as **Love waves**.

5.4 Two elastic half spaces are bonded together along $x_2 = 0$, and waves are made to propagate along the interface between the two materials. The material properties in the lower half space $x_2 < 0$ are denoted by a superscript l, whilst those in the upper half space $x_2 > 0$ have no superscript. By looking for two-dimensional solutions of Navier's equations in the form

$$u_1 = \{A_1 e^{-b_1 x_2} + A_2 e^{-b_2 x_2}\} \exp\{ik(x_1 - ct)\},$$
$$u_2 = \left\{\frac{-b_1}{ik} A_1 e^{-b_1 x_2} + \frac{ik}{b_2} A_2 e^{-b_2 x_2}\right\} \exp\{ik(x_1 - ct)\},$$
$$b_1 = k\left(1 - \frac{c^2}{c_1^2}\right)^{\frac{1}{2}},$$
$$b_2 = k\left(1 - \frac{c^2}{c_2^2}\right)^{\frac{1}{2}},$$

$$u_1^1 = \{A_3 e^{b_3 x_2} + A_4 e^{b_4 x_2}\} \exp\{ik(x_1 - ct)\},$$

$$u_2^1 = \left\{\frac{b_3}{ik} A_3 e^{b_3 x_2} - \frac{ik}{b_4} A_4 e^{b_4 x_2}\right\} \exp\{ik(x_1 - ct)\},$$

$$b_3 = k\left(1 - \frac{c^2}{(c_1)^2}\right)^{\frac{1}{2}},$$

$$b_4 = k\left(1 - \frac{c^2}{(c_2)^2}\right)^{\frac{1}{2}},$$

that satisfy continuity of stress and displacement, show that

$$\begin{vmatrix} 1 & 1 & -1 & -1 \\ \frac{b_1}{k} & \frac{k}{b_2} & \frac{b_3}{k} & \frac{k}{b_4} \\ 2\frac{b_1}{k} & \left(2 - \frac{c^2}{c_2^2}\right)\frac{k}{b_2} & 2\frac{\mu^1}{\mu}\frac{b_3}{k} & \frac{\mu^1}{\mu}\left(2 - \frac{c^2}{(c_2')^2}\right)\frac{k}{b_4} \\ 2 - \frac{c^2}{c_2^2} & 2 & -\frac{\mu^1}{\mu}\left(2 - \frac{c^2}{(c_2')^2}\right) & -2\frac{\mu^1}{\mu} \end{vmatrix} = 0.$$

Is it possible to deduce that this equation has at least one real root? These are known as **Stoneley waves**.

5.5 Torsional waves propagate down the axis of a hollow cylinder, with inner radius a and outer radius b. What is the relationship between their frequency ω and wavelength λ?

If, instead of being hollow, the inner part of the cylinder were to be filled by an elastic material with properties μ' and ρ', would this allow the propagation of torsional waves?

5.6 Show that the speed of Rayleigh waves in an elastic medium with $\lambda = \mu$ is given by $c_R = 2c_2/(3 + 3^{\frac{1}{2}})^{\frac{1}{2}}$.

5.7 Find high frequency approximations to longitudinal waves in a circular elastic rod using the WKBJ method. Appropriate equations are given in subsection 5.4.4.

5.8 Try to predict the structural form of the vertical displacement $w(r, z, t)$ in the case of a suddenly applied point force F_0 on the free surface of an elastic half space with properties λ, μ and ρ using a dimensional argument. Use this form to *attempt* to find a similarity solution of the problem.

5.9 A time harmonic normal line stress, $\sigma_{yy} = -F_0\delta(x)e^{i\omega t}$, is applied to the free surface $y = 0$ of a two-dimensional elastic half space. Find a solution of the wave equations (5.36) with a harmonic time dependence in the form of a Fourier integral. Calculate the displacements on the free surface as $|x| \to \infty$, and show that they consist of a Rayleigh wave and attenuated dilatational and rotational waves.

5.10 A circular cylindrical hole of radius a in an unbounded elastic body is subject to a shear stress around its circumference of the form

$$\begin{aligned} \sigma_{r\theta} &= \sigma_0 \delta(z)\delta(r-a)H(t), \\ \sigma_{rr} &= \sigma_{rz} = 0, \end{aligned}$$

where $H(t)$ is the Heaviside function.

(a) Show that the motion caused by this stress involves a circumferential displacement $v = v(r, z, t)$ only.

(b) State the equation governing $v(r, z, t)$ and solve the initial boundary value problem by an integral transform method.

6

Electromagnetic Waves

In order to understand the propagation of electromagnetic waves, we need to study the equations that govern electromagnetic phenomena – **Maxwell's equations**. Rather than dive straight in by writing the equations down, we begin by giving sufficient background material for a reader new to this area. This will only scratch the surface of the subject, and the interested reader should look elsewhere for an in depth introduction, for example, the books by Jackson (1975) and Clemmow (1973). In particular, we will not consider any relativistic or quantum effects.

6.1 Electric and Magnetic Forces and Fields

Most matter in the universe is thought to be composed of electrons, with a negative charge, protons, with a positive charge, and neutrons, which have no charge. An attractive force is exerted by an electron on a proton and vice versa. A repulsive force is exerted by an electron on another electron and by a proton on another proton. In a static situation, opposite charges attract, like charges repel due to the **electric force** between them. The SI unit of charge is the **coulomb (C)**. An electron has negative charge $e = 1.6 \times 10^{-19}$ C and mass $m_e = 9.1 \times 10^{-31}$ kg. A proton has positive charge e and mass $1839\, m_e$.

It is found experimentally that, in a vacuum, the electric force between two stationary point charges of magnitudes q_1 and q_2 at $\mathbf{x}_1$ and $\mathbf{x}_2$ acts along the line between the charges, and is proportional to $q_1 q_2$ and $|\mathbf{x}_1 - \mathbf{x}_2|^{-2}$, an inverse square law. We can therefore write this force $\mathbf{F}$ as

$$\mathbf{F} = \frac{q_1 q_2 \mathbf{r}}{4\pi\epsilon_0 r^3}, \tag{6.1}$$

where

$$\mathbf{r} = \mathbf{x}_1 - \mathbf{x}_2, \quad r = |\mathbf{r}|,$$

and the constant of proportionality is written in terms of ϵ_0, the **electrical permittivity** of the vacuum, with $\epsilon_0 = 8.854 \times 10^{-12}\,\mathrm{F\,m^{-1}}$ or (F = farad) $\mathrm{C^2\,s^2\,kg^{-1}\,m^{-3}}$. Note that under most circumstances we can treat air at room temperature as if it were a vacuum. The electric force is rather like the gravitational force, except that its magnitude depends upon the product of the particles' charges rather than masses, and it can therefore be repulsive if the particles have charges of opposite sign. Gravity is always attractive.

How does the electric force act? Rather than thinking in terms of one particle acting on another, it is usual to think of the electric force in terms of the **electric field** generated by a charged particle. We define the electric field, $\mathbf{E}(\mathbf{r})$, due to a stationary point charge q_1 at $\mathbf{x}_1$ so that the force, $\mathbf{F}$, on a particle with charge q_2 at $\mathbf{x}_2$ is

$$\mathbf{F} = q_2\mathbf{E}, \tag{6.2}$$

and hence

$$\mathbf{E} = \frac{q_1\mathbf{r}}{4\pi\epsilon_0 r^3}. \tag{6.3}$$

Note that the electric field is unbounded at the point at which it is generated. If there are a number of point charges, $q_1, q_2, \ldots, q_n$ at $\mathbf{x}_1, \mathbf{x}_2, \ldots, \mathbf{x}_n$, which generate electric fields $\mathbf{E}_1, \mathbf{E}_2, \ldots, \mathbf{E}_n$, the total force exerted by these charges on any other point charge will be the sum of the forces, $\mathbf{F}_i$, exerted by each of the n charges individually, so that

$$\mathbf{F} = \sum_{i=1}^{n} \mathbf{F}_i.$$

We therefore have the **principle of superposition for electric fields**,

$$\mathbf{E} = \sum_{i=1}^{n} \mathbf{E}_i. \tag{6.4}$$

It is also found experimentally that a **magnetic force** is exerted by **permanent magnets** and carriers of **electric current** on other permanent magnets and carriers of electric current. An electric current is a flux of electric charge. The most common example is the flux of electrons in a current carrying wire. The magnetic force has very different characteristics from the electric force, both physically and mathematically. There are no magnetic charges, or **magnetic monopoles** as they are usually called, in

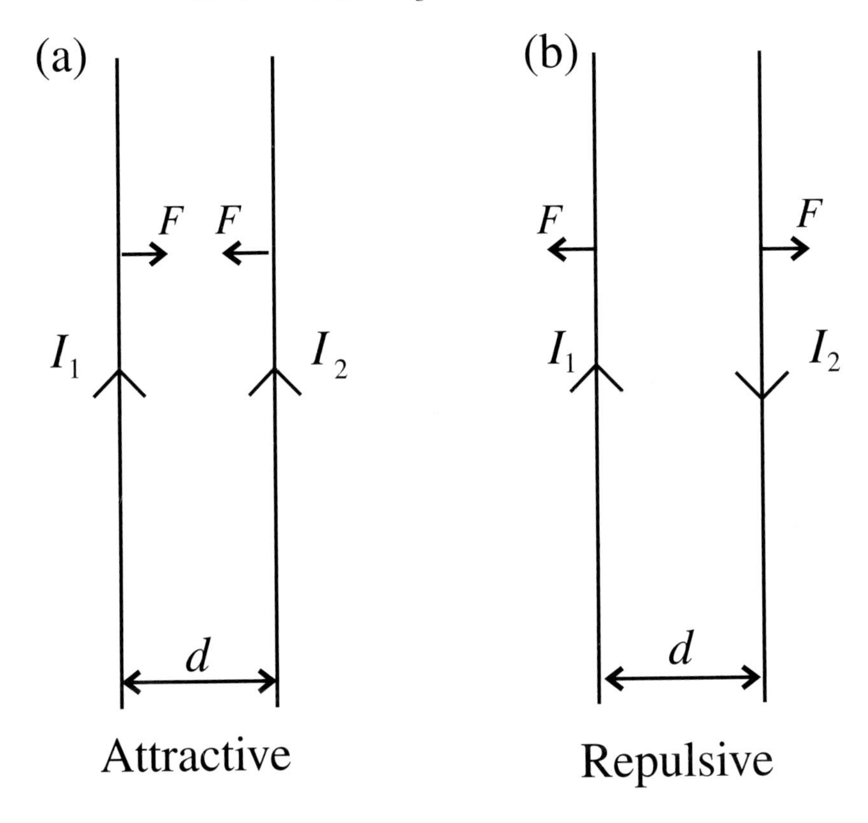

Fig. 6.1. The magnetic forces between two thin, parallel current carrying wires.

Maxwell's theory. Whether magnetic monopoles actually exist remains an open question – none have yet been observed experimentally. The basic unit of magnetism is the **magnetic dipole**, which we can think of as an infinitesimal loop of electric current. Permanent magnets are materials that can support magnetic dipoles in their atomic structure. Rather than go into the detailed mathematics of magnetic dipoles, it is easier to consider the force between two thin, current carrying wires.

Consider two long, thin, parallel, straight wires a perpendicular distance d apart, each carrying a steady current, I_1 and I_2 respectively, as shown in figure 6.1. It is found experimentally that the two wires exert a magnetic force on each other, perpendicular to the wires, proportional to $I_1 I_2$ and inversely proportional to d^2. If the currents are in the same direction the force is attractive. If the currents are in opposite directions the force is repulsive. This is similar to the electric force, but electric currents take the place of static charges, and the direction of the force is perpendicular to the direction of the currents. The generalisation of this to the magnetic force, $\mathbf{F}$, between two arbitrary closed loops of current, C_1 and C_2, is

$$\mathbf{F} = \frac{\mu_0 I_1 I_2}{4\pi} \oint_{C_2} \oint_{C_1} \frac{d\mathbf{x}_2 \times \{d\mathbf{x}_1 \times (\mathbf{x}_1 - \mathbf{x}_2)\}}{|\mathbf{x}_1 - \mathbf{x}_2|^3}, \tag{6.5}$$

where $\mathbf{x}_1$ and $\mathbf{x}_2$ are points on the curves C_1 and C_2 respectively, and μ_0 is the **magnetic permeability** of the vacuum, with $\mu_0 = 4\pi \times 10^{-7}\,\mathrm{H\,m^{-1}}$ (H = henry) or $\mathrm{kg\,m\,C^{-2}}$. Equation (6.5) is the **Biot–Savart law**. The next step is to write the equivalent of (6.2) and (6.3) for the **magnetic field**. Since there are no magnetic monopoles, we have to do this in terms of small elements of current, and the equations are strictly only valid when used in integrals along lines of current. We define the magnetic field, $d\mathbf{B}(\mathbf{r})$, due to a small line element of current, $I_1 d\mathbf{x}_1$, at $\mathbf{x}_1$ so that the force on a line element $I_2 d\mathbf{x}_2$ at $\mathbf{x}_2$ is

$$d\mathbf{F} = I_2 d\mathbf{x}_2 \times d\mathbf{B}, \tag{6.6}$$

and hence

$$d\mathbf{B} = \frac{\mu_0}{4\pi} I_1 d\mathbf{x}_1 \times \frac{\mathbf{r}}{r^3}. \tag{6.7}$$

In addition, we have the **principle of superposition for magnetic fields**,

$$\mathbf{B} = \sum_{i=1}^{n} \mathbf{B}_i. \tag{6.8}$$

We can combine (6.2) and (6.6) to give the **Lorentz force law**,

$$\mathbf{F} = q\,(\mathbf{E} + \mathbf{v} \times \mathbf{B}), \tag{6.9}$$

for a charge q moving with velocity $\mathbf{v}$ in an electric field $\mathbf{E}$ and magnetic field $\mathbf{B}$. The magnetic force exerted on a moving charge acts perpendicular to the line of motion ($\mathbf{v}\times\mathbf{B}$), and tends to deflect the charge from a straight line path.

So far we have been thinking in terms of isolated point charges and lines of current. In order to make any progress with macroscopic problems, we need to define a **charge density**, $\rho(\mathbf{x})$, as

$$\rho(\mathbf{x}) = \frac{1}{V} \sum_{V} q, \tag{6.10}$$

where V is a volume centred on $\mathbf{x}$ which is much bigger than an atom but much smaller than any macroscopic volume. This is just the continuum approximation applied to charges. Similarly we can define a macroscopic **electric current density**, $\mathbf{j}$, as

$$\mathbf{j}(\mathbf{x}) = \frac{1}{V}\sum_{V} q\mathbf{v}, \tag{6.11}$$

where $\mathbf{v}$ is the velocity of the charge q. Conservation of charge can then be written as

$$\frac{\partial \rho}{\partial t} + \nabla \cdot \mathbf{j} = 0. \tag{6.12}$$

Some of the greatest scientists of the nineteenth century showed that there is a close connection between electrical and magnetic phenomena, as we shall see below. Einstein's theory of special relativity (which is beyond the scope of this book) shows that these two fields are manifestations of a single **electromagnetic field**. Perhaps even more surprisingly, electric and magnetic fields, which were originally introduced as abstract concepts to make it easier to think about the forces between charges and currents, are able to propagate away from the charges that generate them. As we shall see, electromagnetic waves are propagating electric and magnetic fields, and account for the full spectrum from radio waves through microwaves and visible light up to X- and gamma rays.

6.2 Electrostatics: Gauss's Law

By the principle of superposition, (6.4), the electric field due to a distribution of charge with density $\rho(\mathbf{x})$ is

$$\mathbf{E}(\mathbf{x}) = \frac{1}{4\pi\epsilon_0}\int \frac{\rho(\mathbf{x}_0)(\mathbf{x}-\mathbf{x}_0)}{|\mathbf{x}-\mathbf{x}_0|^3}\, d^3\mathbf{x}_0, \tag{6.13}$$

where the integral is over all of space. This is rather an inconvenient way to express how electric charges generate an electric field, and we would prefer a formula that expresses this at any given point in space.

Consider a volume V with surface ∂V and outward unit normal $\mathbf{n}$, as illustrated in figure 6.2. Now consider the electric field due to a point charge q at $\mathbf{x}_0$,

$$\mathbf{E}(\mathbf{x}) = \frac{q}{4\pi\epsilon_0}\frac{\mathbf{x}-\mathbf{x}_0}{|\mathbf{x}-\mathbf{x}_0|^3}. \tag{6.14}$$

If $\mathbf{x}_0$ does not lie within V then $\mathbf{E}$ is finite throughout V and we can use the divergence theorem

$$\int_{\partial V} \mathbf{E}\cdot\mathbf{n}\, dS = \int_V \nabla\cdot\mathbf{E}\, dV.$$

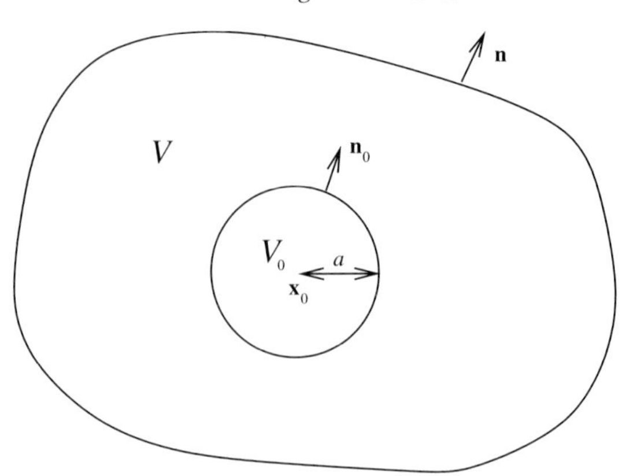

Fig. 6.2. The volumes V and V_0.

If we calculate $\nabla \cdot \mathbf{E}$ from (6.14), we find that it is zero, and hence that

$$\int_{\partial V} \mathbf{E} \cdot \mathbf{n}\, dS = 0, \quad \text{when } \mathbf{x}_0 \notin V.$$

If $\mathbf{x}_0$ does lie within V we must proceed differently, since $\mathbf{E}$ is unbounded at $\mathbf{x}_0$. Define the the volume V_0 to be a sphere of radius a centred on $\mathbf{x_0}$ with outward unit normal $\mathbf{n}_0$, and a small enough that V_0 lies entirely within V, as shown in figure 6.2. The region $V' = V - V_0$ is the volume V with the small sphere V_0 excised from it. Since $\mathbf{E}$ is bounded within V', we can proceed as before and show that

$$\int_{\partial V'} \mathbf{E} \cdot \mathbf{n}\, dS = 0.$$

Since $\partial V' = \partial V + \partial V_0$,

$$\int_{\partial V} \mathbf{E} \cdot \mathbf{n}\, dS = \int_{\partial V_0} \mathbf{E} \cdot \mathbf{n}_0\, dS = \int_0^{\pi} \sin\theta\, d\theta \int_0^{2\pi} d\phi\, a^2 \frac{q}{4\pi\epsilon_0 a^2} = \frac{q}{\epsilon_0},$$

where (r, θ, ϕ) are spherical polar coordinates. Note that this explains why the factor of 4π is included in the definition of the force on a point charge, (6.1). If we now use the principle of superposition, we obtain **Gauss's law**,

$$\int_{\partial V} \mathbf{E} \cdot \mathbf{n}\, dS = \frac{1}{\epsilon_0} \int \rho(\mathbf{x}) d^3\mathbf{x}. \tag{6.15}$$

This relates the flux of electric field out of a volume to the charge contained within it. Since this must hold for all volumes V, the divergence theorem shows that

$$\nabla \cdot \mathbf{E} = \frac{\rho}{\epsilon_0}. \tag{6.16}$$

This is the first of Maxwell's equations, and is also found to hold for moving charges.

6.3 Magnetostatics: Ampère's Law and the Displacement Current

For a distribution of current density $\mathbf{j}(\mathbf{x})$, the principle of superposition, (6.8), shows that the magnetic field generated is

$$\mathbf{B}(\mathbf{x}) = \frac{\mu_0}{4\pi} \int \mathbf{j}(\mathbf{x}_0) \times \frac{\mathbf{x} - \mathbf{x}_0}{|\mathbf{x} - \mathbf{x}_0|^3} d^3\mathbf{x}_0. \tag{6.17}$$

Since

$$\frac{\mathbf{x} - \mathbf{x}_0}{|\mathbf{x} - \mathbf{x}_0|^3} = -\nabla \left(\frac{1}{|\mathbf{x} - \mathbf{x}_0|} \right),$$

we can write (6.17) as

$$\mathbf{B}(\mathbf{x}) = \frac{\mu_0}{4\pi} \nabla \times \int \frac{\mathbf{j}(\mathbf{x}_0)}{|\mathbf{x} - \mathbf{x}_0|} d^3\mathbf{x}_0. \tag{6.18}$$

This immediately shows us that

$$\nabla \cdot \mathbf{B} = 0. \tag{6.19}$$

This is the second of Maxwell's equations, and we show later that it is also true for moving charges. Comparing (6.19) with (6.16), the equivalent equation for the electric field, we can see that the charge density, ρ, acts as a source of the electric field, but that the right hand side of (6.19) is zero. This expresses the fact that there are no sources of magnetic field (magnetic monopoles), only magnetic dipoles, or infinitesimal current loops, in Maxwell's theory.

It is also possible to manipulate (6.18) to obtain $\mathbf{j}$ in terms of $\mathbf{B}$ as

$$\nabla \times \mathbf{B} = \mu_0 \mathbf{j}. \tag{6.20}$$

This is **Ampère's law**, and its derivation is exercise 6.5. Is (6.20) correct for unsteady current distributions? Taking the divergence gives

$$\nabla \cdot (\nabla \times \mathbf{B}) = 0 = \mu_0 \nabla \cdot \mathbf{j},$$

but conservation of charge, (6.12), shows that

$$\nabla \cdot \mathbf{j} = -\frac{\partial \rho}{\partial t} \neq 0.$$

Equation (6.20) cannot, therefore, hold for unsteady currents. It was Maxwell's great insight that (6.20) will be correct if an extra term is added to it. If we write

$$\nabla \times \mathbf{B} = \mu_0 (\mathbf{j} + \mathbf{j}_{\mathrm{D}}),$$

then to be consistent with (6.12) we need

$$\nabla \cdot \mathbf{j}_{\mathrm{D}} = \frac{\partial \rho}{\partial t}.$$

Now (6.16) shows that by choosing

$$\mathbf{j}_{\mathrm{D}} = \epsilon_0 \frac{\partial \mathbf{E}}{\partial t},$$

we can satisfy (6.12). This is not the only possible choice of $\mathbf{j}_{\mathrm{D}}$, but experiments show that it is the right choice. What Maxwell postulated on the grounds of mathematical elegance and simplicity was found to be correct. The modified form of (6.20), correct for unsteady current distributions, is therefore

$$\nabla \times \mathbf{B} = \mu_0 \mathbf{j} + \frac{1}{c_0^2} \frac{\partial \mathbf{E}}{\partial t}, \tag{6.21}$$

where we define $c_0 = 1/\sqrt{\mu_0 \epsilon_0}$ for reasons that will become clear later. This is the third of Maxwell's equations. The quantity $\mathbf{j}_{\mathrm{D}}$, is called the **displacement current**, and is crucial for the existence of electromagnetic waves.

6.4 Electromagnetic Induction: Faraday's Law

If a closed loop of wire is moved through a magnetic field or a magnet is moved through a stationary, closed loop of wire, a current is generated in the loop. This is an example of **electromagnetic induction** and is the basis of the generation of electricity by dynamos. Consider a closed loop of wire, C, in an unsteady magnetic field $\mathbf{B}$. Since a current flows through the loop, there must be a non-zero force on the electrons, and hence

$$\oint_C \mathbf{F} \cdot d\mathbf{x} \neq 0.$$

Equation (6.9) shows that, for an individual electron,

$$\mathbf{F} \propto (\mathbf{E} + \mathbf{v} \times \mathbf{B}).$$

Now $(\mathbf{v} \times \mathbf{B}) \cdot d\mathbf{x} = 0$ since the velocity, $\mathbf{v}$, of the electrons is parallel to the direction, $d\mathbf{x}$, of the wire. This means that the **electromotive force**, $\mathscr{E}$, where

$$\mathscr{E} \equiv \oint_C \mathbf{E} \cdot d\mathbf{x},$$

is non-zero. Experiments show that the electromotive force is given by **Faraday's law**,

$$\mathscr{E} = -\frac{d}{dt}\,[\text{flux of } \mathbf{B} \text{ through } C] = -\frac{d}{dt}\int_S \mathbf{B} \cdot \mathbf{n}\, dS, \tag{6.22}$$

where S is any smooth surface that spans C, with outward unit normal $\mathbf{n}$. This shows, using Stokes' theorem, that the electromotive force is

$$\int_S (\nabla \times \mathbf{E}) \cdot \mathbf{n}\, dS = -\int_S \frac{\partial \mathbf{B}}{\partial t} \cdot \mathbf{n}\, dS,$$

and hence that

$$\nabla \times \mathbf{E} = -\frac{\partial \mathbf{B}}{\partial t}. \tag{6.23}$$

This is the fourth and final Maxwell equation. Note that taking the divergence of (6.23) gives

$$\frac{\partial}{\partial t}(\nabla \cdot \mathbf{B}) = -\nabla \cdot (\nabla \times \mathbf{E}) = 0. \tag{6.24}$$

Since (6.19) states that $\nabla \cdot \mathbf{B} = 0$ for steady currents, (6.24) shows that (6.19) must hold for unsteady currents as well.

To summarise, Maxwell's equations in a vacuum are

$$\nabla \cdot \mathbf{E} = \frac{\rho}{\epsilon_0}, \tag{6.25}$$

$$\nabla \cdot \mathbf{B} = 0, \tag{6.26}$$

$$\nabla \times \mathbf{E} = -\frac{\partial \mathbf{B}}{\partial t}, \tag{6.27}$$

$$\nabla \times \mathbf{B} = \mu_0 \mathbf{j} + \frac{1}{c_0^2}\frac{\partial \mathbf{E}}{\partial t}. \tag{6.28}$$

Note that the equation for conservation of charge, (6.12), is implied by Maxwell's equations. It can be derived by taking the divergence of (6.28) and eliminating $\mathbf{E}$ using (6.25).

6.5 Plane Electromagnetic Waves

We will now consider the **source-free Maxwell's equations**, which are (6.25) to (6.28) with $\rho = 0$ and $\mathbf{j} = 0$, so that

$$\nabla \cdot \mathbf{E} = 0, \tag{6.29}$$

$$\nabla \cdot \mathbf{B} = 0, \tag{6.30}$$

$$\nabla \times \mathbf{E} = -\frac{\partial \mathbf{B}}{\partial t}, \tag{6.31}$$

$$\nabla \times \mathbf{B} = \frac{1}{c_0^2}\frac{\partial \mathbf{E}}{\partial t}. \tag{6.32}$$

These govern the behaviour of the electric and magnetic fields in the absence of any charges and currents in a vacuum, and have the obvious solution $\mathbf{E} = \mathbf{B} = 0$. Are there any other solutions, which would suggest that the electric and magnetic fields can exist even when there are no local charges or currents to act as sources? If we combine (6.31) and (6.32) we obtain

$$\nabla \times (\nabla \times \mathbf{E}) = -\frac{1}{c_0^2}\frac{\partial^2 \mathbf{E}}{\partial t^2}, \quad \nabla \times (\nabla \times \mathbf{B}) = -\frac{1}{c_0^2}\frac{\partial^2 \mathbf{B}}{\partial t^2}.$$

The vector identity

$$\nabla \times (\nabla \times \mathbf{A}) = \nabla(\nabla \cdot \mathbf{A}) - \nabla^2 \mathbf{A},$$

along with (6.29) and (6.30) finally gives us

$$\nabla^2 \mathbf{E} = \frac{1}{c_0^2}\frac{\partial^2 \mathbf{E}}{\partial t^2}, \tag{6.33}$$

$$\nabla^2 \mathbf{B} = \frac{1}{c_0^2}\frac{\partial^2 \mathbf{B}}{\partial t^2}. \tag{6.34}$$

This shows that $\mathbf{E}$ and $\mathbf{B}$ satisfy the **vector wave equation**, with wave speed c_0. We have studied the scalar wave equation extensively in chapters 2 and 3, where we found that, for plane waves, the general solution is a sum of left- and right-propagating waves moving with speed c_0. Solutions of the vector wave equation must be treated somewhat differently, but nevertheless are basically waves that propagate at speed $c_0 \approx 3 \times 10^8\ \mathrm{m\,s^{-1}}$. Since this is so close to the experimentally measured speed of light, Maxwell was led to postulate, correctly, that light is an electromagnetic wave. The wavelength of visible light lies in the narrow range 4×10^{-7} to 8×10^{-7} m. The full spectrum of electromagnetic waves ranges through shorter wavelength gamma rays ($< 10^{-11}$ m) and X-rays (10^{-10} m) to longer wavelength infra-red radiation (10^{-5} m), microwaves (10^{-2} m) and radio waves (> 1 m). None of these can be detected directly by the human eye, but they are of huge practical importance.

We will focus our attention on plane electromagnetic waves. In a plane wave propagating in the z-direction, the electric and magnetic fields are given by

$$\mathbf{E}(\mathbf{r},t) = \mathbf{E}_c(z)e^{i\omega t}, \quad \mathbf{B}(\mathbf{r},t) = \mathbf{B}_c(z)e^{i\omega t}, \tag{6.35}$$

where ω is the constant angular frequency, $\mathbf{r} = (x, y, z)$ is the position vector, $\mathbf{E}_c(z)$ and $\mathbf{B}_c(z)$ are complex vector functions of z, and in all cases it is understood that we are dealing with the real part of $\mathbf{E}$ and $\mathbf{B}$. Substituting these functional forms into (6.33) and (6.34) shows that

$$\mathbf{E}_c(z) = \mathbf{E}_0 e^{\pm i\omega z/c_0}, \quad \mathbf{B}_c(z) = \mathbf{B}_0 e^{\pm i\omega z/c_0},$$

where $\mathbf{E}_0$ and $\mathbf{B}_0$ are constant, complex vectors. Since the fields are simply functions of $z \pm c_0 t$, they are plane waves propagating in the positive or negative z-direction with phase speed c_0. However, we must choose $\mathbf{E}_0$ and $\mathbf{B}_0$ so that (6.29) to (6.32) are satisfied. If

$$\mathbf{E}_0 = (E_{0x}, E_{0y}, E_{0z}) \text{ and } \mathbf{B}_0 = (B_{0x}, B_{0y}, B_{0z}),$$

then substituting for $\mathbf{E}$ in (6.29) shows that

$$E_{0z} = 0.$$

Similarly, (6.30) shows that

$$B_{0z} = 0.$$

The electric and magnetic fields therefore lie in the (x, y)-plane, perpendicular to the direction of propagation. Both (6.31) and (6.32) show that

$$B_{x0} = \pm\frac{1}{c_0}E_{y0}, \quad B_{y0} = \mp\frac{1}{c_0}E_{x0}.$$

If we define the wavenumber vector

$$\mathbf{k} = \left(0, 0, \frac{\omega}{c_0}\right),$$

and consider a wave travelling in the positive $\mathbf{k}$-direction, we can write

$$\mathbf{E}(\mathbf{r},t) = \mathbf{E}_0 e^{i(\omega t - \mathbf{k}\cdot\mathbf{r})}, \quad \mathbf{B}(\mathbf{r},t) = \frac{1}{\omega}(\mathbf{k} \times \mathbf{E}_0)\, e^{i(\omega t - \mathbf{k}\cdot\mathbf{r})}. \tag{6.36}$$

This solution is valid for any complex wavenumber vector $\mathbf{k}$ and complex

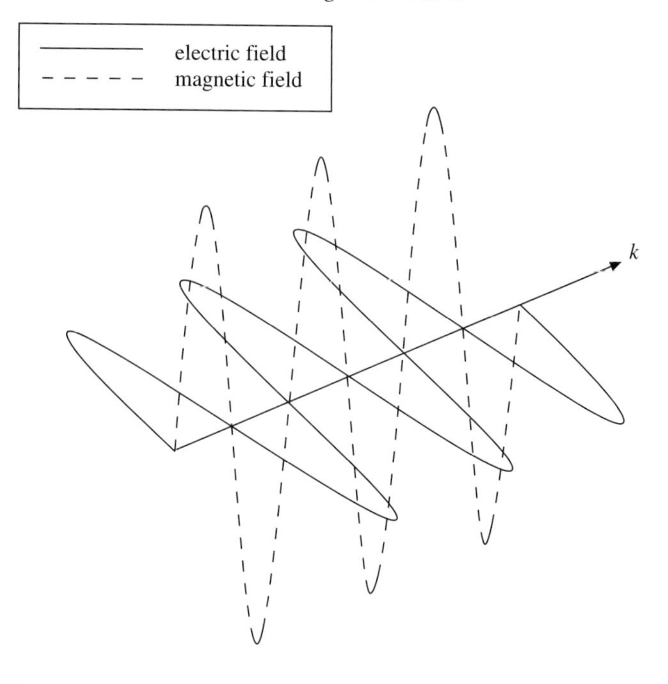

Fig. 6.3. The electric and magnetic fields in a plane polarised wave.

vector $\mathbf{E}_0$, provided that $|\mathbf{k}| = \omega/c_0$ and $\mathbf{E}_0 \cdot \mathbf{k} = 0$. If $\mathbf{k}$ has a finite imaginary part, the wave decays exponentially in the direction of $\mathrm{Im}(\mathbf{k})$. Equations (6.36) show that $\mathbf{B}_0 \propto \mathbf{k} \times \mathbf{E}_0$, and the three vectors $\mathbf{E}_0$, $\mathbf{B}_0$ and $\mathbf{k}$ form a pairwise perpendicular set, with $\mathbf{k}$ in the direction of propagation and $\mathbf{E}_0$ and $\mathbf{B}_0$ perpendicular to each other in the plane transverse to $\mathbf{k}$. We refer to the plane containing $\mathbf{k}$ and $\mathbf{E}_0$ as the **plane of polarisation**, as shown in figure 6.3.

With $\mathbf{k}$ real, if $\mathbf{E}_0$, and hence $\mathbf{B}_0$, is proportional to a real vector, the plane of polarisation is independent of $\omega t - \mathbf{k} \cdot \mathbf{r}$ and the wave is **plane polarised**. Otherwise, the plane of polarisation will rotate as $\omega t - \mathbf{k} \cdot \mathbf{r}$ varies. In particular,

$$\mathbf{E}(\mathbf{r}, t) = \mathrm{Re}(\mathbf{E}_0)\cos(\omega t - \mathbf{k} \cdot \mathbf{r}) - \mathrm{Im}(\mathbf{E}_0)\sin(\omega t - \mathbf{k} \cdot \mathbf{r}).$$

If $|\mathrm{Re}(\mathbf{E}_0)| = |\mathrm{Im}(\mathbf{E}_0)|$, the vector $\mathbf{E}$ traces out a circle as $\omega t - \mathbf{k} \cdot \mathbf{r}$ varies, and the wave is said to be **circularly polarised**. In any other case, $\mathbf{E}$ traces out an ellipse, and the wave is said to be **elliptically polarised**. A plane wave with arbitrary elliptical polarisation can be decomposed into a sum of two linearly polarised waves, simply by noting that

$$\mathbf{E}_0 = \mathrm{Re}(\mathbf{E}_0) + i\mathrm{Im}(\mathbf{E}_0).$$

Alternatively a plane wave can be written as the sum of two waves with circular polarisation, since

$$\mathbf{E}_0 = \frac{1}{2}(1+i)\{\mathrm{Re}(\mathbf{E}_0) + \mathrm{Im}(\mathbf{E}_0)\} + \frac{1}{2}(1-i)\{\mathrm{Re}(\mathbf{E}_0) - \mathrm{Im}(\mathbf{E}_0)\}.$$

The ability to decompose any plane wave into two plane polarised waves with perpendicular electric, and hence magnetic, fields will prove to be very useful in the following sections.

We can now consider how energy is stored in an electromagnetic field, and how it is transported by a plane wave. Consider a space with electric and magnetic fields $\mathbf{E}$ and $\mathbf{B}$ and charge and current densities ρ and $\mathbf{j}$. Since $\mathbf{j} = \rho\mathbf{v}$, the Lorentz force law, (6.9), shows that the electromagnetic force per unit volume is

$$\mathbf{F} = \rho\mathbf{E} + \mathbf{j} \times \mathbf{B}. \tag{6.37}$$

The rate of working against these forces is

$$\mathbf{v} \cdot \mathbf{F} = \rho\mathbf{v} \cdot \mathbf{E} + \mathbf{v} \cdot \mathbf{j} \times \mathbf{B} = \mathbf{E} \cdot \mathbf{j}.$$

Using (6.28), the total rate of working, W_V, in some finite volume V is

$$\begin{aligned} W_V &= \int_V \mathbf{E} \cdot \mathbf{j}\, dV = \int_V \mathbf{E} \cdot \left(\frac{1}{\mu_0} \nabla \times \mathbf{B} - \epsilon_0 \frac{\partial \mathbf{E}}{\partial t} \right) dV \\ &= -\frac{d}{dt} \left(\frac{1}{2} \epsilon_0 \int_V |\mathbf{E}|^2\, dV \right) + \frac{1}{\mu_0} \int_V \mathbf{E} \cdot \nabla \times \mathbf{B}\, dV. \end{aligned}$$

The vector identity

$$\mathbf{E} \cdot \nabla \times \mathbf{B} = \mathbf{B} \cdot \nabla \times \mathbf{E} - \nabla \cdot (\mathbf{E} \times \mathbf{B}),$$

along with (6.27) and the divergence theorem, shows that

$$W_V = -\frac{d}{dt} \left(\frac{1}{2} \epsilon_0 \int_V |\mathbf{E}|^2\, dV + \frac{1}{2\mu_0} \int_V |\mathbf{B}|^2\, dV \right) - \frac{1}{\mu_0} \int_{\partial V} (\mathbf{E} \times \mathbf{B}) \cdot \mathbf{n}\, dS. \tag{6.38}$$

Now suppose that the charges and currents are confined to a finite volume and that there are no electromagnetic waves radiating energy away to infinity. Then, for V sufficiently large, the final term on the right hand side of (6.38) is negligible and, since

$$W_V = -\frac{d}{dt}(\text{energy}),$$

the energy stored in the electromagnetic field must be

$$J = \frac{1}{2}\epsilon_0|\mathbf{E}|^2 + \frac{1}{2\mu_0}|\mathbf{B}|^2. \tag{6.39}$$

This is consistent with more detailed derivations of the energy stored in an electromagnetic field (see for example Jackson (1975)). The neglected term in (6.38) represents the flux of energy through the surface ∂V of V. In particular,

$$\mathbf{P}_0 = \frac{1}{\mu_0}\mathbf{E} \times \mathbf{B}, \tag{6.40}$$

is the aptly named **Poynting vector**, which gives the flux of energy per unit area in any electromagnetic field, in particular in an electromagnetic wave. (Poynting was Professor of Physics at the University of Birmingham from 1900 to 1914.) Note that, since the Poynting vector is a product of two complex vector fields, we must be careful to take the real part of each before forming the product, since $\mathrm{Re}(\mathbf{E} \times \mathbf{B}) \not\equiv \mathrm{Re}(\mathbf{E}) \times \mathrm{Re}(\mathbf{B})$. If we do this and calculate the average energy density in a plane electromagnetic wave over a single period, $T \leq t \leq T + 2\pi/\omega$, we find that there is equipartition of energy between the electric and magnetic fields, and that the average energy density is

$$\langle J \rangle = \frac{1}{2\mu_0}|\mathbf{B}_0|^2 = \frac{1}{2}\epsilon_0|\mathbf{E}_0|^2.$$

It is also straightforward to show that

$$\mathbf{P}_0 = \frac{1}{2\mu_0}\frac{c_0}{|\mathbf{k}|}\mathbf{k}|\mathbf{B}_0|^2,$$

which points in the direction of propagation, as we would expect. The group velocity, which is the velocity at which energy is transported, is therefore equal to

$$\mathbf{c}_g = \frac{\mathbf{P}_0}{\langle J \rangle} = c_0\frac{\mathbf{k}}{|\mathbf{k}|}.$$

This is just the phase velocity of the wave and there is therefore no dispersion.

6.6 Conductors and Insulators

So far we have only considered electromagnetic phenomena in a vacuum. In most situations at room temperature, gases can be treated as vacua. Must Maxwell's equations be modified for electric and magnetic fields in liquids and solids? Since liquids and solids are composed of densely packed atoms, each of which consists of a cloud of negatively charged electrons surrounding a positively charged nucleus, we should not be surprised that solids and liquids are not passive carriers of electromagnetic fields and do not behave in the same way as gases. The major distinction that we have to draw is between **conductors** and **insulators**.

In perfect insulators, the electrons are bound sufficiently tightly to the nuclei of the constituent atoms that they cannot flow and form a current. However, in any applied electric field there will be some deformation of the distribution of electrons around each nucleus. This leads to a modification of the permittivity and magnetic permeability of the material, so that (6.25) to (6.28) become

$$\nabla \cdot \mathbf{D} = \rho, \tag{6.41}$$

$$\nabla \cdot \mathbf{B} = 0, \tag{6.42}$$

$$\nabla \times \mathbf{E} = -\frac{\partial \mathbf{B}}{\partial t}, \tag{6.43}$$

$$\nabla \times \mathbf{H} = \mathbf{j} + \frac{\partial \mathbf{D}}{\partial t}, \tag{6.44}$$

where

$$\mu_0 \mathbf{H} = \mathbf{B} - \mathbf{M}, \quad \mathbf{D} = \epsilon_0 \mathbf{E} + \mathbf{P}, \tag{6.45}$$

where $\mathbf{P}$ is the electric dipole moment per unit volume, or **polarisation**, and $\mathbf{M}$ the magnetic dipole moment per unit volume, or **magnetisation**. The relationships between the fields $\mathbf{H}$ and $\mathbf{B}$, and $\mathbf{D}$ and $\mathbf{E}$, given by (6.45) are constitutive laws for the material. For non-magnetic materials, $\mathbf{M} = 0$, and we will discuss how to model $\mathbf{P}$ in subsection 12.2.1. For isotropic materials and electromagnetic waves that are sufficiently weak that the response of the material is linear (in practice for visible light this is true for anything but laser light), we can write $\mathbf{D} = \epsilon \mathbf{E}$, $\mu \mathbf{H} = \mathbf{B}$, where ϵ is the **dielectric constant** of the material and μ its magnetic permeability. For anisotropic materials, which we will not consider here, μ and ϵ are tensors, since the directionality of the atomic structure can distort the fields generated by charges and currents.

Metals are conductors that have an outer shell of electrons that are so loosely bound to their atoms that, in the presence of an electric field, they are able to move through the conductor as an electric current. This process can also occur in polar liquids, ionic solutions and ionised gases, and we need a constitutive law to describe it. To a good approximation, most isotropic conductors satisfy **Ohm's law**,

$$\mathbf{j} = \sigma \mathbf{E}, \tag{6.46}$$

where σ is the **conductivity**, a scalar for isotropic materials. Conductivity is measured in $\mathrm{S\,m^{-1}}$ (S = siemens) or $\mathrm{C^2\,s\,kg^{-1}\,m^{-4}}$. Good conductors, such as copper and silver, have conductivities of the order of $10^7\,\mathrm{S\,m^{-1}}$, whilst poor conductors, like paraffin, have conductivities of the order of $10^{-11}\,\mathrm{S\,m^{-1}}$.

By substituting (6.46) into (6.41) and using conservation of charge, (6.12), we find that

$$\frac{\partial \rho}{\partial t} = -\frac{\sigma}{\epsilon}\rho. \tag{6.47}$$

If there is a distribution of charge in a conductor, the charge will decay to zero as $t \to \infty$, over a time scale

$$\tau_\sigma = \epsilon/\sigma, \tag{6.48}$$

with

$$\rho(\mathbf{x}, t) = \rho(\mathbf{x}, 0)\exp\left(-\frac{t}{\tau_\sigma}\right).$$

The larger the conductivity, σ, the faster the charge decays. For a good conductor, τ_σ can be as small as 10^{-19} s, whilst poor conductors can hold their charge for hours, or even days.

We can explain the decay of charge in a conductor by noting that the charges generate an electric field. This drives electric currents, as described by Ohm's law, (6.46), which carry the charge away. For a finite volume of conductor, the charge carried by the current must end up at the surface. Conductors are therefore capable of supporting a surface charge with density ρ_s, and also a surface current with density $\mathbf{j}_s$. Since the electric current in the interior also decays to zero, (6.46) shows that $\mathbf{E} \to 0$ as $t \to \infty$. In a static situation, the electric field is therefore zero in a conductor. Moreover, we can define a **perfect conductor** to be a conductor for which $\sigma = \infty$, and hence $\mathbf{E}$ is always zero in its interior.

Real materials are neither perfect conductors nor perfect insulators. Their behaviour under the influence of electromagnetic fields depends strongly on the time scale, τ_0, associated with the imposed fields, but for sufficiently weak fields, not on their intensity (see subsection 12.2 for a discussion of the effect of strong electromagnetic fields in the form of laser light). If the fields change sufficiently rapidly, in particular $\tau_0 \ll \tau_\sigma$, where τ_σ is defined by (6.48), any material will behave as an insulator, since the electrons cannot respond instantaneously (see subsection 12.2.1). We will avoid these issues by considering only perfect conductors and perfect insulators. This is a very reasonable approximation for the electromagnetic wave phenomena that we will consider in the rest of this chapter.

To consider the propagation of plane waves in an isotropic medium, we can simply go through the same analysis as we did in section 6.5 for wave propagation in a vacuum. All of the results we have used so far are unchanged, except that ϵ_0, μ_0 and $c_0 = 1/\sqrt{\epsilon_0\mu_0}$ must be replaced with ϵ, μ and $c = 1/\sqrt{\epsilon\mu}$. There is therefore no dispersion in the propagation of electromagnetic waves in a uniform medium with constant ϵ and μ. Experiments show, however, that real media do not have constant values of ϵ and μ. In particular, ϵ is found to be strongly dependent on the angular frequency, ω, of the wave. Moreover, absorption due to resonances between the wave and vibrational modes of the atoms that make up the medium leads to exponential decay of the wave at certain frequencies. This can be modelled by taking ϵ to be complex, with the imaginary part significant close to these resonances. All of these effects will, as we have seen for linear water waves, lead to dispersion, which we consider briefly at the end of this chapter and in more detail in subsection 12.2.1. As an example of this effect, water absorbs electromagnetic waves at microwave frequencies, as exploited in the design of microwave ovens.

6.7 Reflection and Transmission at Interfaces

By studying the reflection and transmission of plane electromagnetic waves at interfaces between different materials we can explain many experimental observations concerning the reflection and refraction of visible light. First of all, we need to know what boundary conditions should be applied to the electric and magnetic fields at the interface between two different media. Strictly speaking, we should derive the integral form of Maxwell's equations, from which both the partial differential equations (6.25) to (6.28) and the appropriate boundary conditions can be derived. We will proceed less formally here.

6.7.1 Boundary Conditions at Interfaces

Consider a small cylinder, C, with height $2l$ and base area dS, that straddles the interface between two different media with its axis perpendicular

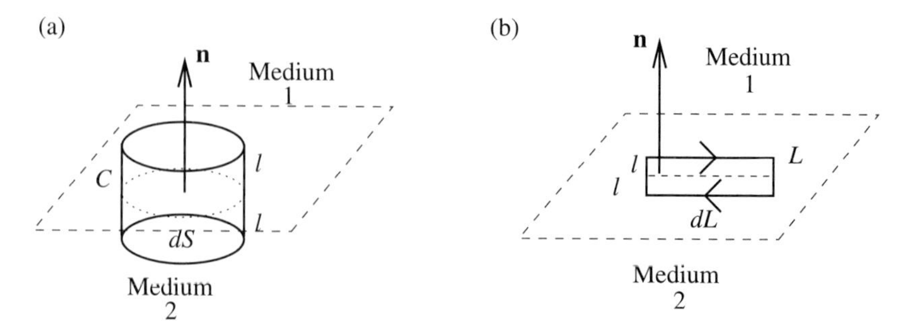

Fig. 6.4. The small cylinder, C, and loop, L, used in the derivation of the boundary conditions at interfaces.

to the interface, as shown in figure 6.4(a). Let $\mathbf{n}$ be the unit normal to the interface pointing into medium 1. Now integrate (6.41) over the cylinder and apply the divergence theorem. If we make l sufficiently small that $l^2 \ll dS$, in other words if we shrink the cylinder down towards the interface, we find that

$$\epsilon_1 \mathbf{E}_1 \cdot \mathbf{n} - \epsilon_2 \mathbf{E}_2 \cdot \mathbf{n} = \rho_s, \tag{6.49}$$

where ϵ_1 and ϵ_2 are the permittivities of the two media and ρ_s is the surface charge density. This shows that $\epsilon \mathbf{E} \cdot \mathbf{n}$ may be discontinuous at an interface if there is a layer of charge confined there. By a similar argument using the simpler equation (6.26), we obtain

$$\mathbf{B}_1 \cdot \mathbf{n} = \mathbf{B}_2 \cdot \mathbf{n}. \tag{6.50}$$

This states that the normal component of the magnetic field is continuous across interfaces.

Now consider a small loop, L, that threads through an interface, with two sides of length dL parallel to the interface and two sides of length $2l$ perpendicular to the interface, as shown in figure 6.4(b). We can integrate (6.44) over a surface, S, bounded by L and use Stokes' theorem to give

$$\oint_L \mathbf{H} \cdot d\mathbf{l} = \int_S \left\{ \mathbf{j} + \frac{\partial \mathbf{D}}{\partial t} \right\} \cdot d\mathbf{S}.$$

For $l \ll dL$, in other words as the loop shrinks down onto the surface,

$$\frac{1}{\mu_1} \mathbf{B}_1 \times \mathbf{n} - \frac{1}{\mu_2} \mathbf{B}_2 \times \mathbf{n} = \mathbf{j}_s, \tag{6.51}$$

where μ_1 and μ_2 are the magnetic permeabilities of the two media and $\mathbf{j}_s$ is the surface current density. Note that we have assumed that $\partial \mathbf{D}/\partial t$ is

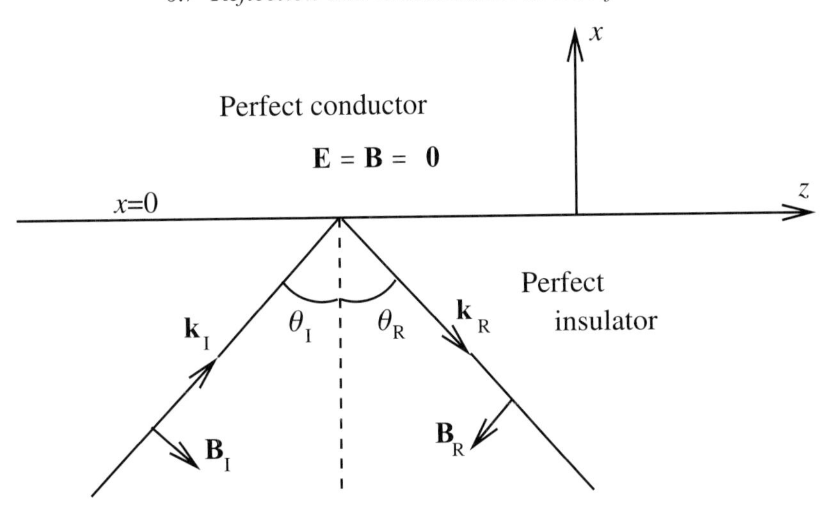

Fig. 6.5. Reflection of a plane polarised wave with electric field parallel to the interface at a perfect conductor.

bounded. This boundary condition states that the tangential components of $\mathbf{B}/\mu$ can be discontinuous at an interface if there is a layer of surface current confined there. By a similar argument using (6.43), we find that

$$\mathbf{E}_1 \times \mathbf{n} = \mathbf{E}_2 \times \mathbf{n}. \tag{6.52}$$

In other words, the tangential components of the electric field are continuous at an interface.

To summarise, the normal component of the magnetic fields, $\mathbf{B} \cdot \mathbf{n}$, and the tangential components of the electric field, $\mathbf{E} \times \mathbf{n}$, are always continuous across an interface between two different media. In addition, if both materials behave as perfect insulators, no surface currents can exist and the tangential components of $\mathbf{B}/\mu$ are continuous across the interface. We also assume below that no surface charge has been deposited at the interface between two insulators, and hence that the normal component of $\epsilon\mathbf{E}$ is continuous.

6.7.2 Reflection by a Perfect Conductor

Consider the situation shown in figure 6.5, where a plane polarised electromagnetic wave is incident from a perfect insulator, which occupies the half space $x < 0$, on a perfect conductor, which occupies the half space $x > 0$. We begin by considering the case where the incident electric field $\mathbf{E}_\mathrm{I}$ is in the y-direction, parallel to the interface. The wave has angular frequency ω_I, wavenumber vector $\mathbf{k}_\mathrm{I}$, angle of incidence θ_I and magnetic field $\mathbf{B}_\mathrm{I}$. More precisely,

$$\left.\begin{aligned}\mathbf{E}_\mathrm{I} &= E_\mathrm{I}(0,1,0)\exp\left[i\left\{\omega_\mathrm{I}t - \frac{\omega_\mathrm{I}}{c}(x\cos\theta_\mathrm{I} + z\sin\theta_\mathrm{I})\right\}\right],\\ \mathbf{B}_\mathrm{I} &= \frac{E_\mathrm{I}}{c}(-\sin\theta_\mathrm{I},0,\cos\theta_\mathrm{I})\exp\left[i\left\{\omega_\mathrm{I}t - \frac{\omega_\mathrm{I}}{c}(x\cos\theta_\mathrm{I} + z\sin\theta_\mathrm{I})\right\}\right],\\ \mathbf{k}_\mathrm{I} &= \frac{\omega_\mathrm{I}}{c}(\cos\theta_\mathrm{I},0,\sin\theta_\mathrm{I}).\end{aligned}\right\} \quad (6.53)$$

Since $\mathbf{E} \equiv 0$ in a perfect conductor, we expect that there will just be a reflected wave, with electric and magnetic fields $\mathbf{E}_\mathrm{R}$ and $\mathbf{B}_\mathrm{R}$, such that

$$\left.\begin{aligned}\mathbf{E}_\mathrm{R} &= (E_{\mathrm{R}x},E_{\mathrm{R}y},E_{\mathrm{R}z})\exp\left[i\left\{\omega_\mathrm{R}t - \frac{\omega_\mathrm{R}}{c}(-x\cos\theta_\mathrm{R} + z\sin\theta_\mathrm{R})\right\}\right],\\ \mathbf{B}_\mathrm{R} &= (B_{\mathrm{R}x},B_{\mathrm{R}y},B_{\mathrm{R}z})\exp\left[i\left\{\omega_\mathrm{R}t - \frac{\omega_\mathrm{R}}{c}(-x\cos\theta_\mathrm{R} + z\sin\theta_\mathrm{R})\right\}\right],\\ \mathbf{k}_\mathrm{R} &= \frac{\omega_\mathrm{R}}{c}(-\cos\theta_\mathrm{R},0,\sin\theta_\mathrm{R}),\end{aligned}\right\} \quad (6.54)$$

with

$$B_{\mathrm{R}x} = -\frac{1}{c}E_{\mathrm{R}y}\sin\theta_\mathrm{R},\quad B_{\mathrm{R}y} = \frac{1}{c}(E_{\mathrm{R}x}\sin\theta_\mathrm{R} + E_{\mathrm{R}z}\cos\theta_\mathrm{R}),$$
$$B_{\mathrm{R}z} = -\frac{1}{c}E_{\mathrm{R}y}\cos\theta_\mathrm{R}.$$

Also, since $\mathbf{k}_\mathrm{R}\cdot\mathbf{E}_\mathrm{R} = 0$,

$$E_{\mathrm{R}x}\cos\theta_\mathrm{R} = E_{\mathrm{R}z}\sin\theta_\mathrm{R}. \quad (6.55)$$

The total electric field at the interface, $x = 0$, is therefore

$$\begin{aligned}\mathbf{E}_0 = {}& E_\mathrm{I}(0,1,0)\exp\left\{i\left(\omega_\mathrm{I}t - \frac{\omega_\mathrm{I}}{c}z\sin\theta_\mathrm{I}\right)\right\}\\ &+(E_{\mathrm{R}x},E_{\mathrm{R}y},E_{\mathrm{R}z})\exp\left\{i\left(\omega_\mathrm{R}t - \frac{\omega_\mathrm{R}}{c}z\sin\theta_\mathrm{R}\right)\right\}.\end{aligned}$$

Since we need this to hold for all values of z and t, we must have $\omega_\mathrm{I} = \omega_\mathrm{R} = \omega$ and $\theta_\mathrm{I} = \theta_\mathrm{R} = \theta$. In other words, the frequency of the wave is unchanged, and the angle of incidence is equal to the angle of reflection. We can now apply the boundary condition that the transverse components of the electric field must be continuous, and hence zero, at $x = 0$. This, along with (6.55), leads to

$$E_{\mathrm{R}x} = E_{\mathrm{R}z} = 0,\quad E_{\mathrm{R}y} = -E_\mathrm{I}.$$

This means that the reflected wave also has its electric field parallel to the interface, but that the sense of the electric and magnetic fields is reversed by the reflection.

The remaining boundary condition is on the normal component of $\mathbf{B}$. Equation (6.43) shows that, in the perfect conductor, where $\mathbf{E} \equiv 0$, $\partial \mathbf{B}/\partial t = 0$, and hence that the normal component of $\mathbf{B}$ is independent of time at the interface. However, the normal component of $\mathbf{B}$ due to the incident and reflected waves at the interface is time harmonic. The only way that these can be equal is if $\mathbf{B} \cdot \mathbf{n} = 0$ at the interface. This is indeed satisfied by the solution above, since

$$B_{\mathrm{R}x} = \frac{1}{c} E_{\mathrm{I}} \sin\theta.$$

Note also that (6.32) in the perfect conductor shows that $\mathbf{B}$ can be written as the gradient of a potential which, by (6.30), satisfies Laplace's equation. If we assume that there is no magnetic field at infinity in the conductor, the condition that $\mathbf{B} \cdot \mathbf{n} = 0$ at the interface shows that the magnetic field is zero everywhere within the perfect conductor.

At the surface, the total magnetic field is

$$\mathbf{B} = \frac{2E_{\mathrm{I}}}{c}(0, 0, \cos\theta) \exp\left[i\left\{\omega t - \frac{\omega}{c} z \sin\theta\right\}\right].$$

The boundary condition (6.51) then shows that there is a surface current density,

$$\mathbf{j}_{\mathrm{s}} = \frac{2\mu_0 E_{\mathrm{I}}}{c}(0, \cos\theta, 0) \exp\left[i\left\{\omega t - \frac{\omega}{c} z \sin\theta\right\}\right], \tag{6.56}$$

which flows in the y-direction along the boundary. It is this sheet of current that allows the transverse component of the magnetic field to be non-zero at the surface of the conductor. An alternative point of view, which will be of use to us in subsection 11.2.2 where we consider the diffraction of electromagnetic waves, is to think of the field for $x < 0$ as being generated by the sheet of current at $x = 0$. In other words, if we specify $\mathbf{j}_{\mathrm{s}}$ as a boundary condition through (6.56), the fields $\mathbf{E}$ and $\mathbf{B}$ are the appropriate solution of Maxwell's equations.

We can now consider the analogous case where the magnetic field of the plane polarised incident wave is in the y-direction, parallel to the interface. The same arguments show that the reflected wave has the same frequency as the incident wave, and that the angle of incidence is equal to the angle of reflection. Then, by applying the boundary conditions at $x = 0$, we find that the reflected wave also has the magnetic field parallel to the interface, but that, in this case, the sense of the wave is preserved by the reflection.

We showed in section 6.5 that any plane electromagnetic wave can be written as the sum of two plane polarised waves. In particular, such a

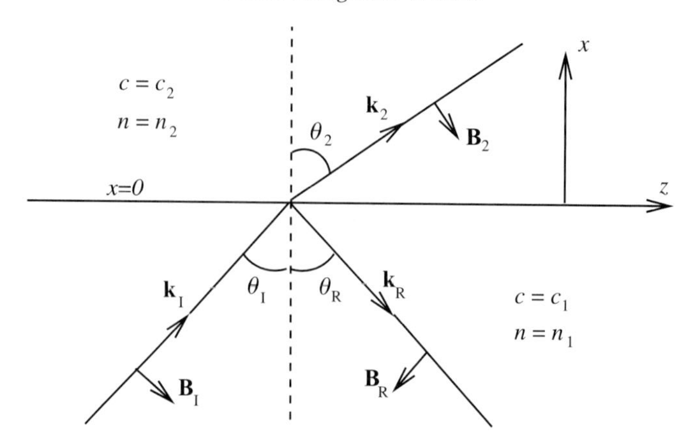

Fig. 6.6. Reflection and refraction of a wave with its electric field parallel to an interface between two perfect insulators.

wave incident on a perfect conductor can be written as the sum of two waves, one with the electric field parallel to the interface, one with the magnetic field parallel to the interface. This means that after reflection, the sense of the component of the electric field parallel to the interface is reversed, whilst the parallel component of the magnetic field is unchanged. These phenomena can be verified by experiments with visible light. The observation that the angle of incidence is equal to the angle of reflection corresponds to our experience of reflections from smooth, metal surfaces.

6.7.3 Reflection and Refraction by Insulators

When a plane polarised electromagnetic wave with electric field in the y-direction is incident on a plane interface between two insulators, the analysis is rather more complicated than that of the previous subsection, since a wave may be transmitted into the second medium. The situation is shown in figure 6.6. The incident and reflected waves are given again by (6.53) and (6.54), whilst the transmitted wave is

$$\left.\begin{aligned}\mathbf{E}_2 &= (E_{2x}, E_{2y}, E_{2z})\exp\left[i\left\{\omega_2 t - \frac{\omega_2}{c}(x\cos\theta_2 + z\sin\theta_2)\right\}\right],\\ \mathbf{B}_2 &= (B_{2x}, B_{2y}, B_{2z})\exp\left[i\left\{\omega_2 t - \frac{\omega_2}{c}(x\cos\theta_2 + z\sin\theta_2)\right\}\right],\\ \mathbf{k}_2 &= \frac{\omega_2}{c}(\cos\theta_2, 0, \sin\theta_2).\end{aligned}\right\} \quad (6.57)$$

The z- and t-dependence of the components of the fields at $x = 0$ must have the same functional form if the boundary conditions are to be satisfied. This means that

$$\omega_{\mathrm{I}} t - \frac{\omega_{\mathrm{I}}}{c_1} z \sin\theta_{\mathrm{I}} = \omega_{\mathrm{R}} t - \frac{\omega_{\mathrm{R}}}{c_1} z \sin\theta_{\mathrm{R}} = \omega_2 t - \frac{\omega_2}{c_2} z \sin\theta_2,$$

for all z and t, where c_1 and c_2 are the wave speeds in the two media. We can immediately see that we need $\omega_{\mathrm{I}} = \omega_{\mathrm{R}} = \omega_2 = \omega$ and $\theta_{\mathrm{I}} = \theta_{\mathrm{R}} = \theta_1$. All three waves have the same frequency and the angle of incidence equals the angle of reflection. We also require

$$\frac{\sin\theta_1}{c_1} = \frac{\sin\theta_2}{c_2}. \tag{6.58}$$

This is known as **Snell's law**, and gives the angle at which the transmitted wave propagates through the second medium. We found a similar relation in subsection 5.3.2, (5.30), for the reflection of an elastic dilatational wave from a free solid surface, but for the reflected rotational wave rather than a transmitted wave. Unless $c_1 = c_2$, $\theta_2 \neq \theta_1$, and the wave is bent or **refracted** by the interface. This phenomenon can be observed with visible light when looking into a pool of water, since refraction makes the water appear shallower than it actually is. Snell's law is usually expressed in terms of the **refractive index**, n, of each material, where

$$n = \frac{c_0}{c} = \sqrt{\frac{\mu\epsilon}{\mu_0\epsilon_0}},$$

so that

$$n_1 \sin\theta_1 = n_2 \sin\theta_2. \tag{6.59}$$

The speed of light in any medium is always less than that in free space, so the refractive index is always greater than one.

If we now assume that there is no charge at the interface, then $\epsilon\mathbf{E}\cdot\mathbf{n}$ must be continuous there, so that

$$\epsilon_1 E_{\mathrm{R}x} = \epsilon_2 E_{2x}. \tag{6.60}$$

Two perfect insulators cannot support a surface current at their interface, so the transverse components of $\mathbf{B}/\mu$ must be continuous there. This leads to

$$\frac{1}{c_1\mu_1}\left(E_{\mathrm{R}x}\sin\theta_1 + E_{\mathrm{R}z}\cos\theta_1\right) = \frac{1}{c_2\mu_2}\left(E_{2x}\sin\theta_2 - E_{2z}\cos\theta_2\right), \tag{6.61}$$

and

$$\frac{\cos\theta_1}{c_1\mu_1}(E_{\mathrm{I}} - E_{\mathrm{R}y}) = \frac{\cos\theta_2}{c_2\mu_2}E_{2y}. \tag{6.62}$$

The condition that the electric field should be perpendicular to the direction of propagation gives

$$E_{Rx}\cos\theta_1 - E_{Rz}\sin\theta_1 = 0, \tag{6.63}$$

$$E_{2x}\cos\theta_2 + E_{2z}\sin\theta_2 = 0. \tag{6.64}$$

The continuity of the transverse components of the electric field at the interface gives

$$E_{I} + E_{Ry} = E_{2y}, \tag{6.65}$$

$$E_{Rz} = E_{2z}, \tag{6.66}$$

whilst continuity of the normal component of **B** is also given by (6.65). Equations (6.60), (6.61), (6.63), (6.64) and (6.66) are five homogeneous equations linear in the x- and z-components of the reflected and transmitted electric fields. Since the determinant of these equations is never zero, the only solution is $E_{Rx} = E_{Rz} = E_{2x} = E_{2z} = 0$, and hence the reflected and transmitted waves also have their electric fields parallel to the interface. For simplicity, we now assume that we are dealing with non-magnetic materials, so that $\mu_1 = \mu_2 = \mu_0$. We can easily solve (6.62) and (6.65) to give

$$\frac{E_{2y}}{E_I} = \frac{2n_1\cos\theta_1}{n_1\cos\theta_1 + n_2\cos\theta_2}, \quad \frac{E_{Ry}}{E_I} = \frac{n_1\cos\theta_1 - n_2\cos\theta_2}{n_1\cos\theta_1 + n_2\cos\theta_2}, \tag{6.67}$$

and by (6.59),

$$\frac{E_{2y}}{E_I} = \frac{2\sin\theta_2\cos\theta_1}{\sin(\theta_2+\theta_1)}, \quad \frac{E_{Ry}}{E_I} = \frac{\sin(\theta_2-\theta_1)}{\sin(\theta_2+\theta_1)}, \tag{6.68}$$

which gives the intensity of the reflected and transmitted/refracted waves.

There is one aspect of this phenomenon that we have yet to address. Consider Snell's law, (6.59), which gives the angle of refraction by

$$\sin\theta_2 = \frac{n_1}{n_2}\sin\theta_1. \tag{6.69}$$

If $n_1 > n_2$, when $\theta_1 = \theta_c$, where the **critical angle** is

$$\theta_c = \sin^{-1}\left(\frac{n_2}{n_1}\right), \tag{6.70}$$

the transmitted wave has $\theta_2 = \pi/2$ and propagates along the interface.

For $\theta_1 > \theta_c$ there is no real solution of (6.69). Since $\cos\theta_2$ is imaginary in this case, with

$$\cos\theta_2 = i\sqrt{\frac{n_1^2}{n_2^2}\sin^2\theta_1 - 1},$$

(6.67) shows that

$$\frac{E_{Ry}}{E_I} = \frac{n_1\cos\theta_1 - in_2\sqrt{\frac{n_1^2}{n_2^2}\sin^2\theta_1 - 1}}{n_1\cos\theta_1 + in_2\sqrt{\frac{n_1^2}{n_2^2}\sin^2\theta_1 - 1}}. \tag{6.71}$$

This means that $|E_{Ry}| = |E_I|$, so that all of the energy in the incident wave is reflected. This is called **total internal reflection**. Equation (6.67) also shows that

$$\frac{E_{2y}}{E_I} = \frac{2n_1\cos\theta_1}{n_1\cos\theta_1 + in_2\sqrt{(n_1^2)\,(n_2^2)\sin^2\theta_1 - 1}}, \tag{6.72}$$

which suggests that there is also a transmitted wave with a finite amplitude. However, since

$$\mathbf{k}_2 = \frac{\omega}{c_2}\left(i\sqrt{\frac{n_1^2}{n_2^2}\sin^2\theta_1 - 1}, 0, \frac{n_1}{n_2}\sin\theta_1\right),$$

the transmitted wave decays exponentially in the x-direction, into the second insulator, and hence does not carry any energy away with it. Note that (6.71) shows that the reflected wave undergoes a change of phase on reflection. This is often exploited to manufacture devices that can manipulate the polarisation of visible light.

We can carry out the same analysis for a wave with its magnetic field parallel to the interface, and find that

$$\frac{B_{2y}}{B_I} = \frac{2\sin 2\theta_1}{\sin 2\theta_1 + \sin 2\theta_2}, \quad \frac{B_{Ry}}{B_I} = \frac{\sin 2\theta_1 - \sin 2\theta_2}{\sin 2\theta_1 + \sin 2\theta_2}. \tag{6.73}$$

The one interesting difference between expressions (6.73) and (6.68) is that when $\sin 2\theta_1 = \sin 2\theta_2$, $B_{Ry} = 0$ and there is no reflected wave. Since

$$\sin 2\theta_1 - \sin 2\theta_2 = 2\cos(\theta_1 + \theta_2)\sin(\theta_1 - \theta_2),$$

the only physically relevant solution occurs when $\theta_1 + \theta_2 = \frac{\pi}{2}$, when Snell's law shows that

$$n_1\sin\theta_1 = n_2\sin\left(\frac{\pi}{2} - \theta_1\right) = n_2\cos\theta_1.$$

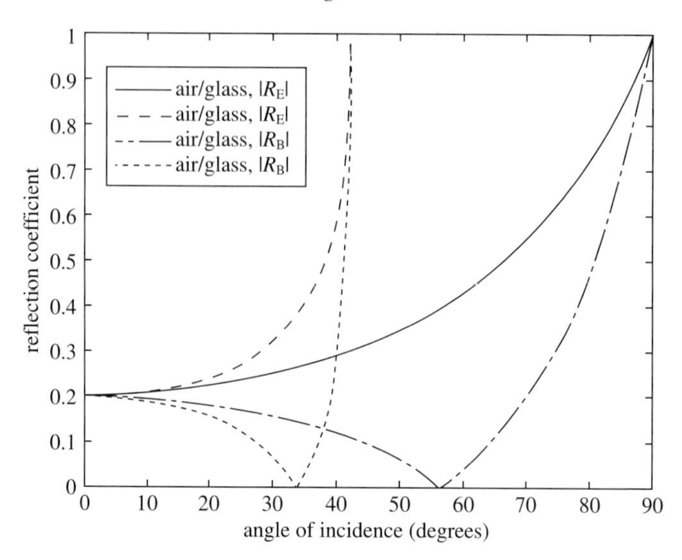

Fig. 6.7. The reflection coefficients for a plane wave incident from air onto glass (air/glass), and from glass onto air (glass/air).

There is therefore no reflected wave when $\theta_1 = \theta_\mathrm{B}$, where **Brewster's angle** is

$$\theta_\mathrm{B} = \tan^{-1}\left(\frac{n_2}{n_1}\right). \tag{6.74}$$

Plots of the magnitudes of the reflection coefficients

$$R_E = \frac{E_{\mathrm{R}y}}{E_\mathrm{I}}, \quad R_B = \frac{B_{\mathrm{R}y}}{B_\mathrm{I}},$$

for the interface between air and a typical glass, with refractive index 1.5, are shown in figure 6.7. The phenomena of total internal reflection and zero reflected component of the magnetic field parallel to the surface at Brewster's angle can be clearly seen.

As we have seen, an arbitrarily polarised plane wave incident on a plane interface between two insulating materials can be written as the sum of two waves, one with its electric field parallel to the interface, one with its magnetic field parallel to the interface. This means that, if the wave is incident at Brewster's angle, only the plane polarised component of the wave with its electric field parallel to the interface is reflected. Even if the angle of incidence is not exactly Brewster's angle, there is still a tendency for the reflected light to be dominated by this component, as shown in figure 6.7. This is the basis of Polaroid sunglasses which do not allow the wave reflected at Brewster's angle from horizontal surfaces to pass through them, so that only the less intense wave component is seen.

We noted in the previous section that the dielectric constant of a material, ϵ, is usually a function of the angular frequency, ω, of an electromagnetic wave passing through it. This means the refractive index is also a function of ω. One consequence of this is that waves of different frequencies are refracted by different amounts. This physical effect can be used to split a beam of light up into its components of different frequencies using a **prism**, as illustrated in figure 6.8. The same effect can also occur when sunlight shines through a sky full of raindrops to produce a rainbow.

6.8 Waveguides

We have seen in chapter 3 how acoustic waves can propagate along a rigid cylinder as a discrete set of axisymmetric wave modes, and in chapter 5 studied similar behaviour in the dynamics of elastic rods. The same phenomenon can occur for electromagnetic waves. Microwaves are usually transmitted using metal waveguides, and we will study these in subsection 6.8.1. Electromagnetic waves can also be transmitted in insulating waveguides. In particular, laser light can be transmitted in **optical fibres**. These are used to convey information rapidly and efficiently over large distances. We will study a model for the simplest possible optical fibre in subsection 6.8.2.

6.8.1 Metal Waveguides

A metal waveguide is typically a circular cylinder within which microwaves can propagate. We will approximate this as a cylindrical region of free space with radius a, surrounded by a perfect conductor. We therefore seek solutions of (6.29) to (6.32) subject to $\mathbf{E} \times \mathbf{n} = 0$ and $\mathbf{B} \cdot \mathbf{n} = 0$ at the interface, where $\mathbf{n}$ is the unit normal, as shown in figure 6.9. We seek solutions of the form

$$\mathbf{E} = \mathbf{E}(r, \theta)e^{i(\omega t - kz)}, \quad \mathbf{B} = \mathbf{B}(r, \theta)e^{i(\omega t - kz)}. \tag{6.75}$$

As for acoustic and elastic waves, we expect that propagating modes will only exist for a discrete, finite set of eigenvalues, k_j.

Since the fields satisfy the vector wave equations, (6.33) and (6.34),

$$\left(\nabla_{\mathrm{t}}^2 + \frac{\omega^2}{c_0^2} - k^2\right)\begin{pmatrix}\mathbf{E}\\ \mathbf{B}\end{pmatrix} = 0, \tag{6.76}$$

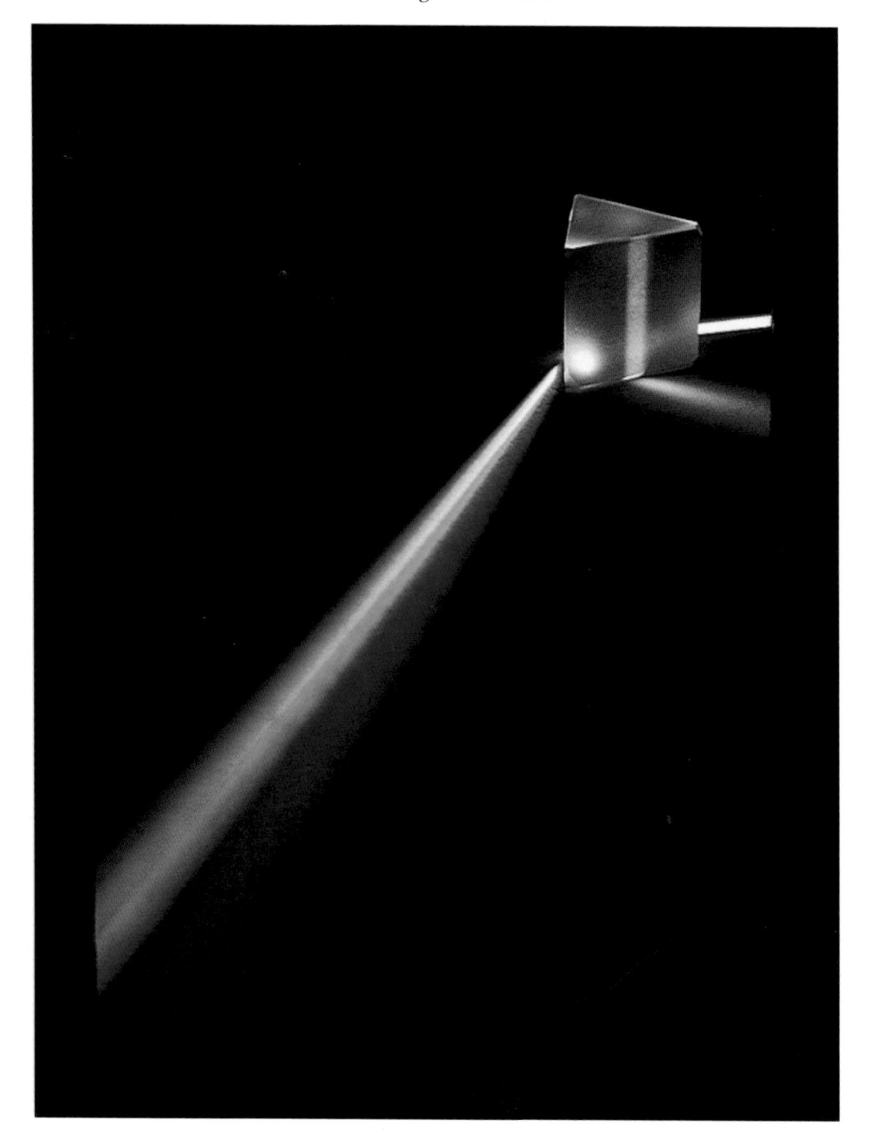

Fig. 6.8. A beam of light is split into its components of different frequencies by a prism.

where $\nabla_{\mathrm{t}}^2 = \nabla^2 - \partial^2/\partial z^2$ is the Laplacian operator in the plane transverse to the axis. We now write **E** and **B** as

$$\mathbf{E} = \mathbf{E}_{\mathrm{t}} + E_z\hat{\mathbf{z}}, \quad \mathbf{B} = \mathbf{B}_{\mathrm{t}} + B_z\hat{\mathbf{z}}, \tag{6.77}$$

where $\hat{\mathbf{z}}$ is the unit vector in the z-direction, which points along the axis of

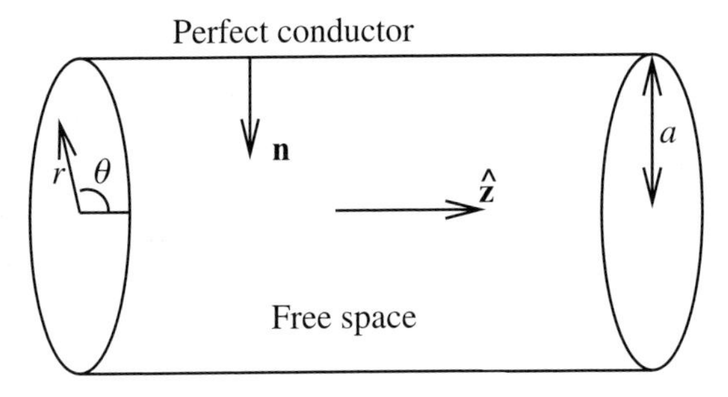

Fig. 6.9. A metal waveguide.

the cylinder. Here, $\mathbf{E}_t$ and $\mathbf{B}_t$ are the components of the fields transverse to the axis, and E_z and B_z the axial components. The axial components satisfy

$$\left(\nabla_t^2 + \frac{\omega^2}{c_0^2} - k^2\right)\begin{pmatrix} E_z \\ B_z \end{pmatrix} = 0. \tag{6.78}$$

Using cylindrical polar coordinates, the boundary conditions that $\mathbf{E} \times \mathbf{n} = 0$ and $\mathbf{B} \cdot \mathbf{n} = 0$ at the interface can be written in terms of the axial components as (see exercise 6.6)

$$\frac{\partial B_z}{\partial r} = E_z = 0 \quad \text{at } r = a. \tag{6.79}$$

We shall see that these can only be satisfied in general if either E_z or B_z is zero throughout the waveguide. This leads us to consider separately **transverse electric (TE) modes**, for which $E_z = 0$, and **transverse magnetic (TM) modes**, for which $B_z = 0$. In geometries that are not simply connected, it is also possible to have **transverse electromagnetic (TEM) modes**, for which $E_z = B_z = 0$, and we will consider these in subsection 6.8.2.

In TE modes, B_z satisfies (6.78) subject to

$$\frac{\partial B_z}{\partial r} = 0 \quad \text{at } r = a.$$

This is precisely the problem satisfied by the acoustic potential in a circular acoustic waveguide (section 3.5.3). The axisymmetric eigensolutions are therefore

$$B_{zj} = B_0 J_0\left(\frac{m_j r}{a}\right), \tag{6.80}$$

with

$$k_j = \sqrt{\frac{\omega^2}{c^2} - \frac{m_j^2}{a^2}}, \tag{6.81}$$

where m_j, $j = 0, 1, \ldots$, satisfy $J_0'(m_j) = 0$, and B_0 is an arbitrary real constant. The solution is propagating rather than evanescent provided $\omega > c_0 m_j / a$.

In TM modes, E_z satisfies (6.78) subject to

$$E_z = 0 \quad \text{at } r = a.$$

The axisymmetric eigensolutions for E_z are therefore of the same form as for B_z in the TE wave, except that now m_j must be the zeros of J_0. Note that the only plane wave solution possible in the circular waveguide is the TE mode with $m_0 = 0$. Also, since the turning points and zeros of J_0 never coincide, it is clear why at least one of the axial field components must be non-zero in any eigensolution, and hence that there are no TEM modes. Now that we have found the axial components of all the axisymmetric wave modes, the transverse components can be determined using Maxwell's equations (see exercise 6.6).

6.8.2 Weakly Guiding Optical Fibres

Most optical fibres consist of a cylindrical insulator, the **core**, surrounded by a layer of another insulator, the **cladding**. Both insulators have $\mu = \mu_0$. Typically their dielectric constants are real, since the fibre is designed to absorb as little light as possible, and hence be optically transparent. In addition the refractive index in the cladding is slightly less than that in the core. If we consider a plane wave incident on a flat interface between these two materials, the results of section 6.7 show that there will be total internal reflection if the wave is incident almost parallel to the interface, with $\frac{\pi}{2} - \theta_1 \ll 1$, but that otherwise, most of the wave is transmitted into the second medium. For the cladding and core of an optical fibre, this suggests that only waves that propagate almost along the axis of the core will be guided by it, with other waves able to escape into the cladding. This suggests in turn that the guided modes will be close to TEM modes, since axially propagating waves have their electric and magnetic fields in the transverse direction. Such a fibre is said to be **weakly guiding**.

We proceed as we did for the metal waveguide, except that we replace the perfect conductor with an insulator with refractive index n_2, and replace the cylindrical region of free space with an insulating core of refractive index n_1. The replacement of the finite cladding of the fibre with an infinite cladding is reasonable, since we are interested in fibres that confine most of the light to the core. We again seek solutions of the form (6.75), so that **E** and **B** satisfy (6.76). (In practice, optical fibres often, but not always, have a core whose refractive index varies across its radius. The resulting waveguide has similar properties to the case of a uniform core, which we analyse here, but the mathematics is less straightforward.)

Focusing our attention on the electric field, the transverse components satisfy

$$\left(\nabla_{\mathrm{t}}^2 + k_0^2 n^2 - k^2\right) \mathbf{E}_{\mathrm{t}} = 0, \tag{6.82}$$

where $k_0 = \omega/c_0$ is the free space wavenumber and $n = n_1$ for $0 \le r < a$ and n_2 for $r > a$. The axial component of the electric field is given by (6.29) as

$$E_z = -\frac{i}{k}\nabla_{\mathrm{t}} \cdot \mathbf{E}_{\mathrm{t}}. \tag{6.83}$$

If we now make the radial coordinate dimensionless using

$$\hat{r} = \frac{r}{a},$$

we find that

$$\left(\hat{\nabla}_{\mathrm{t}}^2 + U^2\right) \mathbf{E}_{\mathrm{t}} = 0 \quad \text{for } 0 \le \hat{r} < 1, \tag{6.84}$$

$$\left(\hat{\nabla}_{\mathrm{t}}^2 - W^2\right) \mathbf{E}_{\mathrm{t}} = 0 \quad \text{for } \hat{r} > 1, \tag{6.85}$$

$$E_z = i\frac{\sqrt{2\Delta}}{V}\hat{\nabla}_{\mathrm{t}} \cdot \mathbf{E}_{\mathrm{t}}, \tag{6.86}$$

where

$$U = a\sqrt{k_0^2 n_1^2 - k^2}, \quad W = a\sqrt{k^2 - k_0^2 n_2^2},$$

$$V = ak_0\sqrt{n_1^2 - n_2^2}, \quad 2\Delta = 1 - \frac{n_2^2}{n_1^2},$$

and

$$\hat{\nabla}_{\mathrm{t}} = \frac{\partial^2}{\partial \hat{r}^2} + \frac{1}{\hat{r}}\frac{\partial}{\partial \hat{r}} + \frac{1}{\hat{r}^2}\frac{\partial^2}{\partial \theta^2}.$$

The parameter Δ is the **relative index difference**, and $\Delta \ll 1$ for a weakly guiding fibre. The parameter $V = ak_0 n_1\sqrt{2\Delta}$ is the **waveguide parameter**. We will assume that $k_0 n_2 < k < k_0 n_1$, so that the parameters U and

W are real. We shall see below that this leads to oscillatory fields for $0 \le \hat{r} < 1$, so that there is a sequence of eigensolutions, and exponentially decaying fields for $\hat{r} > 1$, so that the fields are localised in the core. We will also need to assume that U, W, and hence $V = \sqrt{U^2 + W^2}$, are of $O(1)$ for $\Delta \ll 1$ in order that U and W appear as eigenvalues in (6.84) and (6.85). Equations (6.84) to (6.86) then show that $\mathbf{E}_t = O(1)$ and $E_z = O(\Delta^{1/2})$. A similar calculation for the magnetic field shows that $\mathbf{B}_t = O(1)$ and $B_z = O(\Delta^{1/2})$. Since $V = O(1)$, $ak_0n_1 = O(\Delta^{-1/2})$ and hence the free space wavelength of the wavemode is much smaller than the radius of the core. This is consistent with the simple physical argument with which we began this subsection. This argument was valid provided that the waves could be modelled as approximately plane and incident on a flat interface ($k_0 a \gg 1$), and suggested that we should find approximate TEM modes, for which $E_z \ll 1$ and $B_z \ll 1$.

The leading order TEM mode solution can be decomposed into an x-polarised and a y-polarised wave of the form

$$\mathbf{E}_t = F(\hat{r})\,(A\hat{\mathbf{x}} + B\hat{\mathbf{y}}),$$

for arbitrary complex constants A and B, with $\hat{\mathbf{x}}$ and $\hat{\mathbf{y}}$ the unit vectors in the x- and y-directions in Cartesian coordinates. The function $F(\hat{r})$ satisfies

$$F'' + \frac{1}{\hat{r}}F' + U^2F = 0 \quad \text{for } 0 \le \hat{r} < 1, \tag{6.87}$$

$$F'' + \frac{1}{\hat{r}}F' - W^2F = 0 \quad \text{for } \hat{r} > 1. \tag{6.88}$$

At $\hat{r} = 1$, the usual boundary conditions are satisfied at leading order if $F(\hat{r})$ is continuous there. The solution is therefore

$$F(\hat{r}) = \left\{ \begin{array}{cc} J_0(U\hat{r})/J_0(U) & \text{for } 0 \le \hat{r} \le 1, \\ K_0(W\hat{r})/K_0(W) & \text{for } \hat{r} \ge 1, \end{array} \right\} \tag{6.89}$$

where K_0 is the zeroth order modified Bessel function. For E_z to be small at leading order, we also require that $F'(\hat{r})$ should be continuous at $\hat{r} = 1$. The eigenvalues, k_j, are therefore solutions of the equation

$$U\frac{J_0'(U)}{J_0(U)} = W\frac{K_0'(W)}{K_0(W)},$$

or equivalently

$$U\frac{J_1(U)}{J_0(U)} = W\frac{K_1(W)}{K_0(W)}. \tag{6.90}$$

If we now define the parameter X by

$$k = k_0 n_1 \left(1 - X\sqrt{2\Delta}\right), \tag{6.91}$$

where $0 < X < 1$, (6.90) becomes

$$G(X, V) \equiv \sqrt{X} J_1\left(V\sqrt{X}\right) K_0\left(V\sqrt{1-X}\right) - \sqrt{1-X} J_0\left(V\sqrt{X}\right) K_1\left(V\sqrt{1-X}\right) = 0. \tag{6.92}$$

There is no analytical method for determining the eigenvalues, $X_j(V)$, which satisfy $G(X_j, V) = 0$. Figure 6.10 shows the function $G(X, V)$ for various values of V. For V sufficiently small, there is a unique solution, $X = X_0(V)$, of $G(X, V) = 0$, which gives the wavenumber of the **fundamental mode** of the fibre. As V increases, the number of solutions increases. The defining series expansions for the Bessel functions show that

$$G(X, V) \sim -\frac{1}{2} J_1(V) \log(1 - X) \quad \text{as } X \to 1,$$

and hence that $G(X, V) \to \pm\infty$ as $X \to 1$, with the sign determined by the sign of $J_1(V)$. Although we will not prove it here, this suggests that, as V increases, a new solution appears whenever V increases past a zero of J_1. Since the first three zeros of $J_1(V)$ are at $V \approx 3.83$, 7.02 and 10.2, this is consistent with figure 6.10.

Manufacturers of optical fibres often seek to design **single mode fibres**. These only allow the fundamental mode to propagate. Our analysis shows that such a fibre must have $V < 3.83$ if the fundamental mode is to be the only propagating axisymmetric mode. In fact, an analysis of the non-axisymmetric modes shows that to ensure that none of these can propagate, we need $V < V_c$, where the **critical waveguide parameter** is $V_c \approx 2.41$, the smallest zero of J_0. Figure 6.11 shows the radial distribution of the intensity of the electric field in the fundamental mode, $F(r)$, for various values of V. This is equivalent to fixing the free space wavelength of the light and the refractive indices of the fibre, and varying the radius of the core, $a \propto V$. We can see that, for most of the field to be confined to the core, we need to make V, and hence a, as large as possible. For a single mode fibre, the optimum choice is therefore $V = V_c$.

Optical fibres are used to carry information encoded as pulses of light. Any such pulse should ideally be of a single wavelength, since otherwise dispersion will act to separate the various components of the pulse and smear it out axially. If the pulses spread too far they start to overlap,

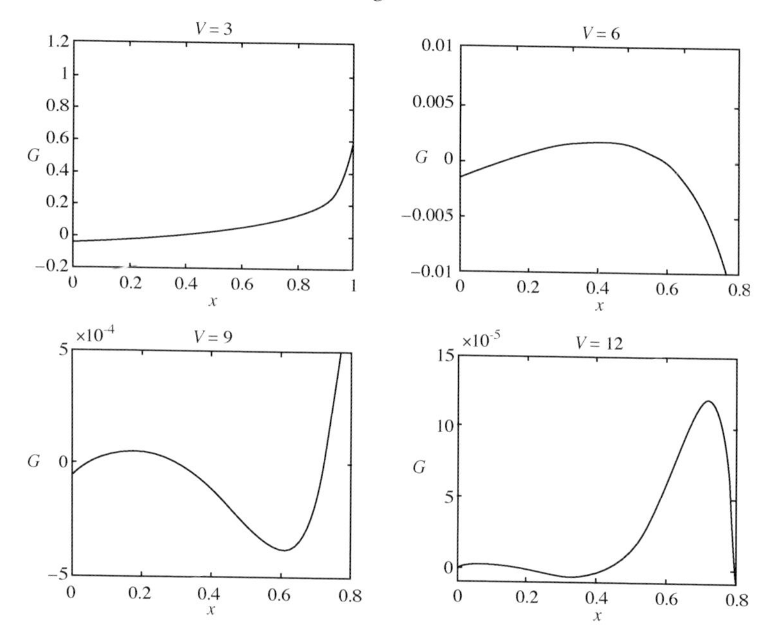

Fig. 6.10. The function $G(X, V)$ for $V = 3$, 6, 9 and 12.

and the information carried may be lost. One way of counteracting this tendency to spread is to note that, as well as the dispersion due to the different group velocities of components of different wavelengths (see subsection 3.5.2), there is dispersion due to the variation of the refractive index with wavelength caused by the properties of the material of the fibre. Specifically, the time for energy to propagate a unit distance along the fibre is

$$\tau = \frac{1}{c_g} = \frac{dk}{d\omega} = -\frac{\lambda^2}{2\pi c}\frac{dk}{d\lambda},$$

and hence

$$\frac{d\tau}{d\lambda} = -\frac{\lambda}{\pi c}\frac{dk}{d\lambda} - \frac{\lambda^2}{2\pi c}\frac{d^2k}{d\lambda^2}.$$

If the small range of wavelengths contained in the pulse is $\delta\lambda$, then the range of transit times due to dispersion is

$$\delta\tau \approx -\frac{\lambda\delta\lambda}{2\pi c}\left(\lambda\frac{d^2k}{d\lambda^2} + 2\frac{dk}{d\lambda}\right) = -\frac{\lambda\delta\lambda}{2\pi c}f(\lambda, n_1(\lambda), n_2(\lambda)). \tag{6.93}$$

In order to minimise the spreading of a pulse we can try to minimise

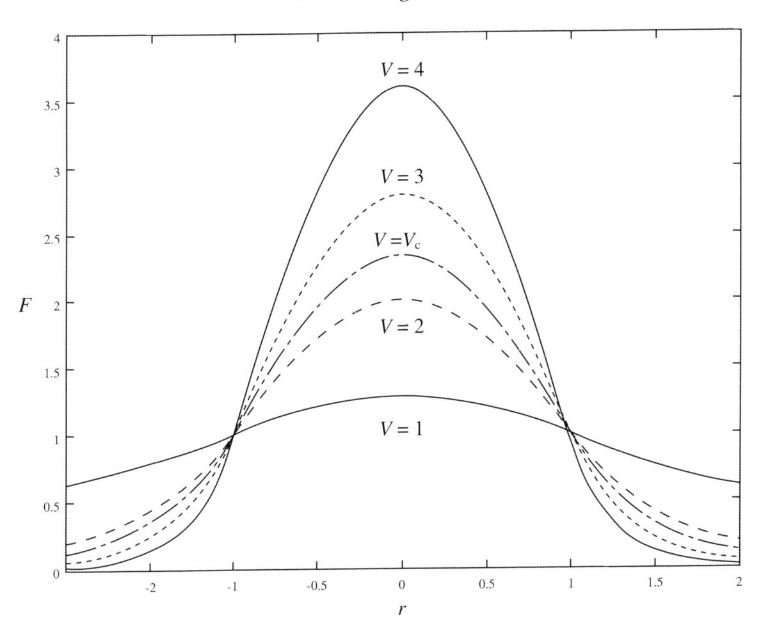

Fig. 6.11. The intensity of the electric field in the fundamental mode for $V = 1$, 2, $V_c \approx 2.41$, 3 and 4.

the function f. This function is dependent on λ both directly (**waveguide dispersion** due to the dependence of the group velocity on λ) and indirectly (**material dispersion** due to the dependence of the group velocity on refractive indices, $n_1(\lambda)$ and $n_2(\lambda)$). For silica fibres, there is a critical wavelength, $\lambda_c \approx 1.3\,\mu$m, such that for $\lambda < \lambda_c$ material dispersion is dominant, and for $\lambda > \lambda_c$ waveguide dispersion is dominant. For waves with wavelength close to λ_c the two effects tend to cancel out, and there is very little dispersion. This phenomenon has been used successfully to manufacture optical fibres in which pulses of around this wavelength propagate with little spreading. For a nice description of the problems involved in designing and modelling optical fibres, see Pask (1992).

This approach to pulse transmission in optical fibres is based entirely on linear theory. Another method makes use of the nonlinear response of an optical fibre for sufficiently intense laser light, the **Kerr effect**. This tends to compress the pulse, whilst dispersion tends to smear it out. When these two competing effects balance, the formation of **optical envelope solitons** is possible. These can be modelled using the **nonlinear Schrödinger equation** (NLS), and form the subject of section 12.2.

6.9 Radiation

The light generated by chemical or nuclear reactions, for example a candle or the sun, can usually be treated as spherically symmetric, with an equal intensity in each direction. However this is not the case when electromagnetic waves, in particular radio waves and microwaves used for communication, are generated by antennas. In this section we study how electromagnetic waves can be generated by localised regions containing time varying currents. The discussion has much in common with that of the generation of sound waves by sources in plane walls (section 3.7). There is, however, a major difference. It is not physically possible to have an isolated region where the amount of charge varies with time, since charge is conserved. Time varying currents are the simplest generators of electromagnetic waves. This leads to what are known as **dipole fields**, which have a strong directional dependence.

6.9.1 Scalar and Vector Potentials

The most convenient way of representing the electric and magnetic fields when considering the waves generated by localised sources is to use a pair of vector and scalar potentials. Since $\nabla \cdot \mathbf{B} = 0$, it can be shown that the magnetic field can always be written as the curl of a **vector potential**, **A**, so that

$$\mathbf{B} = \nabla \times \mathbf{A}.$$

Similarly, (6.27) is

$$\nabla \times \mathbf{E} = -\frac{\partial}{\partial t}\nabla \times \mathbf{A},$$

and the electric field can be written in terms of the gradient of a **scalar potential**, ϕ, with

$$\mathbf{E} = -\nabla\phi - \frac{\partial \mathbf{A}}{\partial t}.$$

On substituting these representations of the fields into the remaining two Maxwell equations, (6.25) and (6.28), we find that

$$\frac{1}{c_0^2}\frac{\partial^2 \mathbf{A}}{\partial t^2} - \nabla^2 \mathbf{A} = \mu_0 \mathbf{j} - \nabla\left(\frac{1}{c_0^2}\frac{\partial \phi}{\partial t} + \nabla \cdot \mathbf{A}\right),$$

$$\frac{1}{c_0^2}\frac{\partial^2 \phi}{\partial t^2} - \nabla^2 \phi = \frac{\rho}{\epsilon_0} + \frac{\partial}{\partial t}\left(\frac{1}{c_0^2}\frac{\partial \phi}{\partial t} + \nabla \cdot \mathbf{A}\right).$$

The final step is to note that we have some freedom in choosing **A** and ϕ. In particular, the mapping

$$\mathbf{A} \mapsto \mathbf{A} + \nabla X, \quad \phi \mapsto \phi - \frac{\partial X}{\partial t},$$

leaves the fields unchanged and

$$\frac{1}{c_0^2}\frac{\partial \phi}{\partial t} + \nabla \cdot \mathbf{A} \mapsto \frac{1}{c_0^2}\frac{\partial \phi}{\partial t} + \nabla \cdot \mathbf{A} + \nabla^2 X - \frac{1}{c_0^2}\frac{\partial^2 X}{\partial t^2}.$$

We can therefore choose **A** and ϕ so that

$$\frac{1}{c_0^2}\frac{\partial \phi}{\partial t} + \nabla \cdot \mathbf{A} = 0.$$

This is known as choosing the **Lorentz gauge**. The potentials now satisfy

$$\frac{1}{c_0^2}\frac{\partial^2 \mathbf{A}}{\partial t^2} - \nabla^2 \mathbf{A} = \mu_0 \mathbf{j}, \tag{6.94}$$

$$\frac{1}{c_0^2}\frac{\partial^2 \phi}{\partial t^2} - \nabla^2 \phi = \frac{\rho}{\epsilon_0}. \tag{6.95}$$

These are forced wave equations, which are very convenient to work with because the forcing is given as simple multiples of the charge and current densities, ρ and **j**.

We can write down the general solution of (6.95) using what we learnt about acoustic sources in section 3.6. When $\rho/\epsilon_0 = f(t)\delta(\mathbf{r})$, the solution is

$$\phi(\mathbf{r}, t) = \frac{1}{4\pi r} f\left(t - \frac{r}{c_0}\right),$$

where $r = |\mathbf{r}|$. This is the potential due to a point source. Since

$$\frac{\rho(\mathbf{r}, t)}{\epsilon_0} = \frac{1}{\epsilon_0}\int_V \rho(\mathbf{r}', t)\delta(\mathbf{r}')\, d^3\mathbf{r}',$$

where ρ is only non-zero within the volume V, the solution of (6.95) is

$$\phi(\mathbf{r}, t) = \frac{1}{4\pi\epsilon_0}\int_V \frac{\rho\left(\mathbf{r}', t - \frac{R}{c_0}\right)}{R} d^3\mathbf{r}', \tag{6.96}$$

where $R = |\mathbf{r} - \mathbf{r}'|$. Similarly,

$$\mathbf{A}(\mathbf{r}, t) = \frac{\mu_0}{4\pi}\int_V \frac{\mathbf{j}\left(\mathbf{r}', t - \frac{R}{c_0}\right)}{R} d^3\mathbf{r}'. \tag{6.97}$$

The appearance of $t - R/c_0$, the **retarded time**, reflects the fact that information propagates at speed c_0, and the potentials at $\mathbf{r}$ respond to changes in the charges and currents at any point $\mathbf{r}'$ a time R/c_0 after they occur.

6.9.2 The Electric Dipole

As we have already noted, charge is conserved, so the simplest time varying source is a localised distribution of currents. The simplest of these is the **time harmonic electric dipole**. Consider a charge $q_0 e^{i\omega t}$ at $\mathbf{r} = 0$ and a charge $-q_0 e^{i\omega t}$ on the z-axis at $\mathbf{r} = \mathbf{l} = (0, 0, l)$. Since these charges vary with time, there must be a line current of magnitude $\mathbf{j} = i\omega q_0 e^{i\omega t}$ confined to the z-axis between the origin and $(0, 0, l)$. We now substitute this into (6.97) and take the limit $q_0 \to \infty$, $l \to 0$ with $\mathbf{p} = q_0 \mathbf{l}$, the **dipole moment**, bounded and non-zero. We obtain

$$\mathbf{A} = \frac{i\omega\mu_0 p}{4\pi r} e^{i(\omega t - kr)} \hat{\mathbf{z}}, \tag{6.98}$$

where $p = |\mathbf{p}|$, $k = \omega/c_0$ and $\hat{\mathbf{z}}$ is the unit vector in the z-direction. This gives a magnetic field

$$\mathbf{B} = \nabla \times \mathbf{A} = \left(\hat{\mathbf{r}}\frac{d}{dr}\right) \times |\mathbf{A}|\hat{\mathbf{z}} = -\frac{i\omega\mu_0 p}{4\pi r}\left(ik + \frac{1}{r}\right) e^{i(\omega t - kr)} \hat{\mathbf{r}} \times \hat{\mathbf{z}}, \tag{6.99}$$

where $\hat{\mathbf{r}}$ is the unit radial vector. This is illustrated in figure 6.12 in terms of spherical polar coordinates, where the unit azimuthal vector, and hence the magnetic field, points into the page and is given by $\hat{\mathbf{z}} \times \hat{\mathbf{r}}/\sin\theta$. In terms of its components,

$$B_r = B_\theta = 0, \quad B_\phi = \frac{i\omega\mu_0 p}{4\pi r}\sin\theta\left(ik + \frac{1}{r}\right) e^{i(\omega t - kr)}. \tag{6.100}$$

The electric field is given by (6.28) as

$$\mathbf{E} = \frac{c_0^2}{i\omega}\nabla \times \mathbf{B} = \frac{1}{r\sin\theta}\frac{\partial}{\partial\theta}\left(\sin\theta B_\phi\right)\hat{\mathbf{r}} - \frac{1}{r}\frac{\partial}{\partial r}\left(rB_\phi\right)\hat{\theta},$$

where $\hat{\theta}$ is the unit vector in the θ-direction. This leads to

$$\left.\begin{aligned} E_r &= \frac{c_0 i\omega p\mu_0}{2\pi r^2}\cos\theta\left(1 - \frac{i}{kr}\right) e^{i(\omega t - kr)}, \\ E_\theta &= -\frac{c_0 p\omega\mu_0 k}{4\pi r}\sin\theta\left(1 - \frac{i}{kr} - \frac{1}{k^2 r^2}\right) e^{i(\omega t - kr)}, \quad E_\phi = 0. \end{aligned}\right\} \tag{6.101}$$

Although the strength of the fields depends upon θ, $\mathbf{E}$ and $\mathbf{B}$ are perpendicular at any point in space.

We can now consider how the fields behave in the far field, when

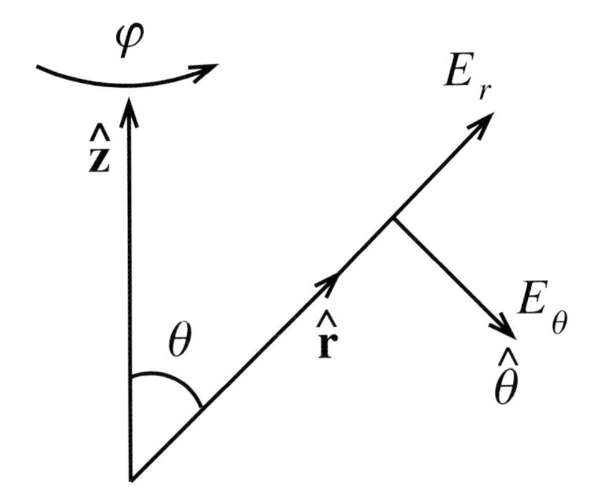

Fig. 6.12. The field of an electric dipole in the $\hat{z}$-direction. The magnetic field points into the page.

$kr \gg 1$. In this case, $E_r = O(r^{-2})$, whilst E_θ and B_ϕ are asymptotically larger, of $O(r^{-1})$, with, using $k = \omega/c_0$,

$$E_\theta \sim -\frac{p\omega^2\mu_0}{4\pi r}\sin\theta\, e^{i(\omega t-kr)}, \quad B_\phi \sim -\frac{p\omega^2\mu_0}{4\pi c_0 r}\sin\theta\, e^{i(\omega t-kr)}. \tag{6.102}$$

At leading order, the fields take the same form as a plane polarised wave in the far field. However, the factor of $\sin\theta$ means that the strength of this wave is greatest in the direction perpendicular to that of the dipole moment. The Poynting vector in the far field is radial, and has magnitude

$$|\mathbf{P}_0| = \frac{1}{\mu_0}|\mathbf{E}\times\mathbf{B}| = \frac{p^2\omega^4\mu_0}{16\pi^2 c_0 r^2}\sin^2\theta\cos^2(\omega t - kr).$$

The total energy flux, or equivalently power output, across a large sphere of radius r, time averaged over a period of the oscillation, is therefore

$$\bar{W}_{\text{tot}} = \frac{p^2\omega^4\mu_0}{16\pi^2 c_0}\int_{t_0}^{t_0+\frac{2\pi}{\omega}}\cos^2(\omega t - kr)\,dt\int_0^{2\pi}\int_0^{\pi}\sin^3\theta\, d\theta\, d\phi = \frac{p^2\omega^4\mu_0}{12\pi c_0}.$$

Remembering that $p = Il/\omega$, where I is the current in the dipole and l its length, we can write this as

$$\bar{W}_{\text{tot}} = \frac{1}{2}R_d I^2,$$

where

$$R_{\text{d}} = \frac{l^2\omega^2\mu_0}{6\pi c_0} \tag{6.103}$$

is the **radiation resistance** of the dipole. The rate at which energy is dissipated when a current I flows through a resistance R is $\frac{1}{2}RI^2$. By analogy, driving a current I with angular frequency ω in a dipole produces a qualitatively similar dissipative effect, since energy is carried away by the radiated wave.

6.9.3 The Far Field of a Localised Current Distribution

By analogy with the results of chapter 3, we might expect that, at leading order, the radiation in the far field of any time harmonic, localised distribution of currents would be equivalent to that of an electric dipole. However, this is not precisely what happens, since there is not, in general, any well-defined direction associated with such an arbitrary current distribution.

In the far field of a distribution of time harmonic currents confined to a finite region,

$$R = |\mathbf{r} - \mathbf{r}'| \sim r\sqrt{1 - 2\frac{\mathbf{r}\cdot\mathbf{r}'}{r^2}} \sim r - \hat{\mathbf{r}}\cdot\mathbf{r}',$$

and hence from (6.97), at leading order

$$\mathbf{A} = \frac{\mu_0 e^{i(\omega t - kr)}}{4\pi r}\int_V \mathbf{j}(\mathbf{r}')e^{ik\hat{\mathbf{r}}\cdot\mathbf{r}'}d^3\mathbf{r}'. \tag{6.104}$$

We can immediately see that this is of $O(r^{-1})$, as for the electric dipole, but that the integral gives a directional dependence to the wave that depends upon the distribution of the current. By comparing (6.98) with (6.104), we can see that we simply need to replace $i\omega p\hat{\mathbf{z}}$, the current in the electric dipole, with

$$\mathbf{F}(\hat{\mathbf{r}}) = \int_V \mathbf{j}(\mathbf{r}')e^{ik\hat{\mathbf{r}}\cdot\mathbf{r}'}d^3\mathbf{r}', \tag{6.105}$$

and arrive at the general far field solution

$$\mathbf{B} = -\frac{ik\mu_0}{4\pi r}e^{i(\omega t - kr)}\mathbf{F}\times\hat{\mathbf{r}}, \quad \mathbf{E} = c_0\mathbf{B}\times\hat{\mathbf{r}}. \tag{6.106}$$

Note that this field is, except in special cases, like that of an elliptically polarised plane wave, since $\mathbf{F}$ is a complex vector. Since $\mathbf{E}$ and $\mathbf{B}$ are perpendicular to each other and to $\hat{\mathbf{r}}$, we can write

$$\mathbf{B} = -\frac{ik\mu_0}{4\pi r}e^{i(\omega t - kr)}(0, F_\theta, F_\phi), \quad \mathbf{E} = \frac{ik\mu_0 c_0}{4\pi r}e^{i(\omega t - kr)}(0, F_\phi, -F_\theta),$$

in terms of the components of $\mathbf{F}$ in spherical polar coordinates. The Poynting vector, which is radial, therefore has magnitude

$$|\mathbf{P}_0| = \frac{k^2\mu_0 c_0}{16\pi^2 r^2}\cos^2(\omega t - kr)\left\{(\mathrm{Re}(F_\theta))^2 + \left(\mathrm{Re}(F_\phi)\right)^2\right\}.$$

The time averaged power output is

$$\bar{W} = \frac{k^2\mu_0 c_0}{32\pi^2 r^2}\left\{(\mathrm{Re}(F_\theta))^2 + \left(\mathrm{Re}(F_\phi)\right)^2\right\}, \tag{6.107}$$

and the total, time averaged power output of the current distribution is

$$\bar{W}_{\mathrm{tot}} = \int_0^{2\pi}\int_0^{\pi} r^2 \bar{W}\sin\theta\, d\theta\, d\phi. \tag{6.108}$$

The directional dependence is now characterised in terms of the **gain**, defined to be

$$G(\theta,\phi) = \frac{4\pi r^2}{\bar{W}_{\mathrm{tot}}}\bar{W}. \tag{6.109}$$

This is normalised so that if the power radiated by the waves were the same in each direction, the gain would be unity, although, since $\mathbf{F}$ cannot be independent of $\hat{\mathbf{r}}$, this is not actually possible.

The electric dipole has

$$\bar{W} = \frac{p^2\omega^4\mu_0}{32\pi^2 c_0 r^2}\sin^2\theta, \quad \bar{W}_{\mathrm{tot}} = \frac{p^2\omega^4\mu_0}{12\pi c_0},$$

and hence

$$G(\theta,\phi) = \frac{3}{2}\sin^2\theta,$$

which is shown in figure 6.13.

6.9.4 The Centre Fed Linear Antenna

A simple, practical method of generating electromagnetic waves is to drive a time harmonic current in a finite length of wire – an **antenna**. If we know the distribution of current in the antenna, we can use (6.105) and (6.106) to determine the far field and (6.108) and (6.109) to obtain the radiation resistance and gain. However, whatever method is used to drive the antenna, the actual determination of the current distribution in the wire is a nasty boundary value problem coupled to the near field of the antenna. Fortunately, for most practical purposes a sensible approximation of the current in the antenna leads to a reasonable estimate of the radiation resistance and gain. One physical constraint is that the current must be zero at the end of each wire.

Consider a straight, thin wire extending from $z = -d/2$ to $z = d/2$, with a small gap at the origin across which an oscillating potential drives a current in the antenna. By considering approximations to the full problem, it can be shown that an antenna has its lowest radiation resistance, and is therefore most efficient, when its length is a half wavelength of the wave that it radiates (see the book by Sander and Reed (1986)). We will not go into this here, but it should not come as a great surprise, considering that the current must be zero at each end of the wire, and bearing in mind what we know about the response of other one-dimensional, linear systems of finite length, such as strings and tubes (see sections 2.2 and 3.4). On this basis, it is reasonable to assume that, for a half wavelength antenna, the current distribution is

$$I(z,t) = I_0 \cos\left(\frac{\pi z}{d}\right) e^{i\omega t}, \tag{6.110}$$

with $kd = \pi$. We shall return to the question of the accuracy of this approximation below. Using this in (6.105) gives

$$\mathbf{F} = I_0 \hat{\mathbf{z}} \int_{-d/2}^{d/2} \cos\left(\frac{\pi z}{d}\right) e^{ikz\cos\theta} dz = \frac{2I_0 d}{\pi} \frac{\cos\left(\frac{1}{2}\pi\cos\theta\right)}{\sin^2\theta} \hat{\mathbf{z}}. \tag{6.111}$$

Since this always points in the z-direction, the far field will look like a plane polarised wave. We also note that the mean current in the wire is

$$\bar{I} = \frac{1}{d} \int_{-d/2}^{d/2} \cos\left(\frac{\pi z}{d}\right) dz = \frac{2}{\pi} I_0. \tag{6.112}$$

In terms of spherical polar coordinates, $\mathbf{F}$ has no component in the azimuthal direction, so that

$$F_\phi = 0, \quad F_\theta = \frac{2I_0 d}{\pi} \frac{\cos\left(\frac{1}{2}\pi\cos\theta\right)}{\sin\theta}, \tag{6.113}$$

and hence energy flux in the far field is, from (6.107),

$$\bar{W} = \frac{\mu_0 c_0 I_0^2}{4\pi} K, \tag{6.114}$$

where

$$K = \int_0^\pi \frac{\cos^2\left(\frac{1}{2}\pi\cos\theta\right)}{\sin\theta} d\theta \approx 1.219. \tag{6.115}$$

The integral K can be written in terms of a special function known as the

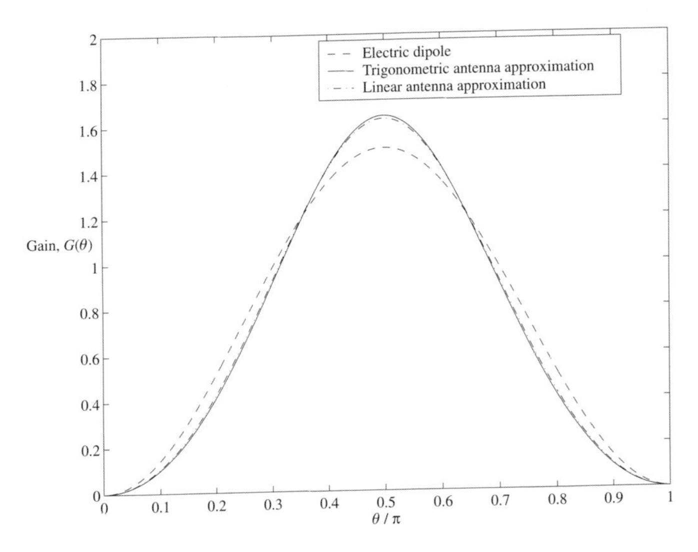

Fig. 6.13. The gain from an electric dipole, a half wavelength antenna with a trigonometric current distribution, (6.110), and with a linear current distribution, (6.118).

cosine integral, but we might as well just evaluate it numerically using the trapezium rule. This gives the gain of the antenna as

$$G_1(\theta, \phi) = \frac{2}{K} \frac{\cos^2\left(\frac{1}{2}\pi \cos\theta\right)}{\sin^2\theta}, \tag{6.116}$$

as shown in figure 6.13. The gain is very similar to that of an electric dipole, with most of the energy transmitted close to the direction perpendicular to the wire.

We can now use (6.112) to determine the radiation resistance, writing $\bar{W}_{\text{tot}} = \frac{1}{2}R_1\bar{I}^2$, where

$$R_1 = \frac{\mu_0 c_0 K \pi}{8} \approx 0.92 R_{\text{d}}, \tag{6.117}$$

where R_{d}, given by (6.103) is the radiation resistance of an electric dipole with length k/π and current $\bar{I}$. The half wavelength antenna therefore has a slightly lower radiation resistance than the equivalent electric dipole.

In order to show that this result is not very sensitive to our assumption about the distribution of current in the antenna, we can repeat the analysis with a piecewise linear current distribution (see exercise 6.11)

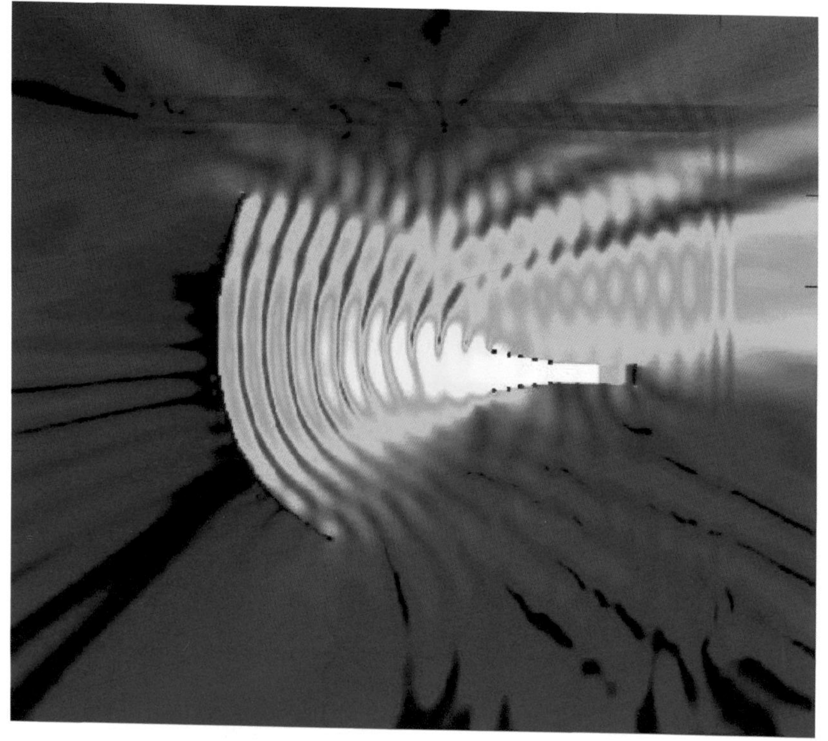

Fig. 6.14. The experimentally measured gain from a half wavelength antenna and a parabolic dish.

given by

$$I(z,t) = I_0 \left(1 - \frac{2|z|}{d}\right) e^{i\omega t}. \tag{6.118}$$

The gain is almost indistinguishable from that which we obtain using (6.110), as shown in figure 6.13. The experimentally measured gain from a half wavelength antenna is shown in figure 6.14.

Exercises

6.1

(a) An electron moves in a constant, spatially uniform electric field, **E**. If the electron is not initially at rest, show that, in general, its subsequent path is a parabola. Under what circumstances does the electron move in a straight line?

(b) An electron moves in a constant, spatially uniform magnetic field, **B**. If the electron is not initially at rest, show that, in general, its subsequent path is a helix. Under what circumstances does the electron move in (i) a straight line, (ii) a circle?

6.2 Each of a pair of infinitely long, parallel, thin straight wires carries a charge q per unit length, which travels along the wire at speed u. Show that the ratio of the magnetic to the electric force per unit length on either line is u^2/c_0^2.

6.3 Use the Biot–Savart law to determine the magnetic field at the centre of a circular loop of wire of radius a, carrying a steady current, I.

6.4 Consider a uniform shell of charge with density ρ lying in $a < r < b$, where r is the distance from the origin. Use Gauss's law to obtain the electric field everywhere. What is the potential? What happens as $b \to a$ and $\rho \to \infty$ with $\rho(b - a) = \rho_s$?

6.5 Deduce from the analysis given in section 6.2 that

$$\nabla^2 \left(\frac{1}{|\mathbf{x} - \mathbf{x}_0|} \right) = -4\pi\delta(\mathbf{x} - \mathbf{x}_0).$$

By taking the curl of (6.18), show that

$$\nabla \times \mathbf{B} = \mu_0 \mathbf{j}(\mathbf{x}) + \frac{\mu_0}{4\pi} \nabla \int \mathbf{j}(\mathbf{x}_0) \nabla_0 \cdot \left(\frac{1}{|\mathbf{x} - \mathbf{x}_0|} \right) d^3\mathbf{x}_0,$$

where ∇_0 is the gradient with respect to $\mathbf{x}_0$. Integrate once by parts, and use the steady form of (6.12) to deduce Ampère's law, (6.20).

6.6 By substituting the definition, (6.77), of the transverse and axial components of the time harmonic fields in a waveguide into Maxwell's equations (6.31) and (6.32), show that

$$\mathbf{E}_t = \frac{1}{k^2 - \frac{\omega^2}{c_0^2}} (ik\nabla_t E_z - i\omega \hat{\mathbf{z}} \times \nabla_t B_z),$$

and find a similar equation for the transverse component of $\mathbf{B}$ in terms of the axial components of the electric and magnetic fields. Using these equations, show that in a circular metal waveguide, boundary conditions (6.79) are appropriate.

6.7 Determine the eigenfrequencies of a square, perfectly conducting waveguide with sides of length a, and calculate the velocity at which energy propagates along the waveguide for each mode.

6.8 A layer of glass with refractive index $n > 1$ and thickness a is sandwiched between a perfect conductor and a vacuum.

(a) Determine the equation satisfied by the eigenfrequencies of TM modes that can propagate along the glass.

(b) A plane polarised wave of frequency ω is incident on the glass at an angle α with its magnetic field parallel to the surface of the glass. Determine the amplitude of the reflected wave.

6.9 Determine the near field of a time harmonic electric dipole, and show that both the electric and magnetic fields are of $O(r^{-3})$. Show that this is equivalent to the field of a static electric dipole due to charges q_0 and $-q_0$ at $(0,0,0)$ and $(0,0,l)$.

6.10 Determine the far field of two identical time harmonic electric dipoles with moments in the same direction, separated by a distance, $d > 0$. Describe the far field when (i) $\omega d/c_0 \ll 1$, (ii) $\omega d/c_0 \gg 1$.

6.11 Consider the half wavelength antenna with a piecewise linear current distribution, given by (6.118). Show that the gain is proportional to

$$\frac{\sin^2\theta}{\cos^4\theta}\left\{1-\cos\left(\frac{1}{2}\pi\cos\theta\right)\right\}^2,$$

and that the ratio of its radiation resistance to that when the current distribution is given by (6.110) is

$$\frac{64}{\pi^4}\left(\int_0^{\pi}\frac{\sin^3\theta}{\cos^4\theta}\left\{1-\cos\left(\frac{1}{2}\pi\cos\theta\right)\right\}^2\right)d\theta\Big/\int_0^{\pi}\frac{\cos^2\left(\frac{1}{2}\pi\cos\theta\right)}{\sin\theta}d\theta.$$

Evaluate this ratio numerically using the trapezium rule, and hence show that the radiation resistance only differs by about 1% between these two approximations.

Part two

Nonlinear Waves

So far we have studied only *linear* wave systems. With the exception of electromagnetic waves, we formulated the governing equations by looking at small amplitude disturbances of steady states – a string in equilibrium, a motionless ideal gas or elastic solid, the flat, undisturbed surface of a fluid or solid. If $\mathbf{y}_1$ and $\mathbf{y}_2$ are solutions of a linear system of equations, then $a_1\mathbf{y}_1 + a_2\mathbf{y}_2$ is also a solution for any constants a_1 and a_2. In particular, this means that separation of variables and integral transform methods allow us to determine the solution. In fact, these are the only techniques we have used. Compare what happens for **nonlinear** systems, for example, disturbances of a steady state that do not have a small amplitude. If $\mathbf{y}_1$ and $\mathbf{y}_2$ is a solution of a nonlinear system of equations, then, in general, neither $\mathbf{y}_1 + \mathbf{y}_2$ nor $k\mathbf{y}_1$, with k a constant, are solutions. Our standard mathematical techniques fail, and we must think again.

We begin by introducing some of the techniques that can be used to study nonlinear systems of equations by looking at some specific examples in chapter 7. In section 7.1 we study in detail a simple model for the flow of traffic. The governing equation determines how the density of cars changes along a road with a single lane. In chapter 3 we studied small amplitude disturbances to a compressible gas. In section 7.2 we investigate finite amplitude disturbances. The system of equations has three dependent variables, velocity, density and pressure. Each of these systems can be studied in terms of **characteristic curves**, which carry information from the initial conditions forward in time in a sense that we shall explain below. Another generic feature of these systems is that **shock waves** can form, and we devote some time to investigating their properties.

We have already derived the governing equations for shallow water waves. The system has two dependent variables, water depth and horizontal velocity, and we begin chapter 8 by considering nonlinear waves on shallow water. After studying Stokes' expansion for weakly nonlinear progressive gravity waves on deep water, we move on to look at how, in shallow water, the nonlinear steepening of water waves can be balanced by the effects of linear dispersion, as expressed by the KdV equation. This is a nonlinear equation, and we examine its wave solutions, known as cnoidal waves. The KdV equation also has a family of solitary wave solutions, which we will consider in detail in section 12.1. We also show how we can use complex variable theory to derive some exact solutions for nonlinear capillary waves.

In chapter 9 we study chemical and electrochemical waves. Chemical waves arise from a self-sustained balance between the tendency of molecular diffusion to smear out distributions of chemicals and limit the maximum concentration, and the tendency of certain reactions to cause the maximum chemical concentration to rise. We end the chapter by considering how we can model the propagation of nerve impulses – an example of an electrochemical wave.

In all of these systems we are interested in how the effect of a nonlinear process (for example, wave steepening or chemical reaction) is modified by that of a linear process (for example, diffusion or dispersion). In general, we find that the result is a coherent, propagating structure (a shock, a soliton or a chemical wave), sustained by a balance between the opposing effects of the linear and nonlinear processes.

7

The Formation and Propagation of Shock Waves

In this chapter we consider various physical systems in which shock waves arise. These systems can be studied in terms of characteristic curves, on which information from the boundary and initial conditions propagates. However, this approach usually only gives a valid solution for a finite time, after which the solution at some points becomes multi-valued. This difficulty can be dealt with by inserting discontinuities in the solution, which represent shock waves.

7.1 Traffic Waves

Traffic flow modelling has developed rapidly over the last forty years, and sophisticated models are used in the planning of new roads and analysis of existing road networks. The type of model that we will discuss is the simplest possible and was one of the first to be postulated. In spite of this, it manages to capture many of the qualitative and quantitative features of real traffic flows. It is an excellent way of introducing the mathematics and physics of shock waves, and the solutions can be readily interpreted in terms of our everyday experience of road travel.

7.1.1 Derivation of the Governing Equation

We begin by stating our main assumptions.

— **There is only one lane of traffic and no overtaking**. This may seem restrictive, but the inclusion of several lanes with traffic switching between lanes, along with a model for overtaking, is a difficult business. Moreover, the model that we will develop has been shown to be in reasonable agreement with observations, even for multi-lane roads

Fig. 7.1. Bunching in lines of traffic.

(see however Kerner (1999) for an introduction to more complex phenomena on multi-lane roads).

— **We can define a local car density, ρ, as the number of cars per unit length of road**. Formally, we are invoking the continuum approximation, which is the basis of fluid mechanics. In fluid mechanics, we do not want to analyse the motion of every fluid molecule, so, at any point, we take a small but finite averaging volume surrounding the point in question, and define the mean density in this volume to be the effective density at the point. In traffic flow modelling, at any point we take a small but finite length of road and define the car density to be proportional to the number of cars within it. Of course, small in this context means something rather different from what it means in fluid mechanics, but the idea is the same. We end up with a function, $\rho(x, t)$, which is defined for every position x and time t.

— **The local car velocity, $v(x, t)$, is a function of the local car density alone ($v \equiv v(\rho)$)**. The underlying idea is that the velocity at which a car is driven is dependent only on the distance from the car in front, and hence on the local car density. This is perhaps the most unrealistic assumption in the model, and we shall see later that we need to modify it.

Now consider a finite length of road, $x_1 \leq x \leq x_2$. The rate of change of the number of cars in this interval is equal to the flux of cars in at x_1

minus the flux of cars out at x_2, or

$$\frac{\partial}{\partial t}\int_{x_1}^{x_2}\rho(x,t)\,dx = q(x_1,t) - q(x_2,t), \tag{7.1}$$

where $q(x,t)$ is the car flowrate. In terms of ρ and v,

$$q = \rho v(\rho). \tag{7.2}$$

Equation (7.1) is the integral expression for conservation of cars, and must hold for any x_1 and x_2. Notice that there are no x-derivatives, a fact that will prove useful later.

For continuous densities, we can take the limit $x_2 \to x_1$ and obtain the more familiar, differential form for conservation of cars as

$$\frac{\partial \rho}{\partial t} + \frac{\partial q}{\partial x} = 0. \tag{7.3}$$

This is just the one-dimensional version of the generic conservation equation

$$\frac{\partial \mathbf{a}}{\partial t} + \nabla \cdot \mathbf{q} = 0,$$

where **a** is a vector whose components are the densities of all the conserved quantities and **q** their fluxes. Equation (7.3) is the canonical form for **kinematic waves**. These are waves that arise purely because of the need to conserve mass, or here cars. No force balance is involved, which distinguishes kinematic waves from **dynamic waves**.

The next issue to be addressed is the functional form of $v(\rho)$. We assume that:

— There is an upper limit, $\rho_{\max}$, on the possible density of cars, corresponding to bumper-to-bumper traffic, so that $v(\rho_{\max}) = 0$.
— As the car density increases, drivers slow down. We assume that $v(\rho)$ decreases monotonically for $0 \le \rho \le \rho_{\max}$. Note that traffic flows in the positive x-direction.

As an example, consider figure 7.2, which shows car density and flowrate plotted against car velocity for a real road. The data comes from the M25 motorway between junctions 13 and 14, Britain's busiest stretch of road. Each data point represents an average over one minute between 7 a.m. and 7 p.m. on a weekday. Also shown is a straight line fit to the density data, and the corresponding fitted flowrate. The maximum flowrate lies at around 45 miles per hour, considerably slower than the speed limit of 70 miles per hour. The highest flowrate of cars can be achieved at this

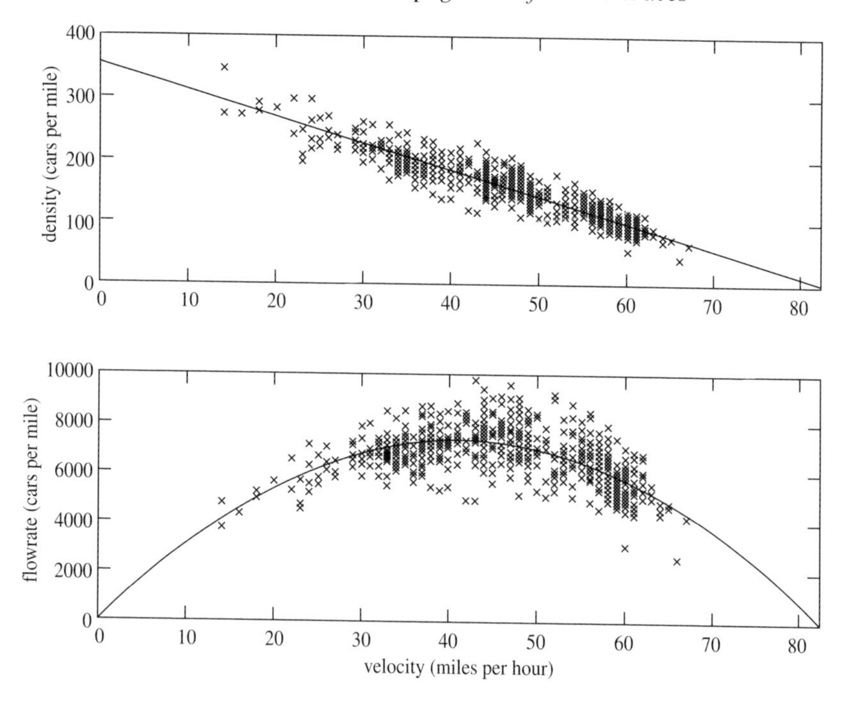

Fig. 7.2. Measured flowrate and density for a stretch of the M25 motorway between 7 a.m. and 7 p.m. on a weekday in 1999. Each data point is an average over one minute.

rather low speed, but with closely packed cars. Indeed, for any density greater than about 200 cars per mile, a higher flowrate can be obtained if everyone slows down and reduces their distance from the car in front.

7.1.2 Small Amplitude Disturbances of a Uniform State

Before studying the full nonlinear problem, it is prudent to consider briefly the linearised model that governs the propagation of small disturbances to a uniform flow of traffic.

By using the chain rule on the x-derivative, we can write (7.3) as

$$\frac{\partial \rho}{\partial t} + c(\rho) \frac{\partial \rho}{\partial x} = 0, \tag{7.4}$$

where

$$c(\rho) = \frac{d}{d\rho}(\rho v) = v(\rho) + \rho v'(\rho) \tag{7.5}$$

is the **kinematic wave speed**. We now look for solutions that are small

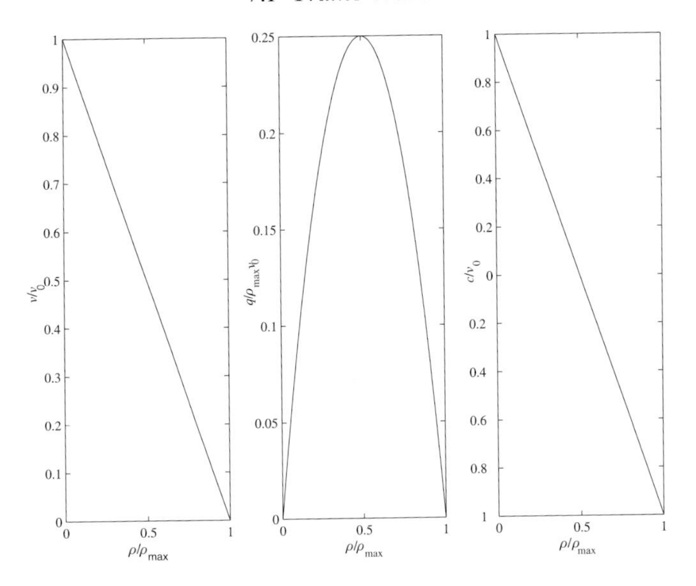

Fig. 7.3. The car velocity, flux function and kinematic wave speed given by (7.14).

amplitude disturbances of a uniform state $\rho = \rho_0$, where ρ_0 is a constant. By writing

$$\rho = \rho_0 + \tilde{\rho}, \quad \text{with } \tilde{\rho} \ll 1, \tag{7.6}$$

and linearising (7.4) we obtain

$$\frac{\partial \tilde{\rho}}{\partial t} + c(\rho_0)\frac{\partial \tilde{\rho}}{\partial x} = 0. \tag{7.7}$$

If we let $\eta = x - c(\rho_0)t$ and look for solutions $\tilde{\rho}(\eta, t)$, we find that (7.7) becomes $\partial\tilde{\rho}/\partial t = 0$, and hence the general solution is

$$\tilde{\rho} = f(x - c(\rho_0)t)$$

for any function, f. This is a **kinematic wave** that propagates in the positive x-direction without change of form at the kinematic wave speed appropriate to the uniform state, $c(\rho_0)$.

If the maximum value of the flux function, $q(\rho)$ is at $\rho = \rho^*$, then $c > 0$ for $\rho < \rho^*$ and $c < 0$ for $\rho > \rho^*$, as illustrated in figure 7.3 for a simple model that we will study later. This means that kinematic waves propagate in the opposite direction to the traffic flow when $\rho > \rho^*$. This demonstrates that what is propagating is not cars, but disturbances in the medium made up of cars; no cars travel backwards when $c < 0$. You may have experienced this phenomenon on a busy road when a sudden

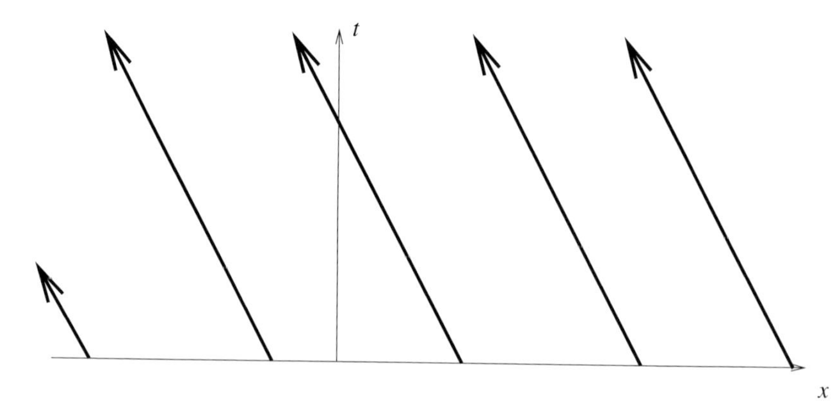

Fig. 7.4. The characteristics for the linearised problem.

increase in car density reaches you from the traffic ahead for no apparent reason.

Kinematic waves occur in many other physical systems. A popular explanation for the dynamics of the spiral arms of galaxies, including our own, is that they are kinematic waves that rotate about the galactic centre, whose underlying medium is interstellar dust and gas. It is not stars that rotate with the arms, but an increased tendency for bright young stars to be born in the high density regions of the wave (see Crosswell (1995)).

Now consider the curves $x = X(t)$ on which ρ is constant. Since the solution propagates at speed $c(\rho_0)$ without change of form, these are just the straight lines

$$x = X(t) = x_0 + c(\rho_0)\,t, \tag{7.8}$$

known as **characteristic curves**, or simply **characteristics**. These are illustrated in figure 7.4 for a case where $c(\rho_0) < 0$. This construction may not seem to be of much use, but we shall see that characteristic curves govern the propagation of the initial car density in the full nonlinear problem.

7.1.3 The Nonlinear Initial Value Problem

We wish to solve

$$\frac{\partial \rho}{\partial t} + c(\rho)\frac{\partial \rho}{\partial x} = 0, \tag{7.9}$$

subject to the initial condition

$$\rho(x, 0) = \rho_0(x). \tag{7.10}$$

Can we construct a set of characteristics, $x = X(t)$, on which ρ is constant, for this nonlinear problem? On these curves

$$\rho\left(X\left(t\right),t\right) = \rho\left(X\left(0\right),0\right) = \rho_0\left(X\left(0\right)\right),$$

and hence

$$\frac{d}{dt}\left\{\rho\left(X\left(t\right),t\right)\right\} = \frac{\partial \rho}{\partial t} + \frac{dX}{dt}\frac{\partial \rho}{\partial x} = 0.$$

If we compare this expression with (7.9), it is clear that we require

$$\frac{dX}{dt} = c\left(\rho\right). \tag{7.11}$$

However, ρ is constant on each characteristic, by definition, so on each characteristic $c(\rho)$ is constant, dX/dt is constant and each characteristic is a straight line given by

$$x = X(t) = x_0 + c\left(\rho_0\left(x_0\right)\right)t \quad \text{for } -\infty < x_0 < \infty. \tag{7.12}$$

The solution is given implicitly by (7.12) and

$$\rho\left(x,t\right) = \rho_0\left(x_0\right). \tag{7.13}$$

This defines the solution, but it is easier to see what is going on by considering a specific problem and examining how the characteristics affect the development of the initial car density profile.

For our example problem, we will use the model

$$v(\rho) = \frac{v_0}{\rho_{\max}}\left(\rho_{\max} - \rho\right). \tag{7.14}$$

This is the simplest possible form for the velocity function, consistent with our earlier assumptions about its behaviour, and, as we have seen in figure 7.2, is in reasonable agreement with real data. In this case,

$$c(\rho) = \frac{v_0}{\rho_{\max}}\left(\rho_{\max} - 2\rho\right), \tag{7.15}$$

and $\rho^* = \frac{1}{2}\rho_{\max}$, as illustrated in figure 7.3. The initial conditions that we will study are

$$\rho_0(x) = \frac{\rho_L + \rho_R e^{x/L}}{1 + e^{x/L}}, \tag{7.16}$$

for positive constants ρ_L, ρ_R and L. We begin by examining what happens when $\rho_L > \rho_R$. Note that $\rho \to \rho_L$ as $x \to -\infty$ and $\rho \to \rho_R$ as $x \to \infty$, with the change between these two states occurring over a distance of $O(L)$, as illustrated in figure 7.5. We study this simple initial condition to

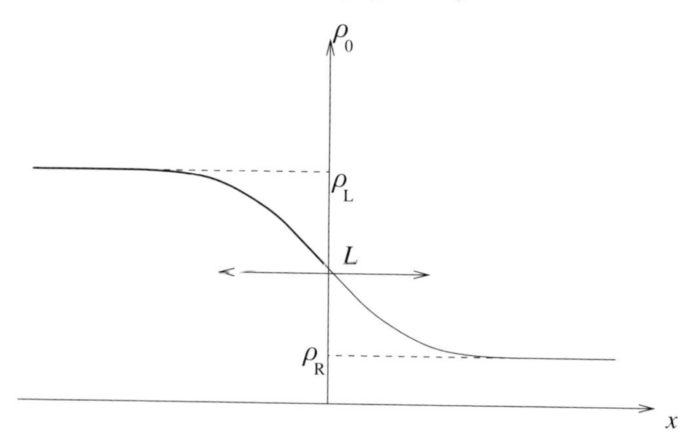

Fig. 7.5. The initial conditions for the initial value problem given by (7.16) with $\rho_L > \rho_R$.

illustrate the fundamental difference between cases where the car density increases with x and those where it decreases.

Since the kinematic wave speed, $c(\rho)$, is a decreasing function of ρ, and the initial conditions have $\rho_0(x)$ a decreasing function of x, $c(\rho_0(x))$ is an increasing function of x, with

$$c(\rho_0(x)) = \frac{v_0}{\rho_{max}}\left\{\frac{\rho_{max} - 2\rho_L + (\rho_{max} - 2\rho_R)\, e^{x/L}}{1 + e^{x/L}}\right\}, \tag{7.17}$$

as illustrated in figure 7.6. This means that the dX/dt increases as x_0 increases, and hence the characteristics are as illustrated in figure 7.7. There is a unique characteristic through every point in the domain of solution. Qualitatively, the spreading out of the characteristics leads to a spreading out of the initial density profile, as shown in figure 7.8. Note that the solution is sketched in a frame of reference moving to the right at speed $c(\rho_0(0))$, so that the density at the point $x - c(\rho_0(0))t = 0$ remains constant. Each point on the initial profile is shifted to the right by a distance $c(\rho_0(x_0))\,t$. Remember, this does not mean that cars are actually moving with speed $c(\rho_0(x_0))$, simply that a disturbance propagates at this speed. Cars accelerate out of the high density region into the low density region, and the deficit in cars travels backwards. Cars are the molecules that make up the medium whose properties we are studying.

Now consider the initial conditions (7.16) in the limit $L \to 0$, so that

$$\rho_0(x) = \left\{\begin{array}{ll} \rho_L & \text{for } x < 0, \\ \rho_R & \text{for } x > 0. \end{array}\right\} \tag{7.18}$$

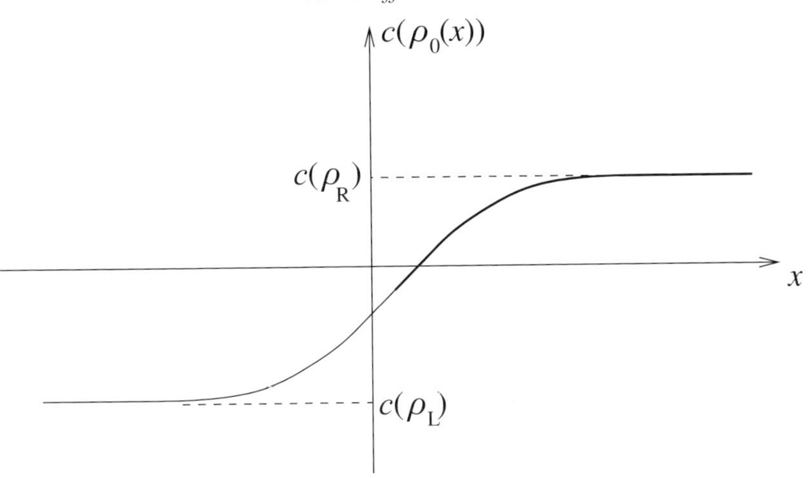

Fig. 7.6. The initial kinematic wave speed for the initial value problem given by (7.16) with $\rho_L > \rho_R$.

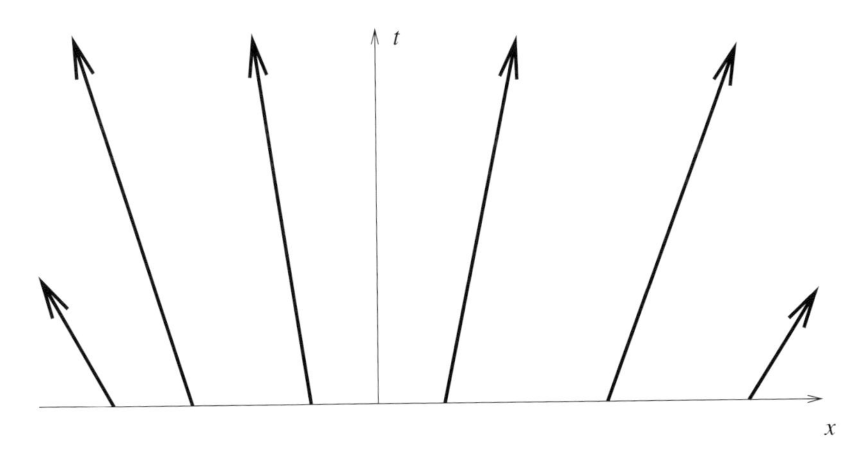

Fig. 7.7. The characteristics for the initial value problem given by (7.16) with $\rho_L > \rho_R$.

The initial value problem given by (7.9) and (7.18) is known as the **Riemann problem**, and is of fundamental importance both for understanding the behaviour of this type of system and for constructing numerical solution schemes (see LeVeque (1990)). The characteristics are given by

$$\left.\begin{aligned} x &= x_0 + c(\rho_L)\,t, \quad \text{for } x_0 < 0, \\ x &= x_0 + c(\rho_R)\,t, \quad \text{for } x_0 > 0. \end{aligned}\right\} \tag{7.19}$$

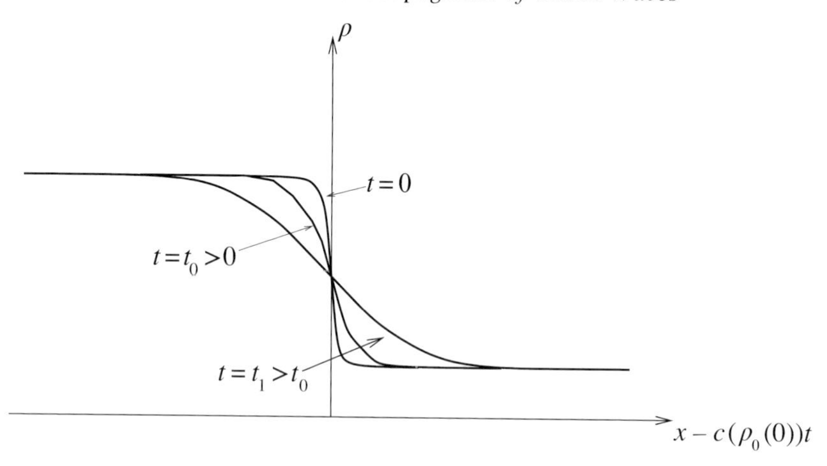

Fig. 7.8. The development of the car density for the initial value problem given by (7.16) with $\rho_L > \rho_R$.

For characteristics that begin at the origin, $x_0 = 0$ and the ratio x_0/L is indeterminate as $L \to 0$. To deal with this we let $x_0 = k_0 L$, where k_0 is a constant. Now $x_0/L = k_0$ for all values of L, and a characteristic beginning at the origin is given by

$$x = \frac{v_0}{\rho_{\max}} \left\{ \frac{\rho_{\max} - 2\rho_L + (\rho_{\max} - 2\rho_R)\, e^{k_0}}{1 + e^{k_0}} \right\} t, \tag{7.20}$$

for any value of k_0. As k_0 varies from $-\infty$ to ∞, this describes a family of straight lines through the origin with $c\,(\rho_L) < x/t < c\,(\rho_R)$. A family of characteristics emanating from a single point in this way is known as an **expansion fan**, **expansion wave** or **rarefaction wave**. We can now sketch all of the characteristics in figure 7.9.

To the left of the characteristic $x = c(\rho_L)t$ there is a uniform density $\rho = \rho_L$, whilst to the right of $x = c(\rho_R)t$ there is a uniform density $\rho = \rho_R$. Between these two characteristics the expansion fan solution is given by the characteristic equation itself, $c\,(\rho) = x/t$. To summarise, the Riemann problem with $\rho_L > \rho_R$ has solution

$$\rho\,(x,t) = \left\{ \begin{array}{cc} \rho_L & \text{for } x \le c\,(\rho_L)\,t, \\ \frac{1}{2}\rho_{\max}\left(1 - x/v_0 t\right) & \text{for } c(\rho_L)t \le x \le c(\rho_R)t, \\ \rho_R & \text{for } x \ge c\,(\rho_R)\,t, \end{array} \right\} \tag{7.21}$$

as shown in figure 7.10. When $\rho_L = \rho_{\max}$ and $\rho_R = 0$, we have the situation that occurs when a traffic light goes green. From an initially stationary line of bumper-to-bumper traffic, the front cars accelerate off

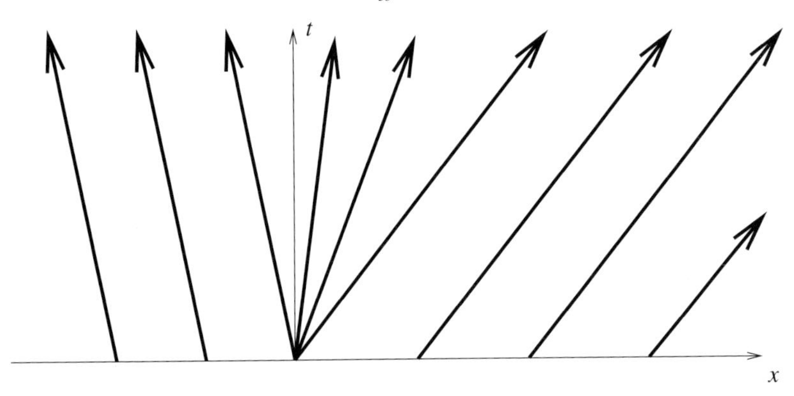

Fig. 7.9. The characteristics for the Riemann problem with $\rho_L > \rho_R$.

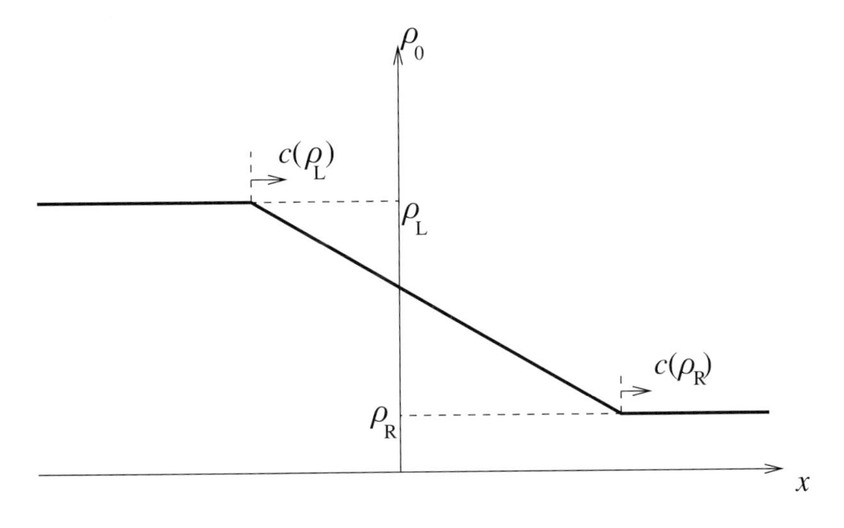

Fig. 7.10. The solution of the Riemann problem with $\rho_L > \rho_R$.

into the clear stretch of road ahead, and the front of the queue spreads out with the effect of the green light propagating back through the traffic at a constant velocity $c(\rho_{max}) = -v_0$.

At this point, it is worth noting the self-similarity of this solution. The only quantities involved in the Riemann problem are ρ, ρ_L, ρ_R, ρ_{max}, v_0, x and t. There is no geometrical length scale in the problem, so we expect the solution to be a function of some dimensionless combination of the above quantities. The only such combinations are ratios of two densities, and the quantity x/v_0t. This leads us to suspect that the appropriate

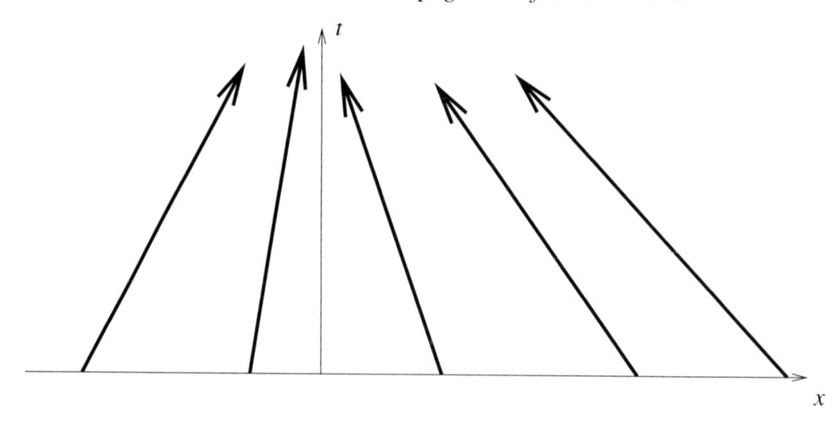

Fig. 7.11. The characteristics for the initial value problem given by (7.16), $\rho_R > \rho_L$.

solution is of the form $\rho = \rho_1 f\left(x/v_0 t\right)$ for some function f and density ρ_1. The solution (7.21) is clearly of this form. In addition, if we substitute this form into (7.9) we find that

$$\left\{\frac{x}{t} - c(\rho)\right\} f'\left(\frac{x}{v_0 t}\right) = 0. \tag{7.22}$$

This means that the solution must consist of spatially uniform sections and expansion fans.

We now consider the apparently equally straightforward case, $\rho_R > \rho_L$. By the converse of the arguments presented above, the kinematic wave speed $c(\rho_0(x))$ is now a decreasing function of x, and hence the characteristics are as shown in figure 7.11. It is clear from the figure that characteristics must eventually intersect. By definition, ρ is constant on each characteristic, so if two intersect, which value of ρ are we to take as the solution? As it stands, our initial value problem becomes ill-posed as soon as any characteristics meet, and we must consider what is missing from our simple model. Before turning to this question we can calculate under what circumstances characteristics will intersect, and where and when this happens.

Consider two characteristics that intersect. Any characteristic that begins at some point between these intersecting characteristics must meet one of them at an earlier time, as shown in figure 7.12, ignoring the unlikely case when *all* the intermediate characteristics intersect at a single point. This means that the earliest intersection must be between neighbouring characteristics. Consider two characteristics, $x = X_1(t)$ and

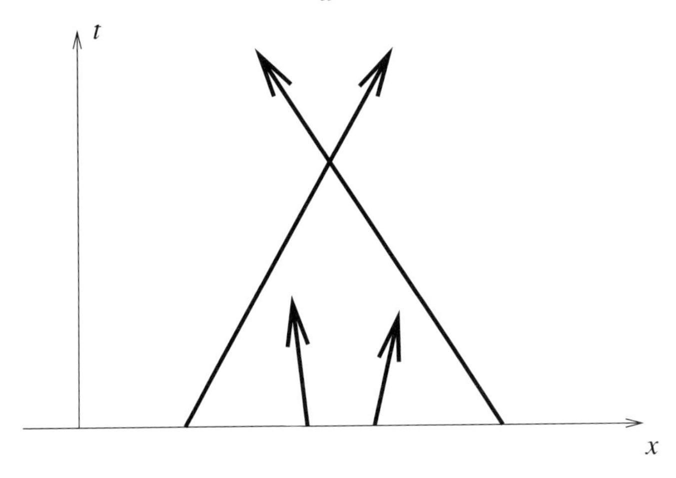

Fig. 7.12. If two characteristics intersect, any enclosed characteristic must meet one of them at an earlier time.

$x = X_2(t)$ given by

$$X_1(t) = x_0 + c(x_0)t,$$
$$X_2(t) = x_0 + \delta x + c(x_0 + \delta x)t.$$

For notational convenience we write $c(x)$ for $c\,(\rho_0\,(x))$ here. As $\delta x \to 0$ we obtain neighbouring characteristics, and

$$X_2(t) \approx x_0 + \delta x + c(x_0)t + \delta x c'(x_0)t,$$

where $c'(x) = dc(\rho_0(x))/dx$. If these characteristics intersect at time $t = T$, $X_1(T) = X_2(T)$ and hence

$$T = -\frac{1}{c'(x_0)}.$$

There are two points to be made here:

— If $c'(x_0) > 0$ for $-\infty < x_0 < \infty$, the characteristics never intersect for $t > 0$, and the solution constructed using characteristics is valid everywhere. This is what we found when $\rho_L > \rho_R$.

— If $c'(x_0) < 0$ at *any* point x_0, a pair of characteristics will intersect and the solution as constructed becomes ill-defined. This first occurs when $t = T_{\min}$, where

$$T_{\min} = \min_{-\infty < x_0 < \infty} \left\{ -\frac{1}{c'(x_0)} \right\}, \tag{7.23}$$

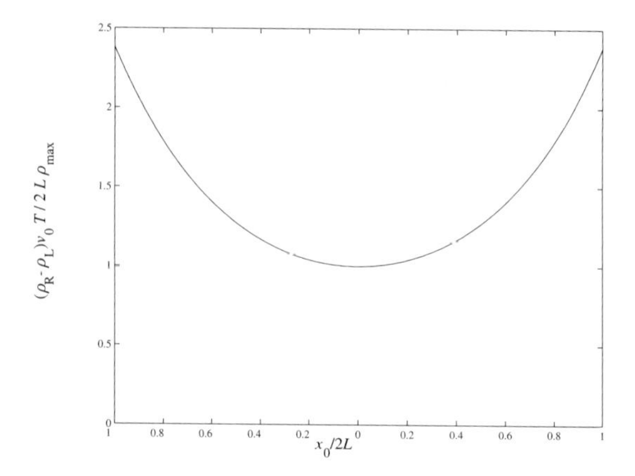

Fig. 7.13. The time, T at which neighbouring characteristics meet for the initial value problem given by (7.16) with $\rho_R > \rho_L$.

in other words, where the initial slope of the car density is most negative. The solution constructed using characteristics is, therefore, valid for $0 \leq t \leq T_{min}$.

For our particular example, we calculate that

$$T = \frac{2L\rho_{max}}{(\rho_R - \rho_L)\,v_0}\cosh^2\left(\frac{x_0}{2L}\right), \tag{7.24}$$

which is sketched in figure 7.13, and hence

$$T_{min} = \frac{2L\rho_{max}}{(\rho_R - \rho_L)\,v_0}, \tag{7.25}$$

with the first intersection of characteristics occurring at $x_0 = 0$. As $t \to T_{min}$ the density profile steepens until, at $t = T_{min}$, an infinite slope develops at $x = c(\rho_0(0))T_{min}$, as shown in figure 7.14. As we have seen, the solution is not valid for $t > T_{min}$, but we can still sketch what we obtain using the characteristics to construct the solution.

Figure 7.15 shows that the profile overturns and becomes multi-valued. Within this region, there are three characteristics through each point, and hence three possible values of ρ. At the boundaries of this multi-valued region, two neighbouring characteristics meet, and hence the location of the multi-valued region for our example problem can be deduced from (7.24). A typical example is sketched in figure 7.13. The appearance of an infinite slope at $x = 0$ when $t = T_{min}$ suggests that a **shock wave** is formed. Mathematically, a shock wave is a discontinuity in one or more

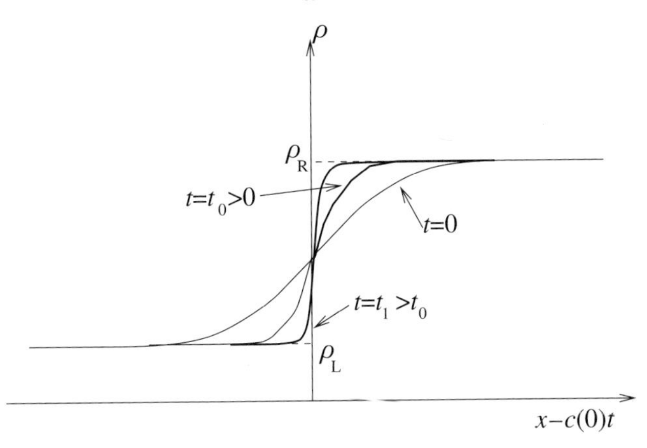

Fig. 7.14. The solution of the initial value problem given by (7.16) with $\rho_R > \rho_L$ for $t \leq T_{\min}$.

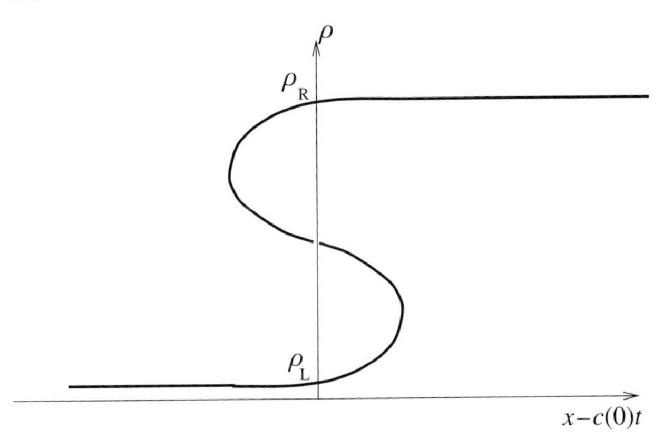

Fig. 7.15. The multi-valued solution of the initial value problem given by (7.16) with $\rho_R > \rho_L$ for $t > T_{\min}$.

of the dependent variables (here there is only one dependent variable). Physically, a shock wave is a thin surface across which one or more of the physical properties changes rapidly and some physical effect, often viscous dissipation, cannot be neglected as it can in the rest of the domain of solution. Before considering what effects we have neglected in our traffic flow model, and whether their inclusion allows us to show that a shock wave actually exists, we can consider where such a shock wave might be located.

Where should we insert a shock wave into the profile shown in figure 7.15? Since our governing equation is an expression of the fact that

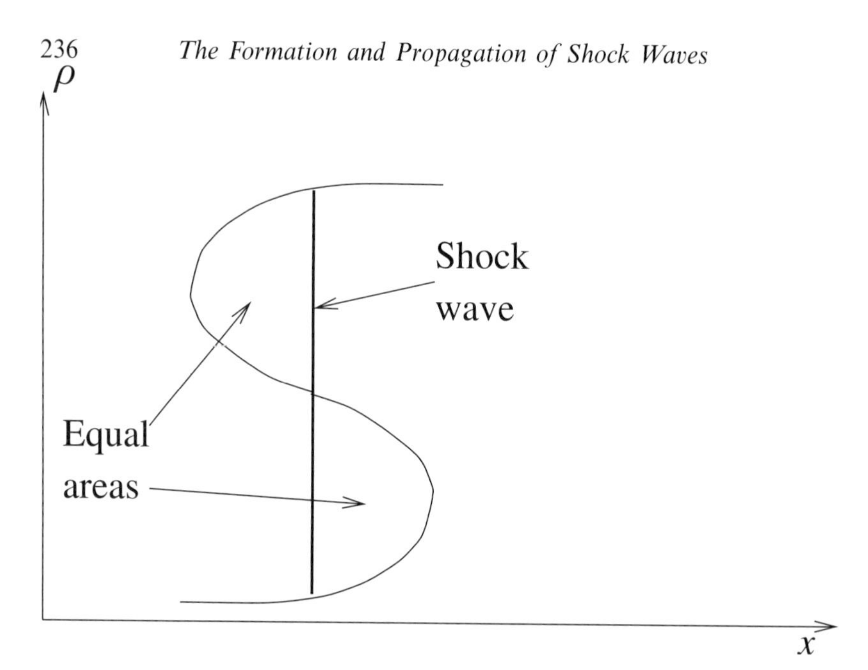

Fig. 7.16. The equal area rule.

the number of cars is preserved, we should insert a discontinuity so as to cut off equal areas in the profile, as shown in figure 7.16. This is known as the **equal area rule**, and using it the number of cars is conserved. Note that this solution satisfies the integral form of car conservation, (7.1). In this form, there are no x-derivatives, and the discontinuity does not cause us a problem. The characteristics and the position of the shock are illustrated in figure 7.17. This procedure is known as **shock fitting**. Note that characteristics enter the shock and then play no further part in the construction of the solution. Consider the case $\rho_R = \rho_{max}$, $\rho_L < \rho_{max}$. This is what happens as cars approach a stationary queue behind a red traffic light. On meeting the queue, cars slow down to a stop, and the lengthening of the queue is achieved through a shock wave propagating backwards.

In section 10.1, we will show how to justify this procedure by introducing some extra physics into the problem, specifically, the reasonable notion that (most) drivers actually look a little further ahead than the bumper of the car in front.

7.1.4 The Speed of the Shock

Now that we have found that shock waves can form, how fast do they move? Consider a shock whose position is given by $x = s(t)$, $\rho = \rho^-$ at

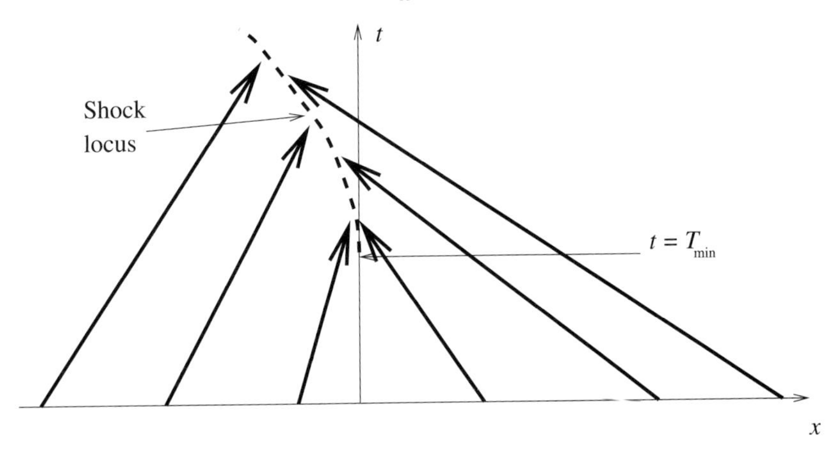

Fig. 7.17. The characteristics and shock locus for the initial value problem given by (7.16) with $\rho_R > \rho_L$.

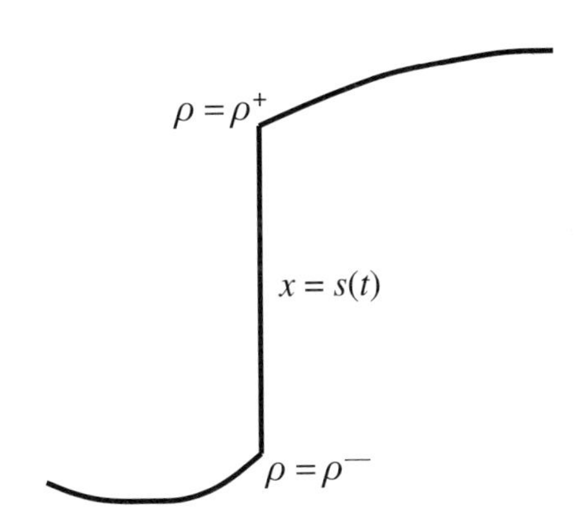

Fig. 7.18. A shock wave.

$x = s^-$, and $\rho = \rho^+$ at $x = s^+$, as sketched in figure 7.18. Conservation of cars in integral form, (7.1), gives

$$\frac{\partial}{\partial t}\left(\int_{x_1}^{s(t)} + \int_{s(t)}^{x_2}\right)\rho(x,t)dx = q(x_1,t) - q(x_2,t),$$

and hence

$$\int_{x_1}^{x_2} \frac{\partial \rho}{\partial t}dx + \frac{ds}{dt}\rho^- - \frac{ds}{dt}\rho^+ = q(x_1,t) - q(x_2,t). \tag{7.26}$$

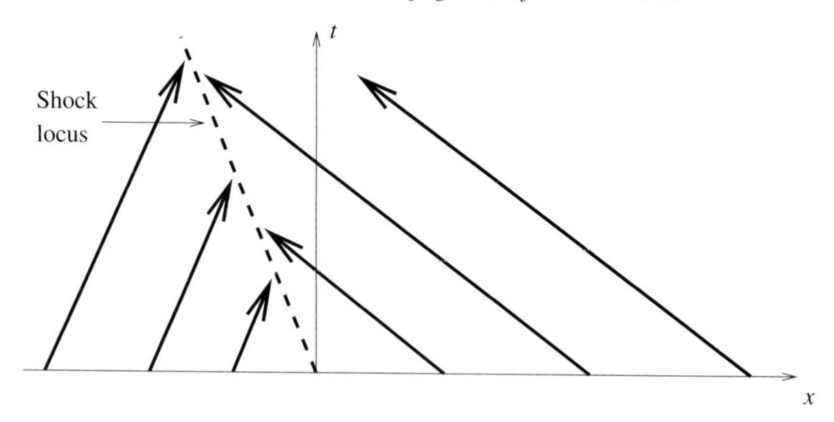

Fig. 7.19. The characteristics and shock locus for the Riemann problem with $\rho_L < \rho_R$.

If we now let $x_2 \to s(t)$ and $x_1 \to s(t)$, the integral vanishes and we are left with

$$q\left(\rho^{-}\right) - \rho^{-}\frac{ds}{dt} = q\left(\rho^{+}\right) - \rho^{+}\frac{ds}{dt},$$

and hence the shock speed is given by

$$\frac{ds}{dt} = \frac{q\left(\rho^{-}\right) - q\left(\rho^{+}\right)}{\rho^{-} - \rho^{+}}. \tag{7.27}$$

For the simple case, with $q(\rho) = \rho v(\rho) = \rho v_0(\rho_{\max} - \rho)/\rho_{\max}$, we find that $ds/dt = (v(\rho^{-}) + v(\rho^{+}))/2$. As we shall see in section 7.2, this procedure can be generalised to systems of conservation laws.

We can now return, after our lengthy diversion, to our example initial value problem with $\rho_L < \rho_R$. What happens as $L \to 0$? In other words, what is the solution of the Riemann problem when $\rho_L < \rho_R$? Equation (7.23) shows that $T_{\min} \to 0$ as $L \to 0$, so in this limit a shock forms immediately. The initial conditions take the form of a shock wave, and this persists for all time, rather than opening out into an expansion wave as was the case for $\rho_L > \rho_R$. The constant speed of the shock can be calculated from (7.27). The characteristics are illustrated in figure 7.19. Note that all of the characteristics terminate in the shock. Looking back to the case $\rho_L > \rho_R$, we found that the characteristics for the appropriate solution were as shown in figure 7.9. However, we can also construct a shock wave solution of this Riemann problem, with characteristics as shown in figure 7.20. We can exclude this solution by considering the possible travelling wave solutions with $\nu > 0$ (see section 10.3). Another

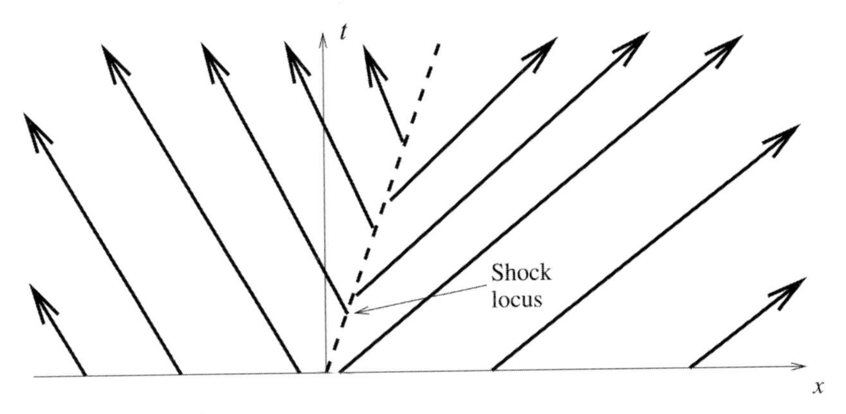

Fig. 7.20. An unphysical shock wave solution of the Riemann problem with $\rho_L > \rho_R$.

way of excluding this solution is by noting that it has characteristics that originate at the shock. This is unphysical, since the solution should depend upon the initial conditions, not on conditions at the shock.

7.2 Compressible Gas Dynamics

In chapter 3 we studied sound waves, which are small amplitude disturbances of a stationary body of compressible gas. In this section we study the dynamics of a compressible gas when the amplitude of the disturbances is not small. In particular, we will find that shock waves can form, and study their properties. Figure 7.21 shows a shadowgraph of the shock waves generated by a fast moving projectile. The pressure, density, velocity and local sound speed in the gas can all change discontinuously across shock waves. An essential preliminary to understanding this is a brief study of the thermodynamics of ideal gases.

7.2.1 Some Essential Thermodynamics

In classical kinetic theory, the **theorem of equipartition of energy** tells us that there is an average internal energy $\frac{1}{2}kT$ associated with every degree of freedom of the molecules in an ideal gas, for which there is no intermolecular attraction, where k is **Boltzmann's constant** and T is the absolute temperature. For an atom in translational motion there are three degrees of freedom, so its internal energy is $\frac{3}{2}kT$. For one mole of these atoms, the energy is $\frac{3}{2}kN_A T$, where N_A is **Avogadro's number**.

Fig. 7.21. The shock waves generated by a projectile at Mach number $M \approx 1.7$.

The **universal gas constant** is $R = kN_A \approx 8.3\,\mathrm{J\,mol^{-1}\,K^{-1}}$, so the internal energy of a mole of atoms is $E = \frac{3}{2}RT$. For a mole of molecules composed of two atoms, in addition to the translational degrees of freedom there are two further rotational degrees of freedom, so that the total internal energy is $E = \frac{5}{2}RT$. This is a reasonable approximation for air, which consists mainly of the diatomic gases nitrogen and oxygen. Note that E is directly proportional to the absolute temperature in an ideal gas.

Let's now consider the changes in temperature and pressure that occur when a gas is heated and its volume allowed to vary. If an amount of heat Q is absorbed (or given up) by a volume of gas, and the work done in any change of volume is W, then the change in internal energy (which we consider to be reversible) is given by the **first law of thermodynamics**,

$$dE = Q - W. \tag{7.28}$$

If we neglect viscous and magnetic effects, we have $W = pdV$, so that $dE = Q - pdV$.

We now introduce the idea of the **entropy** of a gas, S, as the amount of thermal energy that is unavailable for conversion into mechanical energy. This leads us to regard entropy as a measure of the randomness or disorder in the gas. In particular, if an amount of heat Q is absorbed at an absolute temperature T, then the entropy is increased by an amount $dS = Q/T$, and we can state the **second law of thermodynamics** as

$$dE = T\,dS - p\,dV. \tag{7.29}$$

If we now regard E as a function of S and V,

$$dE = \left(\frac{\partial E}{\partial S}\right)_V dS + \left(\frac{\partial E}{\partial V}\right)_S dV,$$

so that we can formally define

$$T = \left(\frac{\partial E}{\partial S}\right)_V, \quad p = -\left(\frac{\partial E}{\partial V}\right)_S.$$

In these equations, the subscript indicates which variable is to be held constant during the differentiation.

There are two further energies that will be of use to us here: the **enthalpy**, $H = E + pV$, and the **free energy**, $F = E - TS$. These are measures of how much energy the gas has available to exchange with its surroundings. Now $dH = dE + p\,dV + V\,dp$, which, using the second law of thermodynamics, (7.29), can be written as

$$dH = T\,dS + V\,dp.$$

Regarding H as a function of S and p, we have

$$T = \left(\frac{\partial H}{\partial S}\right)_p, \quad V = \left(\frac{\partial H}{\partial p}\right)_S.$$

Assuming smoothness of the second partial derivatives of the enthalpy gives

$$\left(\frac{\partial T}{\partial p}\right)_S = \left(\frac{\partial V}{\partial S}\right)_p. \tag{7.30}$$

A similar calculation for the free energy gives

$$\left(\frac{\partial p}{\partial T}\right)_V = \left(\frac{\partial S}{\partial V}\right)_T, \tag{7.31}$$

a result that we will make use of shortly.

If the gas absorbs an amount of heat Q and its temperature rises by dT, we can define the **specific heat** c by the relation $Q = c\,dT$. The first law of thermodynamics states that $dE = cdT - pdV$. If this absorption of heat takes place at constant volume, we have $c = c_V$, the **specific heat at constant volume**, and $dE = c_V dT$, $c_V = (\partial E/\partial T)_V$. As we have seen, for a diatomic ideal gas $c_V = \frac{5}{2}R$. If the absorption of heat takes place at constant pressure, $c = c_p$, the **specific heat at constant pressure**, and $dp = 0$, so we have $c_p dT = dE + pdV$, so that

$$c_p = \left(\frac{\partial E}{\partial T}\right)_p + p\left(\frac{\partial V}{\partial T}\right)_p = \left[\frac{\partial}{\partial T}(E + pV)\right]_p = \left(\frac{\partial H}{\partial T}\right)_p,$$

which we can use to derive a relationship between c_p and c_V. If we regard E as a function of $V(p, T)$ and T rather than V and S, then

$$\begin{aligned} c_p = \left(\frac{\partial H}{\partial T}\right)_p &= \frac{\partial}{\partial T}\{E(V(p,T),T) + pV\} \\ &= \left(\frac{\partial E}{\partial V}\right)_T\left(\frac{\partial V}{\partial T}\right)_p + \left(\frac{\partial E}{\partial T}\right)_V + p\left(\frac{\partial V}{\partial T}\right)_p. \end{aligned}$$

This means that

$$c_p = c_V + \left\{\left(\frac{\partial E}{\partial V}\right)_T + p\right\}\left(\frac{\partial V}{\partial T}\right)_p.$$

Now, from the second law of thermodynamics, (7.29),

$$\left(\frac{\partial E}{\partial V}\right)_T = T\left(\frac{\partial S}{\partial V}\right)_T - p,$$

so that

$$c_p = c_V + T\left(\frac{\partial S}{\partial V}\right)_T\left(\frac{\partial V}{\partial T}\right)_p,$$

and from (7.31)

$$c_p = c_V + T\left(\frac{\partial p}{\partial T}\right)_V\left(\frac{\partial V}{\partial T}\right)_p.$$

For one mole of an ideal gas $pV = RT$. Evaluating the partial derivatives

$$\left(\frac{\partial p}{\partial T}\right)_V = \frac{R}{V}, \quad \left(\frac{\partial V}{\partial T}\right)_V = \frac{R}{p},$$

then gives $c_p = c_V + R$. For a diatomic ideal gas, $c_p = \frac{7}{2}R$.

From the second law of thermodynamics, (7.29),

$$dS = c_V\frac{dT}{T} + p\frac{dV}{T} = c_V\frac{dT}{T} + R\frac{dV}{V},$$

and hence

$$\frac{dS}{c_V} = \frac{dT}{T} + \left(\frac{c_p}{c_V} - 1\right)\frac{dV}{V}.$$

If we define $\gamma = c_p/c_V$, the **ratio of specific heats** of the gas, then this relationship is easily integrated to give $TV^{\gamma-1} = Ae^{S/c_V}$ for some constant A. Using the ideal gas law, this gives

$$\frac{S}{c_V} = \log\left(\frac{p}{\rho^\gamma}\right) + \text{constant}, \tag{7.32}$$

and hence

$$p = \kappa e^{S/c_V}\rho^\gamma \tag{7.33}$$

for some constant, κ. For a diatomic ideal gas, $\gamma = 1.4$. This is the equation we used in section 3.1 when we determined the speed of sound in a gas. As we will show below, the entropy of a gas satisfies a very simple equation, and the thermodynamics that we have studied in this section is enough to allow us to investigate the properties of shock waves. It is now clear that the internal energy E of a unit mass of gas can be written as

$$E = c_V T = \frac{p}{(\gamma-1)\rho}. \tag{7.34}$$

7.2.2 Equations of Motion

We will now assume that the gas is inviscid and that the effect of gravity is negligible, as we did in section 3.1, but we will not assume that the motion of the gas is a small disturbance to a stationary body of gas. It is most convenient to write the equations for the conservation of mass and momentum in the form

$$\frac{D\rho}{Dt} + \rho\nabla\cdot\mathbf{u} = 0, \tag{7.35}$$

$$\frac{D\mathbf{u}}{Dt} + \frac{1}{\rho}\nabla p = 0. \tag{7.36}$$

We now also need the equation for conservation of energy. The energy per unit mass of the gas consists of the internal energy, E, and the kinetic energy, $\frac{1}{2}|\mathbf{u}|^2$. The equation for conservation of energy relates the flux of this energy to the rate of working of the pressure forces and is given by

$$\frac{\partial}{\partial t}\left(\frac{1}{2}\rho|\mathbf{u}|^2 + \rho E\right) + \nabla\cdot\left\{\left(\frac{1}{2}\rho|\mathbf{u}|^2 + \rho E + p\right)\mathbf{u}\right\} = 0. \tag{7.37}$$

This can be rearranged to give

$$\frac{D}{Dt}\left(\frac{1}{2}\rho|\mathbf{u}|^2+\rho E\right)+\left(\frac{1}{2}\rho|\mathbf{u}|^2+\rho E\right)\nabla\cdot\mathbf{u}+\nabla\cdot(p\mathbf{u})=0,$$

and hence

$$\frac{1}{2}|\mathbf{u}|^2\frac{D\rho}{Dt}+\rho\mathbf{u}\cdot\frac{D\mathbf{u}}{Dt}+E\frac{D\rho}{Dt}+\rho\frac{DE}{Dt}+\left(\frac{1}{2}\rho|\mathbf{u}|^2+\rho E\right)\nabla\cdot\mathbf{u}+p\nabla\cdot\mathbf{u}+\mathbf{u}\cdot\nabla p=0.$$

If we now eliminate $\nabla\cdot\mathbf{u}$ and $D\mathbf{u}/Dt$ using (7.35) and (7.36), we obtain

$$\frac{DE}{Dt}-\frac{p}{\rho^2}\frac{D\rho}{Dt}=0.$$

Since we know that

$$T\,dS=dE-pdV=dE-\frac{p}{\rho^2}\,d\rho,$$

we finally arrive at

$$\frac{DS}{Dt}=0.$$

This is the simplest possible way of expressing conservation of energy in an ideal, inviscid flow, and states that entropy is advected with the flow, and hence is constant on streamlines. Physically, this result comes directly from the notion of an ideal gas, where the molecular diffusivity is zero. Consequently, no heat can be transferred between fluid particles, and the entropy must be in thermodynamic equilibrium. This type of flow is said to be **isentropic**. If the entropy is spatially uniform, the flow is said to be **homentropic**. In particular, in a homentropic flow with no shock waves, $S=S_0$, then $p=k\rho^{\gamma}$, where $k=\kappa e^{S_0/c_v}$ is a constant. This is the relationship between pressure and density that we used in chapter 3. Equations (7.35) and (7.36) become

$$\frac{\partial\rho}{\partial t}+\nabla\cdot(\rho\mathbf{u})=0, \tag{7.38}$$

$$\frac{\partial\mathbf{u}}{\partial t}+\mathbf{u}\cdot\nabla\cdot\mathbf{u}+k\gamma\rho^{\gamma-2}\nabla\rho=0. \tag{7.39}$$

For one-dimensional flows these are

$$\frac{\partial\rho}{\partial t}+\frac{\partial(\rho u)}{\partial x}=0, \tag{7.40}$$

$$\frac{\partial u}{\partial t}+u\frac{\partial u}{\partial x}+k\gamma\rho^{\gamma-2}\frac{\partial\rho}{\partial x}=0. \tag{7.41}$$

7.2.3 Construction of the Characteristic Curves

Can we construct characteristic curves for the system (7.40) and (7.41)? The easiest way to answer this question is to consider the system in terms of u and the local sound speed,

$$c = \left(\frac{\partial p}{\partial \rho}\right)_S^{1/2} = \sqrt{\gamma k \rho^{\gamma-1}}.$$

In terms of c, (7.40) and (7.41) become

$$2\frac{\partial c}{\partial t} + 2u\frac{\partial c}{\partial x} + (\gamma - 1)c\frac{\partial u}{\partial x} = 0, \tag{7.42}$$

$$\frac{\partial u}{\partial t} + u\frac{\partial u}{\partial x} + \frac{2c}{\gamma - 1}\frac{\partial c}{\partial x} = 0. \tag{7.43}$$

If we now add or subtract appropriate multiples of these equations we obtain

$$\frac{\partial}{\partial t}\left(u \pm \frac{2c}{\gamma - 1}\right) + (u \pm c)\frac{\partial}{\partial x}\left(u \pm \frac{2c}{\gamma - 1}\right) = 0. \tag{7.44}$$

By analogy with our analysis of the equations for one-dimensional traffic flow, we can see that the functions $R_\pm = u \pm 2c/(\gamma - 1)$ are constant on the two sets of characteristic curves, $X_\pm(t)$, where

$$\frac{dX_\pm}{dt} = u \pm c. \tag{7.45}$$

Note that these are not necessarily straight lines. The functions $R_\pm(u, c)$ are called the **Riemann invariants** of the system. To summarise:

- On the C_+ characteristics, given by $dX_+/dt = u + c$, the C_+ invariant, $R_+ = u + 2c/(\gamma - 1)$, is constant.
- On the C_- characteristics, given by $dX_-/dt = u - c$, the C_- invariant, $R_- = u - 2c/(\gamma - 1)$, is constant.

For the solution of a given initial value problem to be well-defined, a single C_+ characteristic and a single C_- characteristic must pass through each point in the domain of solution. The values of u and ρ at each point can then be determined from the initial values of $R_\pm$ on each characteristic.

We have already seen in section 7.1 how shock waves may develop in this type of system, but that not every mathematically plausible shock solution is physically correct. How can we extend these ideas to shock waves in ideal gases? One approach is to study the effect of viscosity and heat conduction in the neighbourhood of a shock. We will do this in section 10.2 by using the method of multiple scales to determine the behaviour of a small disturbance of a uniform state. For general disturbances, it can be shown that characteristics must enter the shock locus, as we found in section 7.1 for shocks in traffic. An alternative but equivalent constraint is that *the entropy of a fluid particle must increase as it passes through a shock.* Outside the shock, we have seen that $DS/Dt = 0$, but the assumptions involved in the derivation of this equation break down at the shock. In particular, molecular diffusivities cannot be neglected at the shock, and entropy is generated there.

Example: The Generation of a Shock by a Uniformly Accelerating Piston. Consider an ideal gas confined in a long, straight cylinder by a tightly fitting piston, initially at rest at $x = 0$. In chapter 3, we showed how small amplitude oscillations of such a piston generate sound waves that propagate along the tube. We now consider the case where the gas and piston are initially at rest, and the piston moves into the gas with uniform acceleration a. The piston therefore lies at $x = \frac{1}{2}at^2$, and the gas in the region $x \geq \frac{1}{2}at^2$, initially with sound speed c_0, as shown in figure 7.22. We can anticipate some of what will happen here using linear theory. Combining $c_0^2 = \gamma/\rho$, and the equation of state $pV = RT$ we have c_0 proportional to $\sqrt{T}$. When the piston starts to compress the gas the temperature will rise. As this process continues the wave speed c_0 will rise and waves produced at later times will catch up with those produced earlier giving a multivaluedness to the solution. Let us now analyse this more quantitatively. We assume that the C_- characteristics that originate in the gas when $t = 0$ fill the domain. On these characteristics,

$$R_- = u - \frac{2c}{\gamma - 1} = -\frac{2c_0}{\gamma - 1},$$

and hence

$$c = c_0 + \frac{1}{2}(\gamma - 1)u,$$

throughout the gas. This also shows that all of the C_+ characteristics are straight lines, since on these, R_+, and hence u, c and dX_+/dt, are each constant. In particular, the C_+ characteristic that originates at the piston when $t = t_0$ satisfies

$$\frac{dX_+}{dt} = u + c = c_0 + \frac{1}{2}(\gamma + 1)at_0,$$

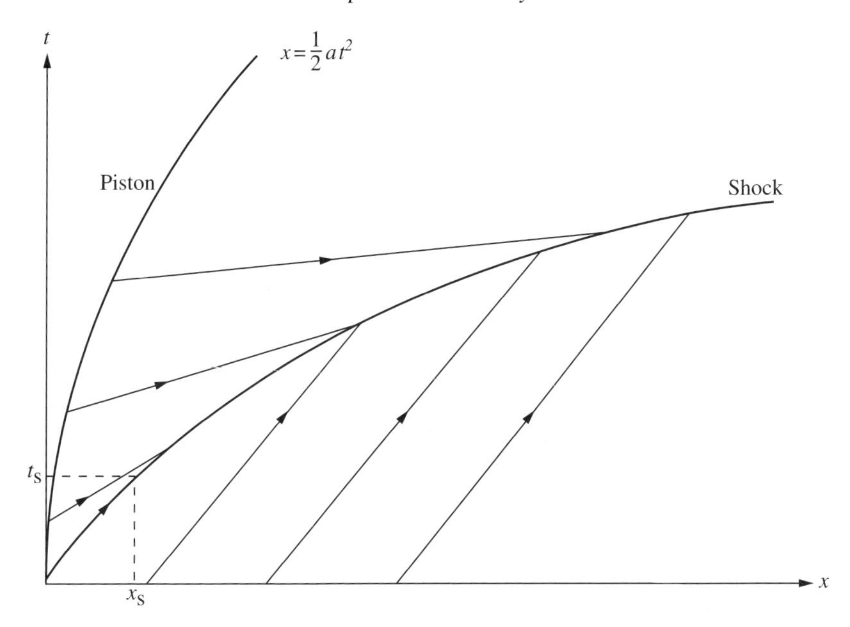

Fig. 7.22. The characteristics and shock path for the uniformly accelerating piston.

and hence

$$X_+(t;t_0) = \frac{1}{2}at_0^2 + \left\{c_0 + \frac{1}{2}(\gamma+1)at_0\right\}(t - t_0).$$

The slopes of these straight lines increase with t_0, and we therefore expect them to intersect at some finite time t_s, when a shock forms, as shown in figure 7.22. As usual, we expect this to occur on neighbouring characteristics. For $\delta t_0 \ll 1$,

$$X_+(t, t_0 + \delta t_0) \approx X_+(t, t_0) + \delta t_0 \frac{\partial X_+}{\partial t_0}(t, t_0).$$

Neighbouring characteristics therefore intersect when $\partial X_+/\partial t_0 = 0$, and hence

$$t = \frac{2c_0}{a(\gamma+1)} + \frac{2\gamma}{\gamma+1}t_0.$$

This first occurs on the characteristic for which $t_0 = 0$, when

$$t = t_s = \frac{2c_0}{a(\gamma+1)}, \quad x = x_s = \frac{2c_0^2}{a(\gamma+1)}.$$

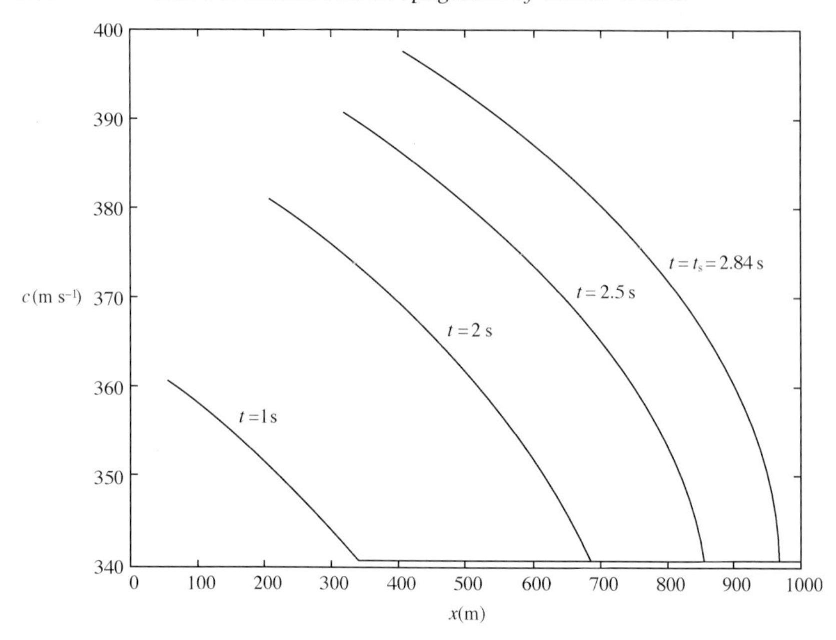

Fig. 7.23. The local sound speed when a piston accelerates at 100 m s^{-2} into a tube of air at atmospheric pressure and room temperature. A shock wave forms when $t = t_s \approx 2.84$ s at $x = x_s \approx 969$ m. Note that the initial sound speed is $c_0 \approx 341$ m s^{-1}.

The gradual acceleration of the piston causes the wave that it generates to steepen until a shock wave forms in the body of the gas at $x = x_S$. This is illustrated in figure 7.23, which shows the local sound speed when the acceleration of the piston is 100 m s^{-2}. The local sound speed increases behind the point $x = c_0 t$, until the gradient becomes infinite and a shock wave forms.

If the motion of the piston is started impulsively when $t = 0$, so that it lies at $x = Vt$ with $V > 0$, a shock forms immediately at the face of the piston (see subsection 8.1.1 for a qualitatively similar solution in the theory of shallow water waves).

For the shock waves that we have encountered so far in traffic flow, a simple application of conservation of cars sufficed to fix the position of the shock for $t > t_s$ through an application of the equal areas rule to the multi-valued solution obtained by the method of characteristics. The significant difference for shocks in gases is that, as we discussed above, the entropy of the gas changes across a shock. This means that once the shock has formed, the solution obtained using the method of characteristics is no longer valid, since it was obtained on the assumption that the entropy of the gas is spatially uniform. A shock generates entropy in its wake.

In the next subsection, we determine what conditions must be satisfied at a shock in an ideal gas, and also demonstrate how the equal area rule can be resurrected when the shock wave is sufficiently weak.

7.2.4 The Rankine–Hugoniot Relations

We can learn a lot about how the various quantities change at a shock by considering the equations for conservation of mass, momentum and energy, (7.35), (7.36) and (7.37), in the neighbourhood of a shock. Assuming that the shock is planar and lies at $x = s(t)$, we can write (7.35) in integral form as

$$\frac{d}{dt}\int_{x_1}^{x_2} \rho dx + [\rho u]_{x_1}^{x_2} = 0. \tag{7.46}$$

We note that

$$\frac{d}{dt}\left(\int_{x_1}^{s(t)} + \int_{s(t)}^{x_2}\right) \rho dx = \left(\int_{x_1}^{s(t)} + \int_{s(t)}^{x_2}\right) \frac{\partial \rho}{\partial t} dx + \rho_{\mathrm{R}}\dot{s} - \rho_{\mathrm{L}}\dot{s},$$

where quantities immediately to the left and right of the shock are denoted by subscripts L and R. Taking the limits $x_1 \to s(t)$ and $x_2 \to s(t)$ in (7.46) then shows that

$$(\rho_{\mathrm{L}} - \rho_{\mathrm{R}})\,\dot{s} = \rho_{\mathrm{L}} u_{\mathrm{L}} - \rho_{\mathrm{R}} u_{\mathrm{R}}.$$

If we define $\bar{u} = u - \dot{s}$, the velocity of the gas relative to the shock, we finally arrive at

$$\rho_{\mathrm{L}}\bar{u}_{\mathrm{L}} = \rho_{\mathrm{R}}\bar{u}_{\mathrm{R}}. \tag{7.47}$$

Similarly, (7.36) and (7.37) show that

$$\rho_{\mathrm{L}}\bar{u}_{\mathrm{L}}^2 + p_{\mathrm{L}} = \rho_{\mathrm{R}}\bar{u}_{\mathrm{R}}^2 + p_{\mathrm{R}}, \tag{7.48}$$

$$\left(\frac{1}{2}\rho_{\mathrm{L}}\bar{u}_{\mathrm{L}}^2 + \rho_{\mathrm{L}}E_{\mathrm{L}} + p_{\mathrm{L}}\right)\bar{u}_{\mathrm{L}} = \left(\frac{1}{2}\rho_{\mathrm{R}}\bar{u}_{\mathrm{R}}^2 + \rho_{\mathrm{R}}E_{\mathrm{R}} + p_{\mathrm{R}}\right)\bar{u}_{\mathrm{R}}. \tag{7.49}$$

These are known as the **Rankine–Hugoniot relations**, and express the fact that the flux of mass, momentum and energy must be continuous at a shock, whilst the pressure, density and internal energy of the gas may not be continuous. Figure 7.24 shows a planar shock in a frame of reference moving with the shock, with spatial coordinate $\bar{x} = x - \dot{s}$.

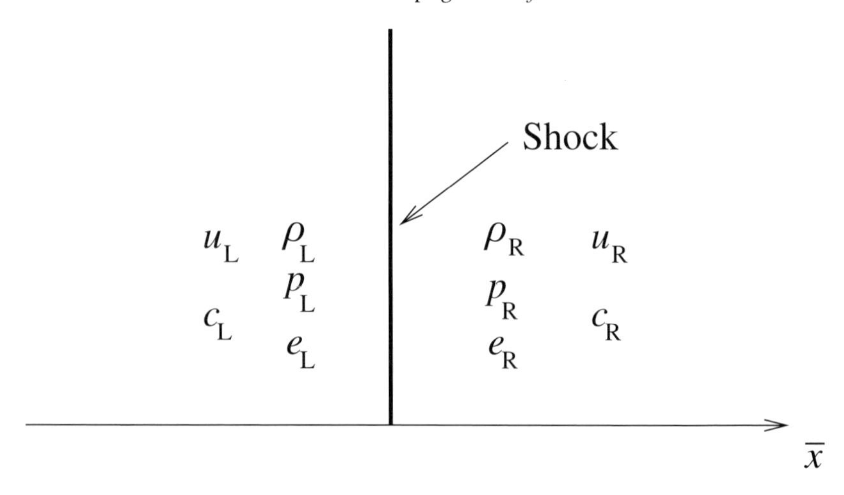

Fig. 7.24. The various physical quantities on either side of a planar shock.

For an ideal gas, $E = p/\rho(\gamma - 1)$, so the equation for continuity of the flux of energy, (7.49), can be written as

$$\left(\frac{1}{2}\bar{u}_L^2 + \frac{\gamma}{\gamma - 1}\frac{p_L}{\rho_L}\right)\rho_L\bar{u}_L = \left(\frac{1}{2}\bar{u}_R^2 + \frac{\gamma}{\gamma - 1}\frac{p_R}{\rho_R}\right)\rho_R\bar{u}_R.$$

Finally, using (7.47),

$$\frac{1}{2}\bar{u}_L^2 + \frac{\gamma}{\gamma - 1}\frac{p_L}{\rho_L} = \frac{1}{2}\bar{u}_R^2 + \frac{\gamma}{\gamma - 1}\frac{p_R}{\rho_R}. \tag{7.50}$$

The Rankine–Hugoniot relations can also be written in terms of the local sound speed, $c = \sqrt{\gamma p/\rho}$, as

$$\rho_L\bar{u}_L = \rho_R\bar{u}_R, \tag{7.51}$$

$$\rho_L\left(\bar{u}_L^2 + \frac{1}{\gamma}c_L^2\right) = \rho_R\left(\bar{u}_R^2 + \frac{1}{\gamma}c_R^2\right), \tag{7.52}$$

$$\frac{1}{2}\bar{u}_L^2 + \frac{1}{\gamma - 1}c_L^2 = \frac{1}{2}\bar{u}_R^2 + \frac{1}{\gamma - 1}c_R^2. \tag{7.53}$$

This is the usual form in which the Rankine–Hugoniot relations are written for an ideal gas.

Now that we have obtained (7.51), (7.52) and (7.53), what can we find out from them? Since these are three equations in six unknowns, $\bar{u}_L$, ρ_L, c_L, $\bar{u}_R$, ρ_R and c_R, we can always eliminate two of these and obtain a single equation that involves any four unknowns. For example, to eliminate $\bar{u}_R$ and c_R, we use equation (7.51) to write

$$\bar{u}_R = \frac{\rho_L \bar{u}_L}{\rho_R}, \tag{7.54}$$

and (7.53) and (7.54) to write

$$c_R^2 = c_L^2 + \frac{1}{2}(\gamma - 1)\left(1 - \frac{\rho_L^2}{\rho_R^2}\right)\bar{u}_L^2, \tag{7.55}$$

If we now use (7.54) and (7.55) to eliminate $\bar{u}_R$ and c_R^2 from (7.52) we find that

$$\rho_L\left(1 - \frac{\rho_L}{\rho_R}\right)\bar{u}_L^2 + \frac{1}{\gamma}(\rho_L - \rho_R)\,c_L^2 = \rho_R\frac{\gamma - 1}{2\gamma}\left(1 - \frac{\rho_L^2}{\rho_R^2}\right)\bar{u}_L^2.$$

If we remove the factor of $(\rho_R - \rho_L)$, we obtain

$$\frac{\rho_L}{\rho_R}\bar{u}_L^2 - \frac{1}{\rho_R}\frac{\gamma - 1}{2\gamma}(\rho_R + \rho_L)\,\bar{u}_L^2 = \frac{1}{\gamma}c_L^2,$$

and a final rearrangement gives

$$(\gamma + 1)\frac{\rho_L}{\rho_R} = \gamma - 1 + \frac{2c_L^2}{\bar{u}_L^2}.$$

If we write this in terms of the local Mach number to the left of the shock, $M_L = \bar{u}_L / c_L$, we obtain

$$\frac{\rho_R}{\rho_L} = \frac{(\gamma + 1)\,M_L^2}{(\gamma - 1)\,M_L^2 + 2}.$$

It is most useful to write this equation in the final form

$$\frac{\rho_R}{\rho_L} = \frac{(\gamma - 1)\,M_L^2 + 2M_L^2}{(\gamma - 1)\,M_L^2 + 2}. \tag{7.56}$$

It is now clear that if $M_L^2 > 1$ then $\rho_R > \rho_L$, and vice versa. In other words, if the flow is locally supersonic ($M_L^2 > 1$), the density on that side of the shock is lower than it is on the other. By symmetry, we can deduce from (7.56) that

$$\frac{\rho_L}{\rho_R} = \frac{(\gamma - 1)\,M_R^2 + 2M_R^2}{(\gamma - 1)\,M_R^2 + 2}, \tag{7.57}$$

and hence that $M_L^2 > 1$ if and only if $M_R^2 < 1$. In other words, the flow must be supersonic on one side of the shock and subsonic on the other.

Many other deductions can be made from the Rankine–Hugoniot

relations using similar manipulations. In particular, the entropies on either side of a shock, S_L and S_R, satisfy (see exercise 7.6)

$$\frac{S_R - S_L}{c_V} = \log\left\{\frac{(1+z)(2\gamma+(\gamma-1)z)^\gamma}{(2\gamma+(\gamma+1)z)^\gamma}\right\}, \tag{7.58}$$

where $z = (p_R - p_L)/p_L$ is the **strength** of the shock. When $z \ll 1$, we say that the shock is **weak**. An expansion of (7.58) for $z \ll 1$ shows that

$$\frac{S_R - S_L}{c_V} \sim \frac{\gamma^2 - 1}{12\gamma^2} z^3. \tag{7.59}$$

This means that for sufficiently weak shocks very little entropy is generated. We will prove a similar result for the loss of energy across a shallow water bore in subsection 8.1.2. Figure 7.25 is a graph of (7.58), which shows that even when $z = 1$, $(S_R - S_L)/c_v \approx 0.01$, and the change in entropy is small. From (7.32), a small change in entropy across the shock leads to a small change in the gas law across the shock. This means that, for sufficiently weak shocks, to a good approximation we can assume that the flow remains isentropic and apply the equal area rule to the density predicted using the method of characteristics. This is the basis of **weak shock theory**, which we shall discuss in more detail in section 10.2.

Figure 7.26 shows a shock wave in air interacting with a sharp edge. The shock wave is generated using a shock tube and the air is at rest ahead of it. Behind the shock the pressure initially increases to 2.4 bar. From these measured quantities, the Rankine–Hugoniot relations give the density behind the shock as $1.8\,\mathrm{kg\,m^{-3}}$, the shock speed as $550\,\mathrm{m\,s^{-1}}$, and the gas velocity behind the shock as $250\,\mathrm{m\,s^{-1}}$.

Example: Reflection of a Shock Wave at a Planar, Solid Wall. As a final example of how the Rankine–Hugoniot relations can be used, let's consider the reflection of a shock wave that is incident normally on a planar, solid wall, as shown in figure 7.27. This is the approximate situation when the shock wave caused by an explosion hits a solid structure. We are particularly interested in the pressure at the wall immediately after the shock is reflected, since it is this that causes the force exerted on the solid structure. If the shock is incident at a significantly oblique angle, the situation is rather more difficult to analyse, and we do not consider this case here.

The first equation that we need, which comes from eliminating p_R and

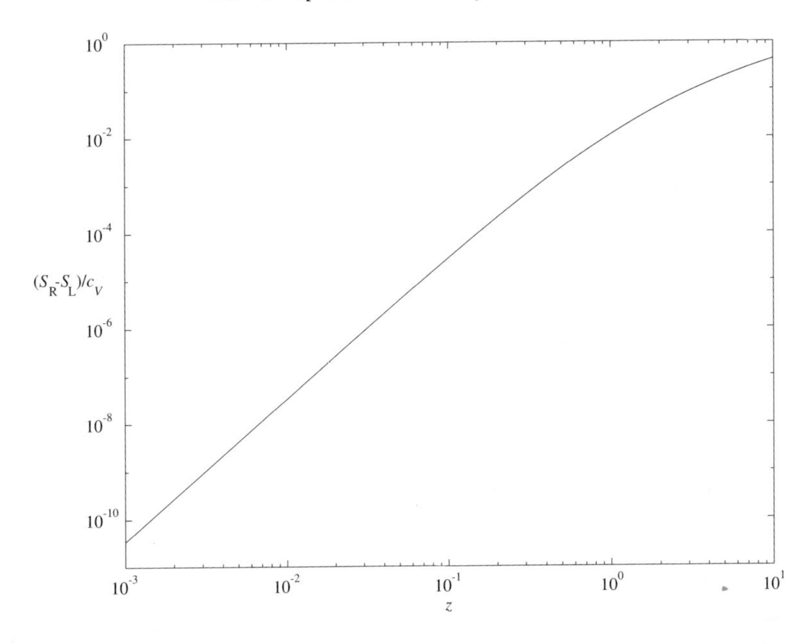

Fig. 7.25. The change in entropy across a shock of strength z.

ρ_R from (7.51) to (7.53), is (see exercise 7.5)

$$\left(\frac{2}{\gamma+1}\right)\bar{u}_L^2 + \bar{u}_L\left(\bar{u}_R - \bar{u}_L\right) - \left(\frac{2\gamma}{\gamma+1}\right)\frac{p_L}{\rho_L} = 0. \tag{7.60}$$

Just before the shock wave reaches the wall, propagating with velocity U_+, the gas pressure and density at the wall take the initial values p_0 and ρ_0. We also know that the normal velocity of the gas is zero at the wall. If the gas pressure, density and normal velocity behind the shock wave are p_s, ρ_s and u_s, we have $\bar{u}_L = u_s - U_+$ and $\bar{u}_R = -U_+$. Substituting this into (7.60) shows that

$$\left(\frac{2}{\gamma+1}\right)(u_s - U_+)^2 - u_s(u_s - U_+) - \left(\frac{2\gamma}{\gamma+1}\right)\frac{p_s}{\rho_s} = 0. \tag{7.61}$$

Immediately after the shock is reflected, its normal velocity is $-U_-$, and the gas pressure and density at the wall are the unknowns, p_1 and ρ_1. However, the normal velocity at the wall must still be zero. In addition, the gas pressure, density and velocity on the other side of the shock are still p_s, ρ_s and u_s, as shown in figure 7.27. This means that $\bar{u}_L = u_s + U_-$

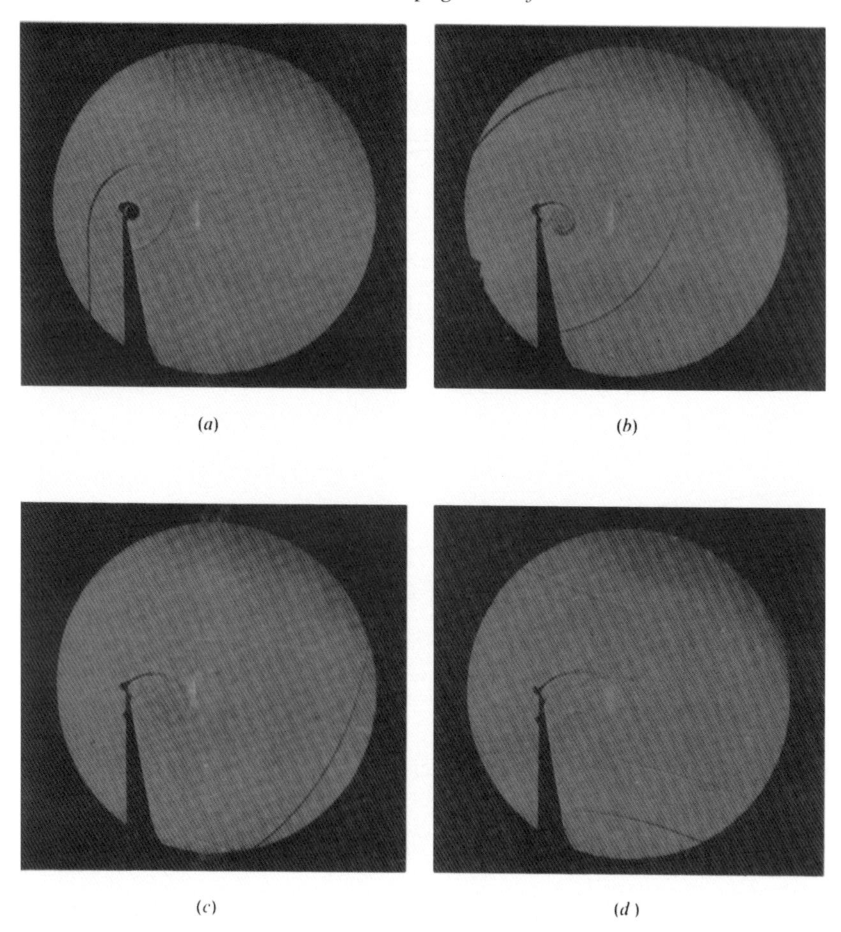

(a) (b) (c) (d)

Fig. 7.26. A shadowgraph photo of a shock interacting with a sharp edge.

and $\bar{u}_{\mathrm{R}} = U_-$, so that

$$\left(\frac{2}{\gamma+1}\right)(u_{\mathrm{s}} + U_-)^2 - u_{\mathrm{s}}(u_{\mathrm{s}} + U_-) - \left(\frac{2\gamma}{\gamma+1}\right)\frac{p_{\mathrm{s}}}{\rho_{\mathrm{s}}} = 0. \tag{7.62}$$

Comparing (7.61) and (7.62), we find that $u_{\mathrm{s}} - U_+$ and $u_{\mathrm{s}} + U_-$ satisfy the same quadratic equation, and must therefore be the two roots, whose product is

$$(u_{\mathrm{s}} - U_+)(u_{\mathrm{s}} + U_-) = -\frac{\gamma p_{\mathrm{s}}}{\rho_{\mathrm{s}}}. \tag{7.63}$$

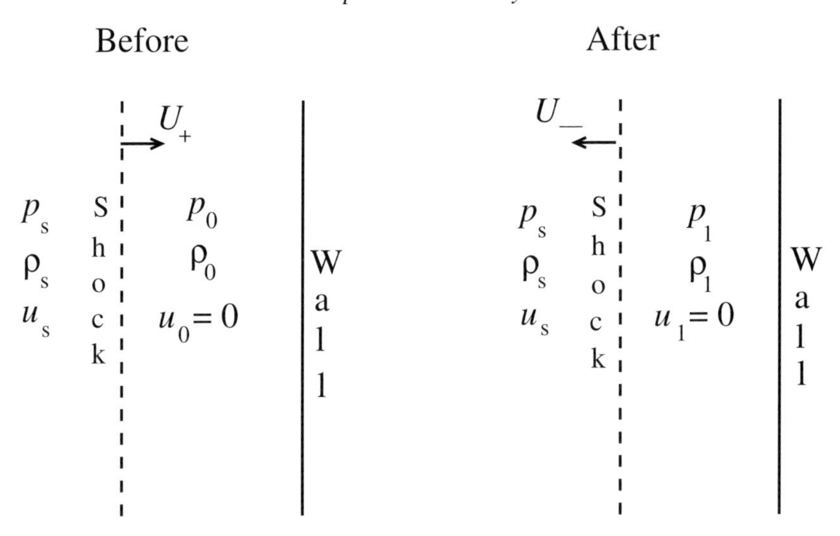

Fig. 7.27. The reflection of a shock wave incident normally on a solid wall.

The other equation that we need is

$$\left(\frac{2}{\gamma+1}\right)\frac{\rho_{\mathrm{L}}\bar{u}_{\mathrm{L}}^2}{p_{\mathrm{L}}} = \frac{p_{\mathrm{R}}}{p_{\mathrm{L}}} + \frac{\gamma-1}{\gamma+1}, \tag{7.64}$$

which comes from (7.51) to (7.53) by eliminating $\bar{u}_{\mathrm{R}}$ and ρ_{R}. From the values on either side of the shock before and after reflection, this implies that

$$\left(\frac{2}{\gamma+1}\right)\frac{\rho_{\mathrm{L}}(u_{\mathrm{s}} - U_+)^2}{p_{\mathrm{L}}} = \frac{p_0}{p_{\mathrm{s}}} + \frac{\gamma-1}{\gamma+1}, \tag{7.65}$$

$$\left(\frac{2}{\gamma+1}\right)\frac{\rho_{\mathrm{L}}(u_{\mathrm{s}} + U_-)^2}{p_{\mathrm{L}}} = \frac{p_1}{p_{\mathrm{s}}} + \frac{\gamma-1}{\gamma+1}. \tag{7.66}$$

If we now multiply these two equations together and use (7.63) to eliminate all the velocities, we find that

$$\left(\frac{p_0}{p_{\mathrm{s}}} + \frac{\gamma-1}{\gamma+1}\right)\left(\frac{p_1}{p_{\mathrm{s}}} + \frac{\gamma-1}{\gamma+1}\right) = \left(\frac{2\gamma}{\gamma+1}\right)^2. \tag{7.67}$$

Finally, when an explosion causes a shock wave to be incident on a wall, we expect that the pressure behind the shock wave will be much greater than that in front of it, so that $p_0 \ll p_{\mathrm{s}}$. Using this approximation in

(7.67), we arrive at the strikingly simple result

$$\frac{p_1}{p_s} \approx \frac{3\gamma - 1}{\gamma - 1}. \tag{7.68}$$

For atmospheric air, with $\gamma \approx 1.4$, this gives $p_1 \approx 8p_s$. Not only does a solid structure, for example a bunker designed to protect its occupants from a conventional or nuclear blast, have to cope with the high pressure p_s incident upon it, the instantaneous pressure is magnified by a factor of eight by the dynamics of the reflection process.

7.2.5 Detonations*

When a shock wave passes through a gas, the temperature behind the shock is greater than that ahead of the shock. We can see this for an ideal gas by eliminating u_R and u_L from the Rankine–Hugoniot relations (7.51) to (7.53). The gas law shows that $p_R\rho_L/p_L\rho_R = T_R/T_L$, and leads to

$$\frac{T_R}{T_L} = \left\{ \frac{(\gamma+1)p_L + (\gamma-1)p_R}{(\gamma-1)p_L + (\gamma+1)p_R} \right\} \frac{p_R}{p_L}. \tag{7.69}$$

Simple calculus shows that the right hand side of this expression is a strictly increasing function of p_R/p_L, and hence that $T_R > T_L$ when $p_R > p_L$. For the shock shown in figure 7.26, the absolute temperature initially increases from about 20 °C to 110 °C. In the limiting case of a strong shock propagating from right to left, so that $p_R \gg p_L$ (see exercise 7.6),

$$\frac{T_R}{T_L} \sim \left(\frac{\gamma - 1}{\gamma + 1} \right) \frac{p_R}{p_L} \gg 1.$$

The stronger the shock, the greater the temperature increase.

If the gas through which the shock travels is combustible, for example a mixture of methane and air or hydrogen and air, the rise in temperature across the shock may be sufficient to initiate a chemical reaction and ignite the gas. Since the chemical reactions involved in combustion are extremely rapid, there are a region where the mixture of gases is burnt, and a region where the mixture is unburnt, separated by a thin **detonation wave**. We will assume that this wave is sufficiently thin relative to any geometrical length scales that we can model it as a surface of discontinuity, just as we did for a shock wave.

We can distinguish two different situations where a detonation wave can exist, in which, as we shall see below, it propagates in slightly

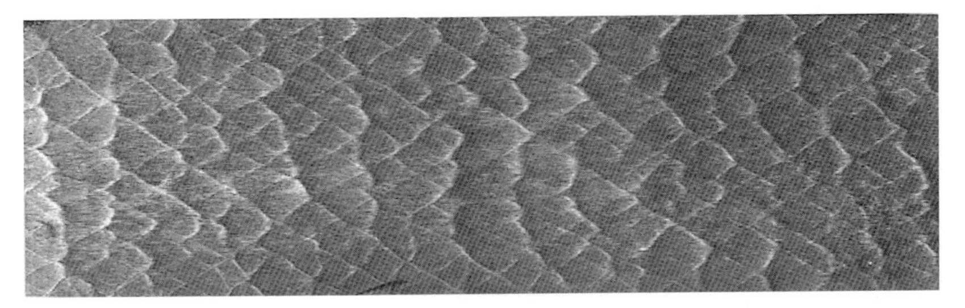

Fig. 7.28. Soot deposited on foil after a CJ detonation wave in a mixture of methane and oxygen has passed down a tube. The cellular pattern indicates that in this case the wave was laterally unstable.

different ways. Firstly, a shock wave may be incident on a combustible mixture and ignite it. Secondly, a combustible mixture may be ignited at a point by a local source of heat, such as a spark, and a detonation wave generated spontaneously by the violence of the chemical reaction. In the second situation, there is also the possibility of a **deflagration wave** propagating. These are rather similar to the chemical waves that we will study in chapter 9, where there is a balance between chemical reaction and diffusion of the heat generated. The dynamics of the compressible gas is of secondary importance in deflagration waves, and there is certainly no shock. The stability of deflagrations and detonations remains the subject of lively debate (for example, Brailovsky and Sivashinsky (1997), Sharpe and Falle (1999)). The effect of a passing detonation wave is shown in figure 7.28.

The Shock and Detonation Adiabatics and the Chapman–Jouguet Point

In order to investigate the propagation of detonation waves, it is first useful to consider the notion of the shock adiabatic for ordinary shock waves. Eliminating the gas temperature from (7.69) in favour of the gas density gives

$$\frac{p_{\mathrm{R}}}{p_{\mathrm{L}}} = \frac{(\gamma+1)\rho_{\mathrm{L}}^{-1} - (\gamma-1)\rho_{\mathrm{R}}^{-1}}{(\gamma+1)\rho_{\mathrm{R}}^{-1} - (\gamma-1)\rho_{\mathrm{L}}^{-1}}. \tag{7.70}$$

For given values of p_{L} and ρ_{L}^{-1}, this equation relates p_{R} to ρ_{R}^{-1} and is known as the **shock adiabatic**. It takes the form of a rectangular hyperbola, as plotted in figure 7.29. From the first two Rankine–Hugoniot relations,

$$j^2 = \frac{p_{\mathrm{L}} - p_{\mathrm{R}}}{\rho_{\mathrm{R}}^{-1} - \rho_{\mathrm{L}}^{-1}}, \tag{7.71}$$

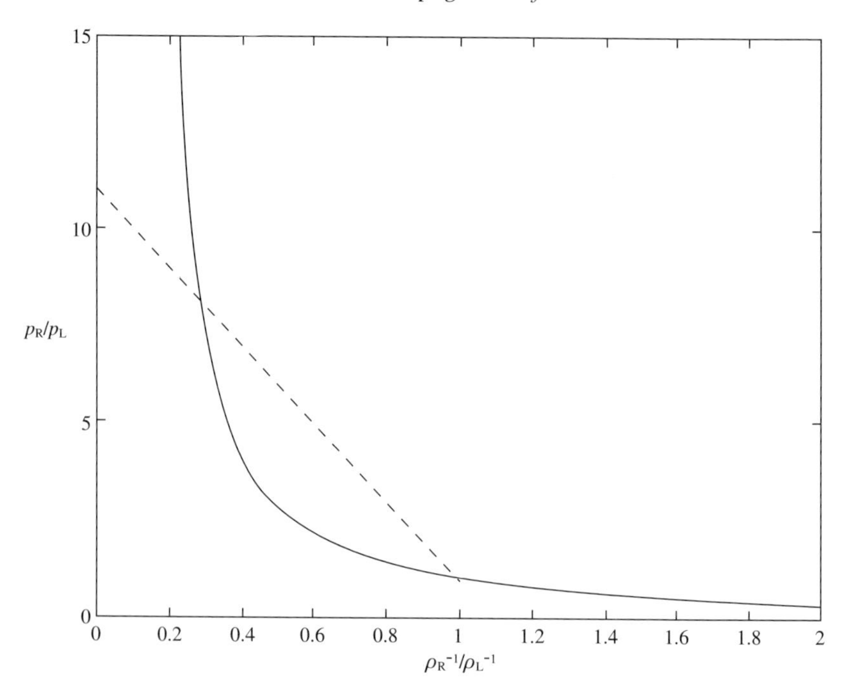

Fig. 7.29. The shock adiabatic for an ideal gas with $\gamma = 1.4$ (solid line), along with a typical straight line given by (7.71), (dashed line).

where $j = \rho_R \bar{u}_R = \rho_L \bar{u}_L$ is the flux of mass through the shock. If we consider the straight line joining the points (ρ_R^{-1}, p_R) and (ρ_L^{-1}, p_L) on the shock adiabatic, as shown in figure 7.29, its slope is therefore $-j^2$. We conclude that, for a given state (ρ_L^{-1}, p_L) ahead of the shock and a given mass flux j through the shock, the state of the gas behind the shock, (ρ_R^{-1}, p_R), can be determined graphically from the shock adiabatic by drawing a line of slope $-j^2$ through the point (ρ_L^{-1}, p_L). The other point of intersection will then give the final state, (ρ_R^{-1}, p_R).

For a detonation wave, the Rankine–Hugoniot relations in the form (7.47) and (7.48) still hold. The first modification to our analysis that we need to make for a detonation wave is to take into account the energy released by the chemical reaction in (7.49). The chemical reaction is confined to a thin region behind the shock wave, which we treat as a discontinuity in the internal energy of the gas. The second is to note that the ratio of specific heats may be different on each side of the detonation, since the chemical composition is different. The third Rankine–Hugoniot relation (7.49) therefore becomes

$$\frac{1}{2}\bar{u}_R^2 + \frac{\gamma_R}{\gamma_R - 1}\frac{p_R}{\rho_R} = \frac{1}{2}\bar{u}_L^2 + \frac{\gamma_L}{\gamma_L - 1}\frac{p_L}{\rho_L} + q, \tag{7.72}$$

where q is the energy per unit mass released by the chemical reaction and γ_R and γ_L are the different specific heat ratios of the burnt and unburnt gases. If we now eliminate $\bar{u}_R$ and $\bar{u}_L$ from the Rankine–Hugoniot relations for the detonation we arrive at the equation of the **detonation adiabatic**,

$$\frac{\gamma_L + 1}{\gamma_L - 1}\frac{p_L}{\rho_L} - \frac{\gamma_R + 1}{\gamma_R - 1}\frac{p_R}{\rho_R} - \frac{p_L}{\rho_R} + \frac{p_R}{\rho_L} = -2q. \tag{7.73}$$

In contrast to the shock adiabatic, the detonation adiabatic does not pass through the initial point (p_L, ρ_L^{-1}). The shock and detonation adiabatics are shown in figure 7.30. The detonation adiabatic lies at a higher pressure for a given value of ρ^{-1} than the shock adiabatic because of the extra heat generated by the chemical reaction. However, (7.71) still holds, since it is derived from conservation of mass and momentum only. This means that the slope of the straight line from the initial state, (p_L, ρ_L^{-1}) to the final state, (p_R, ρ_R^{-1}), which now lies on the detonation adiabatic, is still $-j^2$. It is clear from figure 7.30 that there is now a lower bound on the value of j^2, corresponding to minus the slope of the detonation adiabatic at the point marked CJ, called the **Chapman–Jouguet point**, where the straight line (7.71) is tangent to it. Moreover, for j^2 greater than this minimum value, the line (7.71) meets the detonation adiabatic at two different points, B and C in figure 7.30. Only the state C corresponds to a physically realisable shock. We could deduce this from arguments involving entropy production at the detonation. There is, however, a simpler, physical argument.

The internal structure of the detonation, which we neglect in idealising it as a discontinuity, consists of a shock wave followed by a combustion region. Across the shock, the initial state of the gas changes to that given by the point D on the shock adiabatic, as shown in figure 7.30. Then, across the combustion region, the chemical reaction releases heat and the pressure of the gas decreases until it reaches the equilibrium state given by C on the detonation adiabatic. Thus the state C rather than the state B is reached across the detonation wave, simply because the associated shock heats the gas and precedes the combustion, not the other way around. We conclude that the set of possible states that the gas can reach across a detonation wave is given by the part of the detonation adiabatic lying above the CJ point.

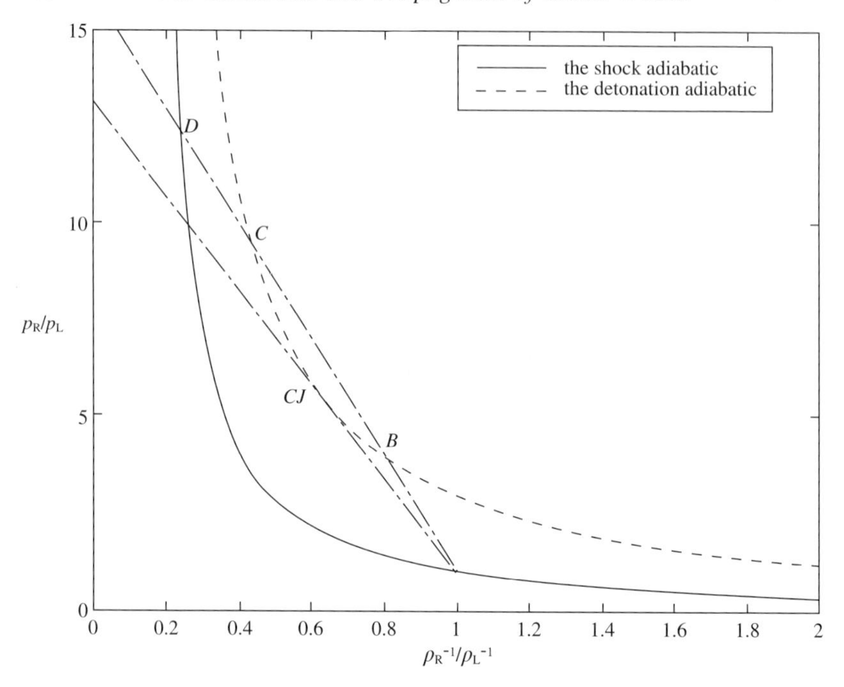

Fig. 7.30. The shock and detonation adiabatics for an ideal gas with $\gamma_R = \gamma_L = 1.4$, along with a typical straight line given by (7.71), which meets the detonation adiabatic at points B and C, and the unique line that meets the detonation adiabatic at the Chapman–Jouguet point CJ (dash–dotted lines).

At the CJ point, $dp_R/d\rho_R^{-1} = -j^2$. Since $dp_R/d\rho_R = c_R^2$ and $j^2 = \rho_R^2 \bar{u}_R^2$, this means that $\bar{u}_R = c_R$ at the CJ point. The detonation wave therefore moves at the local sound speed relative to the velocity of the burnt gas if its burnt state corresponds to the CJ point. This is not possible for a shock wave, since the line given by (7.71) can never be tangent to the shock adiabatic, as is clear from figures 7.29 and 7.30. We can also note that on the part of the detonation adiabatic that lies above the CJ point, for example point C in figure 7.30, $-j^2 > dp_R/d\rho_R^{-1} = -\rho_R^2 c_R^2$, and hence $\bar{u}_R < c_R$. We conclude that, in general, a detonation wave moves at or below the speed of sound relative to the burnt gases behind it, and that the uniquely determined detonation that moves at the local speed of sound corresponds to the Chapman–Jouguet point, CJ, on the detonation adiabatic, and hence the lowest possible pressure and density in the burnt gases.

When a shock wave is incident on a combustible gas and becomes a detonation wave, the strength of the detonation depends on the strength

Fig. 7.31. A spherical detonation wave initiated using high explosives.

of the incident shock, and the state of the burnt gases can lie anywhere on the detonation adiabatic above the CJ point. We say that the detonation may be **over-driven** (see exercise 7.8). However, when a detonation wave is ignited from within a combustible mixture by some local source of heating, it usually corresponds to the CJ point. As an example of this, we will consider the propagation of a spherical detonation wave away from its point of ignition. This has some relevance to the behaviour of supernovas (for example, Wiggins, Sharpe and Falle (1998)).

Example: a Spherical Detonation Wave. Figure 7.31 shows a spherical detonation wave, initiated using high explosives. Let's see if we can describe such a wave mathematically. In spherical polar coordinates, the equations for conservation of mass momentum and energy for a spherically symmetric flow are

$$\frac{\partial \rho}{\partial t} + \frac{\partial(\rho u)}{\partial r} + \frac{2\rho u}{r} = 0, \tag{7.74}$$

$$\frac{\partial u}{\partial t} + u\frac{\partial u}{\partial r} = -\frac{1}{\rho}\frac{\partial p}{\partial r}, \tag{7.75}$$

$$\frac{\partial S}{\partial t} + u\frac{\partial S}{\partial r} = 0. \tag{7.76}$$

For a detonation wave generated at a point, there is no geometrical length scale, and the only parameters in the problem are p_0 and ρ_0, the initial pressure and density in the unburnt gases, and q, the heat generated by the chemical reaction. We can form just two dimensionless groups involving r and t, namely $\rho_0 r^2/p_0 t^2$ and r^2/qt^2. We conclude that the solution will be of similarity form, with all the dependent variables functions of $\eta = r/t$. We need to solve (7.74) to (7.76) for $0 \le \eta < \eta_0$, with a detonation at $\eta = \eta_0$ and the gas at rest in its unburnt state for $\eta > \eta_0$.

If we write $\rho \equiv \rho(\eta)$, $u \equiv u(\eta)$ and $p \equiv p(\eta)$, (7.76) becomes

$$(u - \eta)\, S' = 0, \tag{7.77}$$

where a prime denotes $d/d\eta$. Provided that $u \ne \eta$, which we shall see below does not occur, the entropy S must be a constant behind the spherical detonation wave. We therefore have $p' = c^2\rho'$, and can eliminate p and ρ between (7.74) and (7.75) to obtain

$$u' = \frac{2u}{\eta}\left\{\frac{(u-\eta)^2}{c^2} - 1\right\}^{-1} \quad c' = -(\gamma - 1)\frac{u(u-\eta)}{c\eta}\left\{\frac{(u-\eta)^2}{c^2} - 1\right\}^{-1}. \tag{7.78}$$

These nonlinear ordinary differential equations determine how the gas velocity u and local sound speed c vary behind the detonation wave. We can write (7.78) in dimensionless form by defining

$$X = \frac{\eta}{c_0}, \quad U = \frac{u}{c_0}, \quad C = \frac{c}{c_0}$$

which gives

$$\frac{dU}{dX} = \frac{2U}{X\left\{(U-X)^2 - 1\right\}}, \quad \frac{dC}{dX} = -\frac{(\gamma-1)CU(U-X)}{X\{(U-X)^2 - 1\}}. \tag{7.79}$$

Since there are no sources of mass, the gas velocity must be zero at $X = 0$, and it is helpful to consider the solution when $U \ll 1, C \sim 1$. Provided X is not close to one, (7.79) gives at leading order

$$\frac{dU}{dX} = \frac{2U}{X(X^2 - 1)}. \tag{7.80}$$

We can solve this separable equation and obtain

$$U = k\left(1 - \frac{1}{X^2}\right), \tag{7.81}$$

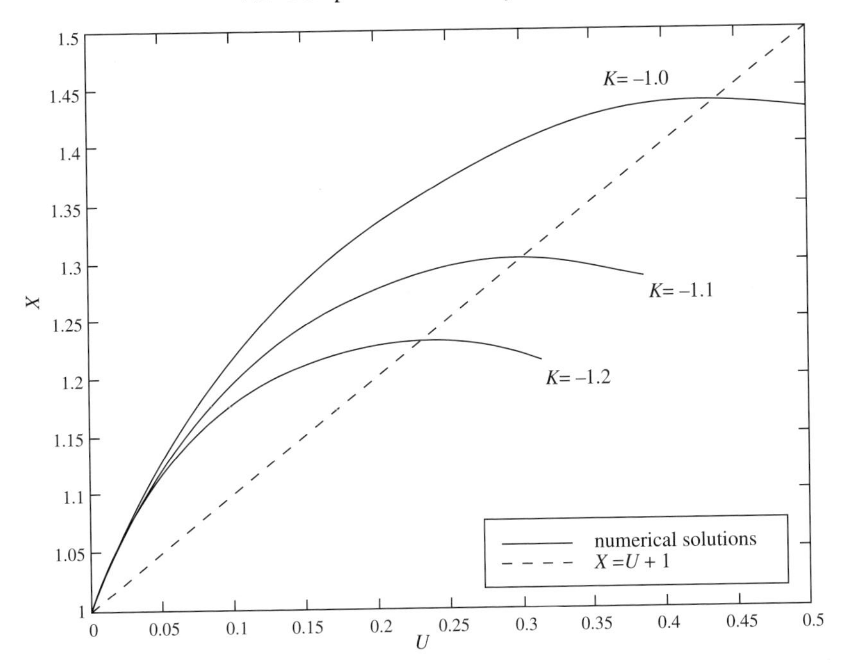

Fig. 7.32. Numerical solutions of (7.83) for various values of K.

where k is a constant. Since $U \to 0$ as $X \to 1$, we cannot in fact assume that X is not close to one, and this solution is not valid. We have therefore shown that U can only be small in the neighbourhood of $X = 1$. To proceed, we define $\hat{X} = X - 1, \hat{C} = C - 1$, and seek a solution for $U, \hat{C}$ and $\hat{X}$ small. At leading order

$$\frac{dU}{d\hat{X}} = \frac{U}{\hat{X} - U}, \quad \frac{d\hat{C}}{d\hat{X}} = \frac{1}{2}(\gamma - 1)\frac{U}{\hat{X} - U - \hat{C}}. \tag{7.82}$$

This equation is linear in $\hat{X}$, and has the implicit solution

$$\hat{X} = KU - \tfrac{1}{2}(\gamma + 1)U \log U, \quad \hat{C} \sim \tfrac{1}{2}(\gamma - 1)U,$$

which gives

$$X \sim 1 - \tfrac{1}{2}(\gamma + 1)U \log U + KU, \quad C \sim 1 + \tfrac{1}{2}(\gamma - 1)U, \quad \text{as } U \to 0, \tag{7.83}$$

for some constant K. This shows that $U \to 0$ and $dU/dX \to 0$ as $X \to 1^+$. We conclude that the solution has $U = 0, C = 1$ for $0 \le X \le 1$, with U and C given implicitly by (7.83) for $(X - 1)$ small and positive. We now need to consider how U and C behave for $X > 1$. Before doing this, let's consider where a detonation wave can exist. For $X > U + C$,

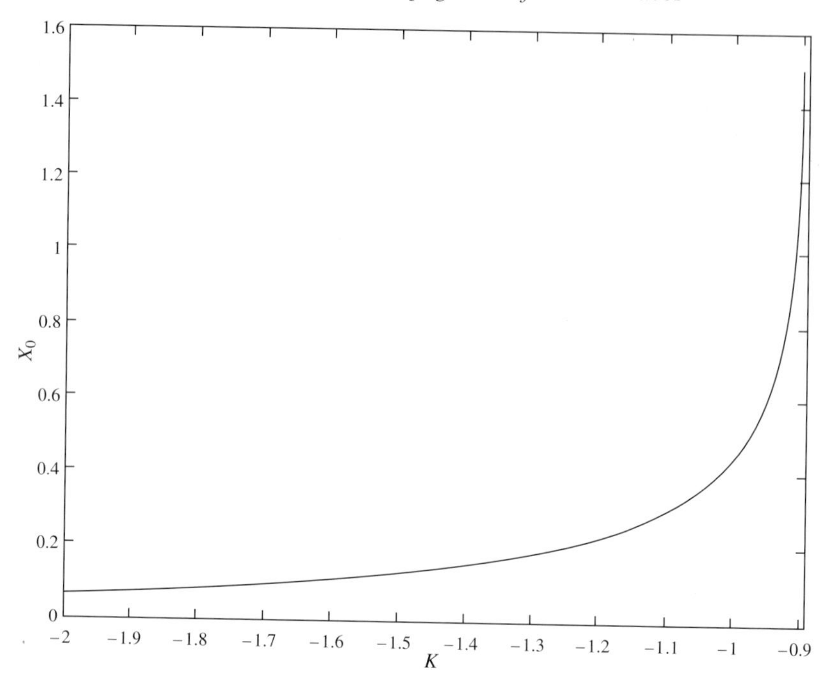

Fig. 7.33. The position at which dU/dX becomes infinite as a function of K.

in terms of the physical variables, $r > ut + ct$. If the detonation wave lies at $r = r_0(t)$ with $r_0(t) > ut + ct$, the local speed of sound is less than the velocity of the burnt gas relative to the speed of the shock. We have already seen that this is not physically possible, and we therefore need to insert the detonation at some point where $X \leq U + C$.

It is rather easier to treat X and C as functions of U. Some numerical solutions of (7.79) are shown in figure 7.32 using (7.83) to begin the integration at $X = 1 + \epsilon$ with $\epsilon \ll 1$ for various values of K. From (7.79) we can see that $dX/dU = 0$ when $X = U \pm C$. Each of the solutions shown in figure 7.32 has $X > U + C$ until it meets the line $X = U + C$ at $X = X_0$, where it has a local maximum. Treating these solutions as giving U as a function of X, they are valid for $1 \leq X \leq X_0$, at which point dU/dX becomes unbounded. Numerical integration of (7.79) suggest that such solutions exist for each $K < K_0$, with $K_0 \approx -2.1$. For $K > K_0$, X becomes unbounded at a finite value of U, and does not provide a solution appropriate to a detonation, since $X > U + C$. The function $X_0(K)$ is plotted in figure 7.33 for $K < K_0$.

To complete the solution, we must insert a detonation wave at $X = X_0$,

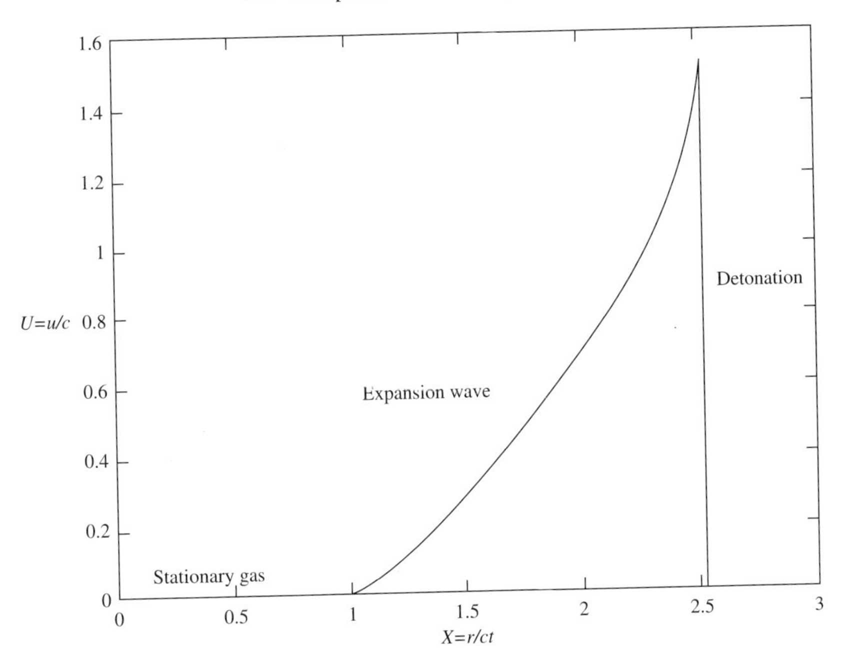

Fig. 7.34. A typical similarity solution for a spherical detonation.

across which U falls to zero, its initial value. This gives a family of solutions with $U = 0$ for $X \leq 1$ and $X \geq X_0(K)$, and U monotonically increasing for $1 \leq X \leq X_0(K)$. A typical solution is shown in figure 7.34. However, at $X = X_0$, $X = U + C$, and hence $u = X_0c_0 - c$. Since, the detonation lies at $r = r_0 = X_0c_0t$ and therefore moves with speed X_0c_0, we conclude that the burnt gases behind the detonation move at the local sound speed relative to the detonation, and hence that the detonation corresponds to the CJ point on the detonation adiabatic. For a mixture of gases, this CJ detonation is associated with a unique value of u, the velocity of the burnt gases behind the detonation, and hence a unique value of X_0. The appropriate similarity solution is therefore selected by the Rankine–Hugoniot conditions at the detonation, in particular, by the amount of heat generated by the chemical reaction.

Physically, we can see that the solution consists of a stationary sphere of gas of radius c_0t centred on the point of ignition, $r = 0$, connected to a spherical Chapman–Jouguet detonation at $r = X_0c_0t$ by an expansion wave across which u increases continuously from zero to $X_0c_0 - c$.

Exercises

7.1 The flow of traffic along a single lane road is modelled using

$$\frac{\partial \rho}{\partial t} + \frac{\partial}{\partial t}(\rho v(\rho)) = 0,$$

where $\rho(x,t)$ is the density of cars and $v(\rho)$ their velocity. If $v(\rho) = v_0 (\rho_{\max} - \rho) / \rho_{\max}$ and

$$\text{(a)}\ \rho(x,0) = \begin{cases} 0 & \text{for } x \le 0, \\ \rho_1 x^2/L^2 & \text{for } 0 \le x \le L, \\ \rho_1 & \text{for } x \ge L, \end{cases}$$

$$\text{(b)}\ \rho(x,0) = \rho_1 \exp(-k|x|),$$

where L, k and $\rho_1 \le \rho_{\max}$ are positive constants, determine when and where the solution first becomes undefined and hence a shock forms in each case. Sketch the solutions up to the development of the shock. Use the equal areas rule to make sketches of the progress of the shock in each case.

7.2 Using the model given in the previous exercise, the initial car density is

$$\rho(x,0) = \rho_0 + H(\pi - |bx|)\, a \sin bx,$$

where ρ_0, a and b are constants, with $\rho_0 > 0$, $|a| < \rho_0$ and $\rho_0 + |a| < \rho_{\max}$, and H is the Heaviside step function.

Show that when ab is positive, a single shock wave forms at time $t = \rho_{\max}/2v_0|ab|$, and that if ab is negative, two shock waves are formed simultaneously, again at time $t = \rho_{\max}/2v_0|ab|$. In each case, where are the shock waves when they form?

By fitting appropriate shocks to the multi-valued solution in each case, show that for $t \gg 1$ the maximum value of ρ is approximately

$$\rho_0 + \frac{\pi \rho_{\max}}{2v_0|b|t}$$

when ab is positive, and

$$\rho_0 + \sqrt{\frac{-2\rho_{\max} a}{v_0 b t}}$$

when ab is negative.

7.3 The flow of cars along a single lane road can be described using a continuous car density, $\rho(x,t)$, with car velocity given by

$$v(\rho) = \frac{v_0}{\rho_{\max}^2} (\rho_{\max} - \rho)^2,$$

for $0 \leq \rho \leq \rho_{\max}$. Write down the equation satisfied by ρ, and determine the kinematic wave speed, $c(\rho)$. Show that $c(\rho)$ is zero at $\rho = \frac{1}{3}\rho_{\max}$ and has a minimum at $\rho = \frac{2}{3}\rho_{\max}$.

At time $t = 0$, the car density is

$$\rho(x, 0) = \frac{\rho_L + \rho_R e^{x/L}}{1 + e^{x/L}},$$

with $0 < \rho_R < \frac{1}{3}\rho_{\max}$ and $\frac{2}{3}\rho_{\max} < \rho_L < \rho_{\max}$. Sketch the function $c(\rho(x, 0))$. Sketch the development of the car density for $t > 0$. How does this solution differ from the solution with $v(\rho) = v_0(\rho_{\max} - \rho)/\rho_{\max}$?

By considering the limit $L \to 0$, show that the car density changes discontinuously from ρ_L to $\rho_{\max} - \rho_L/2$ at a shock wave, which propagates with velocity $c(\rho_{\max} - \rho_L/2)$.

7.4 A piston confines a ideal gas within a semi-infinite tube of uniform cross-section. When $t = 0$ the gas is at rest and has sound speed c_0. For $t \geq 0$:

(a) The piston moves with a constant velocity $-V$ with $V > 0$. Show that the solution takes the form of an expansion fan and determine the solution.

(b) The piston moves with velocity $A\omega \sin \omega t$, where A and ω are positive constants. Show that a shock wave first forms when $t = t_s = 2c_0/A\omega^2(\gamma + 1)$.

7.5 Derive (7.60) and (7.64) from the Rankine–Hugoniot relations.

7.6 A plane shock wave is propagating in an ideal gas with ratio of specific heats γ. In a frame of reference where the shock is stationary at $x = 0$, the sound speed, gas pressure, density and velocity are given by c_R, p_R, ρ_R and $\bar{u}_R$ for $x > 0$, and c_L, p_L, ρ_L and $\bar{u}_L$ for $x < 0$. In addition, $\bar{u}_L > 0$, so the gas passes from left to right through the shock. Use the Rankine–Hugoniot relations to show that

$$z = \frac{2\gamma\left(M_L^2 - 1\right)}{\gamma + 1}, \tag{7.84}$$

$$\frac{\rho_R}{\rho_L} = \frac{2\gamma + (\gamma + 1)z}{2\gamma + (\gamma - 1)z}, \tag{7.85}$$

$$\frac{S_R - S_L}{c_V} = \log\left\{\frac{(1 + z)(2\gamma + (\gamma - 1)z)^{\gamma}}{(2\gamma + (\gamma + 1)z)^{\gamma}}\right\}, \tag{7.86}$$

where $z = (p_R - p_L)/p_L$, $M_L = \bar{u}_L^2/c_L^2$, S_L and S_R are the entropies on either side of the shock, and c_V is the specific heat of the gas at constant volume.

Using (7.86), show that $d(S_R - S_L)/dz > 0$ for $\gamma > 1$ and $z > -1$, and hence that for the entropy of the gas to increase as it passes through the shock, $z > 0$. Now use (7.84) and (7.85) to show that $\rho_R > \rho_L$, and that the flow is subsonic for $x > 0$ and supersonic for $x < 0$.

7.7 A strong detonation wave is normally incident upon a rigid plane wall. What is the pressure at the wall after the detonation is reflected?

7.8 A piston initially at $x = 0$ confines an ideal combustible mixture of gases in a straight, semi-infinite tube lying in $x > 0$. When $t = 0$ the piston moves into the tube at speed $V > 0$. If a detonation wave forms immediately, show that an over-driven detonation is formed if $V > V_0$ and determine V_0.

8

Nonlinear Water Waves

We begin our examination of nonlinear water waves by studying the nonlinear shallow water equations. We looked at the linearised version of these equations in chapter 4. The nonlinear equations are closely related to those that we studied in chapter 7, and there is the possibility of shock, or bore, formation. We then consider the effect of nonlinearity on deep water, progressive gravity waves, determining in particular how the wave speed and waveform depend upon the small amplitude of the wave. When the competing effects of linear dispersion and nonlinear wave steepening act in a shallow water flow, we will show that the **Korteweg–de Vries equation** can control the leading order behaviour of the waves. In the final section, we consider nonlinear capillary waves in deep water, and demonstrate how complex variable theory can be used to derive analytical solutions.

8.1 Nonlinear Shallow Water Waves

In section 4.7 we derived the equations, (4.80) and (4.81), that govern the flow of shallow water, where the horizontal wavelength of disturbances is much greater than the vertical depth. For shallow water flowing over a horizontal bed, with $h_0(x)$ constant, these equations become

$$\frac{\partial h}{\partial t} + u\frac{\partial h}{\partial x} + h\frac{\partial u}{\partial x} = 0, \tag{8.1}$$

$$\frac{\partial u}{\partial t} + u\frac{\partial u}{\partial x} + g\frac{\partial h}{\partial x} = 0, \tag{8.2}$$

where u is the horizontal velocity and h the vertical depth. We have already looked at the linearised version of these equations. We now wish to study the full nonlinear equations.

We can analyse (8.1) and (8.2) using the same methods that we introduced in section 7.2. As before, characteristic curves exist, on which there are Riemann invariants. The local wave speed is $c = \sqrt{gh}$, in terms of which (8.1) and (8.2) become

$$2\frac{\partial c}{\partial t} + 2u\frac{\partial c}{\partial x} + c\frac{\partial u}{\partial x} = 0, \tag{8.3}$$

$$\frac{\partial u}{\partial t} + u\frac{\partial u}{\partial x} + 2c\frac{\partial c}{\partial x} = 0. \tag{8.4}$$

If we now add or subtract these equations we obtain

$$\frac{\partial}{\partial t}(u \pm 2c) + (u \pm c)\frac{\partial}{\partial x}(u \pm 2c) = 0. \tag{8.5}$$

The functions $R_{\pm}(u,h) = u \pm 2c$ are therefore the Riemann invariants of the system. To summarise:

- On the C_+ characteristics, given by $dX_+/dt = u + \sqrt{gh}$, the C_+ invariant, $R_+ = u + 2\sqrt{gh}$, is constant.
- On the C_- characteristics, given by $dX_-/dt = u - \sqrt{gh}$, the C_- invariant, $R_- = u - 2\sqrt{gh}$, is constant.

Note that this is equivalent to the equations of compressible gas dynamics, (7.40) and (7.41) with $\gamma = 2$ and $k = g/2$. However, the conditions that determine the speed of a shock in shallow water are rather different from those that apply for a gas dynamical shock wave.

We study two examples of what can happen in a nonlinear shallow water flow: the dam break problem and a shallow water bore.

8.1.1 The Dam Break Problem

A semi-infinite expanse of shallow water has initial depth h_0 and lies stationary in the domain $x < 0$, held back by a dam at $x = 0$. There is no water on the other side of the dam in the domain $x > 0$. When $t = 0$ the dam breaks. What is the height of the water for $t > 0$ as a function of x? To answer this question, we need to solve the shallow water equations, (8.1) and (8.2), subject to the initial conditions

$$h(x,0) = h_0 H(-x), \qquad u(x,0) = 0, \tag{8.6}$$

which are illustrated in figure 8.1.

On the characteristics that originate at $t = 0$ for $x < 0$, $R_{\pm} = u \pm 2\sqrt{gh} = \pm 2\sqrt{gh_0} = \pm 2c_0$, where $c_0 = \sqrt{gh_0}$ is the initial, linear wave

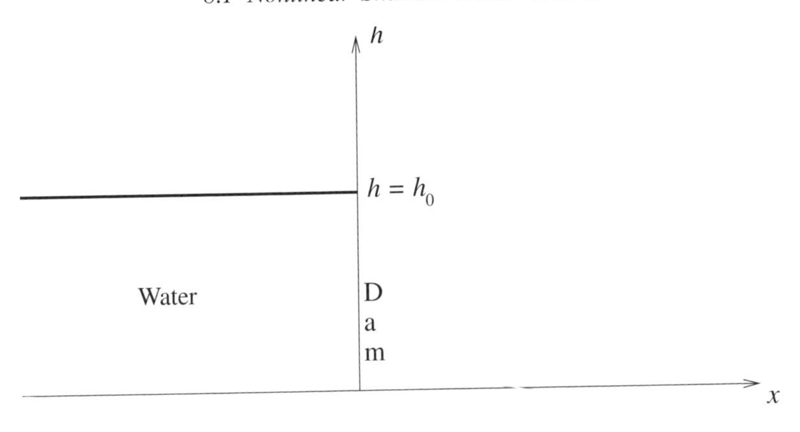

Fig. 8.1. The initial water depth for the dam break problem.

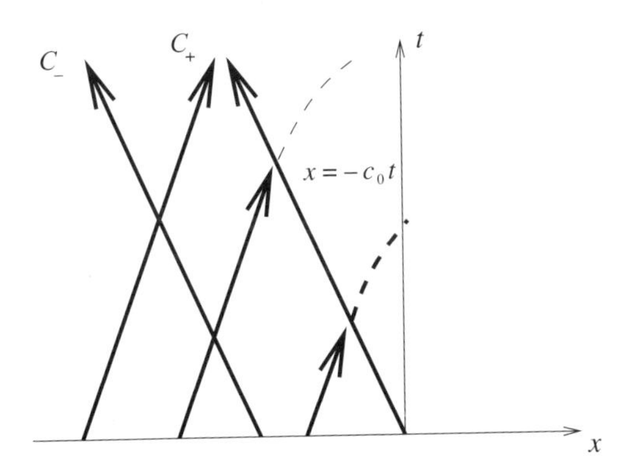

Fig. 8.2. The characteristics in the undisturbed region of the dam break problem.

speed. Therefore, if a C_+ and a C_- characteristic from this region intersect, $u + 2\sqrt{gh} = 2c_0$, $u - 2\sqrt{gh} = -2c_0$, and hence $u = 0$, $h = h_0$. In other words, the water is undisturbed at that point. In addition, $dX_\pm/dt = u \pm \sqrt{gh} = \pm c_0$. In other words, the characteristics are straight lines. Such characteristics must lie in the region $x < -c_0 t$, as shown in figure 8.2, and hence $h = h_0$, $u = 0$ in this region. Notice that the C_+ characteristics leave this region and enter the domain $x > -c_0 t$. We will now assume that these characteristics fill this domain, and justify our assumption later.

For $x > -c_0 t$, the C_- characteristics satisfy

$$\frac{dX_-}{dt} = u - \sqrt{gh}, \tag{8.7}$$

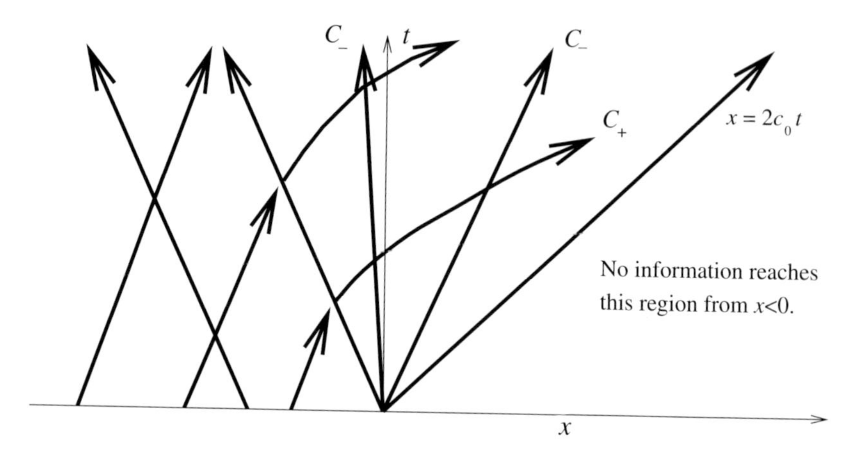

Fig. 8.3. The characteristics for the dam break problem.

and on each curve, $R_- = u - 2\sqrt{gh}$ is constant. However, since we are assuming that this region is filled by C_+ characteristics with $R_+ = u + 2\sqrt{gh} = 2c_0$, u and h must be constant on each C_- characteristic. Equation (8.7) then shows that dX_-/dt is constant on each C_- characteristic, which must therefore be a straight line. Since the fluid only occupies the region $x < 0$ when $t = 0$, these C_- characteristics must start at the origin, as shown in figure 8.3, with $X_-(t) = (u - \sqrt{gh})t$. We have therefore shown that $R_+ = u + 2\sqrt{gh} = 2c_0$ and $u - \sqrt{gh} = x/t$ at each point in the domain $x > -c_0 t$. If we now solve these two equations for u and h we find that

$$h = \frac{h_0}{9}\left(2 - \frac{x}{c_0 t}\right)^2, \qquad u = \frac{2}{3}\left(c_0 + \frac{x}{t}\right). \tag{8.8}$$

Note that this gives $h = 0$ when $x = 2c_0 t$, suggesting that no C_+ characteristics reach the region $x > 2c_0 t$, so that $u = h = 0$ there. The solution is sketched in figure 8.4. Note that the leading edge of the water travels forwards twice as fast as the effect of the dam break travels back through the undisturbed water.

We can now determine the equation of the C_+ characteristics that originate at $t = 0$ in $x < 0$, and show that they fill the domain $x < 2c_0 t$, but do not reach the domain $x > 2c_0 t$. For $x < -c_0 t$ the C_+ characteristics are straight lines with slope c_0, and are given by

$$X_+ = -x_0 + c_0 t, \qquad \text{for } x_0 > 0,\ t < x_0/2c_0. \tag{8.9}$$

When $t = x_0/2c_0$, $X_+ = -c_0 t$, so that for $t > x_0/2c_0$, $dX_+/dt = u + \sqrt{gh}$,

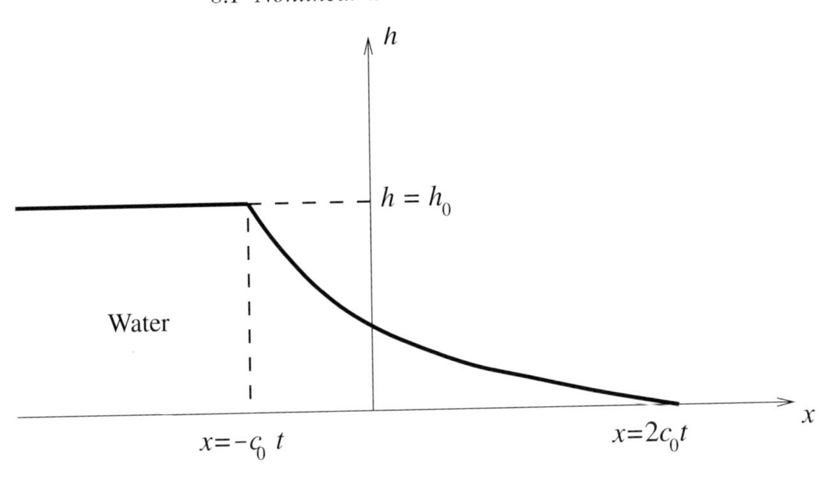

Fig. 8.4. The solution of the dam break problem.

and hence, from the solution (8.8),

$$\frac{dX_+}{dt} = \frac{4}{3}c_0 + \frac{X_+}{3t}.$$

We can easily integrate this linear, first order equation, and the solution that satisfies $X_+ = -x_0/2$ when $t = x_0/2c_0$ is

$$X_+(t) = 2c_0t - 3\left(\frac{1}{2}x_0\right)^{2/3}(c_0t)^{1/3}. \tag{8.10}$$

These curves fill the domain $x < 2c_0t$, and all satisfy $X_+ < 2c_0t$. In other words, our assumptions about these C_+ characteristics are justified. The solution can be summarised as

$$\left.\begin{array}{lr} h = h_0, \quad u = 0, & \text{for } x \le -c_0t, \\ h = \dfrac{h_0}{9}(2 - \dfrac{x}{c_0t})^2, \quad u = \dfrac{2}{3}\left(c_0 + \dfrac{x}{t}\right) & \text{for } -c_0t \le x \le 2c_0t, \\ h = 0, \quad u = 0, & \text{for } x \ge 2c_0t. \end{array}\right\} \tag{8.11}$$

This solution is in reasonable agreement with experimental observations, as shown in figure 8.5.

In this example, one set of characteristics originating in the undisturbed initial state fills the domain of solution. All of the information propagates on the C_- characteristics. This type of solution, where only one set of characteristics is involved in carrying the initial information, is known as a **simple wave**. These are similar to kinematic waves. In this example,

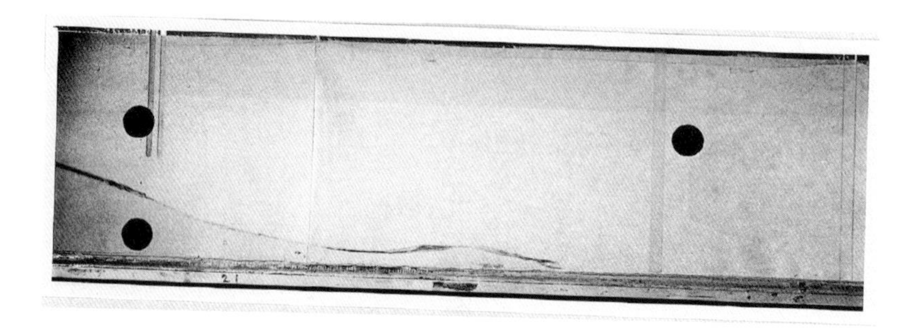

Fig. 8.5. An experimental dam break.

$u = 2(c_0 - \sqrt{gh})$ everywhere, so we can substitute for u in (8.1) and obtain

$$\frac{\partial h}{\partial t} + (2c_0 - 3\sqrt{gh})\frac{\partial h}{\partial x} = 0,$$

which is just a kinematic wave equation with kinematic wave speed $2c_0 - 3\sqrt{gh}$. This explains why all of the C_- characteristics are straight lines in the dam break problem.

More generally, a pair of simple waves is generated by any localised disturbance to a stationary, uniform expanse of shallow water, as illustrated in figure 8.6. After an initial transient, the water close to the initial disturbance returns to the undisturbed state and the simple waves separate, propagating to the right on the C_+ characteristics and to the left on the C_- characteristics.

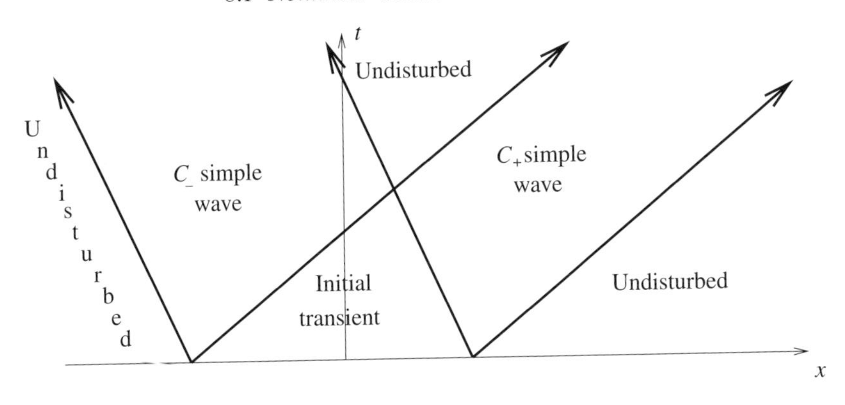

Fig. 8.6. The generation of a pair of simple waves by a localised disturbance in shallow water.

Fig. 8.7. A turbulent bore on the river Severn.

8.1.2 A Shallow Water Bore

A **bore** is just another name for a shock wave when it occurs on the surface of a fluid. These often occur on rivers when an unusually high tide enters a narrowing estuary, the most famous example being the Severn bore (see figure 8.7). We will study a particular mechanism for generating a bore.

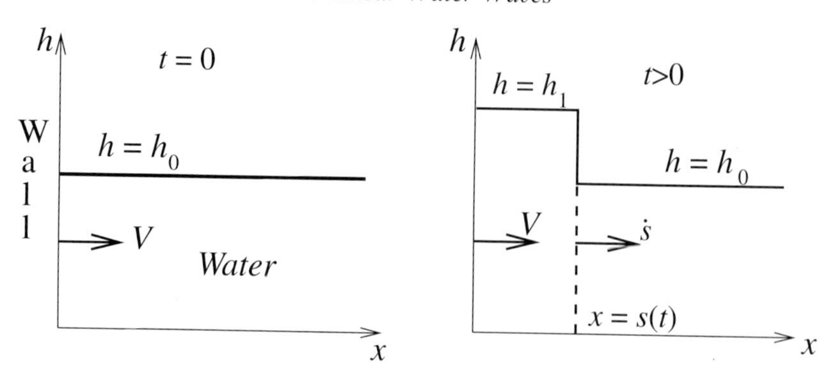

Fig. 8.8. The solution of the moving wall problem.

Consider a semi-infinite expanse of shallow water initially lying stationary in the domain $x > 0$, held back by a wall at $x = 0$. When $t = 0$ the wall starts to move into the water with velocity V. This situation is illustrated in figure 8.8. What happens when $t > 0$ in the domain $x > Vt$? The initial–boundary value problem that we must solve consists of equations (8.1) and (8.2) along with the initial conditions

$$h(x,0) = h_0, \qquad u(x,0) = 0, \qquad \text{for } x > 0, \tag{8.12}$$

and boundary condition

$$u = V, \qquad \text{at } x = Vt. \tag{8.13}$$

When $t = 0$, the slope of the C_+ characteristics that originate in the fluid is $dX_+/dt = u + \sqrt{gh} = \sqrt{gh_0}$, whilst the slope of the C_+ characteristics that originate at the moving wall is $dX_+/dt = V + \sqrt{gh_0} > \sqrt{gh_0}$. This means that the C_+ characteristics from the wall immediately intersect with those from the fluid, and hence, as we have seen for traffic flow, a shock, or bore, must form immediately, moving with velocity $\dot{s}$. The C_+ characteristics and shock locus are sketched in figure 8.9. In order to fit the shock, we must consider the integral conservation laws for mass and momentum at the shock, just as we did in subsection 7.2.4 for compressible gas dynamics. Note that, since equations (8.1) and (8.2) involve only two dependent variables, only two independent quantities can be conserved. As we shall see, energy is dissipated at the bore.

We know that conservation of mass is given by (8.1) or, in conservation form, (4.81). By integrating (4.81) over a finite length of the fluid, we

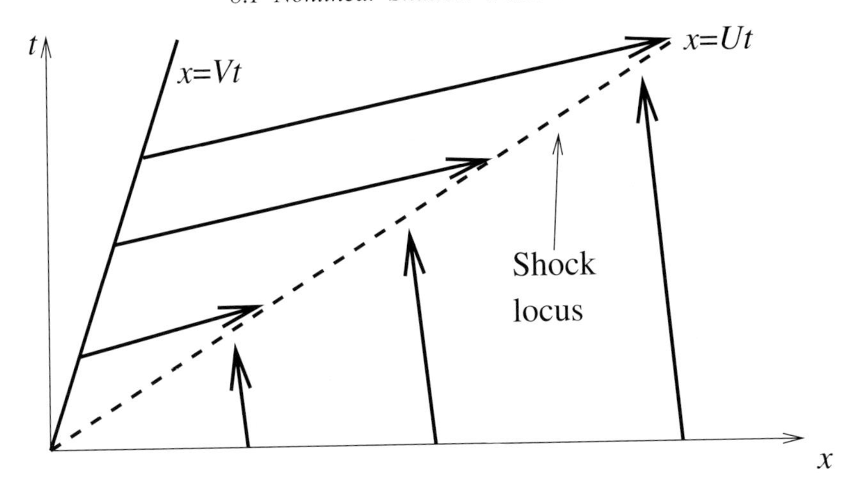

Fig. 8.9. The C_+ characteristics for the moving wall problem.

obtain the integral conservation law

$$\frac{d}{dt}\int_{x_1}^{x_2} \rho h dx + [\rho u h]_{x_1}^{x_2} = 0. \tag{8.14}$$

To obtain a similar law for conservation of momentum, we note that the rate of change of momentum in a finite length of the fluid is equal to the net flux of momentum plus the forces acting on the fluid, which we write as

$$\frac{d}{dt}\int_{x_1}^{x_2} \rho h u dx = -\left[\rho u^2 h\right]_{x_1}^{x_2} + \int_0^{h(x_1)} p(x_1, y, t) dy - \int_0^{h(x_2)} p(x_2, y, t) dy.$$

Since the pressure in a shallow water flow is hydrostatic, we have $p = \rho g(h - y)$, and hence

$$\frac{d}{dt}\int_{x_1}^{x_2} \rho h u dx + \left[\rho u^2 h + \frac{1}{2}\rho g h^2\right]_{x_1}^{x_2} = 0. \tag{8.15}$$

Using the same argument as we did for compressible gas dynamics, the quantities uh and $u^2 h + \frac{1}{2} g h^2$ must not change across a shock, *in a frame of reference moving with the shock*. If the shock is at $x = s(t)$, we can see this explicitly by noting that

$$\begin{aligned}
\frac{d}{dt}\int_{x_1}^{x_2} h dx &= \frac{d}{dt}\left(\int_{x_1}^{s(t)} + \int_{s(t)}^{x_2}\right) h dx \\
&= \left(\int_{x_1}^{s(t)} + \int_{s(t)}^{x_2}\right) \frac{\partial h}{\partial t} dx + h^+ \dot{s} - h^- \dot{s},
\end{aligned}$$

where h^+ and h^- are the values of h to the right and left of the shock. Using this in (8.14), taking the limit $x_1 \to s$ and $x_2 \to s$, leads to

$$\dot{s}\,[h]_-^+ = [uh]_-^+. \tag{8.16}$$

A similar argument for (8.15) leads to

$$\dot{s}\,[uh]_-^+ = \left[u^2h + \frac{1}{2}gh^2\right]_-^+. \tag{8.17}$$

These are the equivalent for shallow water bores of the Rankine–Hugoniot relations (7.47) and (7.48). For the moving wall problem, this means that

$$h_1(V - \dot{s}) = -h_0\dot{s}, \quad h_1(V - \dot{s})^2 + \frac{1}{2}gh_1^2 = h_0\dot{s}^2 + \frac{1}{2}gh_0^2, \tag{8.18}$$

where h_1 is the height of water behind the shock, which must move with the wall at velocity V. If we eliminate the shock speed, $\dot{s}$, we arrive at

$$(1 - H)^2(1 + H) = 2\mathrm{Fr}^2 H, \tag{8.19}$$

where

$$H = \frac{h_1}{h_0}, \quad \mathrm{Fr} = \frac{V}{\sqrt{gh_0}}.$$

When the **Froude number**, Fr, is large, gravitational forces are small, and vice versa. By plotting the left and right hand sides of (8.19), as shown in figure 8.10, we can see that there are two solutions, one with $H > 1$, one with $H < 1$. For the moving wall problem it is clear that we need $H > 1$, since

$$\frac{\dot{s}}{V} = \frac{H}{H - 1}, \tag{8.20}$$

and we need $\dot{s} > 0$. In terms of the physical variables, the shock is therefore at

$$s(t) = \frac{h_1 V}{h_1 - h_0}t > Vt.$$

For $\mathrm{Fr} \ll 1$, $H - 1 \ll 1$, so if the wall has a small velocity relative to the local sound speed, there is only a small change in depth across the shock, whilst for $\mathrm{Fr} \gg 1$, $H \gg 1$, as we might expect.

Finally, we can consider the rate at which energy is dissipated across the shock. For an isolated bore, generated by some mechanism other than a moving wall, it is the requirement that energy is dissipated at the bore that selects the appropriate solution of (8.19). The density of energy in the fluid is the sum of the kinetic and hydrostatic potential energies,

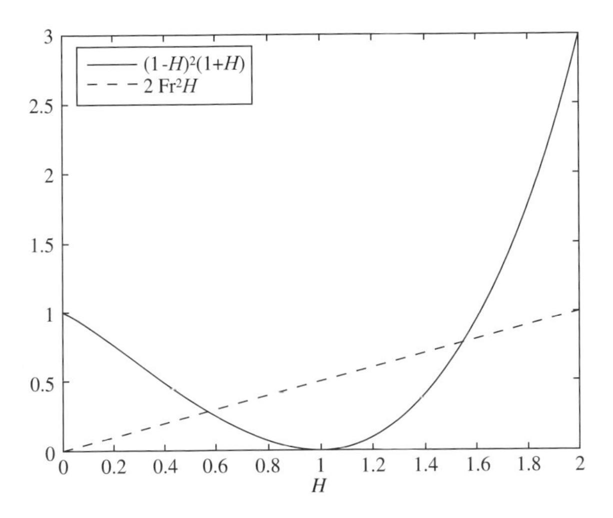

Fig. 8.10. The left and right hand sides of (8.19) for $\text{Fr} = \frac{1}{2}$.

$\frac{1}{2}\rho h u^2 + \frac{1}{2}\rho g h^2$. The rate of change of this energy in a finite length of fluid is equal to the net flux of energy plus the rate of working of the hydrostatic forces acting on the fluid minus the rate of energy dissipation, $L(t)$, assuming that there is a single shock lying within the length of fluid. This leads to

$$\frac{d}{dt}\int_{x_1}^{x_2}\left(\frac{1}{2}\rho h u^2 + \frac{1}{2}\rho g h^2\right)dx = -\left[\frac{1}{2}\rho h u^3 + \frac{1}{2}\rho g h^2 u\right]_{x_1}^{x_2}$$

$$+\int_0^{h(x_1)} p(x_1, y, t)u(x_1, y, t)dy - \int_0^{h(x_2)} p(x_2, y, t)u(x_1, y, t)dy - L,$$

and hence

$$\frac{d}{dt}\int_{x_1}^{x_2}\left(\frac{1}{2}\rho h u^2 + \frac{1}{2}\rho g h^2\right)dx + \left[\frac{1}{2}\rho h u^3 + \rho g h^2 u\right]_{x_1}^{x_2} + L = 0. \tag{8.21}$$

The usual argument then gives

$$L(t) = \dot{s}\left[\frac{1}{2}\rho h u^2 + \frac{1}{2}\rho g h^2\right]_-^+ - \left[\frac{1}{2}\rho u^3 h + g h^2 u\right]_-^+. \tag{8.22}$$

For the moving wall problem,

$$L = \rho\left\{\frac{1}{2}(V - \dot{s})^3 h_1 + (V - \dot{s})g h_1^2 + \frac{1}{2}\dot{s}^3 h_0 + \dot{s} g h_0^2\right\}. \tag{8.23}$$

Fig. 8.11. An undular bore.

By using (8.19) and (8.20) we can reduce this to

$$L = \frac{1}{4}\rho g \dot{s} \frac{(h_1 - h_0)^3}{h_1}. \tag{8.24}$$

For weak bores, where $(h_1 - h_0)/h_0 \ll 1$, the rate of dissipation of the energy is cubically small, echoing the result (7.59) for the rate of production of entropy in gas dynamical shock. For an isolated bore, the requirement that L should be positive means that we need $h_1 > h_0$ for $\dot{s} > 0$. In other words, the depth of fluid must always increase across a bore moving into undisturbed water.

The discontinuous bore that we have considered here is not the only type that is observed. If conditions are right, there is the possibility of a continuous **undular bore**, as shown in figure 8.11. This type of bore, and its relation to nonlinear waves and the discontinuous bore, are discussed by Benjamin and Lighthill (1954).

8.2 The Effect of Nonlinearity on Deep Water Gravity Waves: Stokes' Expansion

In subsection 4.2.1 we determined the leading order solution for progressive gravity waves of small amplitude. By assuming that the amplitude of the wave was much smaller than its wavelength, we were able to linearise the governing equations. In this section, we develop this solution as a formal asymptotic expansion, and show how the waveform and wave speed are affected by the amplitude of the wave. This analysis was first carried out by Sir George Stokes in 1847.

Consider a periodic wave moving from right to left on deep water, with height H (equal to twice the amplitude) and wavelength λ. The small parameter upon which we base our asymptotic expansion is $\epsilon = H/\lambda$. For typical waves generated by the action of the wind, $\epsilon \approx 0.05$, whereas the rather bigger, but still non-breaking, waves generated by a storm may have $\epsilon \approx 0.13$. The maximum value of c for which an irrotational theory is valid is $\epsilon \approx 0.14$. At higher values of ϵ the waves may break, or other physical effects, such as vorticity, could be important (Lamb, 1932).

The flow we consider here can be made steady by choosing a horizontal axis moving with the phase velocity of the wave. This induces a uniform horizontal flow of speed c at great depths. The passage of the wave past any fixed point may induce a change in mean water depth at that point, which is unknown *a priori*. We choose our vertical axis so that the free surface is at $y = 0$ in the absence of wave propagation. We will actually calculate the change in mean depth. Suitable dimensionless variables are

$$\bar{x} = \frac{x}{\lambda}, \quad \bar{y} = \frac{y}{\lambda}, \quad \bar{\phi} = \frac{\phi}{cH}, \quad \bar{\eta} = \frac{\eta}{H}.$$

In terms of these variables, the nonlinear free boundary problem given by (4.3), (4.4) and (4.6) becomes

$$\nabla^2 \bar{\phi} = 0 \quad \text{for } \bar{y} < \epsilon \bar{\eta}, \tag{8.25}$$

$$\frac{1}{2}\epsilon\left(\bar{\phi}_{\bar{x}}^2 + \bar{\phi}_{\bar{y}}^2\right) + K\bar{\eta} = B, \quad \text{at } \bar{y} = \epsilon\bar{\eta}, \tag{8.26}$$

$$\bar{\phi}_{\bar{y}} = \epsilon \bar{\phi}_{\bar{x}} \bar{\eta}_{\bar{x}}, \quad \text{at } \bar{y} = \epsilon\bar{\eta}, \tag{8.27}$$

$$\bar{\phi} \sim \frac{x}{\epsilon}, \quad \text{as } \bar{y} \to -\infty, \tag{8.28}$$

where $K = \lambda g/c^2$. The **Bernoulli constant** B in (8.26) is to be determined, and represents the energy in the fluid. Finally, the solution must be periodic, with period one, whilst the definition of the wave height as the difference in water depth between the peak and the trough gives

$$\bar{\eta}(0) - \bar{\eta}(\tfrac{1}{2}) = -1. \tag{8.29}$$

At leading order, the velocity potential must be $\bar{\phi} = \bar{x}/\epsilon$, which gives $B \sim 1/2\epsilon$ in (8.26). We therefore pose asymptotic expansions

$$\phi = \frac{x}{\epsilon} + \phi_0 + \epsilon\phi_1 + +\epsilon^2\phi_2 + O(\epsilon^3), \quad \eta = \eta_0 + \epsilon\eta_1 + \epsilon^2\eta_2 + O(\epsilon^3),$$

dropping the bars on the variables for notational convenience. We must also pose expansions for the unknown constants B and, more importantly since it is related to the unknown wave speed, K as

$$B = \frac{1}{2\epsilon} + B_0 + \epsilon B_1 + O(\epsilon^2), \quad K = K_0 + \epsilon K_1 + \epsilon^2 K_2 + O(\epsilon^3).$$

Now at leading order,

$$\nabla^2\phi_0 = 0 \quad \text{for } y < 0, \tag{8.30}$$

$$\phi_{0x} + K_0\eta_0 = B_0, \quad \text{at } y = 0, \tag{8.31}$$

$$\phi_{0y} = \eta_{0x}, \quad \text{at } y = 0, \tag{8.32}$$

$$\phi_0 \to 0, \quad \text{as } y \to -\infty, \tag{8.33}$$

$$\eta_0(0) - \eta_0\left(\frac{1}{2}\right) = -1. \tag{8.34}$$

The harmonic solution that decays as $y \to -\infty$ and has period one is

$$\phi_0 = Ae^{2\pi y}\sin 2\pi x.$$

If we eliminate η_0 from the boundary conditions at the free surface, we find that

$$\phi_{0xx} + K_0\phi_{0y} = 0 \quad \text{at } y = 0,$$

and hence $K_0 = 2\pi$. This gives

$$\eta_0 = \frac{B_0}{2\pi} - A\cos 2\pi x.$$

Finally, (8.34) shows that $A = \frac{1}{2}$, and hence

$$\phi_0 = \frac{1}{2}e^{2\pi y}\sin 2\pi x, \quad \eta_0 = \frac{B_0}{2\pi} - \frac{1}{2}\cos 2\pi x, \quad K_0 = 2\pi. \tag{8.35}$$

This is just the linearised solution that we obtained in subsection 4.2.1.

We can now consider the boundary value problem at next order. Note that we must be careful to take into account the small displacement of the free surface by Taylor expanding the terms in the boundary conditions, for example

$$\phi_0(x, \epsilon\eta_0 + O(\epsilon^2)) = \phi_0(x, 0) + \epsilon\eta_0\phi_{0y}(x, 0) + O(\epsilon^2).$$

We find that

$$\nabla^2\phi_1 = 0 \quad \text{for } y < 0, \tag{8.36}$$

$$\phi_{1x} + K_0\eta_1 = B_1 - \frac{1}{2}\left(\phi_{0x}^2 + \phi_{oy}^2\right) - \eta_0\phi_{0xy} - K_1\eta_0, \quad \text{at } y = 0, \tag{8.37}$$

$$\phi_{1y} - \eta_{1x} = \phi_{0x}\eta_{0x} - \eta_0\phi_{0yy}, \quad \text{at } y = 0, \tag{8.38}$$

$$\phi_1 \to 0, \quad \text{as } y \to -\infty, \tag{8.39}$$

$$\eta_1(0) - \eta_1\left(\frac{1}{2}\right) = 0. \tag{8.40}$$

Making use of the leading order solution (8.35) in the boundary conditions leads to

$$\phi_{1x} + 2\pi\eta_1 = B_1 + \frac{1}{2}\pi^2\cos 4\pi x - \frac{1}{2}K_1\cos 2\pi x, \quad \text{at } y = 0, \tag{8.41}$$

$$\phi_{1y} - \eta_{1x} = \pi^2\sin 4\pi x + \pi B_0\sin 2\pi x, \quad \text{at } y = 0. \tag{8.42}$$

The terms proportional to $\cos 2\pi x$ and $\sin 2\pi x$ in (8.41) and (8.42) give rise to terms proportional to $x\cos 2\pi x$ in the solution, which are secular. In order to keep our expansion asymptotic, we must therefore remove them by taking $K_1 = 0$ and $B_0 = 0$. The homogeneous problem has solutions of the same form as the leading order problem, (8.30) to (8.34), but (8.40) shows that these are not appropriate. We must therefore take the solution forced by the $\sin 4\pi x$ and $\cos 4\pi x$ terms in (8.41) and (8.42). These arise from products of $\sin 2\pi x$ and $\cos 2\pi x$ terms. The solution is

$$\phi_1 = 0, \quad \eta_1 = \frac{B_1}{2\pi} + \frac{\pi}{4}\cos 4\pi x, \quad K_1 = 0. \tag{8.43}$$

Although we have found the functional form of the correction to the free surface displacement, we have not yet determined the constant B_1. In addition, since $K_1 = 0$, we can see that the correction to the wave speed, which is determined by K_2, is of $O(\epsilon^2)$ and as yet undetermined. To summarise,

$$\phi = \frac{x}{\epsilon} - \frac{1}{2}e^{2\pi y}\sin 2\pi x + \epsilon^2\phi_2 + O(\epsilon^3),$$

$$\eta = \frac{1}{2}\cos 2\pi x + \epsilon\left(\frac{B_1}{2\pi} + \frac{\pi}{4}\cos 4\pi x\right) + \epsilon^2\eta_2 + O(\epsilon^3),$$

$$B = \frac{1}{2\epsilon} + \epsilon B_1 + \epsilon^2 B_2 + O(\epsilon^3), \quad K = \frac{\lambda g}{c^2} = 2\pi + \epsilon^2 K_2 + O(\epsilon^3).$$

However, we can now see a pattern in the solution procedure. The constants B_0 and K_1 were determined by the condition that no secular terms should appear in the free surface boundary conditions. If we determine the boundary value problem at the next order, we can fix B_1 and K_2 in exactly the same way, without having to actually determine the solution. The coefficient of $\cos 2\pi x$ in the Bernoulli condition, (8.26), at $O(\epsilon^2)$, which must be zero, is

$$-\pi B_2 + \frac{1}{2}K_2 + \frac{3}{8}\pi^3 = 0.$$

The coefficient of $\sin 2\pi x$ in the kinematic condition, (8.27), at $O(\epsilon^2)$, which must also be zero, is

$$\pi B_1 + \frac{5}{8}\pi^3 = 0.$$

We therefore find that

$$K_2 = -2\pi^3, \quad B_1 = -\frac{5}{8}\pi^2.$$

This gives us the corrected solution as

$$\phi = \frac{x}{\epsilon} - \frac{1}{2}e^{2\pi y}\sin 2\pi x + O(\epsilon^2), \tag{8.44}$$

$$\eta = \frac{1}{2}\cos 2\pi x + \epsilon\left(\frac{\pi}{4}\cos 4\pi x - \frac{5\pi}{16}\right) + O(\epsilon^2), \tag{8.45}$$

$$K^{-1} = \frac{c^2}{g\lambda} = \frac{1}{2\pi}\left(1 + \pi^2\epsilon^2\right) + O(\epsilon^3). \tag{8.46}$$

The weakly nonlinear effect of the finite height of the wave is that the wave speed increases, with

$$c^2 \sim \frac{g\lambda}{2\pi}\left(1 + \frac{H^2\pi^2}{\lambda^2}\right), \tag{8.47}$$

whilst the solution for η indicates that the troughs broaden and the crests become narrower as the amplitude increases, as shown in figure 8.12.

We can see from (8.45) that there is a change in mean depth of $\frac{5\pi}{16}\frac{H}{\lambda} + O\left(\frac{H^2}{\lambda^2}\right)$ due to the passage of the wave. If the particle paths are calculated it is found that they are not closed, which indicates that there is a weak mass transport, or **drift**, in the horizontal direction.

In recent times there has been much interest in the use of numerical methods to investigate the properties of nonlinear water waves. Padé

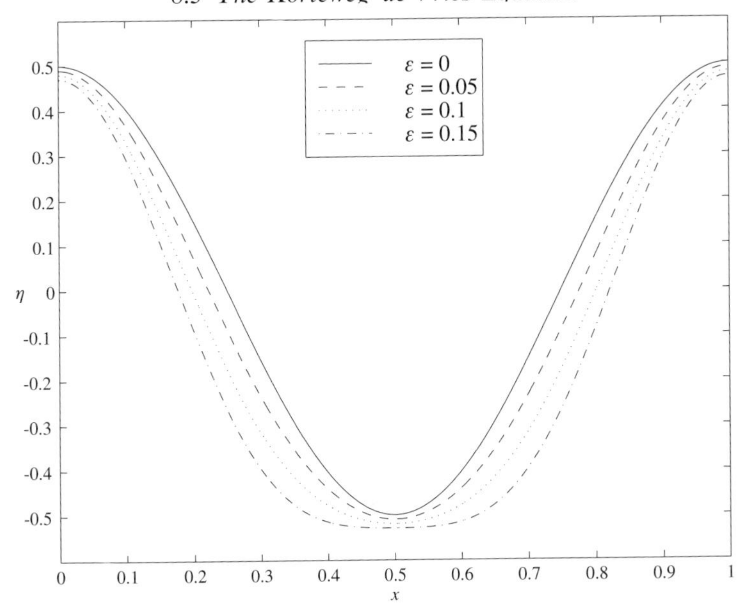

Fig. 8.12. (a) The weakly nonlinear solution (8.45) for various values of ϵ.

approximants have been used to give some interesting results for extreme waves (Longuet-Higgins, 1973, Cokelet, 1977). Bloor (1978) used a conformal mapping technique to investigate capillary–gravity waves in a computationally efficient manner. Integral equation methods have been successfully applied to interfacial waves between fluids of different densities by Turner and Vanden-Broeck (1988) and King and Moni (1995).

The stability of progressive gravity waves to small disturbances has been the subject of many recent studies. The most straightforward results concern the the stability of waves to small amplitude **sideband disturbances**. Using asymptotic methods similar to those used in this section, it can be shown that resonant interactions between modes take energy from the main wave and may eventually cause it to disintegrate. This is known as **Benjamin–Feir instability** (Benjamin and Feir, 1967).

8.3 The Korteweg–de Vries Equation for Shallow Water Waves: the Interaction of Nonlinear Steepening and Linear Dispersion

An analysis of the propagation of waves on the surface of a body of water usually leads to a relationship between wave speed, c, water depth, h,

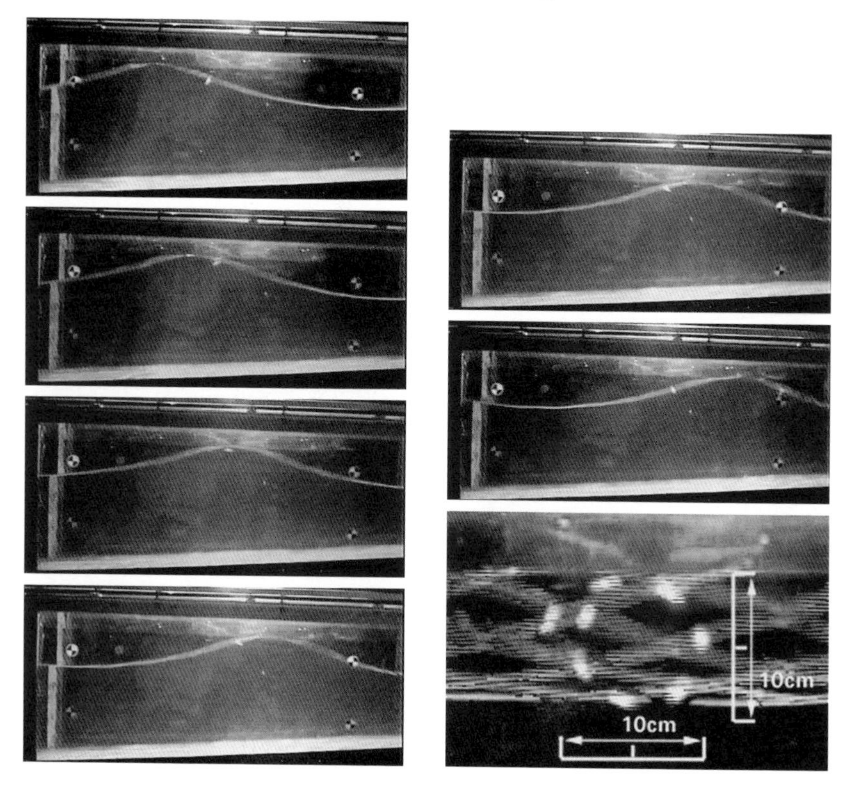

Fig. 8.12. (b) Experimentally observed waves in a channel and their (particle) drift.

wave amplitude, a, wavelength, λ, gravity, g, and possibly surface tension, σ. In the linear theory of water waves, which we studied in chapter 4, two of the main simplifying assumptions are that the wave amplitude is small compared to the mean depth of the water and that the wavelength is large compared to the amplitude. This leads to the dispersion relation

$$c^2 = \frac{g\lambda}{2\pi} \tanh\left(\frac{2\pi h}{\lambda}\right),$$

for gravity waves. We can see that waves with different wavelengths propagate at different speeds, but that the amplitude of the wave does not affect the wave speed – the waves are dispersive but linear. One exception is in the limit of shallow water, $h \ll \lambda$, when $c^2 \sim gh$ and hence no dispersion occurs. In the theory of shallow water waves, we took the amplitude of the disturbance of the surface to be comparable to the depth of the water and neglected vertical accelerations. This leads to a theory in which the governing equations are nonlinear and the wave speed is dependent on the amplitude of the wave (see Peregrine (1967)). However, waves in shallow water are not dispersive. In these two different analyses of the equations that govern the behaviour of water waves there are two important geometrical parameters: $\epsilon = h/\lambda$ and $\alpha = a/h$. By choosing appropriate magnitudes for ϵ and α, we can consider a theory in which dispersion and nonlinearity are in balance, essentially combining the two effects that we studied in the first two sections of this chapter. Nonlinearity tends to steepen wavefronts, whilst dispersion tends to spread them out. As we shall see in chapter 12, when these two effects are in balance, some remarkable waves, known as **solitons**, can be observed.

8.3.1 Derivation of the Korteweg–de Vries Equation

The full equations that govern the propagation of disturbances on the free surface of a body of inviscid water over a flat bed of mean depth h are

$$\phi_{xx} + \phi_{yy} = 0 \quad \text{for } -\infty < x < \infty,\ 0 \le y \le \eta(x,t), \tag{8.48}$$

$$\left.\begin{aligned} &\phi_y = \eta_t + \phi_x\eta_x, \\ &\phi_t + \tfrac{1}{2}(\phi_x^2 + \phi_y^2) + g\eta = B(t) \end{aligned}\right\} \text{ at } y = \eta = h + aH(x,t), \tag{8.49}$$

$$\phi_y = 0 \quad \text{at } y = 0. \tag{8.50}$$

It is now convenient to make these equations dimensionless using the scaled variables

$$\bar{x} = \frac{x}{\lambda},\quad \bar{y} = \frac{y}{h},\quad \bar{\phi} = \frac{h\phi}{\lambda a\sqrt{gh}},\quad \bar{t} = \frac{t\sqrt{gh}}{\lambda}. \tag{8.51}$$

We can gain some insight into why these are appropriate variables by noting that $\sqrt{gh}$ is the linear, non-dispersive wave speed for small-amplitude disturbances. Since the wavelength λ is the typical horizontal length scale, $\lambda/\sqrt{gh}$ is a time scale during which a fluid particle moves one wavelength in the horizontal direction. Using the scaled variables (8.51) in (8.48) to (8.50) we obtain

$$\epsilon^2\bar{\phi}_{\bar{x}\bar{x}} + \bar{\phi}_{\bar{y}\bar{y}} = 0, \quad \text{for } -\infty < \bar{x} < \infty,\ 0 \le \bar{y} \le 1 + \alpha H(\bar{x},\bar{t}), \tag{8.52}$$

$$\bar{\phi}_{\bar{y}} = \epsilon^2 \left(H_{\bar{t}} + \alpha \bar{\phi}_{\bar{x}} H_{\bar{x}} \right) \quad \text{at } y = 1 + \alpha H(\bar{x}, \bar{t}), \tag{8.53}$$

$$\bar{\phi}_{\bar{t}} + \tfrac{1}{2}\alpha(\bar{\phi}_{\bar{x}}^2 + \epsilon^{-2}\bar{\phi}_{\bar{y}}^2) + H = (B(t) - gh)/ag, \quad \text{at } y = 1 + \alpha H(\bar{x}, \bar{t}), \tag{8.54}$$

$$\bar{\phi}_{\bar{y}} = 0 \quad \text{at } \bar{y} = 0, \tag{8.55}$$

where $\epsilon = h/\lambda$ and $\alpha = a/h$. At this point in our analysis it is convenient to make the right hand side of (8.54) zero using the transformation

$$\bar{\phi} \mapsto \bar{\phi} + \int_0^{\bar{t}} \left(\frac{B(s) - gh}{ag} \right) ds,$$

which does not affect the velocity field, $\bar{\mathbf{u}} = \nabla \bar{\phi}$.

We must now examine the double limiting process, $\epsilon \ll 1$, $\alpha \ll 1$. This is always a rather delicate business which can lead to mathematical complications. Fortunately there is a transformation of the equations that completely removes ϵ from the problem so that we can analyse the equations for $\alpha \ll 1$ and forget about ϵ for the moment. Although this transformation can be found systematically in a straightforward way, we will omit the details and define

$$z = \frac{\alpha^{1/2}}{\epsilon}(\bar{x} - \bar{t}), \quad \tau = \frac{\alpha^{3/2}}{\epsilon}\bar{t}, \quad \Phi = \frac{\alpha^{1/2}}{\epsilon}\bar{\phi}, \tag{8.56}$$

so that z is a coordinate centred on a point moving left to right at the linear shallow water wave speed and initially at the origin. The scaled velocity potential, Φ, and scaled time, τ, are simple multiples of $\bar{\phi}$ and $\bar{t}$, whilst the vertical coordinate, $\bar{y}$, is unscaled. Using the scaled variables (8.56), (8.52) to (8.55) become

$$\alpha \Phi_{zz} + \Phi_{\bar{y}\bar{y}} = 0 \quad \text{for } -\infty < z < \infty,\ 0 \le \bar{y} \le 1 + \alpha H(z, \tau), \tag{8.57}$$

$$\Phi_{\bar{y}} = \alpha \left(-H_z + \alpha H_\tau + \alpha \Phi_z H_z \right) \quad \text{at } y = 1 + \alpha H(z, \tau), \tag{8.58}$$

$$H - \Phi_z + \alpha \Phi_\tau + \tfrac{1}{2}\alpha(\Phi_{\bar{y}}^2 + \alpha \Phi_z^2) = 0 \quad \text{at } y = 1 + \alpha H(z, \tau), \tag{8.59}$$

$$\Phi_{\bar{y}} = 0 \quad \text{at } \bar{y} = 0. \tag{8.60}$$

We now pose asymptotic expansions of the form

$$\Phi = \Phi_0 + \alpha \Phi_1 + \alpha^2 \Phi_2 + o(\alpha^2), \tag{8.61}$$

$$H = H_0 + \alpha H_1 + o(\alpha), \tag{8.62}$$

for the velocity potential and position of the free surface. Substituting expansion (8.61) into (8.57) yields at leading order

$$\Phi_{0,\bar{y}\bar{y}} = 0 \Rightarrow \Phi_0 = B_0(z,\tau),$$

at $O(\alpha)$

$$\Phi_{1,\bar{y}\bar{y}} = -\Phi_{0,zz} = -B_{0,zz} \Rightarrow \Phi_1 = -\frac{1}{2}\bar{y}^2 B_{0,zz} + B_1(z,\tau),$$

and at $O(\alpha^2)$

$$\Phi_{2,\bar{y}\bar{y}} = -\Phi_{1,zz} = \frac{1}{2}\bar{y}^2 B_{0,zzzz} + B_{1,zz}$$

$$\Rightarrow \Phi_2 = \frac{1}{24}\bar{y}^4 B_{0,zzzz} - \frac{1}{2}\bar{y}^2 B_{1,zz} + B_2(z,\tau).$$

Note that in solving each of the above equations we have made use of the boundary condition (8.60) at the flat bed, $\bar{y} = 0$. We shall find below that these three terms are sufficient for our purposes, and turn now to the boundary conditions (8.58) and (8.59) at the free surface.

At leading order the Bernoulli equation, (8.59), gives

$$H_0(z,\tau) = \Phi_{0,z} = B_{0,z}, \tag{8.63}$$

a result that also allows the kinematic boundary condition, (8.58), to be satisfied both at leading order and at $O(\alpha)$. From (8.59) at $O(\alpha)$ we obtain

$$H_1 - B_{1,z} + \frac{1}{2}B_{0,zzz} + B_{0,\tau} + \frac{1}{2}B_{0,z}^2 = 0, \tag{8.64}$$

whilst from (8.58) at $O(\alpha^2)$ we obtain

$$-H_0 B_{0,zz} + \frac{1}{6}B_{0,zzzz} - B_{1,zz} = -H_{1,z} + H_{0,\tau} + B_{0,z}H_{0,z}. \tag{8.65}$$

If we now differentiate (8.64) we find that

$$H_{1,z} - B_{1,zz} = -\frac{1}{2}B_{0,zzzz} - B_{0,z\tau} - B_{0,z}B_{0,zz},$$

which allows us to eliminate H_1 and B_1 from (8.65) and arrive at

$$-H_0 B_{0,zz} - \frac{1}{3}B_{0,zzzz} - B_{0,z\tau} - B_{0,z}B_{0,zz} = H_{0,\tau} + B_{0,z}H_{0,z}.$$

Finally, (8.63) gives B_0 in terms of H_0, and hence

$$2H_{0,\tau} + 3H_0 H_{0,z} + \frac{1}{3}H_{0,zzz} = 0. \tag{8.66}$$

This is the celebrated Korteweg–de Vries (KdV) equation, which includes dispersive effects through the term $H_{0,zzz}$ and nonlinear effects through the term $H_0 H_{0,z}$, and governs the behaviour of the small amplitude waves, with $\alpha \ll 1$, that we are considering here.

It is now reasonable to ask when and where the independent variables, z and τ, are of $O(1)$, in order to determine more precisely the region in physical space where the KdV equation is valid as an approximation of the actual flow. If we consider (8.56), which defines z and τ, we find that if $\alpha = O(\epsilon^2)$, then $\bar{t} \gg 1$ and $\bar{x} = \bar{t} + O(1)$. This leads us to interpret any waveform that arises as a solution of the KdV equation as the large time limit of an initial value problem. As we shall see in the next subsection, both nonlinear periodic wave and solitary wave solutions exist. Such waves must travel at a speed that is asymptotically close to the linear shallow water wave speed in order for them to lie in an $O(1)$ region about the point $\bar{x} = \bar{t}$, or equivalently $x = \sqrt{gh}\, t$.

8.3.2 Travelling Wave Solutions of the KdV Equation

The KdV equation has two qualitatively different types of permanent form travelling wave solution. These are referred to as cnoidal waves and solitary waves, for reasons that will become clear below. Before embarking upon an analysis of these waves we must consider what we should expect to find in (z, τ)-space, having made the various transformations that led up to the KdV equation. In terms of the physical variables, a periodic, permanent form travelling wave with wavelength λ has an elevation that satisfies the relationship

$$\eta(x - ct) = \eta(x - ct + \lambda).$$

After the first change of variables, $\bar{x} = x/\lambda$, $\bar{t} = \sqrt{gh}\, t/\lambda$, this leads to

$$H\left(\bar{x} - \frac{c}{\sqrt{gh}}\bar{t}\right) = H\left(\bar{x} - \frac{c}{\sqrt{gh}}\bar{t} + 1\right).$$

In order to be consistent with the requirement that these waves travel at a speed close to the linear shallow water wave speed, we write

$$c = \sqrt{gh}(1 + V),$$

where $V \ll 1$. This gives us

$$H(\bar{x} - \bar{t} - V\bar{t}) = H(\bar{x} - \bar{t} - V\bar{t} + 1).$$

Since we have also assumed that $\epsilon = O(\alpha^2)$, we can write $\alpha^{1/2} = \epsilon_0 \epsilon$, where $\epsilon_0 = O(1)$ for $\epsilon \ll 1$, and hence, in terms of the final scaled variables

$$H\left(z - \frac{V}{\alpha}\tau\right) = H\left(z - \frac{V}{\alpha}\tau + \epsilon_0\right).$$

It is now clear that the wavelength of a periodic solution of the KdV equation is ϵ_0, and for the wave speed to be of $O(1)$ we require $V = O(\alpha)$, and hence can write $V = V_0\alpha$, with $V_0 = O(1)$ for $\alpha \ll 1$. Similar remarks apply to the wave amplitude, and we write this as $\alpha_0 = O(1)$ in the (z, τ)-plane.

If we now substitute $H = H(z - V_0\tau)$ into the KdV equation (8.66), we obtain

$$-2V_0H_\xi + 3HH_\xi + \frac{1}{3}H_{\xi\xi\xi} = 0, \tag{8.67}$$

where $\xi = z - V_0\tau$ is the travelling wave coordinate. This equation can be integrated twice to give

$$\frac{1}{6}H_\xi^2 = V_0H^2 - \frac{1}{2}H^3 + D_1H + D_2 = f(H, V_0, D_1, D_2), \tag{8.68}$$

where D_1 and D_2 are constants of integration. The function f is a cubic polynomial in H, and we will assume that it can be factorised into the form

$$f = \frac{1}{2}(H - H_1)(H - H_2)(H_3 - H). \tag{8.69}$$

Note that

$$V_0 = \frac{1}{2}(H_1 + H_2 + H_3). \tag{8.70}$$

There are two physically important cases that we need to consider, characterised by the nature of the roots H_1, H_2 and H_3 of $f = 0$. Both cases occur when f has three real roots and, if we assume that these are ordered so that $H_3 \geq H_2 \geq H_1$, they are (see exercise 8.5):

(i) $H_3 > H_2 > H_1$, as shown in figure 8.13.
In this case, $f \geq 0$ for $H_2 \leq H \leq H_3$.

(ii) $H_2 = H_1 = 0$, as shown in figure 8.14.
In this case, $f \geq 0$ for $0 \leq H \leq H_3$.

Case (ii) is a subcase of case (i), and we will return to it when we have completed our analysis of case (i). We are only interested in the solution of (8.68) for $H_2 \leq H \leq H_3$, where f is positive, and can sketch the (H_ξ, H)-phase plane in figure 8.15. The choice of both positive and negative square roots establishes the existence of a closed curve in this

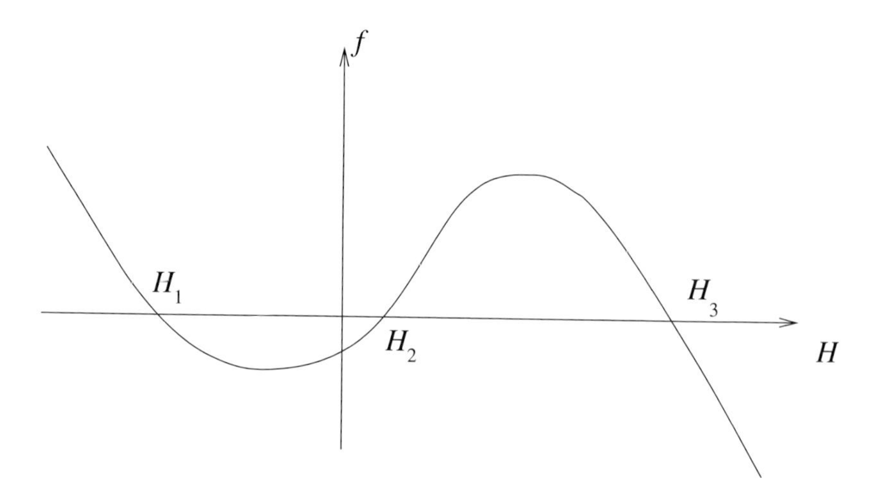

Fig. 8.13. The function f in case (i).

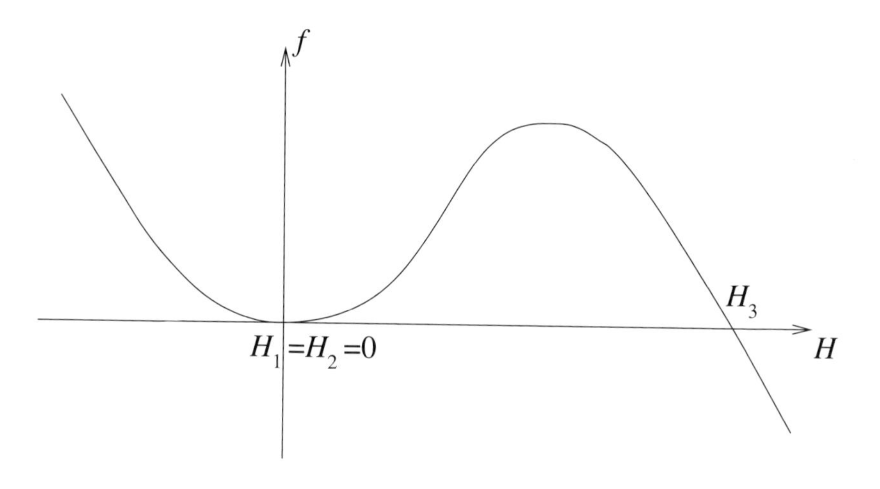

Fig. 8.14. The function f in case (ii).

phase plane, which we can parameterise by the values of V_0, H_2 and H_3. Standard results on the phase plane then allow us to deduce that this represents a periodic solution in (ξ, H)-space, with period or wavelength

$$\epsilon_0 = 2\int_{H_2}^{H_3} \frac{dH}{\sqrt{3(H-H_1)(H-H_2)(H_3-H)}}$$

and wave amplitude

$$\alpha_0 = H_3 - H_2.$$

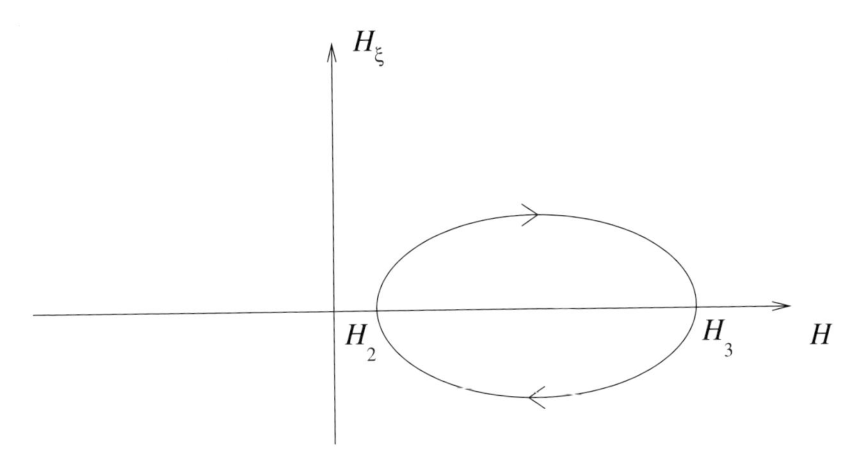

Fig. 8.15. The solution of (8.68) in case (i), plotted in the (H, H_ξ)-phase plane.

We now consider the solution of the differential equation (8.68) in more detail in order to obtain a representation of its solutions in terms of standard functions from which we can deduce more quantitative detail. If we define $X = H/H_3$, (8.68) becomes

$$X_\xi^2 = 3H_3(X - X_1)(X - X_2)(1 - X), \tag{8.71}$$

where $X_i = H_i/H_3$ for $i = 1$ or 2. We take the crest of the wave to be at $\xi = 0$, so that

$$X(0) = 1,$$

which fixes the phase of the wave. If we define a further variable Y via the relation

$$X = 1 + (X_2 - 1)\sin^2 Y,$$

we find that

$$Y_\xi^2 = \frac{3}{4}H_3(1 - X_1)\left\{1 - \left(\frac{1 - X_2}{1 - X_1}\right)\sin^2 Y\right\}, \tag{8.72}$$

and

$$Y(0) = 0. \tag{8.73}$$

In order to get this into a completely standard form we define

$$k^2 = \frac{1 - X_2}{1 - X_1}, \quad l = \frac{3}{4}H_3(1 - X_1).$$

Clearly $0 \le k^2 \le 1$ and $l > 0$. A simple quadrature of equation (8.72) subject to the condition (8.73) then gives us

$$\int_0^Y \frac{ds}{\sqrt{1-k^2\sin^2 s}} = \sqrt{l}\,\xi. \tag{8.74}$$

The real-valued Jacobian elliptic function $y = \mathrm{sn}(x;k)$ can be written in the form

$$x = \int_0^y \frac{dt}{\sqrt{1-t^2}\sqrt{1-k^2t^2}}, \quad \text{for } 0 < k^2 < 1,$$

or equivalently

$$x = \int_0^{\sin^{-1} y} \frac{ds}{\sqrt{1-k^2\sin^2 s}}.$$

We can therefore write (8.74) as

$$\sin Y = \mathrm{sn}(\sqrt{l}\,\xi;k),$$

and hence

$$X = 1 + (X_2 - 1)\mathrm{sn}^2(\sqrt{l}\,\xi;k).$$

It is conventional to write this in terms of the Jacobian elliptic function $\mathrm{cn}(x;k)$, hence the term **cnoidal wave**. Using the result $\mathrm{cn}^2 + \mathrm{sn}^2 = 1$, our final result can be expressed in the form

$$H = H_2 + (H_3 - H_2)\,\mathrm{cn}^2\left(\left\{\frac{3}{4}(H_3 - H_1)\right\}^{1/2}\xi;k\right). \tag{8.75}$$

The values of the function $\mathrm{cn}(x;k)$ have been tabulated, and can also be found using standard computer software. Some typical cnoidal waves are shown in figure 8.16. Note that the wavelength is a function of the amplitude of the wave, and that the greater the amplitude, in other words, the further we are from the small amplitude, linear case, the less sinusoidal the wave appears, with a tendency to narrow peaks and broad troughs, as we found in section 8.2 for weakly nonlinear waves on water of infinite depth (see figure 8.12). The expression for the dimensionless wavelength can also be manipulated into a standard form using the above series of transformations. It is straightforward to show that

$$\epsilon_0 = \frac{4}{\sqrt{3(H_3 - H_1)}}\int_0^{\pi/2} \frac{ds}{\sqrt{1-k^2\sin^2 s}} = \frac{4K(k)}{\sqrt{3(H_3 - H_1)}}, \tag{8.76}$$

where $K(k)$ is the quarter period of the function $\mathrm{sn}(x;k)$, which has also been tabulated, or can be calculated by standard software.

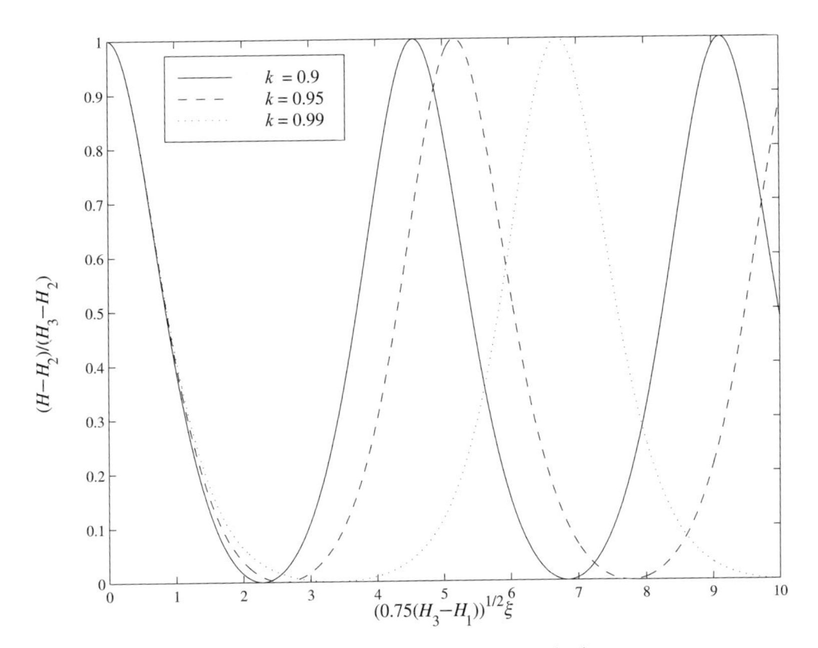

Fig. 8.16. Some typical cnoidal wave solutions.

Our analysis has given us three relationships for the travelling wave system, namely

$$\alpha_0 = H_3 - H_2, \quad \epsilon_0 = \frac{4K(k)}{\sqrt{3(H_3 - H_1)}}, \quad V_0 = \frac{1}{2}(H_1 + H_2 + H_3). \quad (8.77)$$

These are a parametric representation of the dispersion relation for our travelling waves, and can best be understood as follows. For a given wave amplitude, α_0, wavelength, ϵ_0, and uniform perturbation of the depth, H_2, we can see that $H_3 = \alpha_0 + H_2$. H_1 can be calculated from a single transcendental equation involving $K(k)$ to give the relationship

$$V_0 = \frac{1}{2}\left\{2H_2 + H_1\left(\alpha_0, \epsilon_0\right) + \alpha_0\right\}.$$

This is the dispersion relation between phase speed, wavelength, depth and amplitude. The appearance of the wave amplitude in the dispersion relation is in contrast to a linear theory, in which wave speed is independent of amplitude.

In order to compare this analysis with the linear theory, we can consider the limiting case, $(H_3 - H_2) \ll 1$. In this case k is small and positive, and from the definition of $\text{sn}(x;k)$, $\text{sn}(x;0^+) \sim \sin x$ and $K(0^+) \sim \pi/2$. At leading order

$$\epsilon_0 = \frac{2\pi}{\sqrt{3(H_2 - H_1)}}, \qquad V_0 = \frac{1}{2}(H_1 + 2H_2).$$

This leads to the expression

$$H = H_2 + \frac{1}{2}\alpha_0 + \frac{1}{2}\alpha_0 \cos\left(\frac{2\pi}{\epsilon_0}\xi\right).$$

This harmonic wave is the expected solution of the linearised problem. We can also manipulate the dispersion relation into a more standard form. We have

$$H_1 = H_2 - \frac{4\pi^2}{3\epsilon_0^2},$$

so that

$$c = \sqrt{gh}\left\{1 + \frac{1}{2}\alpha\left(3H_2 - \frac{4\pi^2}{3\epsilon_0^2}\right)\right\}.$$

Since, by definition, $\epsilon_0 = \alpha^{1/2}/\epsilon$, we can write this in the form

$$c = \sqrt{gh}\left\{1 - \frac{2\pi^2}{3}\left(\frac{h}{\lambda}\right)^2 + O(\alpha)\right\}.$$

But this is precisely what we get if we take the usual dispersion relation for linear gravity waves,

$$c^2 = \frac{g\lambda}{2\pi}\tanh\left(\frac{2\pi h}{\lambda}\right),$$

and expand it as an asymptotic series for $h \ll \lambda$. In other words, our nonlinear theory is consistent with the linear theory in the limit of small amplitude waves.

We are now in a position to examine case (ii) in detail. As H_1 and H_2 tend to zero through positive values, $k \to 1^-$. The elliptic functions and their periods also simplify in this limit, with $\mathrm{sn}(x; 1^-) \sim \tanh x$ and $K(k) \to \infty$ as $k \to 1^-$. The cnoidal wave therefore loses its periodicity in this limit and becomes a single hump. The expression for the wave profile becomes

$$H = H_3 \,\mathrm{sech}^2\left(\frac{3}{4}H_3^{1/2}\xi\right), \tag{8.78}$$

which is illustrated in figure 8.17, and the wave speed is

$$V_0 = \frac{1}{2}H_3 = \frac{1}{2}\alpha_0. \tag{8.79}$$

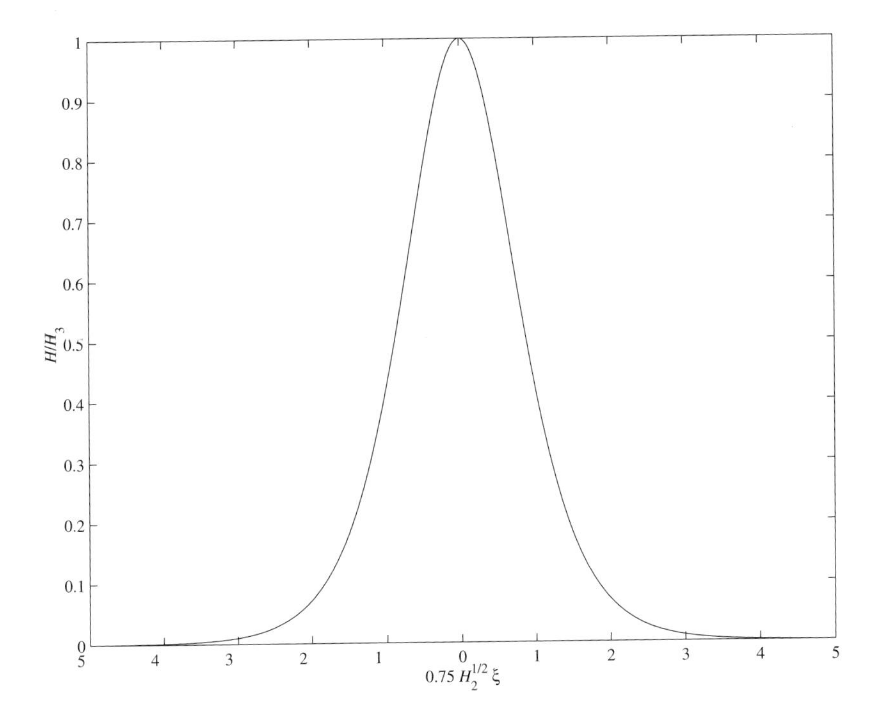

Fig. 8.17. The solitary wave solution given by (8.78).

Alternatively, we could note that when the cubic (8.69) has a repeated root, the direct integration of (8.68) is straightforward. This type of solution is known as a **solitary wave**. It is instructive to write the wave speed in terms of the physical variables, when we find that, to our order of approximation,

$$c = \sqrt{gh}\left(1 + \frac{1}{2}H_3\alpha\right).$$

This shows explicitly how the speed of the solitary wave is related to the height of the hump. A typical solitary wave is shown in figure 8.18. The physical existence of this type of wave was first reported in 1834 by John Scott Russell, who observed a solitary wave with an amplitude of about a foot and a width of about thirty feet on the Edinburgh to Glasgow canal. It was formed when a canal barge stopped suddenly, and he followed it on his horse for a couple of miles.

It is worth noting that although the periodic solutions we have obtained are subject to the Benjamin–Feir instability, which we described in the

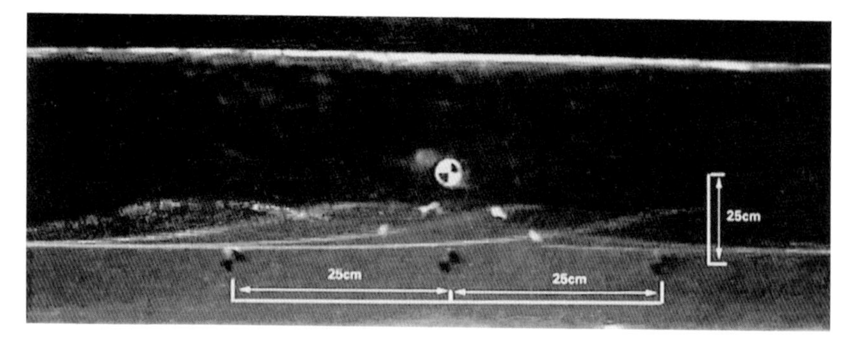

Fig. 8.18. A solitary wave in a channel.

previous section, which will cause them to disintegrate, the solitary wave solution is *not*, and, as Scott Russell's observations suggest, we can expect them to propagate over long distances. As we shall see in chapter 12, it is possible to show that these solitary waves have the remarkable property of being able to collide and pass through each other without changing their form. Solitary wave solutions of a nonlinear equation or system of equations which have this property are known as **solitons**. Although this would not be surprising for a linear equation, for which the principle of superposition holds, remember that the KdV equation is nonlinear. Indeed, we will see in section 12.1 that the interaction of solitons is intrinsically nonlinear, and not simply the result of adding the solutions for individual solitary waves, which would not satisfy the KdV equation. The collision of two solitons is illustrated in figure 8.19.

8.4 Nonlinear Capillary Waves

In this section we again consider two-dimensional nonlinear water waves, with surface tension acting as the *only* restoring force, propagating from right to left with wave height H, wavelength λ and phase speed c. We take the depth of the water to be infinite and suppose that the effect of the passage of the wave is negligible deep down in the water. Again, this is an unsteady flow problem that we can convert into a steady one by using a coordinate system moving from right to left with the phase speed of the waves. In this coordinate system, the fluid down at infinite depth flows horizontally from left to right with speed c. The full nonlinear boundary value problem appropriate to this type of flow was derived in chapter 4, and is given by

$$\nabla^2\phi = 0 \quad \text{for } -\infty < x < \infty,\ -\infty < y < \eta(x), \tag{8.80}$$

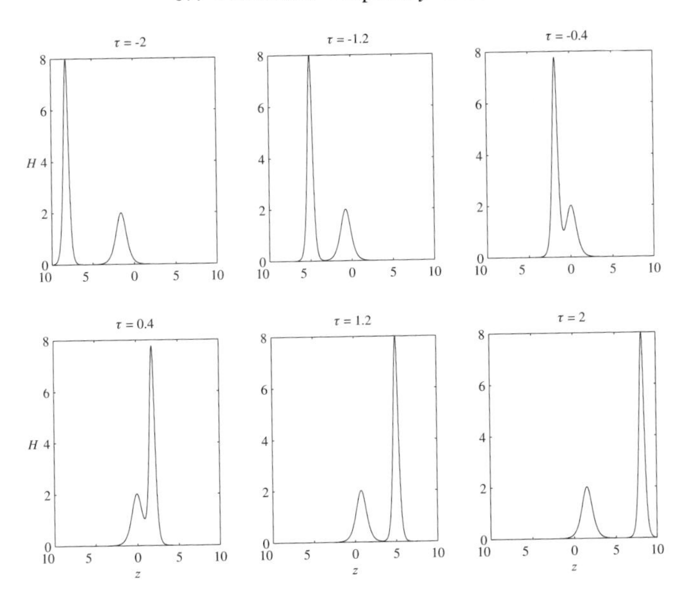

Fig. 8.19. A two soliton solution of the KdV equation.

$$\phi \sim cx \quad \text{as } y \to -\infty, \tag{8.81}$$

$$\phi_y = \eta_x \phi_x, \quad \frac{1}{2}|\nabla\phi|^2 - \frac{\sigma}{\rho}\kappa = \frac{1}{2}c^2 \quad \text{on } y = \eta(x), \tag{8.82}$$

where κ is the curvature of the free surface, which is at $y = \eta(x)$, and we seek a non-trivial periodic solution with wave height $H = \eta_{\text{crest}} - \eta_{\text{trough}} \neq 0$. At first sight it seems impossible to solve this free boundary problem analytically, except by approximate methods appropriate to linear waves ($H \ll \lambda$). However, by reformulating the problem, it is possible to write (8.80) to (8.82) as a boundary value problem on a fixed domain and find an analytical solution. This was first done by G.D. Crapper in 1957 by exploiting an idea due to Stokes in 1847, which exactly locates the position of the free surface.

As the problem is two-dimensional, we can introduce a stream function, $\psi = \psi(x, y)$, which is harmonic and takes the value zero on the free surface and has $\psi \to -\infty$ as $y \to -\infty$. From the definition of the stream function and the velocity potential, we have

$$\phi_x = \psi_y, \quad \phi_y = -\psi_x.$$

Under sensible assumptions about the smoothness of ψ and ϕ, these are just the Cauchy–Riemann equations, which are satisfied by the real and imaginary parts of the analytic function $w = \phi + i\psi$. The quantity w is called the **complex potential**, and is an analytic function of the complex variable $z = x + iy$ in the domain occupied by the fluid.

Let's consider in more detail the properties of this complex potential. We will look for a periodic wave solution, symmetric about $x = \lambda/2$, so that the flow must be horizontal beneath the crests and troughs. This means that $\phi_y = 0$ on straight lines extending vertically down to infinity from a crest or a trough. As these vertical lines head off towards a uniform horizontal flow, $\phi \sim cx$ as $y \to -\infty$, we can see that the change in ϕ over a single wavelength is $c\lambda$. We now concentrate on the single wavelength of the flow labelled $ABCDE$ in figure 8.20(a). This wavelength occupies the vertical strip $A'B'C'D'E'$ in the $w = \phi + i\psi$ plane, as shown in figure 8.20(b). Note that we can choose to take $\phi = 0$ at the point A since the problem is unchanged by the addition of an arbitrary constant to ϕ. The key step in this method is to interchange the role of the variables $z = x + iy$ and $w = \phi + i\psi$ and regard ϕ and ψ as our independent variables, since $w = \phi + i\psi$ provides a one-to-one correspondence between the physical and complex potential planes. This is an example of a **hodograph transformation**. The power of this transformation lies in the fact that it locates the unknown free surface at the known position $\psi = 0$, $0 < \phi < c\lambda$ in the w-plane.

We now define the **complex velocity**,

$$\frac{dw}{dz} = \phi_x + i\psi_x = u - iv, \tag{8.83}$$

where (u, v) are the horizontal and vertical components of the fluid velocity field. If q is the speed of the flow and θ the angle that it makes with the horizontal, $u = q\cos\theta$ and $v = q\sin\theta$, so that

$$\frac{dw}{dz} = qe^{-i\theta}. \tag{8.84}$$

Since the derivative of an analytic function is itself analytic, dw/dz is analytic in the flow domain. If we assume that there are no stagnation points in the flow, then similarly, $\log(dw/dz) = \log q - i\theta \equiv \tau - i\theta$ is analytic. The quantity $\log(dw/dz)$ is an analytic function of $\phi + i\psi$, so $\tau = \log q$ must be harmonic in the strip $-\infty < \psi < 0$, $0 < \phi < c\lambda$. We will now write all of the boundary conditions in terms of τ, so that we can formulate the problem as a boundary value problem for τ in the (ϕ, ψ)-plane.

The kinematic boundary condition, $\phi_y = \eta_x\phi_x$, takes the form $q\sin\theta =$

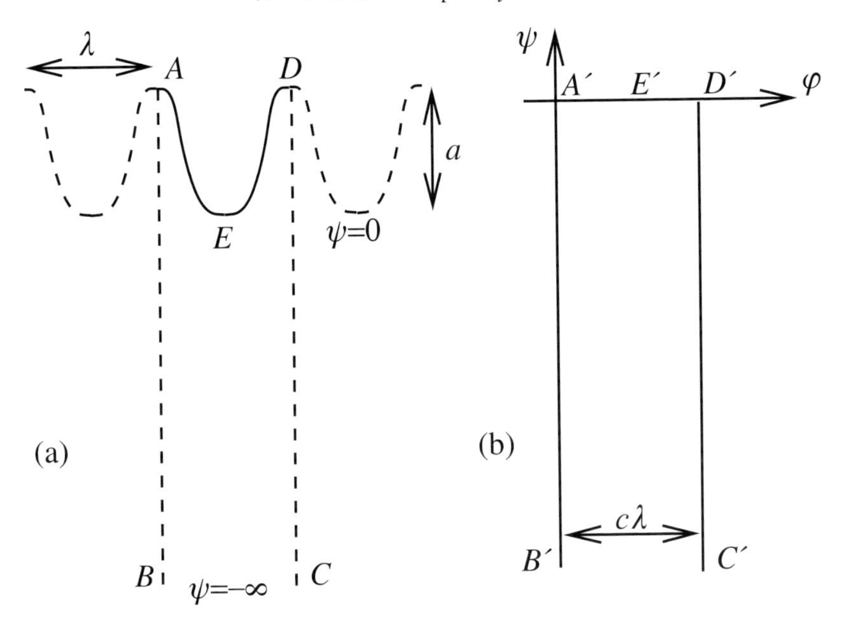

Fig. 8.20. (a) The physical domain occupied by a single wavelength of the wave. (b) The corresponding complex potential plane.

$\eta_x q \cos\theta$, which is identically satisfied in this new formulation, since $\eta_x = \tan\theta$. To deal with the Bernoulli condition, $\frac{1}{2}q^2 - \sigma\kappa/\rho = \frac{1}{2}c^2$, recall that the basic definition of curvature is $\kappa = d\theta/ds$, where s is arc length along the free surface. By the chain rule,

$$\kappa = \frac{\partial\theta}{\partial\phi}\frac{d\phi}{ds} = \frac{\partial\theta}{\partial\phi}\left(\frac{\partial\phi}{\partial x}\frac{dx}{ds} + \frac{\partial\phi}{\partial y}\frac{dy}{ds}\right) = q\frac{\partial\theta}{\partial\phi},$$

using the fact that $\partial\phi/\partial x = q\cos\theta$, $\partial\phi/\partial y = q\sin\theta$, $dx/ds = \cos\theta$ and $dy/ds = \sin\theta$. As τ and θ are the real and imaginary parts of an analytic function they must satisfy the Cauchy–Riemann equations in the variables ϕ and ψ. In particular, $\tau_\psi = \theta_\phi$, so that the Bernoulli condition, applied on $\psi = 0$, is

$$\frac{1}{2}e^{2\tau} - \frac{\sigma}{\rho}e^{\tau}\frac{\partial\tau}{\partial\psi} = \frac{1}{2}c^2,$$

and hence

$$\frac{\sigma}{\rho}\frac{\partial\tau}{\partial\psi} = -\frac{1}{2}\left(c^2 e^{-\tau} - e^{\tau}\right).$$

It is now convenient to write the problem in dimensionless form. The quantity $\sigma/\rho c^2$ has dimensions of length, so we write

$$\left.\begin{aligned} \bar{\tau} = \tau - \log c, \quad \bar{\phi} = \frac{\rho c \phi}{\sigma}, \quad \bar{\psi} = \frac{\rho c \psi}{\sigma}, \quad \bar{\lambda} = \frac{\rho c^2 \lambda}{\sigma}, \\ \bar{w} = \frac{\rho c w}{\sigma}, \quad \bar{z} = \frac{\rho c^2 z}{\sigma}, \quad \bar{\eta} = \frac{\rho c^2 \eta}{\sigma}. \end{aligned}\right\} \tag{8.85}$$

Note that $\bar{q} = q/c$ is the natural way to make the flow speed dimensionless, which leads to $\bar{\tau} = \log \bar{q} = \log q - \log c$, as used above. Having done all of this work, we can now state precisely the problem that we must solve in terms of the new variables as

$$\bar{\tau}_{\bar{\phi}\bar{\phi}} + \bar{\tau}_{\bar{\psi}\bar{\psi}} = 0 \quad \text{for } 0 < \bar{\phi} < \bar{\lambda}, -\infty < \bar{\psi} < 0, \tag{8.86}$$

subject to

$$\bar{\tau}_{\bar{\psi}} = \sinh \bar{\tau} \quad \text{on } \bar{\psi} = 0 \text{ for } 0 < \bar{\phi} < \bar{\lambda}, \tag{8.87}$$

$$\bar{\tau}_{\bar{\phi}} = 0 \quad \text{on } \bar{\phi} = 0, \tfrac{1}{2}\bar{\lambda}, \bar{\lambda} \text{ for } -\infty < \bar{\psi} < 0, \tag{8.88}$$

$$\bar{\tau} \to 0 \quad \text{as } \bar{\psi} \to -\infty \text{ for } 0 < \bar{\phi} < \bar{\lambda}. \tag{8.89}$$

The other Cauchy–Riemann equation, $\bar{\tau}_{\bar{\phi}} = -\theta_{\bar{\psi}}$, leads to (8.88), since $\theta \equiv 0$ beneath the crest and trough of the wave, which we now assume lie at $\bar{\phi} = 0$, $\frac{1}{2}\bar{\lambda}$ and $\bar{\lambda}$.

We now have a linear partial differential equation to solve in a straightforward domain, as shown in figure 8.21. The difficulty lies in the nonlinear boundary condition, $\bar{\tau}_{\bar{\psi}} = \sinh \bar{\tau}$. To deal with this we look for a solution that satisfies

$$\bar{\tau}_{\bar{\psi}} = f(\bar{\psi}) \sinh \bar{\tau} \tag{8.90}$$

throughout the domain, with $f(0) = 1$. Equation (8.90) can be integrated once to give

$$\log \tanh \frac{1}{2}\bar{\tau} = F(\bar{\psi}) + G(\bar{\phi}),$$

where $F'(\bar{\psi}) = f(\bar{\psi})$ and $G(\bar{\phi})$ is a constant of integration (with respect to $\bar{\psi}$). This can be rearranged into the form

$$\bar{\tau} = \log \left\{ \frac{X(\bar{\psi}) + Y(\bar{\phi})}{X(\bar{\psi}) - Y(\bar{\phi})} \right\}, \tag{8.91}$$

where $X(\bar{\psi}) = \exp\{-F(\bar{\psi})\}$ and $Y(\bar{\phi}) = \exp\{G(\bar{\phi})\}$. If we substitute (8.91) into Laplace's equation, we find that

$$X(\bar{\psi})X'(\bar{\psi}) \left(\frac{Y''(\bar{\phi})}{Y(\bar{\phi})} \right)' + Y(\bar{\phi})Y'(\bar{\phi}) \left(\frac{X''(\bar{\psi})}{X(\bar{\psi})} \right)' = 0. \tag{8.92}$$

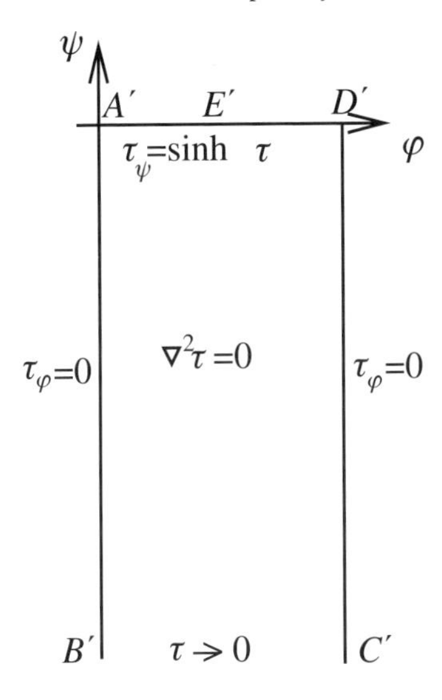

Fig. 8.21. The dimensionless boundary value problem given by (8.86) to (8.88).

It is now sufficient for our purposes to notice that (8.92) can be satisfied by taking $Y'' = -k^2 Y$ and $X'' = k^2 X$, with k a real constant. We find that by using the solution

$$X = A\cosh(k\bar{\psi} + D), \quad Y = A\sin(k\bar{\phi} + B), \tag{8.93}$$

we obtain a three parameter family of solutions that is sufficient to enable us to construct a solution of the nonlinear boundary value problem.

Since the condition $f(0) = 1$ is equivalent to $X'(0) = X(0)$, we need $k = -\coth D$. The condition that $\bar{\tau}_{\bar{\phi}} = 0$ when $\bar{\phi} = 0$ and $\bar{\lambda}/2$ becomes $Y'(0) = Y'(\bar{\lambda}/2) = 0$, which can be satisfied by taking $B = -\pi/2$ and $k = 2\pi/\bar{\lambda}$. The solution then takes the form

$$\bar{\tau} = \log\left\{\frac{\cosh(k\bar{\psi} + D) - \cos k\bar{\phi}}{\cosh(k\bar{\psi} + D) + \cos k\bar{\phi}}\right\}, \tag{8.94}$$

from which we note that $\bar{\tau} \to 0$ as $\bar{\psi} \to -\infty$, as required. We can now calculate θ by integrating $\theta_{\bar{\phi}} = \bar{\tau}_{\bar{\psi}}$, which gives

$$\theta = 2\tan^{-1}\left\{\frac{\sin k\bar{\phi}}{\sinh(k\bar{\psi} + D)}\right\} = i\log\left\{\frac{\sinh(k\bar{\psi} + D) - \sinh ik\bar{\phi}}{\sinh(k\bar{\psi} + D) + \sinh ik\bar{\phi}}\right\}^{1/2}. \tag{8.95}$$

To find the actual shape of the free surface we will need the complex velocity, w. Using $\log(d\bar{w}/d\bar{z}) = \bar{\tau} - i\theta$, we have

$$\log\left(\frac{d\bar{w}}{d\bar{z}}\right) = \log\left[\left\{\frac{\cosh(k\bar{\psi}+D) - \cos k\bar{\phi}}{\cosh(k\bar{\psi}+D) + \cos k\bar{\phi}}\right\} \times \left\{\frac{\sinh(k\bar{\psi}+D) - \sinh ik\bar{\phi}}{\sin(k\bar{\psi}+D) + \sinh ik\bar{\phi}}\right\}^{1/2}\right]$$

Use of the half angle formula and some further simple manipulation, noting that $ik\bar{w} = -k\bar{\psi} + ik\bar{\phi}$, finally allows us to write

$$\log\left(\frac{d\bar{w}}{d\bar{z}}\right) = \log\left\{\tanh^2\left(\frac{ik\bar{w} - D}{2}\right)\right\},$$

and hence

$$\frac{d\bar{w}}{d\bar{z}} = \left(\frac{Ce^{ik\bar{w}} + 1}{Ce^{ik\bar{w}} - 1}\right)^2, \tag{8.96}$$

where we have written $C = -e^{-D}$. This is now easily integrated to give

$$\bar{z} = \bar{w} - \frac{4i}{k}\frac{1}{1 + Ce^{ik\bar{w}}} + \frac{4i}{k}, \tag{8.97}$$

where we have chosen the constant of integration so that $\bar{w} = \bar{z}$ when $C = 0$, which corresponds to a flat surface with no waves. To obtain the equation of the free surface we put $\bar{\psi} = 0$ in (8.97), which gives a parametric representation in terms of the physical coordinates,

$$x + iy = \lambda\left\{\tilde{\phi} + \frac{2iCe^{i2\pi\tilde{\phi}}}{\pi(1 + Ce^{i2\pi\tilde{\phi}})}\right\}, \tag{8.98}$$

where $\tilde{\phi} = \bar{\phi}/\bar{\lambda} = \phi/\lambda c$.

We must now deduce the dispersion relation, and determine any restrictions on the range of values that the parameter C can take. Currently we have

$$\frac{2\pi}{\bar{\lambda}} = k, \quad k = -\coth D, \quad C = -e^{-D}. \tag{8.99}$$

To these we must add a relationship involving the wave height. In the physical plane we have $\eta_{\text{crest}} - \eta_{\text{trough}} = H$. Since our unit of length is $\sigma/\rho c^2$, we need

$$\bar{\eta}\,(0) - \bar{\eta}\left(\frac{1}{2}\bar{\lambda}\right) = \frac{\rho c^2 H}{\sigma}, \tag{8.100}$$

noting that the assumption that the wave has turning points at $x = 0$

and $x = \frac{1}{2}\lambda$ requires that C is real. By taking the imaginary part of (8.98) at $\tilde{\phi} = 0$ and 1 we find that

$$\frac{\rho c^2 H}{\sigma} = \frac{-8C}{k(C^2 - 1)}. \tag{8.101}$$

Now note that (8.99) gives

$$c^2 = \frac{2\pi\sigma}{\rho\lambda}\frac{C^2 - 1}{C^2 + 1}. \tag{8.102}$$

Equations (8.101) and (8.102) show that we need $C < -1$ for H and c^2 to be positive. In addition, eliminating c^2 gives

$$\frac{H}{\lambda} = \frac{-4C}{\pi(C^2 - 1)},$$

and hence

$$C = -\frac{2\lambda}{H\pi}\left(1 + \sqrt{1 + \frac{H^2\pi^2}{4\lambda^2}}\right).$$

Substituting this into (8.102) gives, after some manipulation, the dispersion relation

$$c^2 = \frac{2\pi\sigma}{\rho\lambda}\left(1 + \frac{H^2\pi^2}{4\lambda^2}\right)^{-1/2}. \tag{8.103}$$

We emphasise that this result is exact! When $H \ll \lambda$, (8.103) reduces to $c^2 = 2\pi\sigma/\rho\lambda$, the linear result given by (4.57) in chapter 4.

There is a surprise when we investigate the nonlinear wave profiles, which are shown in figure 8.22. As the waves become steeper, there is a limiting configuration with $H/\lambda = H_c/\lambda \approx 0.73$, such that for $H > H_c$ the free surface is self-intersecting and no longer physical. When $H = H_c$ the free surface just touches itself and entrains a bubble of air, as shown in figure 8.22. This has some application to the mixing of air and water in microgravity environments, but is not particularly important when compared to air entrainment by plunging or spilling breakers under normal gravitational conditions.

Nonlinear capillary waves can also be used to probe the structure of a **Wigner lattice**. Such a lattice is formed when ions are placed just below the surface of superfluid helium, typically at a depth of about 40 nm. Superfluid helium is a substance that (because of quantum effects) has no viscosity, and allows capillary waves to persist on its surface for far longer than in ordinary fluids. The ions organise themselves into a hexagonal lattice, and, by shaking the ions with an external electric

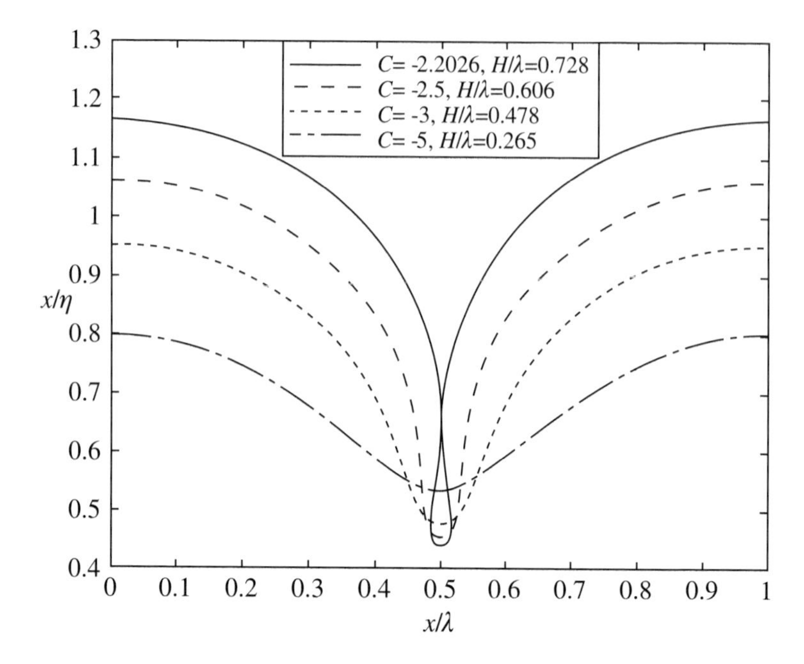

Fig. 8.22. Nonlinear capillary waves of various amplitudes.

field, capillary waves are generated. These are quantised, and known as **ripplons**. The structure of the lattice can then be deduced by studying how it scatters capillary waves. The technique is known as **capillary wave spectroscopy** (see Barenghi, Mellor, Muirhead and Vinen (1986)). Another application of nonlinear capillary waves, to local measurement probes, is given by Billingham and King (1995).

Exercises

8.1 Shallow water of depth $h = h_0$ lies stationary in the region $x < 0$ and a vertical plate with height greater than h_0 is stationary at $x = 0$, holding back the water. There is no water in the region $x > 0$. When $t > 0$ the plate moves into the region $x > 0$ with constant speed V, where $V < 2c_0$ and $c_0^2 = gh_0$.

Show that the height of the water behind the plate is given by

$$h = h_0, \quad \text{for } x < -c_0 t,$$

$$h = \frac{1}{9g}\left(2c_0 - \frac{x}{t}\right)^2, \quad \text{for } -c_0 t < x < \left(\tfrac{3}{2}V - c_0\right) t,$$

$$h = \frac{1}{g}\left(c_0 - \frac{1}{2}V\right)^2, \quad \text{for } \left(\tfrac{3}{2}V - c_0\right)t < x < Vt.$$

8.2 A rigid wall is located at one end of a long tank of water of depth h, with a free surface. When $t = 0$ the wall moves into the water with constant acceleration, A. Use nonlinear shallow water theory to find the form of the wave when $t > 0$.

8.3 For the moving wall problem discussed in subsection 8.1.2 show that the shock speed satisfies

$$\dot{s} \sim c_0 \qquad \text{for } V \ll c_0,$$
$$\dot{s} \sim V + \frac{1}{\sqrt{2}}c_0 \quad \text{for } V \gg c_0,$$

where $c_0 = \sqrt{gh_0}$.

8.4 Periodic waves with finite amplitude a and wavelength λ propagate with phase speed c from right to left in water of depth h. Show that the velocity potential, free surface elevation and phase speed can be written in the form

$$\phi = \frac{\omega a}{k}\frac{\cosh k(y+h)}{\sinh kh}\sin\xi + \omega a^2 Q\frac{\cosh 2k(y+h)}{\sinh 2kh}\sin 2\xi + \cdots,$$

$$\eta = \bar{\eta} + a\cos\xi + ka^2 P\cos 2\xi,$$

$$c^2 = \frac{g}{k}\tanh kh\left[1 + k^2a^2\left\{1 + \frac{1}{\sinh^2 kh} + \frac{9 - 2\tanh^2 kh}{8\sinh^4 kh}\right\} + \cdots\right],$$

where $\xi = kx - \omega t$, $\bar{\eta} = -\frac{1}{2}ka^2\operatorname{cosech} kh$, $P = \frac{1}{4}\coth kh$, $Q = \frac{3}{4}\coth kh \operatorname{cosech}^2 kh$. What is the limiting form of these waves as $h \to \infty$?

8.5 Explain why only cases (i) and (ii) section 8.3.2 are physically relevant.

8.6 We derived the KdV equation in subsection 8.3.1 for gravitational body forces, based upon a balance between wave steepening and dispersion. How is the derivation affected by the inclusion of surface tension?

8.7 By using the exact solution for finite amplitude capillary waves given in section 8.4, find the differential equations satisfied by the particle paths. Determine the asymptotic solution of these equations for $a \ll \lambda$, up to $O(a^3/\lambda^3)$.

9

Chemical and Electrochemical Waves

Chemical waves arise when reaction and diffusion occur simultaneously. We shall see that chemical reaction is represented in the governing equations by nonlinear source terms. Molecular diffusion tends to smear out chemical concentrations, whilst suitable chemical reactions tend to sharpen them. We will concentrate here on chemical reactions occurring in solution, which are not strongly exothermic. We have already seen in subsection 7.2.5 that exothermic reactions can generate detonation waves.

We begin by examining the effects of molecular diffusion and chemical reaction in isolation. Even without molecular diffusion, many chemical reactions exhibit surprising nonlinear behaviour, including limit cycles and chaos. The most famous of these is the **Belousov–Zhabotinskii reaction** or BZ reaction. When the reactants are mixed in a beaker, the colour of the solution changes repeatedly from red to blue with a period of the order of minutes. We then consider the mathematics of diffusion, before combining these two processes to obtain systems of reaction–diffusion equations. Travelling wave solutions of this type of equation describe the propagation of many chemical waves, and form the basis for the analysis of more geometrically complex chemical waves. From these travelling wave solutions we can determine the speed and stability of the waves. A typical chemical wave is shown in figure 9.1. We also consider an example of the use of the method of matched asymptotic expansions to construct an approximate solution. Finally, we consider the propagation of waves in the **nervous system**. After a brief introduction to the biology of nerves, we use distributed circuit theory, based upon Ohm's law and Kirchoff's laws, to derive the **Fitzhugh–Nagumo equations**. These equations represent a caricature of electrochemical activity in a nerve. Their solutions exhibit **threshold effects**, travelling waves and limit cycles which we analyse in some detail.

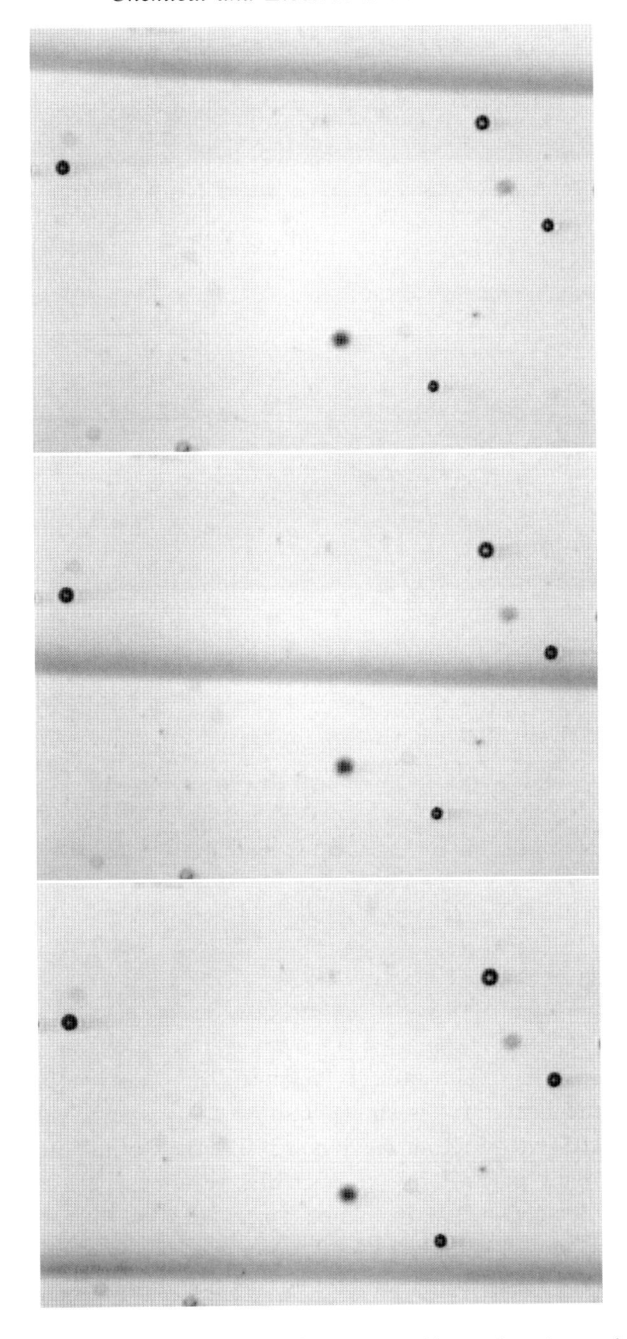

Fig. 9.1. The wavefront in the iodate/arsenous acid reaction. A starch indicator turns the solution black behind the wavefront which is moving at $37\,\mu m\,s^{-1}$.

9.1 The Law of Mass Action

Consider a chemically reacting system that consists of m chemical species, $C_1, C_2, \ldots, C_m$. The **law of mass action**, which can be derived from statistical mechanics (Mandl, 1971), states that if these species react together through

$$\lambda_1 C_1 + \lambda_2 C_2 + \cdots + \lambda_m C_m \to \nu_1 C_1 + \nu_2 C_2 + \cdots + \nu_m C_m, \tag{9.1}$$

the rate at which the reaction proceeds is

$$\dot{r} = k[C_1]^{\lambda_1}[C_2]^{\lambda_2}\ldots[C_m]^{\lambda_m}, \tag{9.2}$$

where $[C_i]$ is the concentration of C_i. The reaction rate $\dot{r}$ has dimensions of mol $m^{-3}\,s^{-1}$, k is a dimensional constant and λ_i and ν_i are the **stoichiometric coefficients**. In a typical solution phase reaction, $\dot{r}$ lies between about 10^{-7} and 10^{-10} mol $m^{-3}\,s^{-1}$. Conservation of total mass can be stated as

$$\sum_{i=1}^{m} \lambda_i m_i = \sum_{i=1}^{m} \nu_i m_i, \tag{9.3}$$

where m_i is the **molar mass** of the ith chemical species. This is an intrinsic measure of the mass of the substance and is the mass of one **mole** of the chemical. A mole is the chemist's preferred unit for amount of substance, and is equal to N_A molecules, where $N_A \approx 6.02 \times 10^{23}$ is **Avogadro's number**.

Conservation of mass for each individual chemical can be written as

$$\frac{d[C_i]}{dt} = (\nu_i - \lambda_i)\,\dot{r}. \tag{9.4}$$

For example, the combustion of hydrogen follows

$$2H_2 + O_2 \to 2H_2O, \tag{9.5}$$

with reaction rate proportional to $[H_2]^2[O_2]$. In reality, most seemingly simple chemical reactions, such as (9.5), take place via a large number of intermediate steps, involving ionised and fragmented states of the chemical species, particularly for gas phase reactions. The law of mass action can, however, often give some insight into the rate at which the reaction proceeds. Moreover, in solution phase reactions, particularly those involving only two species per reaction step, the law of mass action is an excellent model for the rate of reaction.

Example 1: Decay of a Single Reactant. Consider a chemical P that decays

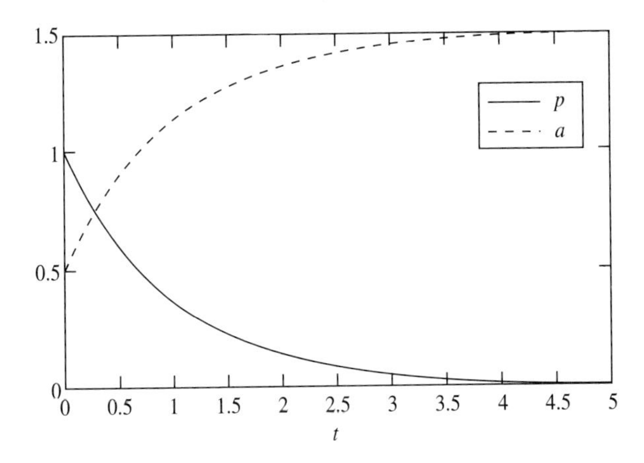

Fig. 9.2. The solution (9.7) of equations (9.6) when $a_0 = 0.5$, $p_0 = 1$ and $k = 1$.

to a product A via the simple reaction step

$$P \to A, \quad \text{rate } k[P].$$

If we define $p = [P]$ and $a = [A]$ to be the concentrations of the two chemicals, then in a well-stirred, spatially uniform reaction vessel the **reaction rate equations** are

$$\frac{dp}{dt} = -kp, \quad \frac{da}{dt} = kp. \tag{9.6}$$

If we know that $p = p_0$ and $a = a_0$ when $t = 0$, we can solve these simple, first order, separable, ordinary differential equations to give

$$p = p_0 e^{-kt}, \quad a = a_0 + p_0 \left(1 - e^{-kt}\right). \tag{9.7}$$

A typical solution is shown in figure 9.2. Note that $p \to 0$ and $a \to a_0 + p_0$ as $t \to \infty$. All of the chemical species P is consumed by the reaction and converted into A.

Example 2: Quadratic Autocatalysis. Consider a chemical species A that reacts with a chemical species B via the single reaction step

$$A + B \to 2B, \quad \text{rate } k[A][B].$$

This is known as an **autocatalytic** reaction step, and B is an **autocatalyst**. The greater the concentration of B, the faster B is produced. We say that B **catalyses** its own production. If we let $a = [A]$ and $b = [B]$, we can

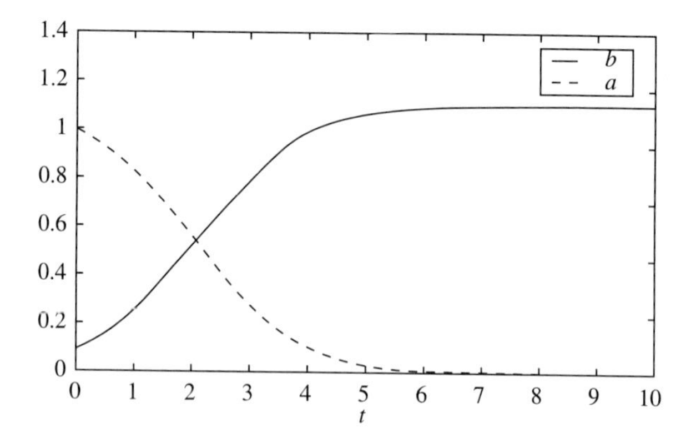

Fig. 9.3. The solution (9.9) of equations (9.8) when $a_0 = 1$, $b_0 = 0.1$ and $k = 1$.

write down the reaction rate equations for a spatially uniform reaction as

$$\frac{da}{dt} = -kab, \quad \frac{db}{dt} = kab. \tag{9.8}$$

Adding these two equations shows that $d(a+b)/dt = 0$, and hence, if $a = a_0$ and $b = b_0$ when $t = 0$, that $a + b = a_0 + b_0$. Eliminating a then gives

$$\frac{db}{dt} = kb(a_0 + b_0 - b).$$

We can solve this separable equation to give the solution

$$b = \frac{b_0(a_0 + b_0)e^{k(a_0+b_0)t}}{a_0 + b_0 e^{k(a_0+b_0)t}}, \quad a = \frac{a_0(a_0 + b_0)}{a_0 + b_0 e^{k(a_0+b_0)t}}. \tag{9.9}$$

Note that $a \to 0$ and $b \to a_0 + b_0$ as $t \to \infty$. A typical solution is shown in figure 9.3. The chemical A is completely converted into B. Since this is true for all $b_0 > 0$, even a tiny initial amount of the autocatalyst, B, sets off the autocatalytic reaction and eventually converts all of the species A into B.

Example 3: A Clock Reaction. As an example of what can happen when there is more than one reaction step involved, consider the system

$$\begin{aligned} P &\to A, &\quad &\text{rate } k_0 p, \\ A + B &\to 2B, &\quad &\text{rate } k_1 ab, \\ B + C &\to D, &\quad &\text{rate } k_2 bc, \end{aligned}$$

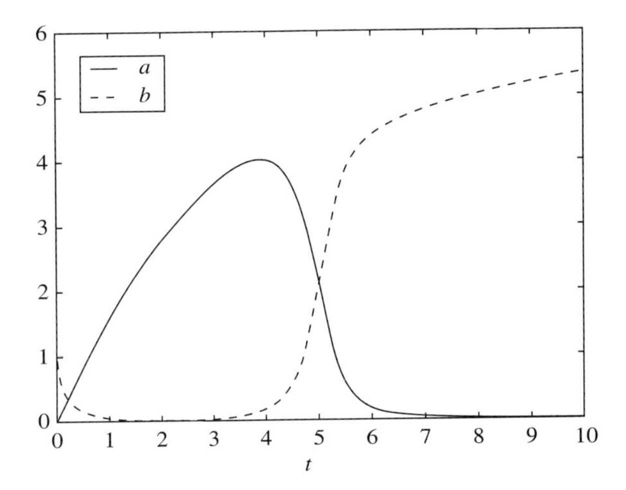

Fig. 9.4. A typical solution of equations (9.10).

where $p = [P]$, $a = [A]$, $b = [B]$ and $c = [C]$. Since the species P is only involved in the simple decay step, $P \to A$,

$$\frac{dp}{dt} = -k_0 p,$$

and hence

$$p = p_0 e^{-k_0 t},$$

where $p = p_0$ when $t = 0$. The concentrations of the species A, B and C are therefore governed by

$$\frac{da}{dt} = k_0 p_0 e^{-k_0 t} - k_1 ab, \quad \frac{db}{dt} = k_1 ab - k_2 bc, \quad \frac{dc}{dt} = -k_2 bc. \tag{9.10}$$

It can be shown that when $k_2 \gg k_1$, the concentration of the autocatalyst B remains small for a well-defined time, before suddenly growing rapidly whilst the concentration of A falls rapidly to become small. This type of reaction is called a **clock reaction**, because of the long **induction period** before the autocatalyst is produced. A typical solution is shown in figure 9.4.

Other spatially uniform chemical systems can show all of the complex behaviour associated with nonlinear ordinary differential equations, from nonlinear oscillations through to chaos. We will not go into this here, since we are interested in the chemical waves that can arise when chemical reaction and diffusion interact in an unstirred, spatially non-uniform chemical system, but the interested reader can consult Murray (1993).

9.2 Molecular Diffusion

Let $c(\mathbf{x}, t)$ be the concentration of a chemical, measured in mol m^{-3}. Let $\mathbf{q}(\mathbf{x}, t)$ be the flux of the chemical, measured in mol s^{-1}. The equation for conservation of the chemical is

$$\frac{\partial c}{\partial t} + \nabla \cdot \mathbf{q} = 0. \tag{9.11}$$

This is derived in exactly the same way as the equation for conservation of mass in fluid mechanics, current in electrodynamics, and cars in traffic modelling.

Fick's law of diffusion states that gradients in c cause a diffusive flux, with

$$\mathbf{q} = -D\nabla c, \tag{9.12}$$

where D is the diffusion coefficient. For the chemical to diffuse away from regions where c is large, D must be positive. Combining (9.11) and (9.12) leads to

$$\frac{\partial c}{\partial t} = \nabla \cdot (D\nabla c). \tag{9.13}$$

If $D = D(c)$, this is a **nonlinear diffusion equation**. We will only consider the spatially homogeneous, linear case, where D is constant, and hence

$$\frac{\partial c}{\partial t} = D\nabla^2 c. \tag{9.14}$$

This is known as the **linear diffusion equation**. It is also a good model for the flow of heat in a uniform medium, with c replaced by temperature, T.

In order to get a feel for how solutions of the diffusion equation, (9.14), behave, we consider some typical solutions that depend on just one spatial variable, x. We can find the solution of the general initial value problem of solving (9.14) subject to

$$c(x, 0) = c_0(x) \text{ for } -\infty < x < \infty, \tag{9.15}$$

using Fourier transforms, as

$$c(x, t) = \frac{1}{\sqrt{4\pi D t}} \int_{-\infty}^{\infty} c_0(X) \exp\left\{\frac{(x - X)^2}{4Dt}\right\} dX. \tag{9.16}$$

When the initial condition is a localised source of chemical, $c_0(x) = \delta(x)$, (9.16) becomes

$$c(x, t) = \frac{e^{-x^2/4Dt}}{\sqrt{4\pi D t}}. \tag{9.17}$$

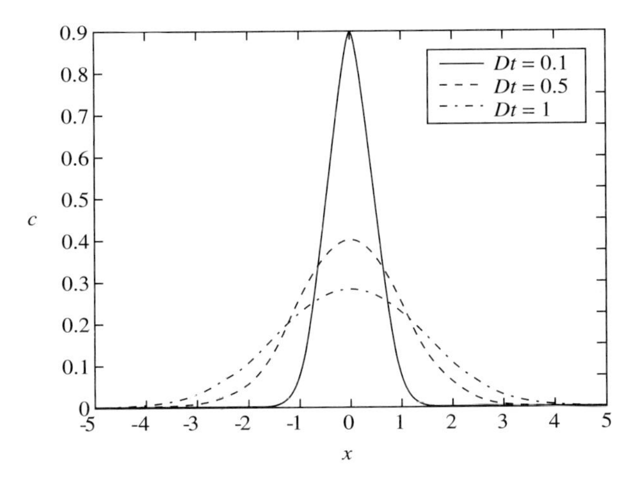

Fig. 9.5. The initial source solution of the one-dimensional diffusion equation at various times.

The solution is Gaussian, and spreads out as time increases, having a typical width of $O(\sqrt{4\pi Dt})$ and maximum height $1/\sqrt{4\pi Dt}$. Note also that diffusion transports the chemical to infinity instantaneously, since $c(x,t) > 0$ for all x when $t > 0$. For $|x| \gg 1$ and $t \ll 1$, the corresponding chemical concentrations are negligibly small, and we should not regard this feature of the solution as unphysical. Typical solutions are shown in figure 9.5.

If $c_0(x) = H(-x)$, the solution is

$$c(x,t) = \frac{1}{\sqrt{\pi}} \int_{x/\sqrt{4Dt}}^{\infty} e^{-s^2}\, ds. \tag{9.18}$$

The initial discontinuity in c at $x = 0$ is smoothed out by the diffusion for $t > 0$, as shown in figure 9.6. As with the initial source solution, (9.17), diffusion transports matter to infinity instantaneously.

9.3 Reaction–Diffusion Systems

Consider a fixed volume V within which chemical species C_i, $i = 1, 2, \ldots, m$, with concentrations $c_i(\mathbf{x}, t)$ are created by chemical reaction at rate $R_i(\mathbf{x}, t)$ whilst being transported by molecular diffusion. Conservation of mass shows that

$$\frac{d}{dt} \int_V c_i(\mathbf{x}, t) dV = \int_{\partial V} D\mathbf{n} \cdot \nabla c_i(\mathbf{x}, t) dS + \int_V R_i(\mathbf{x}, t) dV, \tag{9.19}$$

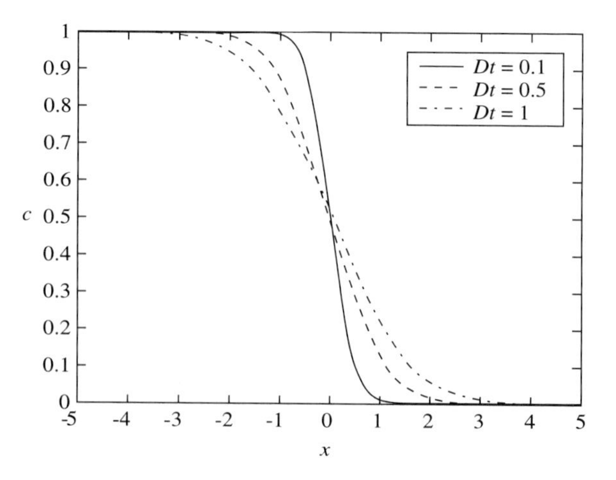

Fig. 9.6. The initial step function solution of the one-dimensional diffusion equation at various times.

where D is the diffusion coefficient and $\mathbf{n}$ is the outward unit normal at ∂V, the surface of V. The divergence theorem then gives

$$\int_V \left\{ \frac{\partial c_i}{\partial t} - \nabla \cdot (D \nabla c_i) - R_i \right\} dV = 0. \tag{9.20}$$

Since this holds for any fixed volume V, the integrand must be zero, so that

$$\frac{\partial c_i}{\partial t} = \nabla \cdot (D \nabla c_i) + R_i. \tag{9.21}$$

This is a system of **reaction–diffusion equations** for the m chemical species. We begin by considering the simplest example that leads to chemical waves.

Let A and B be two chemical species that react via the single autocatalytic step

$$A + B \rightarrow 2B, \quad \text{rate } kab, \tag{9.22}$$

where $a = [A]$ and $b = [B]$. In addition, we assume that molecular diffusion acts on A and B with constant diffusion coefficient, D. Finally, we assume that spatial variations only occur in the x-direction. This is a good model for a reaction proceeding in a long, thin tube. The governing equations for the one-dimensional chemical system are

$$\frac{\partial a}{\partial t} = D \frac{\partial^2 a}{\partial x^2} - kab, \quad \frac{\partial b}{\partial t} = D \frac{\partial^2 b}{\partial x^2} + kab. \tag{9.23}$$

It is now convenient to make these equations dimensionless. We define dimensionless variables

$$\alpha = a/a_0, \ \beta = b/a_0, \ z = x/x^*, \ \tau = t/t^*,$$

where x^* and t^* are quantities with the dimensions of length and time respectively, which we must determine, and a_0 is the spatially uniform initial concentration of the reactant, A. The governing equations, (9.23), become

$$\frac{\partial \alpha}{\partial \tau} = \frac{Dt^*}{x^{*2}} \frac{\partial^2 \alpha}{\partial z^2} - ka_0 t^* \alpha\beta, \quad \frac{\partial \beta}{\partial \tau} = \frac{Dt^*}{x^{*2}} \frac{\partial^2 \beta}{\partial z^2} + ka_0 t^* \alpha\beta.$$

In order to obtain a pair of equations that contain no parameters we choose

$$t^* = \frac{1}{ka_0}, \quad x^* = \sqrt{\frac{D}{ka_0}},$$

so that

$$\tau = ka_0 t, \quad z = x\sqrt{\frac{ka_0}{D}},$$

and hence

$$\frac{\partial \alpha}{\partial \tau} = \frac{\partial^2 \alpha}{\partial z^2} - \alpha\beta, \quad \frac{\partial \beta}{\partial \tau} = \frac{\partial^2 \beta}{\partial z^2} + \alpha\beta. \tag{9.24}$$

Note that the length scale x^* is an order of magnitude estimate of the distance that a chemical diffuses during a period of time over which the reaction (9.22) consumes a significant quantity of the chemical species.

We will also assume that the concentration of B is initially only non-zero for $-L < x < L$, where L is a non-zero positive constant. This localised initial input of the autocatalyst seeds the reaction when $t = 0$. Numerical solutions of this initial value problem show that a pair of travelling waves is always initiated by the presence of the autocatalyst, B. These waves propagate away from the origin at a constant speed, as shown in figure 9.7 for the initial condition $\beta(x, 0) = 1$ for $-1 < x < 1$. Physically, diffusion transports a small amount of B into the unreacted expanse of A, and the autocatalytic chemical reaction then causes this small amount to increase. At the wavefront the spreading caused by diffusion is balanced by the steepening caused by the chemical reaction.

We can now try to find out whether an analysis of (9.24) confirms the existence of permanent form travelling wave solutions. We define $y = z - v\tau$ and seek solutions $\alpha = \alpha(y)$, $\beta = \beta(y)$. This will be a right-propagating travelling wave solution, which we expect to describe the

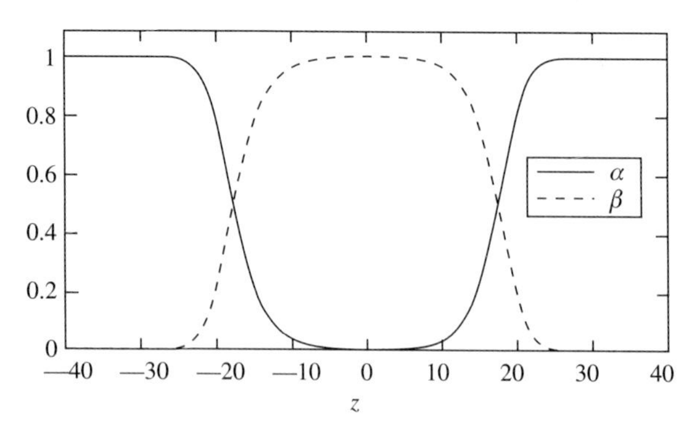

Fig. 9.7. Solution of equations (9.23) when $\beta(x, 0) = 1$ for $-1 < x < 1$ and $\tau = 12$.

solution for $z \gg 1$ after a pair of waves have been initiated by the localised input of B. The constant v is the wave speed, which we will try to determine. Using the chain rule,

$$\frac{\partial \alpha}{\partial \tau} = -v\frac{d\alpha}{dy}, \quad \frac{\partial \alpha}{\partial z} = \frac{d\alpha}{dy},$$

and hence

$$\frac{d^2\alpha}{dy^2} + v\frac{d\alpha}{dy} - \alpha\beta = 0, \quad \frac{d^2\beta}{dy^2} + v\frac{d\beta}{dy} + \alpha\beta = 0.$$

Adding these equations gives

$$\frac{d^2(\alpha+\beta)}{dy^2} + v\frac{d(\alpha+\beta)}{dy} = 0,$$

and therefore

$$\alpha + \beta = c_0 + c_1 e^{-vy},$$

where c_0 and c_1 are constants of integration. We expect that the chemicals are unreacted far ahead of the wave so that $a + b \to a_0$, and hence $\alpha + \beta \to 1$ as $y \to \infty$. We also expect the solution to be bounded as $y \to -\infty$. We therefore take $c_0 = 1$ and $c_1 = 0$, so that $\alpha + \beta = 1$ and hence

$$\frac{d^2\beta}{dy^2} + v\frac{d\beta}{dy} + \beta(1-\beta) = 0. \tag{9.25}$$

If we now define $\gamma = d\beta/dy$, we can write this nonlinear, second order

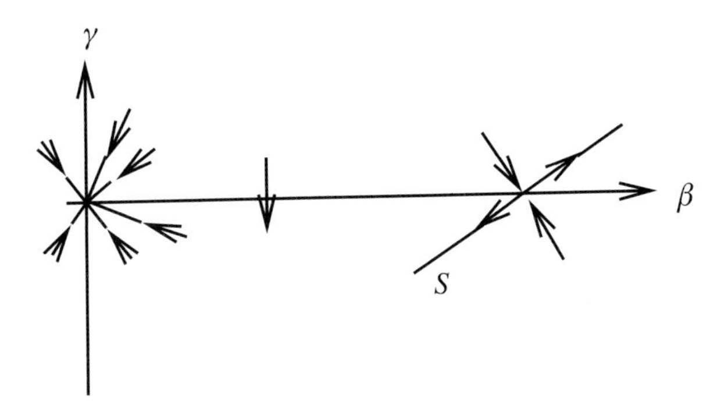

Fig. 9.8. Local behaviour in the phase portrait of the system (9.26).

differential equation as the second order system

$$\frac{d\beta}{dy} = \gamma, \quad \frac{d\gamma}{dy} = -v\gamma - \beta(1-\beta). \tag{9.26}$$

For a physically meaningful solution we need $\alpha \geq 0$ and $\beta \geq 0$, and hence, since $\alpha + \beta = 1$,

$$0 \leq \beta \leq 1. \tag{9.27}$$

We can now consider (9.26) in the phase plane, and try to determine whether solutions that satisfy (9.27) exist.

The system (9.26) has equilibrium points at $(0,0)$ and $(1,0)$ in the (β,γ)-phase plane. A local analysis reveals that $(0,0)$ is a stable node for $v \geq 2$ and a stable spiral for $v < 2$, whilst $(1,0)$ is a saddle point with stable and unstable separatrices leaving the equilibrium point in the directions $\mathbf{e}_\pm = (1, \mu_\pm)^T$, where

$$\mu_\pm = \frac{1}{2}\left(-v \pm \sqrt{v^2+4}\right), \tag{9.28}$$

are the eigenvalues. This local analysis shows that the only integral paths on which β is bounded as $y \to -\infty$ are the unstable separatrices of the saddle point. Moreover, only the unstable separatrix that enters the region where $\beta < 1$, which we will refer to as S, is physically meaningful because of the constraint (9.27). This local behaviour is illustrated in figure 9.8.

If S remains in the region $0 \leq \beta \leq 1$ for $-\infty < y < \infty$, it represents a physically meaningful solution. Now, since $d\gamma/dy < 0$ at $\gamma = 0$ for $\beta > 0$, as shown in figure 9.8, S cannot pass through the positive β-axis, and must therefore either pass through the negative γ-axis, which violates

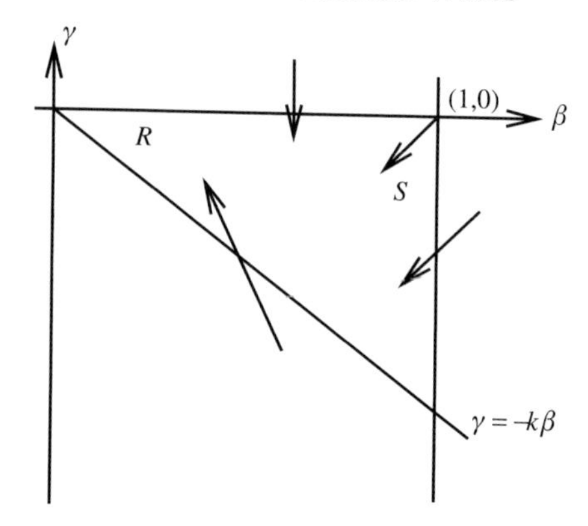

Fig. 9.9. Definition of the region R, along with the behaviour of integral paths on its boundary for the system (9.26).

(9.27), or terminate at the equilibrium point at $(0, 0)$. When $v < 2$, the point $(0, 0)$ is a stable focus, and hence β locally becomes negative. We conclude that *there is no physically meaningful permanent form travelling wave solution for $v < 2$.*

For $v \geq 2$ the point $(0, 0)$ is a stable node. Does the integral path S enter $(0, 0)$ without β becoming negative? Consider the domain R defined by

$$R = \{(\beta, \gamma) \mid \beta < 1, \ -k\beta < \gamma < 0\},$$

where k is a positive constant, which is illustrated in figure 9.9.

On the line $\beta = 1$, $d\beta/dy < 0$ for $\gamma < 0$, and hence integral paths enter R there. On the line $\gamma = 0$, $d\gamma/dy < 0$ for $0 < \beta < 1$, and hence integral paths enter R there. On the line $\gamma = -k\beta$, integral paths enter R if and only if $d\gamma/d\beta < \gamma/\beta$, since $d\beta/dy < 0$ for $\gamma < 0$. However,

$$\frac{d\gamma}{d\beta} - \frac{\gamma}{\beta} = -v - \frac{\beta(1-\beta)}{\gamma} - \frac{\gamma}{\beta} = \frac{1}{k}\left(k^2 - vk + 1 - \beta\right)$$

when $\gamma = -k\beta$. Integral paths will therefore enter the domain R if $k^2 - vk + 1 < 0$. Since $k^2 - vk + 1 = 0$ has real roots for $v \geq 2$, we can choose such a value of k. For example, if $k = \frac{1}{2}v$,

$$\frac{d\gamma}{d\beta} - \frac{\gamma}{\beta} = -\frac{2}{v}\left(\frac{1}{4}v^2 - 1 + \beta\right) < 0 \ \text{ at } \gamma = -\tfrac{1}{2}v\beta.$$

All integral paths therefore enter R. Since S enters R and cannot leave R, by the Poincaré–Bendixson theorem it must terminate at the stable node, $(0, 0)$, and hence has $\beta \to 0$, $\alpha \to 1$ as $y \to \infty$, and $\beta \to 1$, $\alpha \to 0$

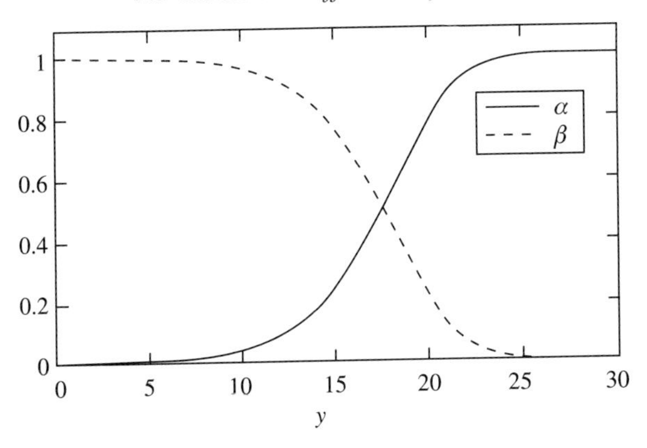

Fig. 9.10. The permanent form travelling wave solution of (9.26) when $v = 2$.

as $y \to -\infty$. There is therefore a physically acceptable permanent form travelling wave solution for all $v \geq 2$. The solution when $v = 2$ is shown in figure 9.10.

We can get some further insight into the form of the travelling wave by considering the solution for $v \gg 1$. In terms of the scaled variable $Y = y/v$, (9.25) becomes

$$\frac{1}{v^2}\frac{d^2\beta}{dY^2} + \frac{d\beta}{dY} + \beta(1-\beta) = 0, \tag{9.29}$$

subject to

$$\beta \to 0 \text{ as } Y \to \infty, \quad \beta \to 1 \text{ as } Y \to -\infty. \tag{9.30}$$

We expand β as

$$\beta = \beta_0 + \frac{1}{v^2}\beta_1 + O\left(\frac{1}{v^4}\right),$$

and substitute into (9.29). At leading order, $\beta_0' = -\beta_0(1-\beta_0)$, which has the simple solution

$$\beta_0 = \frac{1}{1+e^{Y-Y_0}}, \tag{9.31}$$

with Y_0 an arbitrary constant. Without loss of generality, we choose the origin so that $Y_0 = 0$. At first sight this appears to be a regular perturbation problem, with (9.31) satisfying the boundary conditions (9.30). However, on solving the equation satisfied by β_1, we find that

$$\beta_1 = \frac{e^Y}{\left(1+e^Y\right)^2}\log\left\{\frac{e^Y}{\left(1+e^Y\right)^2}\right\}, \tag{9.32}$$

again setting the constant of integration to zero. This means that

$$\beta \sim 1 - e^{Y}\left(1 - \frac{Y}{v^2}\right) + O\left(\frac{1}{v^4}\right) \quad \text{as } Y \to -\infty,$$

$$\beta \sim e^{-Y}\left(1 - \frac{Y}{v^2}\right) + O\left(\frac{1}{v^4}\right) \quad \text{as } Y \to \infty. \tag{9.33}$$

There is therefore a non-uniformity in the expansion when $|Y| = O(v^2)$. To deal with this non-uniformity when $Y > 0$ we define

$$\hat{Y} = \frac{Y}{v^2}, \quad \beta = e^{-v^2\phi(\hat{Y})},$$

with $\hat{Y}$ and ϕ of $O(1)$ as $v \to \infty$. Equation (9.29) becomes

$$1 - \phi' + \frac{1}{v^2}\phi'^2 - \frac{1}{v^4}\phi'' = O(e^{-v^2\phi}). \tag{9.34}$$

An asymptotic expansion for ϕ that matches with the solution for $Y = O(1)$ as $\hat{Y} \to 0$ is easily found to be

$$\phi = \hat{Y} + \frac{1}{v^2}\hat{Y} + O\left(\frac{1}{v^4}\right),$$

and hence

$$\beta = \exp\left\{-\hat{Y}\left(v^2 + 1 + o(1)\right)\right\}. \tag{9.35}$$

This expansion remains uniform as $\hat{Y} \to \infty$, and reproduces the behaviour that we would expect from our earlier analysis in the neighbourhood of the equilibrium point $(1, 0)$. A similar analysis sorts out the non-uniformity as $Y \to -\infty$. In terms of the original variable, for $y = o(v^3)$, the solution tends to a constant value as $y \to \pm\infty$ over a length scale v, with the solution for $|y| = O(v^3)$ representing an adjustment to the far field behaviour. Our main conclusion is that the width of the wavefront is of $O(v)$ when $v \gg 1$.

Having shown that a family of travelling wave solutions exists, we can now ask whether any members of the family are stable. In order to determine this, we proceed in the usual manner, by considering whether a small perturbation to a travelling wave solution grow or decays. If it grows, the solution is unstable, if it decays it is linearly stable. Before we perform the analysis, let's consider what we might expect to find. We have already seen that numerical solutions of initial value problems where the initial input of the autocatalyst is localised show the evolution of a pair of diverging, minimum speed, travelling waves. This strongly suggests that the minimum speed wave is stable, since the numerical error inherent in any such calculation would surely destroy any unstable solution. However, we must be careful with this line of argument. We have seen that there is a family of travelling wave solutions, with a member for each $v \geq 2$. In particular, the local phase plane analysis in the neighbourhood of the stable node of the system (9.26) shows that

$$\beta_v(y) = O(e^{-\lambda y}) \quad \text{as } y \to \infty, \tag{9.36}$$

where

$$\lambda(v) = \frac{1}{2}\left(v - \sqrt{v^2 - 4}\right) \tag{9.37}$$

and $\beta_v(y)$ is the permanent form travelling wave with speed $v \geq 2$. How might one of these faster waves be generated in an initial value problem? This can occur when the initial input of the autocatalyst has

$$\beta(z, 0) = O(e^{-\lambda z}) \quad \text{for } z \gg 1,$$

which generates a travelling wave with, from (9.37), speed $v = (\lambda^2 + 1)/\lambda$, provided that $\lambda < 1$. This is borne out by numerical solutions, and is consistent with solutions obtained by Billingham and Needham (1992) using the method of matched asymptotic expansions. Going back to the stability problem, this suggests that small perturbations of $O(e^{-\lambda(v_0)y})$ as $y \to \infty$ with $v_0 > v$ will destabilise the wave and transform it into a wave of speed v_0. Numerical error will not generate such perturbations in solutions with localised initial inputs of autocatalyst, so we cannot deduce that the minimum speed travelling wave is stable to *all* small perturbations.

Let's now see if we can deduce any of this from a stability analysis of the travelling wave solutions. In order to simplify the analysis, we will assume that $\alpha + \beta \equiv 1$, so that (9.24) reduces to the single reaction–diffusion equation

$$\frac{\partial \beta}{\partial \tau} = \frac{\partial^2 \beta}{\partial z^2} + \beta(1 - \beta), \tag{9.38}$$

known as **Fisher's equation**. In terms of $y = z - v\tau$, (9.38) becomes

$$\frac{\partial \beta}{\partial \tau} = \frac{\partial^2 \beta}{\partial y^2} + v\frac{\partial \beta}{\partial y} + \beta(1 - \beta). \tag{9.39}$$

In order to examine the behaviour of a small perturbation to a travelling wave solution, we write

$$\beta(y, \tau) = \beta_v(y) + b(y, \tau).$$

The function $b(y, \tau)$ is the small perturbation, and we need to determine whether b grows or decays as τ increases. At leading order for $|b| \ll 1$,

$$\frac{\partial b}{\partial \tau} = \frac{\partial^2 b}{\partial y^2} + v\frac{\partial b}{\partial y} + \{1 - 2\beta_v(y)\}\, b, \tag{9.40}$$

and we wish to solve this linear partial differential equation subject to the initial condition

$$b(y, 0) = b_0(y),$$

where $b_0 \to 0$ as $y \to \pm\infty$.

Guided by our discussion of what we expect to discover from this analysis, we now define a new variable, $\hat{b}(y, \tau)$, through

$$b(y, \tau) = e^{-\lambda y}\hat{b}(y, \tau).$$

In terms of $\hat{b}$, (9.40) becomes

$$\frac{\partial \hat{b}}{\partial \tau} = \frac{\partial^2 \hat{b}}{\partial y^2} + \sqrt{v^2 - 4}\frac{\partial \hat{b}}{\partial y} - 2\beta_v(y)\hat{b}, \tag{9.41}$$

to be solved subject to

$$\hat{b}(y, 0) = \hat{b}_0(y) = e^{\lambda y} b_0(y).$$

This helps us because we know that when $v \geq 2$, $\beta_v(y)$ is strictly positive for $-\infty < y < \infty$. Equation (9.41) is therefore a linear advection–diffusion equation with a term, $-2\beta_v\hat{b}$, that represents consumption of $\hat{b}$. Although physical intuition tells us that the solution of such an equation will tend to zero as $\tau \to \infty$, we must prove it!

The next step in this type of stability analysis is usually to look for a solution of the form $\hat{b}(y, t) = e^{-\mu\tau}\bar{b}(y)$, and show that μ must be positive, and hence that $\hat{b} \to 0$ as $\tau \to \infty$. The ordinary differential equation satisfied by $\bar{b}$ is

$$\frac{d^2\bar{b}}{dy^2} + \sqrt{v^2 - 4}\frac{d\bar{b}}{dy} + \{\mu - 2\beta_v(y)\}\, \bar{b} = 0. \tag{9.42}$$

If we were solving this problem on a finite domain, Sturm–Liouville theory would immediately tell us that μ must be positive for non-zero solutions. However, on the infinite domain that we are interested in, the problem is not so straightforward.

Instead of taking this approach, we work directly with (9.41) and make use of a lovely theorem due to Kolodner and Pederson (1966), which is extremely useful for analysing many different reaction–diffusion systems, including nonlinear systems. The idea is very simple. Equation (9.41) models an advective–diffusive process, as exemplified by the equation

$$\frac{\partial A}{\partial \tau} = \frac{\partial^2 A}{\partial y^2} + \sqrt{v^2 - 4}\frac{\partial A}{\partial y}, \tag{9.43}$$

and a reactive process,

$$\frac{\partial R}{\partial \tau} = -2\beta_v(y)R. \tag{9.44}$$

In (9.41), both of these processes occur simultaneously. What happens if we allow the initial conditions firstly to be advected and diffused, and secondly to react, or vice versa? In other words, what if the processes are applied consecutively rather than simultaneously? Under certain rather mild technical conditions, which are satisfied by our system, we have

$$\hat{b}_{\text{RD}}(y,\tau) \leq \hat{b}(y,\tau) \leq \hat{b}_{\text{DR}}(y,\tau), \tag{9.45}$$

where $\hat{b}_{\text{RD}}$ is the solution at time τ when firstly reaction, and secondly diffusion, are applied consecutively, and $\hat{b}_{\text{DR}}$ the solution when these processes are reversed. We therefore have a method for obtaining upper and lower bounds on the perturbation. If we can show that both of these tend to zero as $\tau \to \infty$, we will have demonstrated the stability of the travelling wave. We will not consider the details of the technical conditions for which the theorem holds, but, crucially, we do need $\hat{b}_0(y)$ to be bounded for $-\infty < y < \infty$.

To obtain the bounding solutions, we must begin by solving (9.43) and (9.44) subject to initial conditions

$$A(y,0) = A_0(y), \quad R(y,0) = R_0(y).$$

We can easily solve (9.43) using Fourier transforms to give

$$A(y,\tau) = \frac{1}{\sqrt{4\pi\tau}}\int_{-\infty}^{\infty} A_0(Y)\exp\left[-\frac{\left\{y - Y + (v^2-4)^{1/2}\tau\right\}^2}{4\tau}\right] dY, \tag{9.46}$$

whilst (9.44) is separable, with solution

$$R(y,\tau) = R_0(y)\exp\left(-2\beta_v(y)\tau\right). \tag{9.47}$$

This means that $\hat{b}_{\text{RD}}$ satisfies (9.43) subject to the initial conditions

$$\hat{b}_{\text{RD}}(y,0) = \hat{b}_0(y)\exp\left(-2\beta_v(y)\tau\right), \tag{9.48}$$

treating τ *as it appears in* (9.48) *as a parameter*. This leads to

$$\hat{b}_{\mathrm{RD}}(y,\tau) = \frac{1}{\sqrt{4\pi\tau}} \int_{-\infty}^{\infty} \hat{b}_0(Y) \times \exp\left[-\frac{\{y - Y + (v^2-4)^{1/2}\tau\}^2}{4\tau} - 2\beta_v(Y)\tau \right] dY. \quad (9.49)$$

Similarly,

$$\hat{b}_{\mathrm{DR}}(y,\tau) = \frac{1}{\sqrt{4\pi\tau}} \exp(-2\beta_v(y)\tau) \int_{-\infty}^{\infty} \hat{b}_0(Y) \times \exp\left[-\frac{\{y - Y + (v^2-4)^{1/2}\tau\}^2}{4\tau} \right] dY. \quad (9.50)$$

These solutions are governed by spreading due to diffusion and exponential decay due to chemical reaction, and both $\hat{b}_{\mathrm{RD}}$ and $\hat{b}_{\mathrm{DR}}$ tend to zero as $\tau \to \infty$.

Finally, (9.45) shows that $\hat{b} \to 0$ as $\tau \to \infty$. This means that the travelling wave solution is stable provided that $\hat{b}_0$ is bounded as $y \to \pm\infty$, or equivalently, provided $b_0(y) = o(e^{-\lambda y})$ as $y \to \infty$. Of course, we have not proved that the wave is unstable if this does not hold, only that it is stable if it does. In fact when $b_0(y) = O(e^{-ky})$ as $y \to \infty$ with $k > \lambda$, the travelling wave *is* unstable. We can prove using a variation on this method (see exercise 9.6). For an alternative proof that the wave is unstable to any small perturbation of $O(e^{-ky})$, see Larson (1978). In these unstable cases, provided the perturbation decays exponentially as $z \to \infty$, the wave is transformed into one with a larger wave speed, consistent with our earlier discussion concerning the generation of travelling waves by initial inputs of autocatalyst that decay exponentially as $z \to \infty$.

9.4 Autocatalytic Chemical Waves with Unequal Diffusion Coefficients*

In section 9.3, we studied the chemical waves generated by the addition of a small amount of autocatalyst, B, into a uniform expanse of a reactant, A, when each chemical diffuses with the constant diffusion coefficient, D. We now consider what happens when the diffusion coefficients are different, D_A and D_B, for A and B. We use almost the same dimensionless variables as in section 9.3, with

$$x^* = \sqrt{\frac{D_A}{ka_0}}.$$

The governing equations are now

$$\frac{\partial \alpha}{\partial \tau} = \frac{\partial^2 \alpha}{\partial z^2} - \alpha\beta, \quad \frac{\partial \beta}{\partial \tau} = \delta \frac{\partial^2 \beta}{\partial z^2} + \alpha\beta, \tag{9.51}$$

where

$$\delta = \frac{D_B}{D_A}$$

is the ratio of the diffusion coefficients.

9.4.1 Existence of Travelling Wave Solutions

We now define a travelling wave coordinate, $y = z - v\tau$, and seek solutions $\alpha = \alpha(y)$, $\beta = \beta(y)$. This leads to

$$\frac{d^2\alpha}{dy^2} + v\frac{d\alpha}{dy} - \alpha\beta = 0, \quad \delta\frac{d^2\beta}{dy^2} + v\frac{d\beta}{dy} + \alpha\beta = 0. \tag{9.52}$$

When $\delta = 1$ we recover the problem studied in section 9.3, where we were able to add the two equations and show that $\alpha + \beta = 1$. This is no longer possible, and the problem is rather more difficult.

We know that the chemicals are unreacted ahead of the wave, so that

$$\alpha \to 1, \ \beta \to 0 \text{ as } y \to \infty,$$

whilst behind the wave the chemicals are fully reacted so that

$$\alpha \to 0, \ \beta \to 1 \text{ as } y \to -\infty.$$

We are still able to add equations (9.52) and integrate once to obtain

$$\frac{d\alpha}{dy} + \delta\frac{d\beta}{dy} + v(\alpha + \beta) = v,$$

but a further integration is not possible. If we now define

$$\gamma = \frac{d\beta}{dy},$$

we can write the boundary value problem for α, β and γ as

$$\left.\begin{aligned} \alpha_y &= v(1 - \alpha - \beta) - \delta\gamma, \\ \beta_y &= \gamma, \\ \gamma_y &= -\delta^{-1}(\alpha\beta + v\gamma), \end{aligned}\right\} \tag{9.53}$$

subject to

$$\left.\begin{aligned} \alpha \to 1, \ \beta \to 0, \ \gamma \to 0 \text{ as } y \to \infty, \\ \alpha \to 0, \ \beta \to 1, \ \gamma \to 0 \text{ as } y \to -\infty. \end{aligned}\right\} \tag{9.54}$$

For each positive value of δ we want to know for what values of v a solution exists with $\alpha > 0$ and $\beta > 0$ for $-\infty < y < \infty$. Such a solution represents a physically acceptable travelling wave. When $\delta = 1$ we were able to use phase plane methods to address this problem, firstly by determining the nature of the two equilibrium points, and secondly by constructing a trapping region to show that an integral path connects the equilibrium points for $v \geq 2$. When $\delta \neq 1$, we must analyse the boundary value problem given by (9.53) and (9.54) in (α, β, γ)-phase space. Fortunately, we find that exactly the same techniques work.

Equations (9.53) have two equilibrium points, $(1, 0, 0)$ and $(0, 1, 0)$. In order to determine their type, we must calculate the Jacobian matrix at each point, just as we do for an equilibrium point in the phase plane. The Jacobian is

$$J = \begin{pmatrix} -v & -v & -\delta \\ 0 & 0 & 1 \\ -\delta^{-1}\beta & -\delta^{-1}\alpha & -\delta^{-1}v \end{pmatrix}. \tag{9.55}$$

At the equilibrium point $(0, 1, 0)$ the eigenvalues of J are

$$\lambda_1 = -\frac{v}{\delta},\ \lambda_2 = -\frac{1}{2}\left(v + \sqrt{v^2+4}\right),\ \lambda_3 = -\frac{1}{2}\left(v - \sqrt{v^2+4}\right).$$

There is therefore a two-dimensional stable manifold associated with the two negative eigenvalues, λ_1 and λ_2, composed of integral paths that asymptote to $(0, 1, 0)$ as $y \to \infty$, and a one-dimensional unstable manifold, associated with the positive eigenvalue, λ_3, that consists of two integral paths that asymptote to $(0, 1, 0)$ as $y \to -\infty$. The eigenvector that gives the direction of these unstable integral paths close to the equilibrium point is

$$\mathbf{e}_3 = (\lambda_3(\delta\lambda_3 + v), -1, -\lambda_3)^T.$$

The only integral path that satisfies the boundary condition (9.54) as $y \to -\infty$ and has $\alpha \geq 0$ is therefore the unstable manifold of $(0, 1, 0)$ that lies with $\alpha \geq 0$ in the neighbourhood of the equilibrium point, which we label as S. This fixes S as the only integral path that can be a solution of the boundary value problem.

At the equilibrium point $(1, 0, 0)$ the eigenvalues of J are

$$\mu_1 = -v,\ \mu_2 = -\frac{1}{2\delta}\left(v + \sqrt{v^2 - 4\delta}\right),\ \mu_3 = -\frac{1}{2\delta}\left(v - \sqrt{v^2 - 4\delta}\right).$$

All of these have a negative real part, and hence the equilibrium point is stable. For S to represent a solution of the boundary value problem, it must enter the stable equilibrium point $(1,0,0)$ as $y \to \infty$. We now note that μ_2 and μ_3 are complex when $v < 2\delta^{1/2}$. This means that integral paths in the neighbourhood of $(1,0,0)$ approach the equilibrium point in an oscillatory manner, with β becoming negative. We conclude that *there is no physically acceptable travelling wave solution for* $v < 2\delta^{1/2}$.

We now need to construct a trapping region to show that a travelling wave solution exists for each $v \geq 2\delta^{1/2}$. By analogy with the region of the phase plane used for the case $\delta = 1$ in section 9.3, we define the wedge shaped region R to be

$$R = \{(\alpha, \beta, \gamma) \mid 0 \leq \alpha \leq 1,\ 0 \leq \beta \leq 1,\ -v\beta/2\delta \leq \gamma \leq 0\}.$$

The governing equations (9.53) show that all integral paths at the five bounding planar surfaces of R are directed into R. The integral path S lies within R in the neighbourhood of $(0,1,0)$ and the only stable equilibrium point within R is $(1,0,0)$. Although there is no equivalent of the Poincaré–Bendixson theorem for third order systems, $\gamma = \beta_z < 0$ in R, so that β is monotone decreasing. Therefore, S must asymptote to $(1,0,0)$ as $y \to \infty$. We have now shown that *a unique, physically acceptable travelling wave solution exists for each* $v \geq 2\delta^{1/2}$. We expect that a pair of minimum speed travelling waves, with $v = v_{\min} = 2\delta^{1/2}$, will be generated by a localised input of B into a uniform expanse of A. Note that this is consistent with the results we obtained in section 9.3 when $\delta = 1$. When we use this result to obtain the dimensional minimum travelling wave speed, $v^*_{\min} = v_{\min} x^*/t^*$, we find that

$$v^*_{\min} = 2\left(D_B k a_0\right)^{1/2},$$

independent of D_A. The rate of diffusion of A does not affect the minimum wave speed, although it will affect the form of the travelling wave solution through the parameter δ. This means that it is the rate at which the autocatalyst is transported that determines the wave speed.

Figure 9.11 shows the minimum speed travelling wave for various values of δ. As δ becomes small, the concentration profiles become increasingly asymmetric, with α varying over a much longer length scale than β. As δ becomes large, the concentration profiles remain symmetric, but both vary over a long length scale. The solution for $\delta \ll 1$ is the more interesting, and we will construct an asymptotic solution using the method of matched asymptotic expansions. This limit is physically relevant to enzyme reactions, where the autocatalyst may have larger molecules than the other reactant, and therefore diffuse more slowly. It is also possible to immobilise enzymes in a gel or membrane, so that

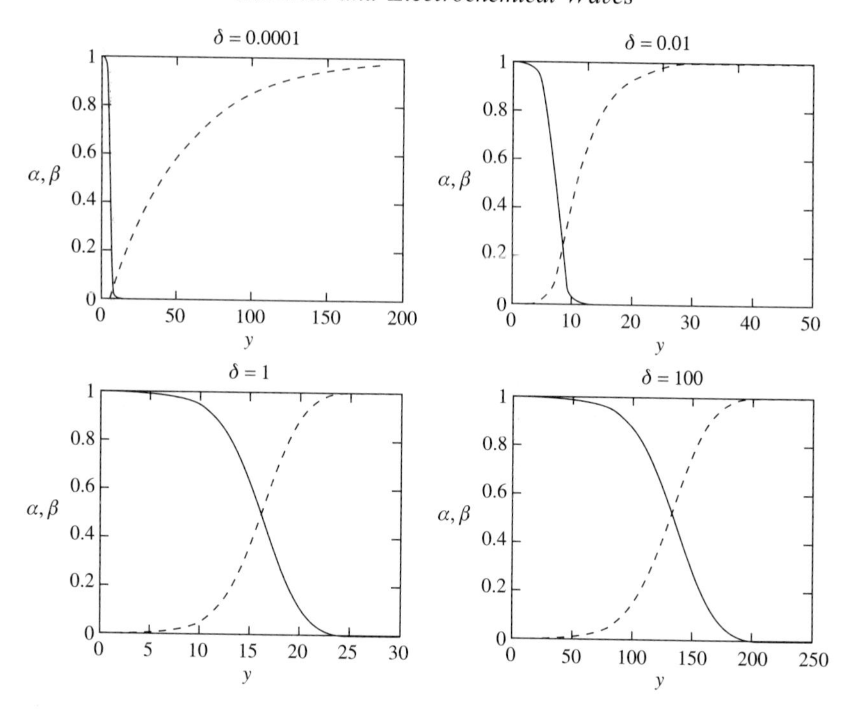

Fig. 9.11. The minimum speed travelling wave for $\delta = 0.0001$, 0.01, 1 and 100. The solid line is β and the dashed line is α.

$D_B = 0$, and hence $\delta = 0$. In this case the minimum wave speed is zero, consistent with the fact that no wave can be generated if the autocatalyst cannot diffuse.

9.4.2 Asymptotic Solution for $\delta \ll 1$

When $\delta \ll 1$, the minimum wave speed, $v_{\min} = 2\delta^{1/2}$, is small. We therefore write

$$v = \delta^{1/2} V,$$

and look for travelling wave solutions with $V = O(1)$. In this case, the positive eigenvalue of the equilibrium point $(0, 1, 0)$ is $\lambda_3 \sim 1$, and the associated eigenvector is $\mathbf{e}_3 \sim (V\delta^{1/2}, -1, -1)^T$. This suggests that we should rescale α so that

$$\alpha = \delta^{1/2} A,$$

with $A = O(1)$. The governing equations (9.53) are now

$$\left.\begin{aligned} A_y &= V(1 - \delta^{1/2}A - \beta) - \delta^{1/2}\gamma, \\ \beta_y &= \gamma, \\ \delta^{1/2}\gamma_y &= -(A\beta + V\gamma). \end{aligned}\right\} \tag{9.56}$$

We will refer to the region where z, A, β, $\gamma = O(1)$, as region 1. At leading order for $\delta \ll 1$ in region 1,

$$\left.\begin{aligned} \gamma &= -A\beta/V, \\ A_y &= V(1 - \beta), \\ \beta_y &= -A\beta/V. \end{aligned}\right\} \tag{9.57}$$

The leading order problem is one that can be analysed in the (A, β)-phase plane. There is a unique equilibrium point at $(0, 1)$, which is a saddle point. We are therefore interested in the appropriate unstable separatrix of this point, so that $A \to 0$, $\beta \to 1$, $\gamma \to 0$ as $z \to -\infty$. This is not surprising since we determined the appropriate scaled variables by considering precisely this limit. We can determine analytical expressions for A, β and γ by noting that

$$\frac{dA}{d\beta} = -\frac{V^2}{A\beta}(1 - \beta).$$

We can integrate this separable ordinary differential equation subject to $A = 0$ when $\beta = 1$ and $A \geq 0$ to obtain

$$A = V\sqrt{2(\beta - 1 - \log\beta)}, \tag{9.58}$$

and hence

$$\frac{d\beta}{dy} = -2^{1/2}\beta\sqrt{\beta - 1 - \log\beta}.$$

This gives us

$$-2^{1/2}(y - y_0) = \int_{1/2}^{\beta} \frac{ds}{s\,(s - 1 - \log s)^{1/2}}, \tag{9.59}$$

where y_0 is the constant of integration. Note that we have arbitrarily chosen $s = \frac{1}{2}$ as the lower limit of our integration, since the integrand has a non-integrable singularity at the more natural choices, 0 or 1. We can rewrite the right hand side of this equation as

$$\begin{aligned} \int_{1/2}^{\beta} \frac{ds}{s\,(s - 1 - \log s)^{1/2}} &= \int_{1/2}^{\beta} \frac{1}{s}\left\{\frac{1}{(s - 1 - \log s)^{1/2}} - \frac{1}{(-\log s)^{1/2}}\right\} ds \\ + \int_{1/2}^{\beta} \frac{ds}{s\,(-\log s)^{1/2}} &= \int_{1/2}^{\beta} \frac{1}{s}\left\{\frac{1}{(s - 1 - \log s)^{1/2}} - \frac{1}{(-\log s)^{1/2}}\right\} ds \\ &\quad -2\,(-\log\beta)^{1/2} + 2\,(\log 2)^{1/2}. \end{aligned}$$

This integral can now be evaluated at $\beta = 0$, since we have subtracted out the singularity, which now appears as the term $2(-\log\beta)^{1/2}$. This means that we can rewrite (9.59) as

$$-2^{1/2}(y - y_1) = \int_0^\beta \frac{1}{s} \times \left\{ \frac{1}{(s-1-\log s)^{1/2}} - \frac{1}{(-\log s)^{1/2}} \right\} ds - 2(-\log\beta)^{1/2}, \tag{9.60}$$

where y_1 is a new constant which absorbs the various constants that emerge during the course of the above manipulations. This constant appears because the system of equations has a translational invariance. In other words, the travelling wave solution remains a solution, however far we shift it to the left or right. This means that we can choose to put the origin of our coordinates wherever we want to, and hence y_1 is arbitrary, so for convenience we let $y_1 = 0$. This completes the leading order solution in region 1. The solution with $V = 2$, the minimum speed travelling wave solution, is illustrated in figure 9.12. In this region the concentration, β, of the autocatalyst falls to zero over an $O(1)$ length scale, independent of δ.

Equation (9.59) shows that $y \to \infty$ as $\beta \to 0$, and hence $A \to \infty$, $\gamma \to 0$ as $y \to \infty$. More precisely, (9.60) shows that

$$A \sim Vy, \quad \beta \sim e^{-\frac{1}{2}y^2}, \quad \gamma \sim -ye^{-\frac{1}{2}y^2} \quad \text{as } y \to \infty. \tag{9.61}$$

Since we have assumed that $A = O(1)$, and hence that $\alpha = O(\delta^{1/2})$, there must be a non-uniformity in the solution as $y \to \infty$. Since we need $\alpha \to 1$, $\beta \to 0$, $\gamma \to 0$ as $y \to \infty$, this suggests that we need to introduce another asymptotic region, within which $y = O(\delta^{-1/2})$, $\alpha = O(1)$, $\log\beta = O(\delta^{-1})$ and $\log(\delta^{1/2}\gamma) = O(\delta^{-1})$. We therefore define scaled variables

$$y = \delta^{-1/2}Y, \ \beta = \exp\{-\phi(Y)/\delta\}, \ \gamma = \delta^{-1/2}\exp\{-\psi(Y)/\delta\},$$

and let region 2 be the region where α, ϕ, ψ, Y are of $O(1)$. In terms of these new variables, (9.53) becomes

$$\left.\begin{aligned} \alpha_Y &= V(1 - \alpha - e^{-\phi/\delta}) - e^{-\psi/\delta}, \\ -\phi_Y e^{-\phi/\delta} &= e^{-\psi/\delta}, \\ -\psi_Y e^{-\psi/\delta} &= -\alpha e^{-\phi/\delta} - Ve^{-\psi/\delta}. \end{aligned}\right\} \tag{9.62}$$

The solution of these equations must match with the solution in region 1 as $Y \to 0$. By rewriting (9.61) in terms of the region 2 variables, we

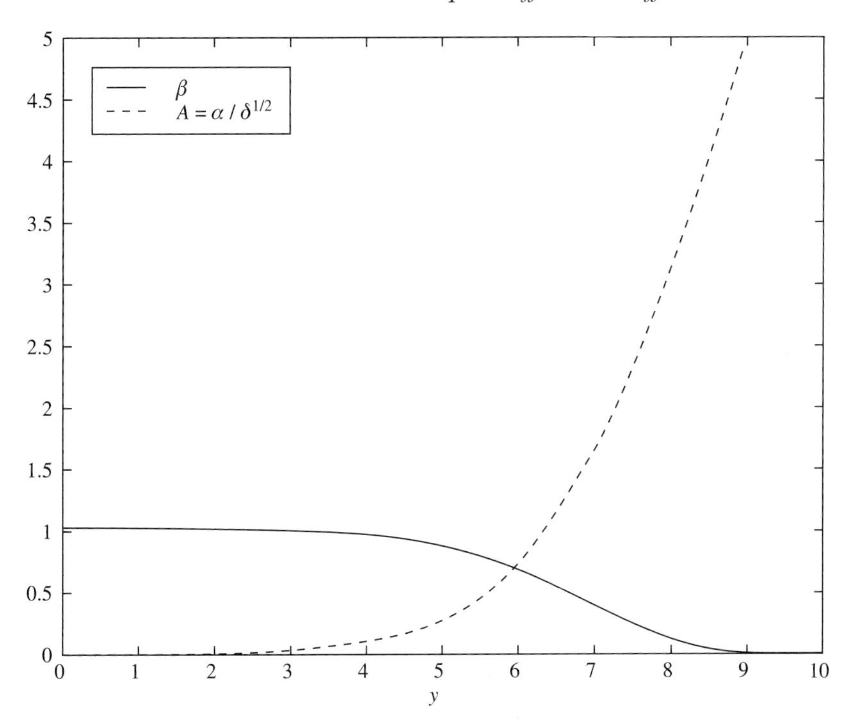

Fig. 9.12. The asymptotic solution in region 1 for $V = 2$.

find that

$$\alpha \sim VY, \quad \phi \sim \frac{1}{2}Y^2, \quad \psi \sim \frac{1}{2}Y^2, \quad \text{as } Y \to 0. \tag{9.63}$$

The leading order equation for α is

$$\frac{d\alpha}{dY} = V(1-\alpha),$$

and the solution that satisfies (9.63) is

$$\alpha = 1 - e^{-VY}. \tag{9.64}$$

This also satisfies the boundary condition, $\alpha \to 1$ as $Y \to \infty$.

Turning now to the equations for ϕ and ψ, these show that $\phi = \psi$ at leading order. We therefore expand ϕ and ψ as

$$\phi = \phi_0 + \delta\phi_1 + O(\delta^2), \quad \psi = \phi_0 + \delta\psi_1 + O(\delta^2).$$

On substituting these and (9.64) into (9.57) we obtain

$$-\phi_{0Y}e^{-\phi_1} = e^{-\psi_1}, \quad -\phi_{0Y}e^{-\psi_1} = -(1-e^{-VY})e^{-\phi_1} - Ve^{-\psi_1}.$$

We can eliminate ϕ_1 and ψ_1 between these equations and arrive at

$$\phi_{0Y}^2 - V\phi_{0Y} + 1 - e^{-VY} = 0.$$

Solving this quadratic equation for ϕ_{0Y}, making use of (9.63) to determine the appropriate root, gives

$$\phi_{0Y} = \frac{1}{2}\left\{V - \sqrt{V^2 - 4\left(1 - e^{-VY}\right)}\right\}. \qquad (9.65)$$

Note that the quantity under the square root is positive for all $Y > 0$, provided $V \geq 2$, which we know is precisely the condition for a physically acceptable travelling wave to exist. We can integrate (9.65) subject to (9.63) to obtain

$$\phi_0 = \psi_0 = \frac{1}{2}VY - \frac{1}{2}\int_0^Y \sqrt{V^2 - 4\left(1 - e^{-Vs}\right)}\,ds. \qquad (9.66)$$

Note that $\phi_0 \to \infty$, $\psi_0 \to \infty$ as $Y \to \infty$, and hence $\beta \to 0$, $\gamma \to 0$ as $Y \to \infty$. The solution in region 2 therefore completes the asymptotic solution for $\delta \ll 1$. For the minimum speed travelling wave, $V = 2$, we can calculate the integral in (9.66) analytically, so that

$$\phi_0 = \psi_0 = Y + e^{-Y} - 1. \qquad (9.67)$$

In region 2 the concentration of the autocatalyst, β, is exponentially small, and the behaviour of ϕ simply adjusts this from $O(e^{-y^2/2})$ as $y \to \infty$ out of region 1, to $O(e^{-Y/\delta})$ as $Y \to \infty$ out of region 2, consistent with the local analysis at the equilibrium point $(1, 0, 0)$. The concentration, α, of the reactant increases from $O(\delta^{1/2})$ leaving region 1 and has $\alpha \to 1$ as $Y \to \infty$ over a length scale where $Y = O(1)$, and hence $y = O(\delta^{-1/2})$. All of these features are consistent with the numerical solutions with $\delta \ll 1$ shown in figure 9.11, and we conclude that our asymptotic solution correctly captures all the features of the travelling wave solution in this limit.

9.5 The Transmission of Nerve Impulses: the Fitzhugh–Nagumo Equations

The brain and central nervous system (c.n.s.) control the limbs and organs of living creatures using two distinct methods. The **hormonal system** consists of glands, which release chemicals into the bloodstream. These chemicals then influence particular organs, transported by the circulation of the blood. The **nervous system**, which we will be studying here, is electrochemical, and works by the transmission of electrochemical pulses down specific paths between the brain and limb or organ to be controlled. These pathways are referred to as **nerves** and, when examined microscopically, are found to consist of thousands of sub-units called **neurons**. **Motor neurons**, which act upon signals from the c.n.s., are attached to muscles or glands. **Sensory neurons** relay a signal from the sensory receptors back into the c.n.s. and hence the brain. **Interneurons** connect together the c.n.s. and motor or sensory neurons. For each sensory neuron in the body there are roughly ten motor neurons and one hundred interneurons.

A typical neuron consists of a long, thin tube called the **axon**, which has a **dendritic cell body** at one end, and is branched at the other, as shown in figure 9.15. The dimensions of a neuron vary depending on the creature and the location in the body. Typical lengths lie in the range 0.01–0.1 mm, with diameter 1–20 μm for vertebrates, so these neurons are long and thin. The axon is surrounded by a chemically permeable membrane of thickness 50 Å (ångström). This can be further enveloped and isolated from the surrounding tissue by a segmented myelin sheath. In general, unmyelinated neurons have a larger diameter than myelinated neurons. The discovery by the biologist Young in 1921 of a very large axon in the giant squid, on which experimentation was relatively easy, marked the beginning of the development of the theories that we will discuss in this section. The difference in size between myelinated and unmyelinated axons is illustrated in figure 9.13. The gross anatomy of the giant squid is outlined in figure 9.14, and schematic diagrams of myelinated and unmyelinated neurons are shown in figure 9.15. Any electrical signal that travels down the axon to the branched end is relayed across the **synaptic gap** to the next neuron by chemicals called **neurotransmitters**. The speed of these signals is astonishingly fast, and has been measured at 120 m s^{-1}.

Experimental measurements of the chemical and electrical activity that occurs when a signal passes down the axon reveal that:

- In the absence of a signal, there is a **resting potential** of about −70 mV in the **axoplasmic material** relative to the material outside the membrane.
- A nervous signal consists of a succession of voltage pulses. Each of these raises the potential very quickly to about +25 mV. There is a somewhat slower return to about −75 mV, and then a very slow

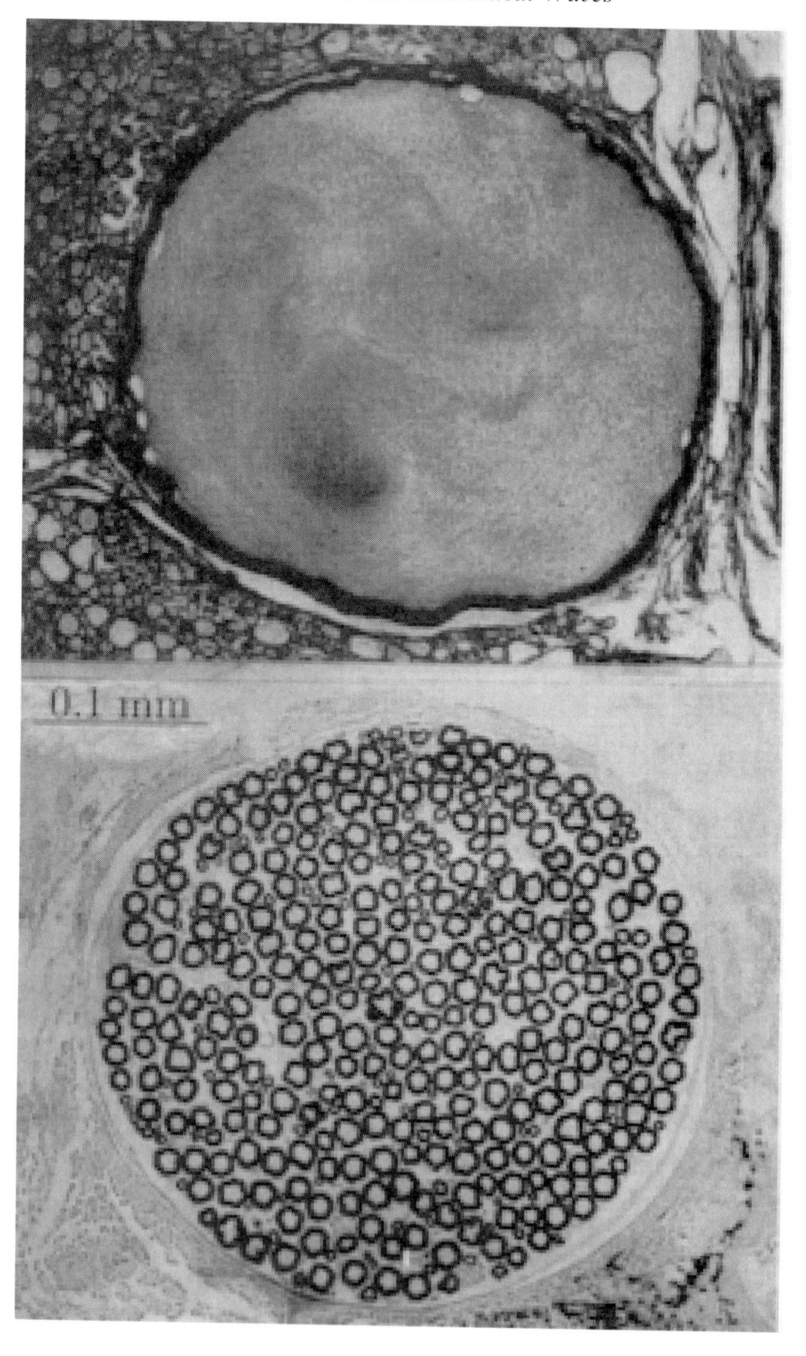

Fig. 9.13. Unmyelinated and myelinated neurons.

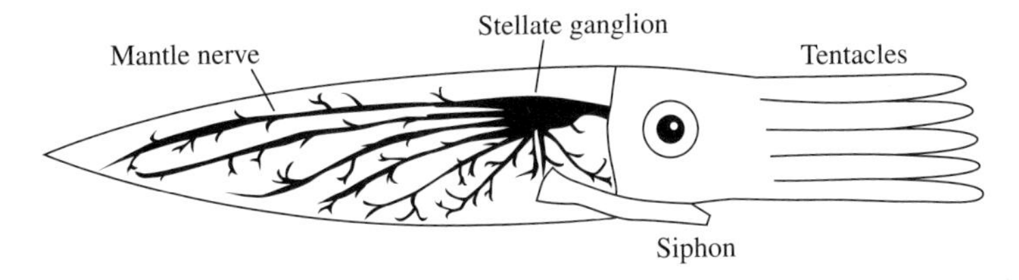

Fig. 9.14. A giant squid.

relaxation to the resting potential. A typical nervous signal is shown in figure 9.16

— The stage at which the voltage is raised rapidly corresponds to an influx of positively–charged sodium ions (Na^+) across the membrane. The return to −75 mV corresponds to a transfer of positively charged potassium ions (K^+) out from the axoplasm to the exterior of the membrane. The final relaxation phase is associated with the passage of negatively charged ions (for example chlorine ions, Cl^-) into the exterior of the membrane. As there is a flow of charge during these processes, there is a flow of current, which will change sign according to the type of charge on the ions and the direction in which they flow. A typical current–voltage relation is sketched in figure 9.17.

Our picture so far of the propagation of signals along the nerves is of an electrochemical process, during which there are a 'firing' of the signal at a cell body, chemical changes across the axon membrane and propagation of a waveform down the axon. This is actually a rather simplistic picture, but it will suffice for our introductory treatment of the subject. There is a further feature that is important in the assessment of the firing mechanism that initiates wave propagation. If a current is made to flow across the membrane in the laboratory, this mimics the flow of ions in a working neuron. It is found that there is a firing of the neuron only when the imposed current is greater than a threshold value. Below this value, the stimulus just dies away and fails to propagate as a wave. It is found that the current caused by the greatest possible flux of Na^+ ions that could occur during the time that the voltage rises to 25 mV is below this threshold value. For this reason, there must be a feedback effect on the membrane whereby the passage of Na^+ ions increases the permeability of the membrane, or equivalently, decreases the resistance to the flow of Na^+ ions, allowing even more to flow in until a sufficient current has built up.

Unmyelinated axon

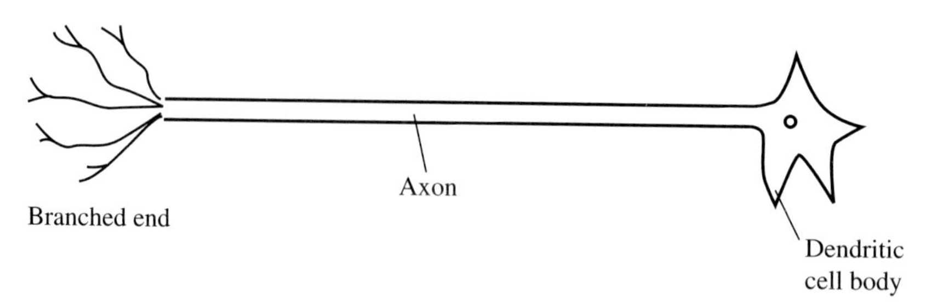

Myelinated axon

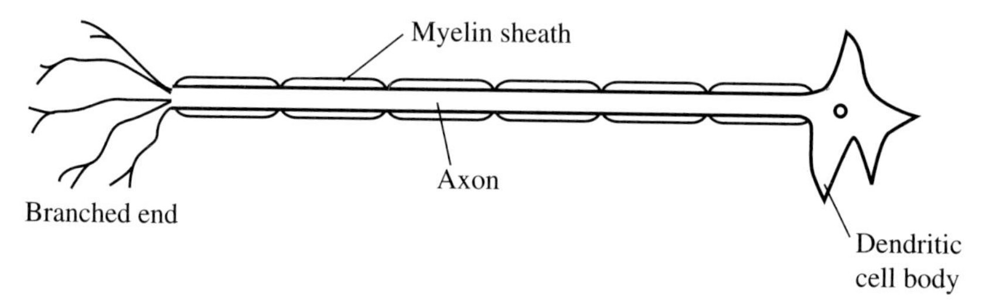

Cross–section (unmyelinated)

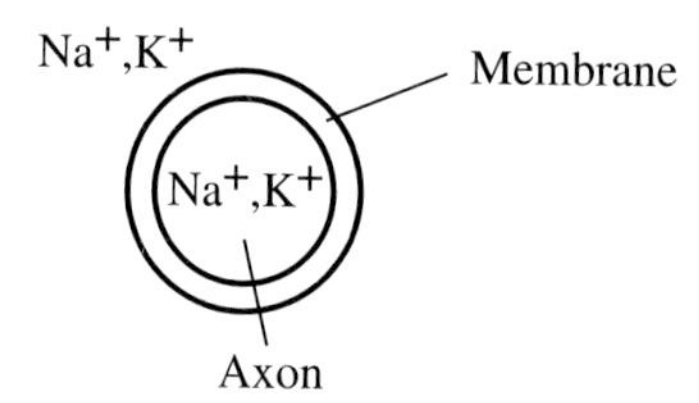

Fig. 9.15. Schematic diagrams of myelinated and unmyelinated neurons.

Before we consider the propagation of waves in the nervous system, it is worth considering the gross anatomy of the giant squid in more detail. In figure 9.14 it can be seen that the longest nerves are the thickest, and vice versa. The reason for this lies in the fact that wave propagation speeds are directly proportional to nerve diameter. This is easily deduced by considering an electrically conducting nerve of radius a with a resistance R and capacitance C. If we assume a relationship between wave speed and the other properties of the nerve in the form $\mu \equiv \mu(a, R, C)$, a dimensional analysis immediately gives $\mu = \mu_0 a/RC$. The giant squid moves through the water by using a form of jet propulsion. It does this by taking in sea water and ejecting it through a siphon by contracting

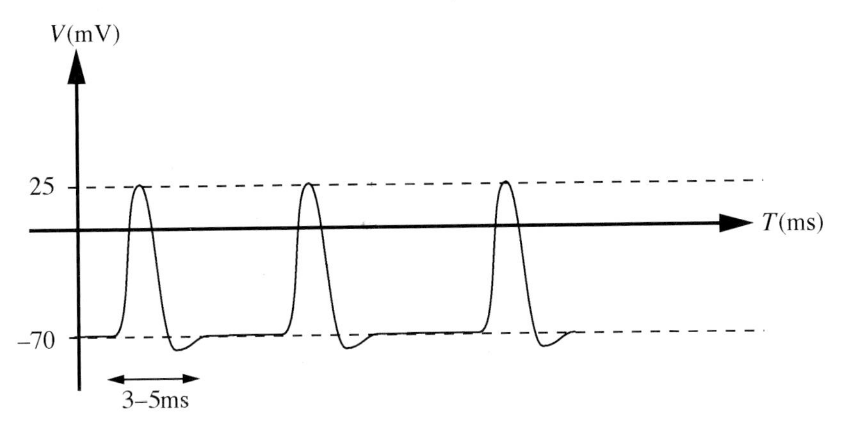

Fig. 9.16. A schematic of measured voltage as a function of time during the transmission of a nervous signal. The frequency of the pulses is usually related to the intensity of the stimulus to the sensor, or the urgency and complexity of the command from the brain.

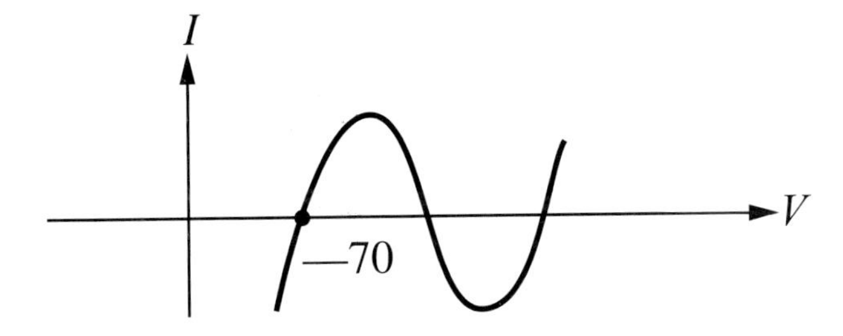

Fig. 9.17. A typical current–voltage relationship due to the motion of ions across the membrane.

its mantle muscles very rapidly. This process is at its most efficient if all of the mantle muscle contracts simultaneously. It can be seen from figure 9.14 that the short and long nerves, with their different radii and hence wave speeds, are arranged to allow this coordinated contraction.

9.5.1 The Fitzhugh–Nagumo Model

The first mathematical models of these phenomena were made by Hodgkin and Huxley in 1952. They considered in detail the different ionic flows across the membrane and the associated voltage changes. We will consider a somewhat simpler model due to Fitzhugh (1961) and Nagumo *et al.* (1962), which should be regarded as a caricature of the electrochemical processes that we have described. It is based upon standard electric circuit theory for the axon, and assumes that we can model the membrane as an imperfect insulator across which current can flow. When considering travelling waves later on, we will concentrate on single pulse solutions only. We will be considering unmyelinated axons, so we ignore the effects of the myelin sheath. This in itself is a further simplification, as the nodes in the sheath lead to a modified form of electrical impulse conduction called saltatory conduction (see Weiss (1996)).

We consider the axon to be a long, thin, cylindrical fibre of radius a, which is electrically conductive, as shown in figure 9.18(a). The outer surface of the fibre is covered by an imperfect insulator, through which charge can move. We denote the density of current flowing along the fibre by $i(x, t)$, and its resistivity by r. If the potential is $v(x, t)$, Ohm's law applied to a slice of the fibre of thickness δx, shown in figure 9.18(b), gives

$$v(x + \delta x, t) - v(x, t) = -i(x, t)\pi a^2 \delta x \, r.$$

On taking the limit $\delta x \to 0$, this gives

$$\frac{\partial v}{\partial x} = -\pi a^2 r i. \tag{9.68}$$

If there is also a loss of current through the imperfect insulator, $j(x, t)$ per unit area, shown in figure 9.18(c), conservation of charge in the same slice of the fibre gives

$$\pi a^2 \{i(x, t) - i(x + \delta x, t)\} - 2\pi a \delta x j(x, t) = 0.$$

In the limit $\delta x \to 0$, we find that

$$\frac{\partial i}{\partial x} = -\frac{2}{a} j. \tag{9.69}$$

We will regard the loss of current through the imperfect insulator to the surrounding tissue as due to both capacitive losses, j_C, and resistive losses, j_R, as shown in figure 9.18(d). By Kirchhoff's first law, $j = j_R + j_C$. By definition,

$$j_C = C \frac{\partial v}{\partial t},$$

where C is the constant capacitance.

In order to model the resistive current loss, j_R, we must account for the experimental observations described earlier. The transport of sodium and potassium ions across the insulator and the recovery phase will make

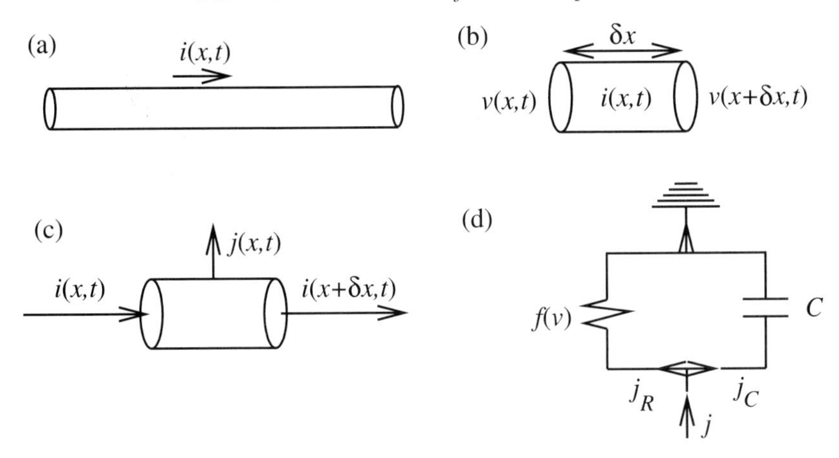

Fig. 9.18. (a) A cylindrical axon of radius a and resistance r per unit length, (b) a slice of the axon showing the changes in potential, (c) a slice of the axon showing the flow of current, (d) the circuit equivalent of the imperfect insulator.

the resistive current a function of the voltage. We tentatively write

$$j_R = \frac{1}{R} f(v) + w,$$

where R is the resistance of the insulator, $f(v)$ is a function that must account for the observed voltage variations due to ion transport and w is a recovery variable that allows the axon to return to its rest state. At this stage it is useful to eliminate i between (9.68) and (9.69) to give

$$\frac{\partial^2 v}{\partial x^2} = 2\pi ar \left\{ C\frac{\partial v}{\partial t} + \frac{1}{R} f(v) + w \right\}.$$

Since we initially expect an inward flow of current, to coincide with the flux of sodium ions, then a reversal to an outward current, to coincide with the flux of potassium ions, and no current at all in the rest state, we will take $f(v)$ to be a cubic polynomial of the form

$$f(v) = \alpha v^3 - \beta v.$$

We assume that the **recovery variable**, w, varies slowly compared with the axon voltage, v, and that it does not vary along the length of the axon. We also choose the simplest possible linear form for its evolution. This has $w = 0$ when v is at its rest state, $v = -\gamma$, so that

$$\frac{\partial w}{\partial t} = \epsilon \left(v + \gamma - \delta w \right),$$

where α, β, γ, δ and ϵ are constants, with $\epsilon \ll 1$. Further justification of this choice can be found in the original paper by Nagumo *et al.* (1962).

By taking a different origin and scales for the v and w variables, it is possible to write the **Fitzhugh–Nagumo equations** in the more convenient form

$$\frac{\partial u}{\partial t} = \frac{\partial^2 u}{\partial x^2} + u(1-u)(u-a) + w, \tag{9.70}$$

$$\frac{\partial w}{\partial t} = bu - cw, \tag{9.71}$$

where $0 < a < 1$ and $b, c > 0$. There are some other restrictions on the parameters a, b and c that we will investigate as and when we encounter them. Note that the rest state is $u = w = 0$.

9.5.2 The Existence of a Threshold

We now consider an experimental situation in which the neuron has a fine, conducting wire inserted along its axis so that there are no axial variations in the voltage. This is referred to as the **space-clamp** by experimentalists, and illustrates the need for the large diameter neuron of the giant squid. In this situation, the Fitzhugh–Nagumo equations reduce to the pair of nonlinear ordinary differential equations

$$\frac{du}{dt} = u(1-u)(u-a) - w, \tag{9.72}$$

$$\frac{dw}{dt} = bu - cw. \tag{9.73}$$

The equilibrium points of this system are the solutions of the cubic equation

$$u\left\{u^2 - (1+a)u + a + \frac{b}{c}\right\} = 0, \quad w = \frac{b}{c}u. \tag{9.74}$$

Since $u = w = 0$ represents the unique rest state of the axon, we require that the quadratic equation $u^2 - (1+a)u + a + b/c = 0$ has no real roots, and hence that

$$(1-a)^2 < \frac{4b}{c}. \tag{9.75}$$

The linear approximation close to the rest state ($|u| \ll 1$, $|w| \ll 1$) is

$$\frac{du}{dt} = -au - w, \quad \frac{dw}{dt} = bu - cw. \tag{9.76}$$

The eigenvalues associated with this system are

$$\lambda_{\pm} = \frac{1}{2}\left\{-(a+c) \pm \sqrt{(a-c)^2 - 4b}\right\},$$

which have negative real part for $b > 0$. The origin is therefore stable, and for $4b < (a-c)^2$ is a stable node, for $4b > (a-c)^2$ a stable focus. By considering the direction of the trajectories of the full nonlinear system, (9.72) and (9.73), on the lines $w = 0$, $u = a$ and $u = 1$, we can sketch the phase portrait, shown in figure 9.19. Trajectories that start between the resting potential $u = 0$ and $u = a$ enter the equilibrium point with u initially decreasing monotonically. Those that start between $u = a$ and $u = 1$ make an excursion with u increasing before finally returning to the rest state, $u = w = 0$. This means that $u = a$ is a threshold, above which the voltage increases, as observed experimentally. It is worth pointing out that there remains the possibility that there are one or more closed orbits surrounding the equilibrium point. However, this can be ruled out by a simple application of Dulac's extension to Bendixson's negative criterion (see exercise 9.8). It is also possible to simplify the model further by replacing the cubic polynomial in (9.72) with a piecewise linear function of u (see exercise 9.9).

9.5.3 Travelling Waves

As we have seen, neurons are typically long and thin. By taking the radius of the axon as our unit of length, we can regard the length of the axon as very large. Accordingly, we now consider a semi-infinite axon, with $-\infty < x \leq 0$, and the origin at the cell body. If a firing event occurs at the cell body, a change in electrical potential travels down the axon. After a suitably long time this may asymptote to a permanent form travelling wave. We consider here a waveform with a single local maximum, which asymptotes to the rest state far ahead of and far behind the wave.

We introduce the travelling wave coordinate, $y = x + \mu t$, where μ is the wave speed, and seek a permanent form travelling wave solution, $u = u(y)$, $w = w(y)$. The Fitzhugh–Nagumo equations show that

$$\mu\frac{du}{dy} = \frac{d^2u}{dy^2} + u(1-u)(u-a) - w, \tag{9.77}$$

$$\mu\frac{dw}{dy} = bu - cw, \tag{9.78}$$

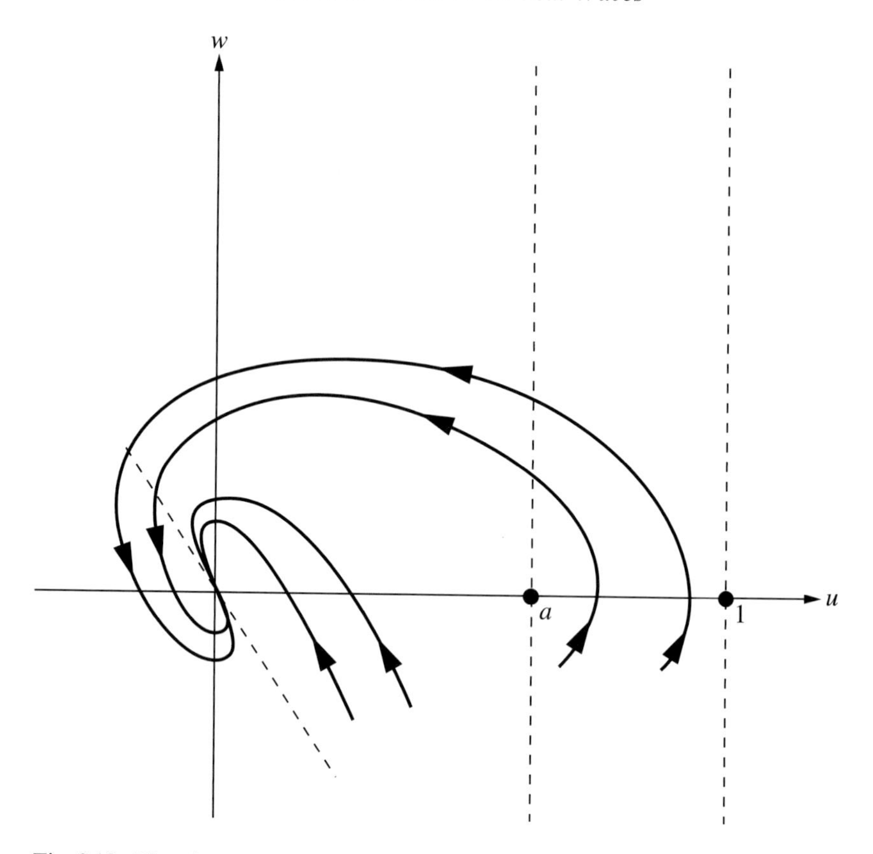

Fig. 9.19. The phase portrait of the space-clamped system, showing the existence of a threshold at $u = a$.

to be solved subject to

$$u \to 0, \quad w \to 0 \quad \text{as } y \to \pm\infty. \tag{9.79}$$

For some of the following analysis, it is convenient to write this as a system of first order ordinary differential equations,

$$\dot{x}_1 = x_2, \tag{9.80}$$

$$\dot{x}_2 = \mu x_2 - x_1(1 - x_1)(x_1 - a) + x_3, \tag{9.81}$$

$$\dot{x}_3 = \frac{b}{\mu} x_1 - \frac{c}{\mu} x_3, \tag{9.82}$$

with

$$(x_1, x_2, x_3) \to 0 \quad \text{as } y \to \pm\infty, \tag{9.83}$$

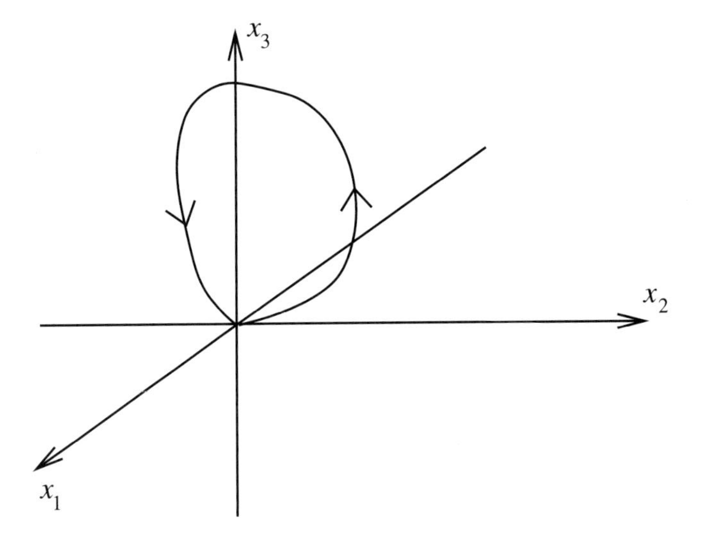

Fig. 9.20. A homoclinic orbit of the space-clamped system.

where $(x_1, x_2, x_3) = (u, du/dy, w)$. The solution that we are looking for is a 'loop' in the three-dimensional phase space, which begins and ends at the origin, as shown in figure 9.20. This is referred to as a **homoclinic orbit** of the dynamical system, and we are faced with the task of proving that one actually exists for some value of μ. This is difficult, and we will not attempt it here. Instead we will content ourselves with finding a lower bound on the wave speed, finding a simple region of the (a, b)-parameter space where a homoclinic orbit does not exist and finally giving some references to numerical calculations that show the evolution of initial data to a travelling wave.

(a) A Lower Bound for the Wave Speed

In the preceding sections on chemical waves, we found the minimum wave speed by insisting that the solution should be positive, since the independent variables were chemical concentrations. In the Fitzhugh–Nagumo model we are working with voltages, which can be positive or negative. To obtain a bound on the wave speed for this system is therefore rather more difficult. We start by multiplying (9.77) and (9.78) by u and w respectively, and integrating between $y = \pm\infty$. This gives

$$\int_{-\infty}^{\infty} \left(\frac{du}{dy}\right)^2 dy = \int_{-\infty}^{\infty} u^2(1-u)(u-a)dy - \int_{-\infty}^{\infty} uwdy, \tag{9.84}$$

$$c \int_{-\infty}^{\infty} w^2 dy = b \int_{-\infty}^{\infty} uw dy. \tag{9.85}$$

This process can be repeated using du/dy and dw/dy to multiply (9.77) and (9.78), which gives

$$\mu \int_{-\infty}^{\infty} \left(\frac{du}{dy} \right)^2 dy + \int_{-\infty}^{\infty} w \frac{du}{dy} dy = 0, \tag{9.86}$$

$$\mu \int_{-\infty}^{\infty} \left(\frac{dw}{dy} \right)^2 dy = b \int_{-\infty}^{\infty} u \frac{dw}{dy} dy. \tag{9.87}$$

Now since

$$\int_{-\infty}^{\infty} u \frac{dw}{dy} dy + \int_{-\infty}^{\infty} w \frac{du}{dy} dy = 0,$$

(9.86) and (9.87) show that

$$b \int_{-\infty}^{\infty} \left(\frac{du}{dy} \right)^2 dy = \int_{-\infty}^{\infty} \left(\frac{dw}{dy} \right)^2 dy. \tag{9.88}$$

Finally, multiplying (9.78) by u and integrating gives

$$b \int_{-\infty}^{\infty} u^2 dy = \mu \int_{-\infty}^{\infty} u \frac{dw}{dy} dy + c \int_{-\infty}^{\infty} uw dy. \tag{9.89}$$

We can now use (9.89) and eliminate the integrals of uw and udw/dy using (9.84), (9.87) and (9.88), to give

$$b \int_{-\infty}^{\infty} u^2 dy = \frac{\mu^2 - c}{b} \int_{-\infty}^{\infty} \left(\frac{dw}{dy} \right)^2 dy + c \int_{-\infty}^{\infty} u^2 (1-u)(u-a) dy. \tag{9.90}$$

This can be further rearranged, using

$$\frac{b}{c} - (1-u)(u-a) = \left\{ u - \frac{1}{2}(1+a) \right\}^2 + \frac{b}{c} - \frac{1}{4}(1-a)^2,$$

to give

$$\int_{-\infty}^{\infty} u^2 \left[\left\{ u - \frac{1}{2}(1+a) \right\}^2 + \frac{b}{c} - \frac{1}{4}(1-a)^2 \right] dy = \frac{\mu^2 - c}{bc} \int_{-\infty}^{\infty} \left(\frac{dw}{dy} \right)^2 dy. \tag{9.91}$$

Now, since we saw in the previous section that (9.75) must hold for the equilibrium state to be unique, all of the integrands in (9.91) are positive. We conclude that $\mu \geq \sqrt{c}$. This is a lower bound on the size of the wave speed. Of course, unlike the chemical waves we have studied, we have not shown that a travelling wave exists for all $\mu \geq \sqrt{c}$, simply that, if a travelling wave does exist, it must travel more quickly than $\sqrt{c}$.

(b) The Nonexistence of a Homoclinic Orbit

For simplicity, we will consider the case $c = 0$. The origin is an equilibrium point of the system (9.80) to (9.82), and the linear approximation close to the origin is

$$\frac{d}{dy}\begin{pmatrix} x_1 \\ x_2 \\ x_3 \end{pmatrix} = \begin{pmatrix} 0 & 1 & 0 \\ a & \mu & 1 \\ b/\mu & 0 & 0 \end{pmatrix}\begin{pmatrix} x_1 \\ x_2 \\ x_3 \end{pmatrix}. \tag{9.92}$$

The eigenvalues of the equilibrium point are therefore given by the roots of the cubic equation

$$\lambda^3 - \mu\lambda^2 - a\lambda - \frac{b}{\mu} = 0. \tag{9.93}$$

As a, b and μ are positive, it is straightforward to show that there is one positive real eigenvalue, λ_1, and two eigenvalues with negative real parts. The eigenvector, $\mathbf{n}_1$, associated with λ_1 has three components of the same sign, and therefore points into either the positive or the negative octant ($x_1, x_2, x_3 > 0$ or $x_1, x_2, x_3 < 0$). The straight line through the origin in the direction of $\mathbf{n}_1$ is the local approximation to the unstable manifold (solutions leave the equilibrium point at a tangent to this), whilst the plane perpendicular to $\mathbf{n}_1$ is the local stable manifold. If a homoclinic orbit exists, it must leave the origin on the unstable manifold and return on the stable manifold. Now the part of the unstable manifold that leaves the origin and enters the negative octant must remain within it, since, when $c = 0$, $\dot{x}_1$, $\dot{x}_2$ and $\dot{x}_3$ are always negative there. The only possible candidate for a homoclinic orbit is therefore the part of the unstable manifold that enters the positive octant.

We can now demonstrate a simple nonexistence result for the homoclinic orbit. If we define

$$L(y) = \frac{1}{2}x_2^2 + \int_0^{x_1(y)} x(1-x)(x-a)\,dx, \tag{9.94}$$

then $L \to 0$ as $y \to \pm\infty$ on the homoclinic orbit, since $x_1 \to 0$ and $x_2 \to 0$. By differentiating (9.94) and using (9.80) to (9.82), we find that

$$\frac{dL}{dy} = \mu x_2^2 + x_2 x_3. \tag{9.95}$$

As the homoclinic orbit leaves the origin and enters the positive octant, L is therefore a positive, increasing function of y. If L is to return to zero, as it must on a homoclinic orbit, dL/dy must vanish at some point $y = y^*$ and become negative, so that L starts to decrease. This can happen if

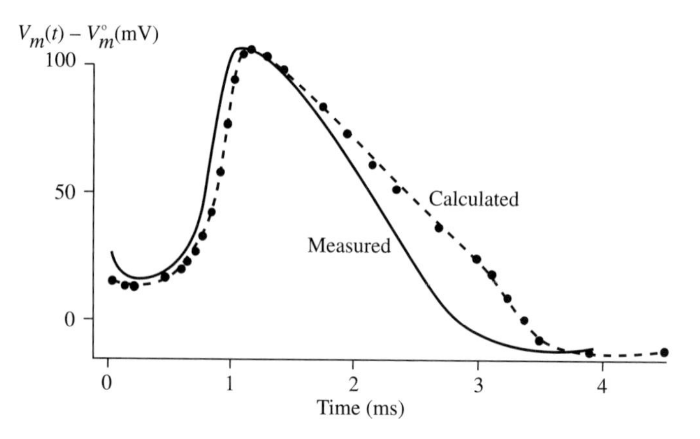

Fig. 9.21. A comparison between measured and predicted waves in an axon from a giant squid.

$x_2(y^*) = 0$ or $x_3(y^*) = -\mu x_2(y^*)$. If we follow the homoclinic orbit into the positive octant, then, directly from the equations for $\dot{x}_1$ and $\dot{x}_3$, (9.80) and (9.82), x_1 and x_3 remain positive while x_2 does. In particular, x_1 will only start to decrease when x_2 changes sign. Similarly, x_3 will only start to decrease when x_1 changes sign, so the variable x_2 will be the first to become negative. In this case, with $x_2 \to 0$ as $y \to y^*$, x_3 is still positive, so we cannot have $x_3 = -\mu x_2$, and hence

$$L(y^*) = \int_0^{x_1(y^*)} x(1-x)(x-a)dx \geq 0. \tag{9.96}$$

For $0 < x < a$ or $x > 1$, the integrand is negative, whilst for $a < x < 1$ it is positive. This means that

$$L(y^*) \leq \int_0^1 x(1-x)(x-a)dx = \frac{1}{12}(1-2a).$$

For a homoclinic orbit to exist we therefore require $1-2a \geq 0$, otherwise the trajectory that we have been considering heads off to infinity. There can therefore only be a homoclinic orbit, and hence a travelling wave solution, if $0 < a < \frac{1}{2}$. For $a > \frac{1}{2}$ there cannot be a homoclinic orbit.

At this stage you may be wondering whether travelling wave solutions do exist. They do, but the details of the calculations are beyond the scope of this textbook. We content ourselves with showing a typical travelling wave and the corresponding experimental measurements in figure 9.21 (see also exercise 9.11).

Electrochemical mechanisms are very common in physiological studies. A description of some applications of these to the function of the **kidney** and **heart** can be found in the books by Keener and Sneyd (1998) and Murray (1993). A more comprehensive biophysical description of the material in this section can be found in the book by Weiss (1998).

Exercises

9.1 A thin tube extends from $x = 0$ to $x = L$. The concentration, $c(x, t)$, of a chemical is held constant at the value c_0 at $x = 0$, and there is no flux of the chemical through the closed end at $x = L$. Initially, there is no chemical in the tube. Assuming that the behaviour of the chemical can be modelled using a one-dimensional diffusion equation with diffusion rate D, write down the appropriate initial–boundary value problem. By writing $c = c_0 + f(x, t)$ and looking for separable solutions of the equation for f, determine the solution in terms of a Fourier series in the x-direction. Show that $c \to c_0$ as $t \to \infty$.

9.2 For well-stirred, spatially uniform systems, write down the governing ordinary differential equations for the chemical concentrations given by the following chemical reaction schemes.

(a)

$$A + 2B \to 3B, \qquad \text{rate constant, } k,$$

(b)

$$\begin{aligned} 2A &\to B, \qquad \text{rate constant, } k_1, \\ B &\to 2C, \qquad \text{rate constant, } k_2, \end{aligned}$$

(c)

$$\begin{aligned} P &\to A, \qquad \text{rate constant, } k_1, \\ P + C &\to A + C, \qquad \text{rate constant, } k_2, \\ A &\to B, \qquad \text{rate constant, } k_3, \\ A + 2B &\to 3B, \qquad \text{rate constant, } k_4, \\ B &\to C, \qquad \text{rate constant, } k_5, \\ C &\to D, \qquad \text{rate constant, } k_6. \end{aligned}$$

Solve, as far as possible, the governing equations for reaction schemes (a) and (b) for general initial conditions. Reaction scheme (c) is known as the cubic cross-catalator and has solutions that exhibit bistability, mixed mode oscillations and chaos.

9.3 Write down the governing equations for the well-stirred system

$$
\begin{aligned}
P &\to A, && \text{rate constant, } k_1,\\
A + 2B &\to 3B, && \text{rate constant, } k_2,\\
B &\to C, && \text{rate constant, } k_3.
\end{aligned}
$$

If $p = p_0$ when $t = 0$, solve the equation for $p = [P]$. By finding suitable dimensionless variables, show that the equations for α and β, the dimensionless concentrations of A and B, can be written as

$$
\begin{aligned}
\frac{d\alpha}{d\tau} &= \mu e^{-\epsilon\tau} - \alpha\beta^2,\\
\frac{d\beta}{d\tau} &= \alpha\beta^2 - \beta,
\end{aligned}
$$

where τ is the dimensionless time, and μ and ϵ are constants to be determined.

When $\epsilon \ll 1$, $e^{-\epsilon\tau} \approx 1$, and this system of equations is autonomous. Show that there is a unique equilibrium point, and determine the nature of this point for each $\mu > 0$. What would you expect to happen as μ slowly decreases?

9.4 For the quadratic autocatalytic system (9.24), ahead of the wavefront, $\alpha \approx 1$ and $\beta \ll 1$, so that the equation for β is, at leading order,

$$
\frac{\partial \beta}{\partial \tau} = \frac{\partial^2 \beta}{\partial z^2} + \beta. \tag{9.97}
$$

Using the transformation $\beta = e^{\tau} B$, show that B satisfies the diffusion equation. Hence, write down the general solution of equation (9.97). If the initial distribution of β is zero outside some finite region of the real line, show that, when z/τ is a positive constant and $\tau \gg 1$, $\log \beta \propto \tau - z^2/4\tau$ at leading order. At which velocity does this suggest that the wavefront will propagate in the full problem?

9.5 A reaction–diffusion system is governed by the equation

$$
\frac{\partial \beta}{\partial \tau} = \frac{\partial^2 \beta}{\partial z^2} + \beta(1 - \beta)(\beta - \mu),
$$

where μ is a constant with $0 < \mu < \frac{1}{2}$. By looking for solutions that are a function of a suitable variable, y, write down a system of first order ordinary differential equations satisfied by permanent form travelling wave solutions that propagate at speed $v > 0$. Show that this system has three equilibrium points, and determine the local behaviour at each. Show that there is an exact solution of the form

$$\beta(y) = \left(1 + e^{ky}\right)^{-1},$$

for a particular pair of values $k = k_0(\mu)$ and $v = v_0(\mu)$. Sketch the phase portrait in this case, and show that there is another solution with $\beta \to 0$ as $y \to -\infty$, and $\beta \to \mu$ as $y \to \infty$.

9.6 (a) Make the substitution $b(y, t) = e^{-ky}\tilde{b}(y, \tau)$ in (9.40), and determine the initial value problem satisfied by $\tilde{b}$.

(b) Obtain upper and lower bounds for $\tilde{b}$, assuming that $\tilde{b}(y, 0)$ is bounded as $y \to \pm\infty$, and show that when $k < \lambda$

$$\sup_{-\infty<y<\infty} |\tilde{b}(y)| \to \infty \quad \text{as } \tau \to \infty.$$

What does this tell you about the stability of the travelling wave solution, $\beta_v(y)$?

9.7 Consider the reaction–diffusion system that we studied in section 9.4, augmented with the reaction step

$$B \to C, \quad \text{rate } k_2 b,$$

which represents linear decay of the autocatalyst.

(a) Show that the equations satisfied by permanent form travelling wave solutions are

$$\frac{d^2\alpha}{dy^2} + v\frac{d\alpha}{dy} - \alpha\beta = 0, \quad \frac{d^2\beta}{dy^2} + v\frac{d\beta}{dy} + \alpha\beta - \epsilon\beta = 0,$$

where ϵ is the dimensionless decay rate, which you should determine.

(b) Appropriate boundary conditions are

$$\alpha \to 1, \quad \beta \to 0 \quad \text{as } y \to \infty, \quad \alpha \to \alpha^*, \quad \beta \to 0 \quad \text{as } y \to -\infty,$$

where α^* is an unknown constant. Make a sketch of what you would expect the solution to look like for $\epsilon \ll 1$.

(c) We now write expansions

$$\alpha = \alpha_0 + \epsilon\alpha_1 + O(\epsilon^2), \quad \beta = \beta_0 + \epsilon\beta_1 + O(\epsilon^2) \quad \text{for } \epsilon \ll 1.$$

When $v = 2$, so that the leading order solution is the

minimum speed travelling wave that we studied in section 9.4, determine the equations satisfied by α_1 and β_1. Hence show that

$$\alpha \sim \alpha_\infty \exp\left\{(\sqrt{2}-1)y\right\}\{1+O(\epsilon)\},$$

$$\beta \sim 1+\frac{1}{2}\epsilon y + O(\epsilon^2) \text{ as } y \to -\infty,$$

where α_∞ is a positive constant.

(d) This suggests that there is a non-uniformity in the expansion for β when $-y = O(\epsilon^{-1})$. If $Y = \epsilon y$, $\phi(Y) = -\epsilon \log \alpha$ and $\hat{\beta}(Y) = \beta$, with Y, ϕ, $\hat{\beta} = O(1)$ for $\epsilon \ll 1$, determine the leading order equations for ϕ and β. Solve these equations, matching your solution carefully with the solution in the previous asymptotic region as $Y \to 0$. Hence show that

$$\log \alpha^* = -\frac{1}{\epsilon}\int_{-\infty}^{0}\left(\sqrt{1+e^{\frac{1}{2}s}}-1\right)ds + O(1).$$

9.8 Use Dulac's extension to Bendixson's negative criterion, with weighting function $\rho = e^{\sigma w}$ for some constant σ, to show that there are no limit cycle solutions of the space-clamped system (9.72) and (9.73).

9.9 When an external current, I_a, is applied to a space-clamped axon, the Fitzhugh–Nagumo model becomes

$$\frac{du}{dt} = u(a-u)(u-1) - w + I_a,$$

$$\frac{dw}{dt} = bu - cw,$$

where $0 < a < 1$ and b, c and I_a are positive constants. The u nullcline has a local maximum at $u = u_{\max}$ and a local minimum at $u = u_{\min}$. Determine $u_{\max}$ and $u_{\min}$ in terms of a and I_a.

We can define a new model with

$$\frac{du}{dt} = w_1(u) - w,$$

$$\frac{dw}{dt} = bu - cw,$$

where the term $u(a-u)(u-1) + I_a$ in the original model has been replaced by a continuous, three section, piecewise linear function, $w_1(u)$, which you should determine, that has the same local maximum and minimum as the u nullcline and has $w_1(0) = w_1(1) = I_a$.

By considering the slopes of the u and w nullclines in this new model, show that for there to be just one equilibrium state for all $I_a \geq 0$ we require that

$$\frac{b}{c} > \frac{2}{9}(a^2 - a + 1).$$

If this condition holds, show that this state is unstable if

$$\frac{2}{9}(a^2 - a + 1) > c,$$

and

$$u_{\min}\left(\frac{b}{c} + a - (a+1)u_{\min} + u_{\min}^2\right) < I_a$$
$$< u_{\max}\left(\frac{b}{c} + a - (a+1)u_{\max} + u_{\max}^2\right).$$

Use Bendixson's negative criterion to show that if a limit cycle solution surrounds the unstable equilibrium point then, at some point on this solution, either $u > u_{\max}$ or $u < u_{\min}$. Note that it can be shown that there is a range of values of the external current, I_a, for which a limit cycle solution does exist.

9.10 Taking the travelling wave equation for the Fitzhugh–Nagumo model,

$$\mu\frac{du}{dy} = \frac{d^2u}{dy^2} + u(1-u)(u-a) - w,$$
$$\mu\frac{dw}{dy} = bu - cw,$$

show that the wave speed μ satisfies the inequality

$$\mu^2 > \frac{4b}{(1-a)^2} > c.$$

9.11 By writing the time derivatives as forward differences and spatial derivatives as central differences, show that a discrete form of the Fitzhugh–Nagumo equations is

$$u_{i,j+1} = u_{i,j} + \frac{\Delta t}{(\Delta x)^2}\left(u_{i+1,j} - 2u_{i,j} + u_{i-1,j}\right)$$
$$+ \left\{u_{i,j}(1-u_{i,j})(u_{i,j} - a) + w_{i,j}\right\}\Delta t,$$
$$v_{i,j+1} = v_{i,j} + (bu_{i,j} - cw_{i,j})\Delta t,$$

where $u_{i,j} = u(i\Delta x, j\Delta t)$. For the initial–boundary value problem

$$u(x,0) = w(x,0) = 0 \quad \text{for } x \geq 0,$$
$$u \to 0, \quad w \to 0 \quad \text{as } x \to \infty \text{ for } t \geq 0,$$
$$u(0,x) = f(t) \quad \text{for } t \geq 0$$

try to discover what 'impulse functions' $f(t)$ give rise to a wave that propagates into $x > 0$ from the initiation site at $x = 0$. (Note that it is a good idea to choose $\Delta t = \frac{1}{4}(\Delta x)^2$ to get a stable numerical scheme, and that you will need a sufficiently long numerical domain to capture the development of any waves.)

Part three
Advanced Topics

Now that we have studied the basic ideas that underlie the generation and propagation of waves in the most commonly occurring situations, we will focus our attention on some more advanced techniques and more difficult problems than those that we have encountered so far in parts one and two. In chapter 10 we study Burgers' equation, which arises in many different areas of applied mathematics. It is the generic example of what happens when there is competition between nonlinear steepening and linear diffusion. As we shall see, Burgers' equation often controls the leading order behaviour of systems in which a linear dissipative process modifies the effect of nonlinear steepening. The examples that we study come from traffic flow, and weakly nonlinear, compressible gas dynamics.

Chapter 11 is concerned with scattering and diffraction – the interaction of waves with solid objects. We show how the solution of linear diffraction and scattering problems, involving either scalar or vector waves, can be approximated using integral representations. For the diffraction of a scalar wave by a single straight edge, we can use the Wiener–Hopf technique to construct the exact solution. We also examine a series solution for the scattering of water waves by a cylinder. In the final chapter we study the remarkable inverse scattering transform as it applies to the Korteweg–de Vries equation, which we met in chapter 8, and the nonlinear Schrödinger equation (NLS), which we will derive in the context of pulse propagation in weakly nonlinear optical fibres. In all of the other examples that we study in this part of the book (with the exception of the Cole–Hopf transformation for Burgers' equation) no analytical solution technique is available for the governing nonlinear partial differential equations. However, such a technique is available for the KdV and NLS equations, along with a wide variety of other important nonlinear equations. By associating it with a scattering problem and determining the evolution of the scattering data, the solution of the KdV equation can be reduced to the solution of a *linear* integral equation. The methods used are far from obvious, and seem almost magical when first encountered. In particular they show that the solitary wave solutions that we studied in chapter 8 are actually solitons. Solitons arise in many different physical systems, and perhaps most importantly today, are often used to transmit information along optical fibres. They have some remarkable, particle-like properties. In particular they can retain their individual identities after collisions and persist over long distances.

10

Burgers' Equation: Competition between Wave Steepening and Wave Spreading

By making some dramatic assumptions about the flow field and pressure gradients in a particular class of fluid flow, Burgers (1948) derived a simple equation with which he hoped to model some aspects of turbulence. Burgers' equation is

$$u_t + uu_x = \nu u_{xx}. \tag{10.1}$$

This is an evolutionary partial differential equation that models the competing effects of wave steepening (uu_x) and diffusion (νu_{xx}). Rather similar systems of equations had been studied prior to this by those working in one-dimensional gas dynamics, see for example the collected works of G.I. Taylor (1963). That turbulence itself is still an open problem puts the Burgers model into historical perspective. The equation is, however, still important in its own right and crops up in various branches of applied mathematics. We begin by considering how Burgers' equation arises in a modified version of the simple model for the flow of traffic that we studied in section 7.1. In particular, by introducing an extra physical process into our model, we are able to justify the shock fitting method. We will also demonstrate a transformation that exactly linearises Burgers' equation. Next, we turn to a classic fluid flow problem and examine how Burgers' equation arises in the study of weak, gas dynamical shock waves. Finally, we examine the asymptotic solution of Burgers' equation as $\nu \to 0$, and use the method of matched asymptotic expansions to construct the solution of a simple initial value problem.

10.1 Burgers' Equation for Traffic Flow

When we introduced the notion of shock fitting in section 7.1, the main assumption that we made was that the solution given by just one particular characteristic curve is the actual solution in the multi-valued region. We can justify this for our example problem by considering an appropriate change of our model that modifies the solution when large gradients in density start to be formed. We have assumed the $q(\rho) = \rho v(\rho)$. However, it seems reasonable to assume that, if a driver sees a large positive gradient in density, in other words she sees that cars are starting to become more closely packed ahead, she will slow down. This suggests that we should change the model for the flowrate of cars to $q(\rho) = \rho v(\rho) - \nu \rho_x$, where $\rho_x = \partial \rho / \partial x$. We will consider the case where $\nu \ll 1$, so that this gradient effect only becomes significant when $\rho_x \gg 1$, in other words, when a shock wave starts to form. If we substitute this new model for $q(\rho)$ into (7.3) we obtain

$$\frac{\partial \rho}{\partial t} + c(\rho)\frac{\partial \rho}{\partial x} = \nu \frac{\partial^2 \rho}{\partial x^2}. \tag{10.2}$$

This is an **advection–diffusion equation**, with $c(\rho)\rho_x$ the advective part and $\nu \rho_{xx}$ the diffusive part. For $\nu \ll 1$, the diffusive term will only be significant where there are large density gradients. For our example problem, where $c(\rho)$ is a linear function of ρ, we can easily rewrite (10.2) in terms of c, and arrive at Burgers' equation,

$$\frac{\partial c}{\partial t} + c\frac{\partial c}{\partial x} = \nu \frac{\partial^2 c}{\partial x^2}. \tag{10.3}$$

This is the generic example of a combination of steepening due to nonlinear advection and dissipation due to linear diffusion. Even more importantly, it is one of the few nonlinear equations for which an exact, analytical solution is available, and we will now derive it.

Define a function $\phi(x, t)$ via $c = -2\nu \partial(\log \phi)/\partial x$. This is known as the **Cole–Hopf transformation**. On substituting this into (10.3) we find that ϕ satisfies the linear diffusion equation

$$\frac{\partial \phi}{\partial t} = \nu \frac{\partial^2 \phi}{\partial x^2}. \tag{10.4}$$

The Cole–Hopf transformation therefore maps Burgers' equation to an equation that we can easily solve. The initial conditions are

$$c(x, 0) = c_0(x), \tag{10.5}$$

which maps to the initial condition

$$\phi(x, 0) = \phi_0(x) = \exp\left(-\frac{1}{2\nu}\int_0^x c_0(s)\, ds\right). \tag{10.6}$$

The solution of the diffusion equation subject to the initial condition (10.6) is

$$\phi(x,t) = \frac{1}{\sqrt{4\pi\nu t}} \int_{-\infty}^{\infty} \phi_0(X) \exp\left\{-\frac{(x-X)^2}{4\nu t}\right\} dX, \tag{10.7}$$

which can be derived using Fourier transforms. We can write this as

$$\phi(x,t) = \frac{1}{\sqrt{4\pi\nu t}} \int_{-\infty}^{\infty} \exp\{-G(x,X,t)/2\nu\}\, dX, \tag{10.8}$$

where G is given by

$$G(x,X,t) = \frac{(x-X)^2}{2t} + \int_0^X c_0(s)\, ds. \tag{10.9}$$

We can now use the fact that $c = -2\nu\partial(\log\phi)/\partial x$ to show that the general solution of Burgers' equation is

$$c(x,t) = \frac{\int_{-\infty}^{\infty} \left(\frac{x-X}{t}\right) \exp\{-G(x,X,t)/2\nu\}\, dX}{\int_{-\infty}^{\infty} \exp\{-G(x,X,t)/2\nu\}\, dX}. \tag{10.10}$$

Now we want to know what this solution looks like for $\nu \ll 1$. Do we recover the solution that we have already obtained using characteristics and shock fitting when $\nu = 0$? In order to proceed, we must estimate each integral in the solution (10.10) for $\nu \ll 1$. In this limit, each integral is dominated by the contributions from the local maxima of $-G(x,X,t)$. If we can calculate these, we can estimate the integral by approximating G using the first couple of terms in its Taylor series. This is known as **Laplace's method**, and can be justified rigorously (Nayfeh, 1993). So where are the local maxima of $-G$? Since

$$\frac{\partial G}{\partial X} = -\left(\frac{x-X}{t}\right) + c_0(X),$$

the turning points of G occur at $X = X_0$, where X_0 satisfies

$$x = X_0 + c_0(X_0)t. \tag{10.11}$$

But this is just the equation of a characteristic. To determine whether this point is a maximum of $-G$, note that

$$\frac{\partial^2 G}{\partial X^2} = \frac{1}{t} + c_0'(X). \tag{10.12}$$

If $c_0'(X_0) > 0$, then each turning point must be a maximum of $-G$ and hence there can only be a single maximum, in agreement with the fact that a single characteristic passes through each point in this case. If $c_0'(X_0) < 0$ a turning point is only a maximum if $t < -1/c'(X_0)$, corresponding to the well-defined solution that we obtain using characteristics for $t < T_{\text{min}}$. So, for a single maximum, what is the approximate solution?

In the neighbourhood of a maximum at $X = X_0$ we can approximate G using its Taylor series as

$$G(x, X, t) \approx G(x, X_0, t) + \frac{1}{2}(X - X_0)^2 \left\{ \frac{1}{t} + c_0'(X_0) \right\},$$

and hence

$$\int_{-\infty}^{\infty} \exp\{-G(x, X, t)/2v\}\, dX \approx \exp\{-G(x, X_0, t)/2v\}$$
$$\times \int_{-\infty}^{\infty} \exp\left\{ -\frac{1}{4v}\left(\frac{1}{t} + c_0'(X_0)\right)(X - X_0)^2 \right\} dX,$$

and

$$\int_{-\infty}^{\infty} \left(\frac{x - X}{t}\right) \exp\{-G(x, X, t)/2v\}\, dX$$
$$\approx \left(\frac{x - X_0}{t}\right) \exp\{-G(x, X_0, t)/2v\}$$
$$\times \int_{-\infty}^{\infty} \exp\left\{ -\frac{1}{4v}\left(\frac{1}{t} + c_0'(X_0)\right)(X - X_0)^2 \right\} dX.$$

Substituting these into (10.10) gives

$$c(x, t) \approx \frac{x - X_0}{t} \approx c_0(X_0). \tag{10.13}$$

So, if a single characteristic passes through a given point, (x, t), there is a single dominant contribution to each integral for $v \ll 1$, and

$$c(x, t) \approx c_0(X_0),$$

where X_0 satisfies

$$x = X_0 + c_0(X_0)t.$$

This is precisely the characteristic based solution that we have already constructed for the case $v = 0$. However, what we are really interested in is what happens when the solution becomes multi-valued, and *three* characteristics pass through each point in the multi-valued region. This corresponds to three turning points of $-G$. The turning point that corresponds to the middle branch of the multi-valued solution, shown in figure 7.15, has $t > 1/c'(X)$, since this is just the condition given earlier for neighbouring characteristics to have intersected, and hence is a minimum of $-G$. The other two characteristics, at $X = X_1$ and $X = X_2$, correspond to maxima of $-G$, and we must consider the contribution of each to the integrals.

Using Laplace's method for each of these maxima we find that

$$c(x,t) \approx \left[K_1\left(\frac{x-X_1}{t}\right)\exp\{-G(x,X_1,t)/2v\}\right.$$
$$\left.+K_2\left(\frac{x-X_2}{t}\right)\exp\{-G(x,X_2,t)/2v\}\right]$$
$$\times\left[K_1\exp\{-G(x,X_1,t)/2v\}+K_2\exp\{-G(x,X_2,t)/2v\}\right] \tag{10.14}$$

where

$$K_1 = \left(\frac{1}{t}+c_0'(X_1)\right)^{-\frac{1}{2}}, \quad K_2 = \left(\frac{1}{t}+c_0'(X_2)\right)^{-\frac{1}{2}}.$$

This is a well-defined, single valued solution. When $v \ll 1$, if $G(x,X_1,t) > G(x,X_2,t)$ then $c \approx (x-X_2)/t \approx c_0(X_2)$, whilst if $G(x,X_2,t) > G(x,X_1,t)$ then $c \approx (x-X_1)/t \approx c_0(X_1)$. In other words, the solution takes the value given by one or other of the characteristics through the point, and the changeover between the two states occurs rapidly in the neighbourhood of the point where $G(x,X_2,t) = G(x,X_1,t)$. This is the position of the shock wave when $v = 0$, which is represented by a discontinuity in ρ and hence c. For v small but non-zero it becomes a rapid change over a thin region. But where is this shock? The condition $G(x,X_2,t) = G(x,X_1,t)$ gives

$$\frac{(x-X_1)^2}{2t}+\int_0^{X_1} c_0(s)\,ds = \frac{(x-X_2)^2}{2t}+\int_0^{X_2} c_0(s)\,ds,$$

and hence, using $c_0(X_1) = (x-X_1)/t$ and $c_0(X_2) = (x-X_2)/t$,

$$\int_{X_1}^{X_2} c_0(s)\,ds = \frac{1}{2}(X_2-X_1)\{c_0(X_1)+c_0(X_2)\}. \tag{10.15}$$

If we can show that this condition corresponds to the equal area rule that we discussed earlier, we have justified our shock fitting procedure.

In order to achieve this, we firstly note that, since c is a linear function of ρ for the specific case that we are considering, the equal area rule for the behaviour of ρ is equivalent to the equal area rule for the behaviour of c. Now let's assume that the equal area rule is correct and see if we can deduce (10.15). Consider the initial profile, $c_0(x)$, and the profile after a shock has formed, as illustrated in figure 10.1. Let A' and B' be the points in the (c,x)-plane at the top and bottom of the shock, as shown in

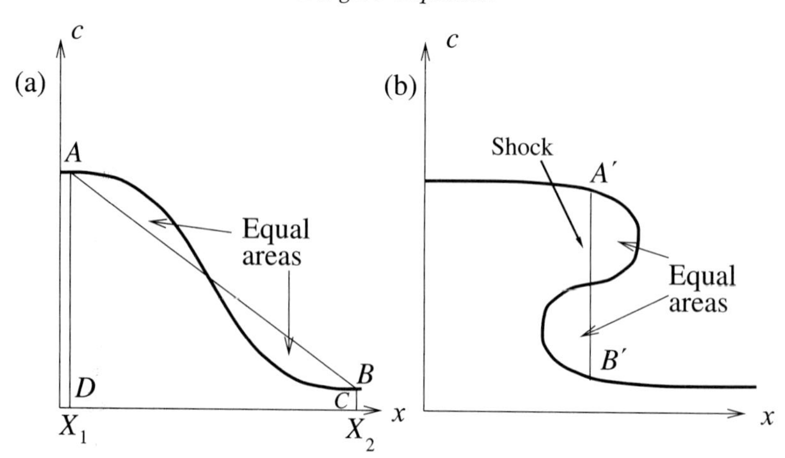

Fig. 10.1. (a) An initial profile, $c_0(x)$. (b) The profile and fitted shock some time later. Equal areas are indicated, and the line AB maps to the line $A'B'$.

figure 10.1(b). These points correspond to A and B in the initial profile, $c_0(x)$, shown in figure 10.1(a). Since the line $A'B'$ can be obtained from AB by translating it a distance $c_0(x)t$ to the right in the same way that $c(x,t)$ is obtained from $c_0(x)$, the areas indicated in figure 10.1(a) map to those indicated in figure 10.1(b), and are therefore equal. This means that the area of the trapezium $ABCD$ is equal to the area under the initial profile between X_1 and X_2. This is, however, precisely what (10.15) says. We conclude that the equal area rule can be deduced from our analysis of Burgers' equation in the limit $\nu \ll 1$. The detailed structure of the solution for $\nu \ll 1$ is the subject of section 10.3.

10.2 The Effect of Dissipation on Weak Shock Waves in an Ideal Gas

In this section, we consider the development of a small, one-dimensional disturbance of a uniform, ideal gas initially in equilibrium. At leading order, such disturbances are the linear sound waves that we studied in chapter 3. However, in section 7.2 we showed that nonlinear steepening can lead to the formation of shock waves, even for small disturbances, if they propagate for a long enough time. We therefore include in the model the main dissipative processes that are important in the neighbourhood of the shock – viscosity and heat conduction. We shall see that Burgers' equation appears again at leading order.

The governing equations for this flow are

$$\frac{\partial \rho}{\partial t} + \rho\frac{\partial u}{\partial x} + u\frac{\partial \rho}{\partial x} = 0, \tag{10.16}$$

$$\rho\frac{\partial u}{\partial t} + \rho u\frac{\partial u}{\partial x} + \frac{\partial p}{\partial x} = \frac{4}{3}\mu\frac{\partial^2 u}{\partial x^2}, \tag{10.17}$$

$$\frac{\partial S}{\partial t} + u\frac{\partial S}{\partial x} = \frac{2}{3}\frac{\mu}{\rho T}\left(\frac{\partial u}{\partial x}\right)^2 + \frac{\lambda}{\rho T}\frac{\partial^2 T}{\partial x^2}. \tag{10.18}$$

All of these variables are as defined in section 7.2. In particular, T is the absolute temperature and S the entropy. The additional terms on the right hand sides of these equations model, in (10.17), the effect of viscosity on the momentum in the x-direction, and in (10.18), the effect of viscosity and heat conduction on the entropy, with λ the thermal diffusivity of the gas. Viscosity acts to diffuse the momentum and generate entropy. In addition, the local sound speed, the gas law and the equation for the entropy are

$$c^2 = \frac{\gamma p}{\rho}, \quad p = \rho RT/m, \quad S = c_V \log\left(\frac{p}{\rho^\gamma}\right). \tag{10.19}$$

We wish to solve these equations subject to the initial conditions

$$p(x,0) = p_0\{1+\epsilon f_0(x)\}, \quad \rho(x,0) = \rho_0\{1+\epsilon g_0(x)\}, \quad u = \epsilon h_0(x), \tag{10.20}$$

where $\epsilon \ll 1$ is dimensionless and the $O(1)$ functions f_0, g_0 and h_0 are zero outside the range $-L < x < L$. These initial conditions define the orders of magnitude of the variables in the uniform state and of their initial perturbations. The initial temperature, entropy and sound speed are determined by (10.19). We therefore define the dimensionless variables

$$\left.\begin{array}{c} \bar{p} = \dfrac{p}{p_0}, \quad \bar{\rho} = \dfrac{\rho}{\rho_0}, \quad \bar{S} = \dfrac{S}{c_V}, \\ \bar{T} = \dfrac{T}{T_0}, \quad \bar{u} = \dfrac{u}{c_0}, \quad \bar{x} = \dfrac{x}{L}, \quad \bar{t} = \dfrac{c_0 t}{L}, \end{array}\right\} \tag{10.21}$$

where $c_0^2 = \gamma p_0/\rho_0$ and $T_0 = p_0/R\rho_0$. In terms of these variables (10.16) to (10.19) become

$$\frac{\partial \bar{\rho}}{\partial \bar{t}} + \bar{\rho}\frac{\partial \bar{u}}{\partial \bar{x}} + \bar{u}\frac{\partial \bar{\rho}}{\partial \bar{x}} = 0, \tag{10.22}$$

$$\bar{\rho}\frac{\partial \bar{u}}{\partial \bar{t}} + \bar{\rho}\bar{u}\frac{\partial \bar{u}}{\partial \bar{x}} + \frac{1}{\gamma}\frac{\partial \bar{p}}{\partial \bar{x}} = \frac{4}{3\mathrm{Re}}\frac{\partial^2 \bar{u}}{\partial \bar{x}^2}, \tag{10.23}$$

$$\frac{\partial \bar{S}}{\partial \bar{t}} + \bar{u}\frac{\partial \bar{S}}{\partial \bar{x}} = \frac{2}{3}\frac{\gamma(\gamma-1)}{\text{Re}}\left(\frac{\partial \bar{u}}{\partial \bar{x}}\right)^2 + \frac{1}{\text{Re Pr}}\frac{1}{\bar{p}}\frac{\partial^2 \bar{T}}{\partial \bar{x}^2}, \tag{10.24}$$

$$\bar{p} = \bar{\rho}\bar{T}, \quad \bar{S} = \log\left(\frac{\bar{p}}{\bar{\rho}^{\gamma}}\right) + \log\left(\frac{p_0}{\rho_0^{\gamma}}\right). \tag{10.25}$$

The dimensionless groups in these equations, apart from the ratio of specific heats, γ, are

$$\text{Re} = \frac{\rho_0 c_0 L}{\mu}, \quad \text{Pr} = \frac{\mu c_V}{\lambda}.$$

The Reynolds number, Re, is a measure of the size of typical viscous forces relative to inertia, whilst the Prandtl number, Pr, is a measure of the rate of diffusion of momentum relative to the rate of diffusion of heat.

We can now restrict our attention to the three independent variables $\bar{u}$, $\bar{p}$ and $\bar{S}$, using (10.25) to eliminate $\bar{\rho}$ and $\bar{T}$ from (10.22) to (10.24). After some simple manipulations of the governing equations we arrive at

$$\frac{\partial \bar{u}}{\partial \bar{t}} + \bar{u}\frac{\partial \bar{u}}{\partial \bar{x}} + \frac{\exp(\bar{S}/\gamma)}{\gamma \bar{p}^{1/\gamma}}\frac{\partial \bar{p}}{\partial \bar{x}} = \frac{4}{3\text{Re}}\frac{\partial^2 \bar{u}}{\partial \bar{x}^2}, \tag{10.26}$$

$$\frac{\partial \bar{p}}{\partial \bar{t}} + \bar{u}\frac{\partial \bar{p}}{\partial \bar{x}} + \gamma\bar{p}\frac{\partial \bar{u}}{\partial \bar{x}} = M, \tag{10.27}$$

$$\frac{\partial \bar{S}}{\partial \bar{t}} + \bar{u}\frac{\partial \bar{S}}{\partial \bar{x}} = \frac{1}{\bar{p}}M, \tag{10.28}$$

where

$$M = \frac{2\gamma(\gamma-1)}{3\text{Re}}\left(\frac{\partial \bar{u}}{\partial \bar{x}}\right)^2 + \frac{1}{\text{Re Pr}}\frac{\partial^2}{\partial \bar{x}^2}\left(\frac{\bar{p}}{\bar{\rho}}\right). \tag{10.29}$$

In order to investigate the behaviour of the perturbations to the initially uniform state, given by (10.20), we define

$$\bar{u} = \epsilon\hat{u}, \quad \bar{p} = 1 + \epsilon\hat{p}, \quad \bar{S} = \log\left(\frac{p_0}{\rho_0^{\gamma}}\right) + \epsilon\hat{S},$$

with $\hat{u}$, $\hat{p}$ and $\hat{S}$ of $O(1)$ as $\epsilon \to 0$. In addition, we assume that $\text{Pr} = O(1)$ and $\text{Re} = 1/\sigma\epsilon$, with $\sigma = O(1)$. Physically, this means that the effects of viscosity and heat conduction appear at the same order, both being small, with the initial perturbation small enough that these effects are still significant.

On substituting these new variables into (10.26) to (10.29), we find that

$$\frac{\partial \hat{u}}{\partial \bar{t}} + \frac{1}{\gamma}\frac{\partial \hat{p}}{\partial \bar{x}} = \epsilon\left[-\hat{u}\frac{\partial \hat{u}}{\partial \bar{x}} + \frac{1}{\gamma^2}\left(\hat{p} - \hat{S}\right)\frac{\partial \hat{p}}{\partial \hat{x}} + \frac{4}{3}\sigma\frac{\partial^2 \hat{u}}{\partial \bar{x}^2}\right] + O(\epsilon^2), \quad (10.30)$$

$$\frac{\partial \hat{p}}{\partial \bar{t}} + \gamma\frac{\partial \hat{u}}{\partial \bar{x}} = \epsilon\left[-\hat{u}\frac{\partial \hat{p}}{\partial \bar{x}} - \gamma\hat{p}\frac{\partial \hat{u}}{\partial \bar{x}} + \frac{\sigma}{\mathrm{Pr}}\left\{\frac{\partial^2 \hat{S}}{\partial \bar{x}^2} + (\gamma-1)\frac{\partial^2 \hat{p}}{\partial \bar{x}^2}\right\}\right] + O(\epsilon^2), \quad (10.31)$$

$$\frac{\partial \hat{S}}{\partial \bar{t}} = \epsilon\left[-\hat{u}\frac{\partial \hat{S}}{\partial \bar{x}} + \frac{\sigma}{\mathrm{Pr}}\left\{\frac{\partial^2 \hat{S}}{\partial \bar{x}^2} + (\gamma-1)\frac{\partial^2 \hat{p}}{\partial \bar{x}^2}\right\}\right] + O(\epsilon^2). \quad (10.32)$$

Note that the term proportional to $(\partial \bar{u}/\partial \bar{x})^2$ that appears in (10.29), which is due to the generation of entropy by viscous dissipation, is of $O(\epsilon^2)$ and does not appear in the leading order or correction terms of these equations. This is related to the observations that we made in section 7.2 concerning the very small amount of entropy generated by weak shocks. Viscous dissipation is a very ineffective generator of entropy in weak flows such as this.

It is now convenient to write the governing equations in terms of the linearised Riemann invariants,

$$R_1 = \gamma\hat{u} + \hat{p}, \quad R_2 = \gamma\hat{u} - \hat{p},$$

so that

$$\frac{\partial R_1}{\partial \bar{t}} + \frac{\partial R_1}{\partial \bar{x}} = \epsilon\left[\left\{-\frac{\gamma+1}{4\gamma}R_1 + \frac{\gamma-3}{4\gamma}R_2 - \frac{1}{2\gamma}\hat{S}\right\}\frac{\partial R_1}{\partial \bar{x}}\right.$$
$$+\left\{-\frac{\gamma+1}{4\gamma}R_1 + \frac{\gamma+1}{4\gamma}R_2 + \frac{1}{2\gamma}\hat{S}\right\}\frac{\partial R_2}{\partial \bar{x}} + \sigma\left\{\frac{2}{3} + \frac{1}{2\mathrm{Pr}}(\gamma-1)\right\}\frac{\partial^2 R_1}{\partial \bar{x}^2}$$
$$\left. +\sigma\left\{\frac{2}{3} - \frac{1}{2\mathrm{Pr}}(\gamma-1)\right\}\frac{\partial^2 R_2}{\partial \bar{x}^2} + \frac{\sigma}{\mathrm{Pr}}\frac{\partial^2 \hat{S}}{\partial \bar{x}^2}\right] + O(\epsilon^2), \quad (10.33)$$

$$\frac{\partial R_2}{\partial \bar{t}} - \frac{\partial R_2}{\partial \bar{x}} = \epsilon\left[\left\{\frac{\gamma+1}{4\gamma}R_1 - \frac{\gamma+1}{4\gamma}R_2 - \frac{1}{2\gamma}\hat{S}\right\}\frac{\partial R_1}{\partial \bar{x}}\right.$$
$$+\left\{\frac{\gamma-3}{4\gamma}R_1 - \frac{\gamma+1}{4\gamma}R_2 + \frac{1}{2\gamma}\hat{S}\right\}\frac{\partial R_2}{\partial \bar{x}} + \sigma\left\{\frac{2}{3} - \frac{1}{2\mathrm{Pr}}(\gamma-1)\right\}\frac{\partial^2 R_1}{\partial \bar{x}^2}$$
$$\left. +\sigma\left\{\frac{2}{3} + \frac{1}{2\mathrm{Pr}}(\gamma-1)\right\}\frac{\partial^2 R_2}{\partial \bar{x}^2} - \frac{\sigma}{\mathrm{Pr}}\frac{\partial^2 \hat{S}}{\partial \bar{x}^2}\right] + O(\epsilon^2), \quad (10.34)$$

$$\frac{\partial \hat{S}}{\partial \bar{t}} = \epsilon\left[-\frac{1}{2\gamma}(R_1 + R_2)\frac{\partial \hat{S}}{\partial \bar{x}}\right.$$

$$+\frac{\sigma}{\text{Pr}}\left\{\frac{1}{2}(\gamma-1)\frac{\partial^2 R_1}{\partial \bar{x}^2}-\frac{1}{2}(\gamma-1)\frac{\partial^2 R_2}{\partial \bar{x}^2}+\frac{\partial^2 \hat{S}}{\partial \bar{x}^2}\right\}\Bigg]+O(\epsilon^2). \quad (10.35)$$

At leading order

$$\frac{\partial R_1}{\partial \bar{t}}+\frac{\partial R_1}{\partial \bar{x}}=0, \quad \frac{\partial R_2}{\partial \bar{t}}-\frac{\partial R_2}{\partial \bar{x}}=0, \quad \frac{\partial \hat{S}}{\partial \bar{t}}=0.$$

This means that R_1 is a function of $\bar{x}-\bar{t}$ only and R_2 is a function of $\bar{x}+\bar{t}$ only, whilst the entropy is independent of $\bar{t}$. This just represents left- and right-travelling simple waves, as we would expect considering that these are just linear sound waves.

In order to study the effect of the dissipative processes, we use the method of multiple scales. We define characteristic coordinates, $X_1=\bar{x}+\bar{t}$ and $X_2=\bar{x}-\bar{t}$, and a slow time scale, $\tau=\epsilon\bar{t}$, and expand the independent variables as

$$R_1=R_{10}(X_1,X_2,\tau)+\epsilon R_{11}(X_1,X_2,\tau)+O(\epsilon^2),$$
$$R_2=R_{20}(X_1,X_2,\tau)+\epsilon R_{21}(X_1,X_2,\tau)+O(\epsilon^2),$$
$$\hat{S}=\hat{S}_0(X_1,X_2,\tau)+\epsilon\hat{S}_1(X_1,X_2,\tau)+O(\epsilon^2).$$

As we have seen, the governing equations at leading order show that

$$R_{10}\equiv R_{10}(X_2,\tau), \quad R_{20}\equiv R_{20}(X_1,\tau), \quad \hat{S}_0\equiv\hat{S}_0(x,\tau).$$

At $O(\epsilon)$, we find that

$$2\frac{\partial R_{11}}{\partial X_1}=-\frac{\partial R_{10}}{\partial \tau}+\left\{-\frac{\gamma+1}{4\gamma}R_{10}+\frac{\gamma-3}{4\gamma}R_{20}-\frac{1}{2\gamma}\hat{S}_0\right\}\frac{\partial R_{10}}{\partial X_2}$$
$$+\left\{-\frac{\gamma+1}{4\gamma}R_{10}+\frac{\gamma+1}{4\gamma}R_{20}+\frac{1}{2\gamma}\hat{S}_0\right\}\frac{\partial R_{20}}{\partial X_1}+\sigma\left\{\frac{2}{3}+\frac{1}{2\text{Pr}}(\gamma-1)\right\}\frac{\partial^2 R_{10}}{\partial X_2^2}$$
$$+\sigma\left\{\frac{2}{3}-\frac{1}{2\text{Pr}}(\gamma-1)\right\}\frac{\partial^2 R_{20}}{\partial X_1^2}+\frac{\sigma}{\text{Pr}}\frac{\partial^2 \hat{S}_0}{\partial \bar{x}^2}, \quad (10.36)$$

$$-2\frac{\partial R_{21}}{\partial X_2}=-\frac{\partial R_{20}}{\partial \tau}+\left\{\frac{\gamma+1}{4\gamma}R_{10}-\frac{\gamma+1}{4\gamma}R_{20}-\frac{1}{2\gamma}\hat{S}_0\right\}\frac{\partial R_{10}}{\partial X_2}$$
$$+\left\{\frac{\gamma-3}{4\gamma}R_{10}-\frac{\gamma+1}{4\gamma}R_{20}+\frac{1}{2\gamma}\hat{S}_0\right\}\frac{\partial R_{20}}{\partial X_1}+\sigma\left\{\frac{2}{3}-\frac{1}{2\text{Pr}}(\gamma-1)\right\}\frac{\partial^2 R_{10}}{\partial X_2^2}$$
$$+\sigma\left\{\frac{2}{3}+\frac{1}{2\text{Pr}}(\gamma-1)\right\}\frac{\partial^2 R_{20}}{\partial X_1^2}-\frac{\sigma}{\text{Pr}}\frac{\partial^2 \hat{S}_0}{\partial \bar{x}^2}, \quad (10.37)$$

$$\frac{\partial \hat{S}_1}{\partial \bar{t}}=-\frac{\partial \hat{S}_0}{\partial \tau}-\frac{1}{2\gamma}(R_{10}+R_{20})\frac{\partial \hat{S}_0}{\partial \bar{x}}$$

$$+\frac{\sigma}{\text{Pr}}\left\{\frac{1}{2}(\gamma-1)\frac{\partial^2 R_{10}}{\partial X_2^2}-\frac{1}{2}(\gamma-1)\frac{\partial^2 R_{20}}{\partial X_1^2}+\frac{\partial^2 \hat{S}_0}{\partial \bar{x}^2}\right\}. \tag{10.38}$$

Although the right hand sides of these equations are a ghastly mess, in order to solve for R_{11}, R_{21} and $\hat{S}_1$ we simply have to integrate once, since R_{10}, R_{20} and $\hat{S}_0$ are the known simple wave solutions. Consider equation (10.36). When we integrate with respect to X_1, any terms on the right hand side that are independent of X_1 will integrate up to a term linear in X_1, which will be unbounded as $X_1 \to \pm\infty$. These are therefore the secular terms. All the other terms involve derivatives with respect to X_1, and integrate up to functions that are bounded at infinity. The behaviour of R_{10} is therefore fixed by the condition that the secular terms must be zero. Similar considerations apply to the other equations, and we finally arrive at

$$\frac{\partial R_{10}}{\partial \tau}+\frac{\gamma+1}{4\gamma}R_{10}\frac{\partial R_{10}}{\partial X_2}=\sigma\left\{\frac{2}{3}+\frac{1}{2\text{Pr}}(\gamma-1)\right\}\frac{\partial^2 R_{10}}{\partial X_2^2}, \tag{10.39}$$

$$\frac{\partial R_{20}}{\partial \tau}+\frac{\gamma+1}{4\gamma}R_{20}\frac{\partial R_{20}}{\partial X_1}=\sigma\left\{\frac{2}{3}+\frac{1}{2\text{Pr}}(\gamma-1)\right\}\frac{\partial^2 R_{20}}{\partial X_1^2}, \tag{10.40}$$

$$\frac{\partial \hat{S}_0}{\partial \tau}=\frac{\sigma}{\text{Pr}}\frac{\partial^2 \hat{S}_0}{\partial x^2}, \tag{10.41}$$

to be solved subject to the initial conditions

$$\left.\begin{array}{c}R_{10}(X_2,0)=\gamma f_0(X_2)+h_0(X_1),\quad R_{20}(X_1,0)=\gamma f_0(X_2)-h_0(X_1),\\ \hat{S}_0(\bar{x},0)=f_0(\bar{x})-\gamma g_0(\bar{x}).\end{array}\right\} \tag{10.42}$$

The entropy is governed at leading order by a simple linear diffusion equation. The two Riemann invariants, which determine the velocity and pressure perturbations, are each governed by Burgers' equation, which explicitly displays the effect of competition between nonlinear steepening and thermal and viscous dissipation. There is no coupling between the three dependent variables R_{10}, R_{20} and $\hat{S}_0$. In addition, although we chose the richest limit in the governing equations by assuming that $\sigma = O(1)$, if σ is asymptotically smaller than this, the same equations govern the leading order development. In this case, as we shall see section 10.3, the leading order solution can be found using the method of characteristics with shock fitting, and dissipation is active only in shocks and corner regions at leading order.

However, we can say rather more about these equations. Since the initial conditions are such that these perturbation are confined to the region $-1 < \bar{x} < 1$ and the solution develops on an $O(\epsilon^{-1})$ time scale, characterised by the slow time variable, τ, the initial conditions lead to the development of two simple waves, as shown in figure 8.6, with the initial input of entropy essentially confined to the region of the initial disturbance. More precisely, for $\tau = O(1)$, $\bar{t} = O(\epsilon^{-1})$. Therefore, when $X_1 = O(1)$, $\bar{x} \sim -\epsilon^{-1}\tau$, so when $X_2 = O(1)$, $\bar{x} \sim \epsilon^{-1}\tau$. Far from each simple wave region, the equations for R_{10} and R_{20} are dominated by diffusion, and these Riemann invariants are exponentially small. Similarly for the entropy far from the initial region, where $\bar{x} = O(1)$. This means that at leading order for $\tau = O(1)$, the perturbation to the entropy is exponentially small in each of the simple wave regions. All of this confirms that the picture that emerged from our analysis of d'Alembert's solution in section 2.3, of an initially localised disturbance splitting into a left- and a right-propagating wave, is correct for one-dimensional sound waves, but that over a long, $O(\epsilon^{-1})$ time scale, the simple waves steepen and form shocks, governed by Burgers' equation through (10.39) and (10.40). In addition, the assumption of isentropic flow, which we made throughout chapter 3, is shown to be appropriate, since any initial disturbance of the entropy is left behind by the diverging waves.

Focusing our attention on the right-propagating simple wave, we can write (10.39) in terms of the physical variables, noting that $R_{20} \ll 1$ at leading order, and hence that $\hat{p} \sim \gamma \hat{u}$, as

$$\frac{\partial p}{\partial t} + \frac{\gamma + 1}{2\rho_0 c_0}(p - p_0)\frac{\partial p}{\partial \eta} = \frac{1}{\rho_0}\left\{\frac{2}{3}\mu + \frac{1}{2}(\gamma - 1)\frac{\lambda}{c_v}\right\}\frac{\partial^2 p}{\partial \eta^2}, \tag{10.43}$$

where $\eta = x - c_0 t$. This equation can be mapped onto the canonical form of Burgers' equation, (10.1), by the transformation

$$\nu = \frac{1}{\rho_0}\left\{\frac{2}{3}\mu + \frac{1}{2}(\gamma - 1)\frac{\lambda}{c_v}\right\}, \quad u = \frac{\gamma + 1}{2\rho_0 c_0}(p - p_0), \quad x = \eta. \tag{10.44}$$

Looking ahead a little, the analysis of the travelling wave solutions of Burgers' equation, which is the subject of the following section, shows that the thickness of the shock is of order

$$x_s = \frac{2\nu}{u_R - u_L} = \frac{4c_0}{(\gamma + 1)p_1}\left\{\frac{2}{3}\mu + \frac{1}{2}(\gamma - 1)\frac{\lambda}{c_v}\right\}, \tag{10.45}$$

where p_1 is the magnitude of the initial disturbance of the pressure. A discussion of the thickness of shocks using a more intuitive, physical approach is given by Lighthill (1978), and reproduces this estimate. In addition, Lighthill discusses a third contribution to shock dissipation, a time lag in the redistribution of the energy in a gas between its various modes of molecular vibration and rotation, which we have not studied here.

10.3 Simple Solutions of Burgers' Equation

10.3.1 Travelling Waves

Consider the solution of (10.1) subject to the boundary co

$$u \to u_{\rm L} \quad \text{as } x \to -\infty, \quad u \tag{10.46}$$

and initial conditions compatible with ... We can now consider what the large time development of this solution is, after any initial transient behaviour has decayed. If we assume that the solution takes the form of a travelling wave with constant speed, c, and $u(x,t) = u(x-ct)$, we find that

$$\nu u_{\zeta\zeta} + u_{\zeta}(c-u) = 0,$$

subject to

$$u \to u_{\rm L} \quad \text{as } \zeta \to -\infty, \quad u \to u_{\rm R} \quad \text{as } \zeta \to \infty, \tag{10.47}$$

where $\zeta = x - ct$ is the travelling wave coordinate. Integrating this once leads to

$$\nu u_{\zeta} + cu - \frac{1}{2}u^2 = C, \tag{10.48}$$

where C is a constant that we can fix in two different ways. Firstly, $u_{\zeta} \to 0$ as $u \to u_{\rm R}$ gives $C = cu_{\rm R} - \frac{1}{2}u_{\rm R}^2$. Secondly, $u_{\zeta} \to 0$ as $u \to u_{\rm L}$ gives $C = cu_{\rm L} - \frac{1}{2}u_{\rm L}^2$. For consistency we must have $cu_{\rm R} - \frac{1}{2}u_{\rm R}^2 = cu_{\rm L} - \frac{1}{2}u_{\rm L}^2$ or $c = \frac{1}{2}(u_{\rm R} + u_{\rm L})$, which is precisely the shock speed that we determined in subsection 7.1.4. Equation (10.48) can now be written as

$$\begin{aligned}\nu u_{\zeta} &= \frac{1}{2}u^2 - \frac{1}{2}(u_{\rm R} + u_{\rm L}) + \frac{1}{2}u_{\rm R}u_{\rm L} \\ &= \frac{1}{2}(u - u_{\rm L})(u - u_{\rm R}).\end{aligned} \tag{10.49}$$

If $u_{\rm R} > u_{\rm L}$, then u is a decreasing function of ζ for $u_{\rm R} > u > u_{\rm L}$. Since this is not compatible with the boundary conditions as $\zeta \to \pm\infty$, we need $u_{\rm L} > u_{\rm R}$ for a travelling wave solution to exist. This is consistent with the results of section 7.1, where we showed that the solution of the Riemann problem when $\nu = 0$ for $u_{\rm R} > u_{\rm L}$ is an expansion fan, whilst the solution for $u_{\rm L} > u_{\rm R}$ is a shock. When $u_{\rm L} > u_{\rm R}$, (10.49) is separable, with solution

$$u = \frac{u_{\rm L} + \exp\left\{\frac{1}{2\nu}(u_{\rm L} - u_{\rm R})(\zeta + \zeta_0)\right\} u_{\rm R}}{1 + \exp\left\{\frac{1}{2\nu}(u_{\rm L} - u_{\rm R})(\zeta + \zeta_0)\right\}}, \tag{10.50}$$

where ζ_0 is a constant of integration.

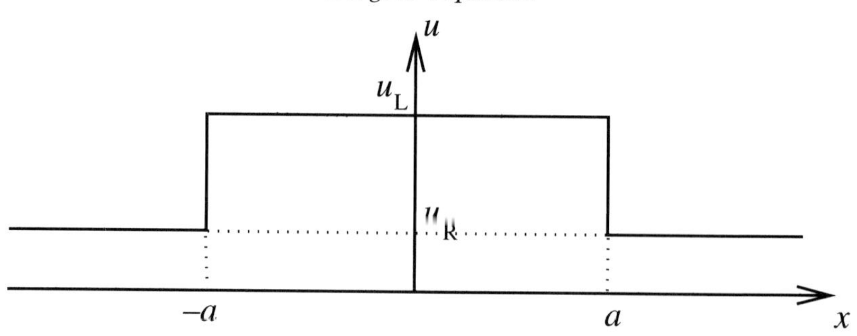

Fig. 10.2. The initial conditions given by (10.51).

The above analysis, informative as it is, does not tell us

(i) whether a smooth monotone decreasing travelling wave is the only structure that can emerge from the initial data,
(ii) what the role of ν is in the solution structure,
(iii) the significance of the constant ζ_0,
(iv) why the particular wave speed $\frac{1}{2}(u_L + u_R)$ is selected.

10.3.2 Asymptotic Solutions for $\nu \ll 1$

In order to shed some light on points (i) to (iv), it is sufficient to consider the initial value problem

$$u_t + uu_x = \nu u_{xx} \qquad \text{on } -\infty < x < \infty, \text{ for } t > 0, \text{ with } \nu \ll 1,$$

$$u(x,0) = \left\{ \begin{array}{lll} u_R & \text{for} & x < -a, \\ u_L & \text{for} & -a < x < a, \\ u_R & \text{for} & x > a, \end{array} \right\} \qquad (10.51)$$

and $u_L > u_R$, as shown in figure 10.2.

The outer solutions

If x, t, u_L, u_R and a are of $O(1)$, we can start our perturbation procedure with the expansion

$$u = u_0(x,t) + O(\nu),$$

so that $u_{0,t} + u_0 u_{0,x} = 0$ at leading order. The solution of this first order hyperbolic system with the top hat initial conditions given above is readily obtained by the method of characteristics as

$$u(x,t) = \left\{ \begin{array}{cc} u_{\mathrm{R}} & \text{for } -\infty < x < -a + u_{\mathrm{R}}t, \\ (x+a)/t & \text{for } -a + u_{\mathrm{R}}t < x < -a + u_{\mathrm{L}}t, \\ u_{\mathrm{L}} & \text{for } -a + u_{\mathrm{L}}t < x < a + \frac{1}{2}(u_{\mathrm{L}} + u_{\mathrm{R}})t, \\ u_{\mathrm{R}} & \text{for } a + \frac{1}{2}(u_{\mathrm{L}} + u_{\mathrm{R}})t < x < \infty. \end{array} \right\} \quad (10.52)$$

It is clear from figure 10.3(a) that there are diverging characteristics spreading out from $x = -a$, $t = 0$. The solution for $-a + u_{\mathrm{R}}t < x < -a + u_{\mathrm{L}}t$ is therefore an expansion fan which connects two uniform regions. The characteristics on either side of $x = a$ converge so that a shock wave is formed when $t = 0$. Using the shock speed that we determined in subsection 7.1.4, the path of the shock is given by $x_{\mathrm{s}} = \frac{1}{2}(u_{\mathrm{L}} + u_{\mathrm{R}})t$, and separates two uniform regions. A sketch of this solution is given in figure 10.3(b). We conclude that we have determined, in the usual nomenclature of matched asymptotic expansions, the outer solution. There is a need for further consideration of the perturbation process in the neighbourhood of the shock, where u_x and u_{xx} are large, and also at the two corners of the expansion fan, where u_x and u_t are discontinuous. It should also be pointed out that when $t = t_{\mathrm{I}} = 4a/(u_{\mathrm{L}} - u_{\mathrm{R}})$, the expansion fan catches up with the shock and modifies its path for $t > t_{\mathrm{I}}$. We give further details of the large time behaviour later in this section. For now, we restrict attention to times $t < t_{\mathrm{I}}$.

The shock structure

To examine more closely the solution in a region close to the shock, it is convenient to introduce the new coordinate $\bar{x}$ defined by

$$x = a + \frac{1}{2}(u_{\mathrm{L}} + u_{\mathrm{R}})t + v^{\alpha}\bar{x}, \quad (10.53)$$

so that $\bar{x} = O(1)$ when $x - a - \frac{1}{2}(u_{\mathrm{L}} + u_{\mathrm{R}})t = O(v^{\alpha})$. The thickness of the shock region, determined by the value of α, is to be found by balancing terms in the transformed version of (10.51). Under the transformation (10.53), with $u = \bar{u}(y,t)$, Burgers' equation becomes

$$v^{\alpha}\bar{u}_t - \frac{1}{2}(u_{\mathrm{L}} + u_{\mathrm{R}})\bar{u}_{\bar{x}} + \bar{u}\bar{u}_{\bar{x}} = v^{1-\alpha}\bar{u}_{\bar{x}\bar{x}}. \quad (10.54)$$

The choice $\alpha = 1$ gives the richest balance in (10.54), so the shock has thickness of $O(v)$ and we must solve

$$v\bar{u}_t + \bar{u}_{\bar{x}}\left\{\bar{u} - \frac{1}{2}(u_{\mathrm{L}} + u_{\mathrm{R}})\right\} = \bar{u}_{\bar{x}\bar{x}} \quad (10.55)$$

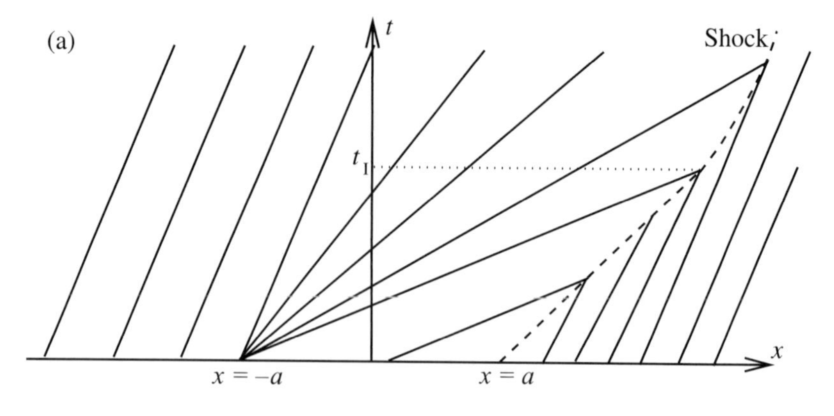

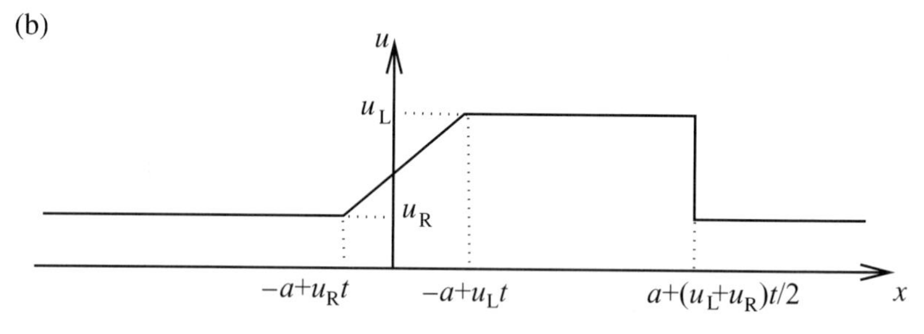

Fig. 10.3. (a) The characteristics and (b) the solution, when $t < t_{\rm I}$, of (10.51) with $\nu = 0$.

subject to the matching conditions $\bar{u} \to u_{\rm L}$ as $\bar{x} \to -\infty$, $\bar{u} \to u_{\rm R}$ as $\bar{x} \to \infty$. As these conditions indicate that $\bar{u} = O(1)$ throughout the region we try to solve (10.55) by an asymptotic expansion of the form

$$\bar{u}(\bar{x}, t) = \bar{u}_0(\bar{x}, t) + O(\nu), \tag{10.56}$$

which, upon substitution into (10.55), gives at leading order

$$u_{0,\bar{x}} \left\{ \bar{u}_0 - \frac{1}{2}(u_{\rm L} + u_{\rm R}) \right\} = \bar{u}_{0,\bar{x}\bar{x}}. \tag{10.57}$$

This is essentially the same as (10.47), and has solution

$$\bar{u}_0 = \frac{u_{\rm L} + \exp\left\{\frac{1}{2}(u_{\rm L} - u_{\rm R})(\bar{x} + \bar{x}_0(t))\right\} u_{\rm R}}{1 + \exp\left\{\frac{1}{2}(u_{\rm L} - u_{\rm R})(\bar{x} + \bar{x}_0(t))\right\}}, \tag{10.58}$$

where $\bar{x}_0(t)$ is an unknown function of time which arises in the integration of (10.57).

To determine the value of $\bar{x}_0(t)$, we first integrate (10.55) with respect to $\bar{x}$ between $-\infty$ and ∞. This leads to

$$\int_{-\infty}^{\infty} \bar{u}_t \, d\bar{x} = 0, \tag{10.59}$$

which is only possible if $\bar{x}_0(t)$ is a constant, say $\bar{x}_0$. To fix the value of $\bar{x}_0$ we must consider what happens to the initial data for small times since, in the current scalings, we have lost the temporal evolution of the initial data. To examine the small time behaviour, we define $\bar{t} = t/\nu$ which, upon substitution into (10.55), gives

$$\hat{u}_{\bar{t}} + \hat{u}_{\bar{x}} \left\{ \hat{u} - \frac{1}{2}(u_L + u_R) \right\} = \hat{u}_{\bar{x}\bar{x}},$$

subject to the initial conditions

$$\hat{u}(\bar{x}, 0) = \begin{cases} u_L & \text{for } \bar{x} < 0, \\ u_R & \text{for } \bar{x} > 0, \end{cases}$$

and boundary conditions

$$\hat{u} \to u_L \quad \text{as } \bar{x} \to -\infty, \quad \hat{u} \to u_R \quad \text{as } \bar{x} \to \infty, \tag{10.60}$$

for all $\bar{t} > 0$, where $\bar{u} = \hat{u}(\bar{x}, \bar{t})$. This initial value problem has a solution that evolves to a steady state as $\bar{t} \to \infty$, which will be $\bar{u}_0$, as given by (10.58). The solution of this problem must be found in general by numerical methods. However, this can be avoided here as we are really only interested in $\hat{u}(0, \bar{t})$. As $\bar{t} \to \infty$ this must match with $\bar{u}_0(0, t)$ as $t \to 0$ and hence fix the value of $\bar{x}_0$. We now find $\hat{u}(0, \hat{t})$ by developing a symmetry argument.

Define $\hat{v} = \hat{u} - \frac{1}{2}(u_L + u_R)$ so that

$$\hat{v}_{\bar{t}} + \hat{v}\hat{v}_{\bar{x}} = \hat{v}_{\bar{x}\bar{x}},$$

subject to initial conditions

$$\hat{v}(\bar{x}, 0) = \begin{cases} \frac{1}{2}(u_L - u_R) & \text{for } \bar{x} < 0, \\ \frac{1}{2}(u_R - u_L) & \text{for } \bar{x} > 0, \end{cases}$$

and boundary conditions

$$\hat{v} \to \frac{1}{2}(u_L - u_R) \quad \text{as } \bar{x} \to -\infty, \quad \hat{v} \to \frac{1}{2}(u_R - u_L) \quad \text{as } \bar{x} \to \infty. \tag{10.61}$$

As the equation and boundary conditions are invariant under the transformations $\hat{v} \to -\hat{v}$, $\bar{x} \to -\bar{x}$, we can deduce that $\hat{v}$ is an odd function of

$\bar{x}$ so that $\hat{v}(0,\bar{t}) = 0$ for all $\hat{t} > 0$. Hence $\hat{u}(0,\hat{t}) = \frac{1}{2}(u_L + u_R)$, matching requires

$$\frac{1}{2}(u_L + u_R) = \frac{u_L + \exp\left\{\frac{1}{2}(u_L - u_R)\bar{x}_0\right\} u_R}{1 + \exp\left\{\frac{1}{2}(u_L - u_R)\bar{x}_0\right\}},$$

and therefore $\bar{x}_0 = 0$. Our solution in the neighbourhood of the shock is now complete.

The corner region

We now consider the right-most corner of the expansion fan in detail, as we currently have a solution that is not differentiable there. The situation at the left corner of the expansion fan is very similar to this. Again, we will resolve this by constructing an inner solution with sufficient differentiability in the neighbourhood of the corner. We define x^* by

$$x = -a + u_L t + \nu^\beta x^*, \tag{10.62}$$

and consider the transformed equation when $x^* = O(1)$. Working with $u = u^*(x^*, t)$, Burgers' equation transforms to

$$\nu^\beta u_t^* - u_L u_{x^*}^* + u^* u_{x^*}^* = \nu^{1-\beta} u_{x^* x^*}^*. \tag{10.63}$$

If we write an expansion of the form

$$u^*(x^*, t) = u_L + \nu^\beta u_1^* + O(\nu^\beta),$$

guided by the form of the expansion fan solution, a balance occurs in the equation if $\beta = 1 - \beta$, and hence $\beta = \frac{1}{2}$. At leading order we obtain

$$u_{1,t}^* + u_1^* u_{1,x^*}^* = u_{1,x^* x^*}^* \tag{10.64}$$

subject to the matching conditions $u_1 \to x^*/t$ as $x^* \to -\infty$ and $u_1 \to 0$ as $x^* \to \infty$. Thus we have the full Burgers equation and two boundary conditions, but without the parameter ν. Rather surprisingly, the form of these boundary conditions actually allows us to solve the problem by using a similarity reduction of the partial differential equation to an ordinary differential equation. To perform this reduction we first note that the partial differential equation and the boundary conditions are invariant under the one parameter (k) group of transforms $(u, x, t) \mapsto (k^{-\frac{1}{2}}\hat{u}, k^{\frac{1}{2}}\hat{x}, k\hat{t})$. The quantities $x/t^{\frac{1}{2}}$ and $ut^{\frac{1}{2}}$ are therefore invariants of the transformation so that standard theory indicates that we can find a solution of the form $u_1^* = u_1(x^*/t^{\frac{1}{2}})/t^{\frac{1}{2}}$ (see, for example, Bluman and Cole (1974)). Substitution into Burgers' equation gives

$$-\frac{1}{2}\eta u_{1,\eta} - \frac{1}{2}u_1 + u_1 u_{1,\eta} = u_{1,\eta\eta}, \tag{10.65}$$

subject to $u_1 \to 0$ as $\eta \to \infty$ and $u_1 \sim \eta$ as $\eta \to -\infty$, where $\eta = x^*/t^{\frac{1}{2}}$. This ordinary differential equation is readily integrated once to give

$$u_{1,\eta} = \frac{1}{2}u_1(u_1 - \eta).$$

This can be linearised by putting $v = 1/u_1$ to give

$$v_\eta = \frac{1}{2}\eta v - \frac{1}{2},$$

with solution

$$v = -\frac{1}{2}e^{\frac{1}{4}\eta^2}\int_{-\infty}^{\eta} e^{-\frac{1}{4}s^2}\,ds. \tag{10.66}$$

Returning to the original variables, we have constructed a solution, valid in a region of thickness $O(v^{\frac{1}{2}})$ about the corner, in the form

$$u^*(x^*,t) = u_L - \frac{2v^{\frac{1}{2}}e^{-x^{*2}/4t}}{t^{\frac{1}{2}}\int_{-\infty}^{x^*/t^{\frac{1}{2}}} e^{-\frac{1}{4}s^2}\,ds} + O(v^{\frac{1}{2}}), \tag{10.67}$$

which possesses the smoothness properties that the outer solution lacked.

Finally, we consider what happens to our solution for $t > 4a/(u_L - u_R) = t_I$, after the expansion fan collides with the shock. Using the methods of section 7.1, we can see that the shock path curves to accommodate this collision. In fact, it takes the form $x = u_R t - a + 4a\left(t/t_I\right)^{\frac{1}{2}}$ and the velocity to the left of the shock is $u = u_R + 4a/(t_I t)^{\frac{1}{2}}$. This indicates that the solution structure we can expect as $t \to \infty$ is a weakening shock whose path approaches that of the u_R characteristics. Behind the shock there is an expansion fan whose slope becomes flat as $t \to \infty$. Rather similar methods to those above can be adopted to match together the different parts of the large time development of this initial data. Further details on this and other issues associated with initial value problems for Burgers' equation can be found in Crighton and Scott (1979).

Exercises

10.1 Use the Cole–Hopf transformation to show that the solution of Burgers' equation with initial conditions

$$u(x,0) = \begin{cases} u_1 & \text{for } x < 0, \\ u_2 & \text{for } x > 0, \end{cases}$$

with $u_1 > u_2$, can be written in the form

$$u(x,t) = \int_{-\infty}^{\infty} \frac{x-\eta}{t}\theta(\eta)\exp\left\{-\frac{(x-\eta)^2}{4\nu t}\right\} d\eta \Big/ \int_{-\infty}^{\infty} \theta(\eta)\exp\left\{-\frac{(x-\eta)^2}{4\nu t}\right\} d\eta,$$

where

$$\theta(x) = \begin{cases} \exp\left(-\frac{u_1 x}{2\nu}\right) & \text{for } x < 0, \\ \exp\left(-\frac{u_2 x}{2\nu}\right) & \text{for } x > 0. \end{cases}$$

Show by manipulation of these integrals that this can be recast in the form

$$u(x,t) = (u_1 - u_2) \times \left[1 + \frac{\operatorname{erfc}\left(-\frac{x-u_2 t}{2\sqrt{\nu t}}\right)}{\operatorname{erfc}\left(-\frac{x-u_1 t}{2\sqrt{\nu t}}\right)} \exp\left\{\frac{(u_1-u_2)(x-ct)}{2\nu}\right\}\right]^{-1},$$

where $c = (u_1 + u_2)/2$. Verify that this satisfies the initial condition, and show that for fixed values of $x - ct$,

$$\lim_{t\to\infty} u(x,t) = u_2 + (u_1 - u_2)\left[1 + \exp\left\{\frac{(u_1-u_2)(x-ct)}{2\nu}\right\}\right]^{-1}.$$

Interpret this result.

10.2 Verify by substitution that the function

$$\phi(x,t) = 1 + \sqrt{\frac{a}{t}}e^{-x^2/4\nu t},$$

with a a positive constant, is a solution of the diffusion equation

$$\frac{\partial \phi}{\partial t} = \nu\frac{\partial^2 \phi}{\partial t^2}.$$

Use the Cole–Hopf transformation to obtain the corresponding solution, $c(x,t)$, of Burgers' equation,

$$\frac{\partial c}{\partial t} + c\frac{\partial c}{\partial x} = \nu\frac{\partial^2 c}{\partial x^2}.$$

Show that this solution of Burgers' equation is anti-symmetric, and that the area, A, under the solution for $x > 0$ is given by

$$A(t) = 2\nu \log\left(1 + \sqrt{\frac{a}{t}}\right).$$

Use this expression to eliminate a and show that the solution can be rewritten as

$$c(x,t) = \frac{x}{t}\left(1 + \frac{e^{x^2/4\nu t}}{e^{A/2\nu} - 1}\right)^{-1}.$$

Determine and sketch the approximate solution when $\nu \ll 1$ and $A(t)$ is bounded.

10.3 Construct an asymptotic solution of Burgers' equation, valid as $\nu \to 0$, for the initial conditions

$$u(x,0) = \begin{cases} 1 & \text{for } |x| > 1, \\ |x| - 1 & \text{for } |x| < 1. \end{cases}$$

Hint: The form of the shock changes twice.

10.4 Find uniformly valid asymptotic solutions of Burgers' equation $u_t + uu_x = \nu u_{xx}$ in the domain $0 < x < \infty$, $0 < t < \infty$, as $\nu \to 0$ subject to the initial condition $u(x,0) = a$ and boundary conditions $u(0,t) = b$, $u(\infty,t) = a$. You should consider the cases $0 < a < b$, $a = b$ and $b > a$ separately, and be sure to determine the leading order solution in all of the shock and corner regions.

11

Diffraction and Scattering

We have already considered various different situations where waves are reflected (sound waves incident on a rigid wall, elastic waves incident on a free surface and electromagnetic waves incident on perfect conductors and insulators). Whilst pure reflection is characterised by the presence of a uniform, infinitely long barrier, at which the wave changes its direction of propagation, scattering is characterised by the presence of a smooth, finite body, which leads to a non-uniformly reflected wave field. In this chapter, we consider barriers to wave propagation that are semi-infinite or finite. We will find that, as well as reflection and scattering, there is another effect caused by the presence of an edge to a barrier. When the incident wave impinges upon the edge, it is scattered in all directions. This is called **diffraction** and can produce some interesting effects. If there are two or more edges, the scattered waves from each edge can interfere with each other to form a **diffraction pattern**. Although reflection, scattering and diffraction are treated using slightly different mathematical methods, the underlying idea is that of the interaction of an incident wave with a boundary. The reflected/scattered/diffracted field is the difference between the unaffected incident wave and the actual solution.

Situations where these effects are important are widespread in the natural world, in the laboratory, in engineering applications and in the home. The sky is blue because of the way sunlight is scattered by air molecules. The technique of **X-ray diffraction** uses diffraction patterns to probe the three-dimensional structure of crystals and certain chemicals. The detection of aircraft using radar relies on their tendency to reflect and diffract electromagnetic waves and an important element of the design of military aircraft is to minimise the amplitude of this diffraction pattern. Rather more domestically, the reflection and diffraction of a plane wave incident at a grating are a basic feature of compact disc players. The information on the disc is stored in a spiral track containing a sequence of densely packed pits. The optical pick-up consists of a laser beam, which is imaged on the disc surface, and a detector, which converts the variation in reflected and diffracted laser light caused by the pit pattern into an electrical signal. This signal is then processed and amplified to produce the sound.

In this chapter, we consider the diffraction and scattering of plane waves. Much of the theory is based on solutions of the two- or three-dimensional wave equation, and is therefore applicable to many different physical situations. We will concentrate on three types of waves. We begin by considering the scalar diffraction problem of a plane acoustic wave incident on a semi-infinite rigid wall. We then examine the propagation and diffraction of plane, linear, sound waves through a hole in a rigid wall – a **scalar diffraction problem**. Next, we consider the same situation for electromagnetic waves. Although the solution we obtain is very similar to that for acoustic waves, it is not the same. It is a **vector diffraction problem**, and the polarisation of the incident wave has an effect on the diffracted wave. Finally, we look at the scattering of water waves, for which the boundary value problem that we must solve is rather different.

11.1 Diffraction of Acoustic Waves by a Semi-infinite Barrier

Let's consider what happens when a plane, harmonic, acoustic wave, with frequency ω and wavenumber k, is incident at an angle α on a semi-infinite barrier $x \geq 0$, $y = 0$, as shown in figure 11.1. This is commonly referred to as **Sommerfeld diffraction**. The potential ϕ satisfies the wave equation. We now write $\phi = \Phi(\mathbf{r})e^{-i\omega t}$ so that Φ satisfies the **Helmholtz equation**,

$$(\nabla^2 + k^2)\Phi = 0 \quad \text{for } -\infty < x < \infty,\ -\infty < y < \infty, \tag{11.1}$$

where $k^2 = |\mathbf{k}|^2 = \omega^2/c^2$. This is now to be solved subject to the no flux boundary condition,

$$\Phi_y(x,0) = 0 \quad \text{for } 0 < x < \infty,$$

together with appropriate conditions at infinity. From the physics of the problem, what type of solution would we expect? The incident wave will be reflected where there is a barrier to reflect it, with the angle of incidence equal to the angle of reflection. This defines region I, which, in polar coordinates with origin at the edge, is $0 < r < \infty$, $0 < \theta < \pi - \alpha$, as shown in figure 11.1. There will also be a diffracted wave in this region.

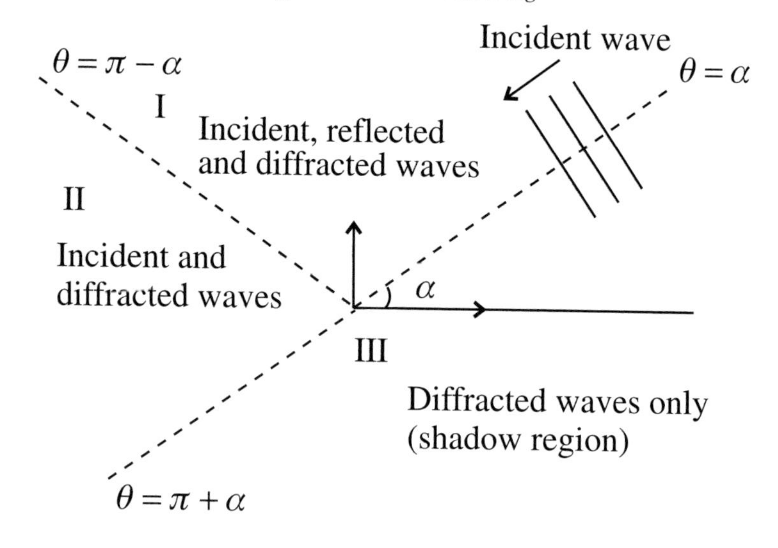

Fig. 11.1. The regions I, II and III, and the coordinate system for half plane diffraction of a plane sound wave.

The parts of the incident wave that miss the barrier but are not in its shadow define region II as $0 < r < \infty$, $\pi - \alpha < \theta < \pi + \alpha$. There are no reflected waves here but there is a diffracted component. Region III, defined as $0 < r < \infty$, $\pi + \alpha < \theta < 2\pi$, is in the shadow of the barrier and only contains diffracted waves. The diffracted waves must propagate outwards from the edge of the barrier.

11.1.1 Preliminary Estimates of the Potential

In order to solve this diffraction problem, we need to have some estimates of the magnitude of the potential in regions I, II and III when r is large and when r is small. We will use an artificial small parameter, δ, to do this, purely to formalise the asymptotics. It also provides a check on the results, since the parameter δ cannot appear when the solutions are written back in terms of the physical variables. (We could also try to find separable solutions but the results we obtain would then need expansion for large r.)

The Far Field: $r \gg 1$

To estimate Φ when $r \gg 1$, we introduce the artificial small parameter, δ, and scaled coordinate $\bar{r} = \delta r$, so that $\bar{r} = O(1)$ when $r = O\left(\delta^{-1}\right) \gg 1$.

Equation (11.1) becomes

$$\delta^2\left(\Phi_{\bar r\bar r}+\frac{1}{\bar r}\Phi_{\bar r}+\frac{1}{\bar r^2}\Phi_{\theta\theta}\right)+k^2\Phi=0. \tag{11.2}$$

To solve this when $\delta \ll 1$, we pose a WKB expansion of the form

$$\Phi=\exp\left\{\frac{A(\bar r,\theta)}{\delta}+B(\bar r,\theta)\log\delta+C(\bar r,\theta)+O(\delta)\right\}. \tag{11.3}$$

After substituting this into (11.1) and collecting like terms, we obtain the hierarchy of equations

$$A_{\bar r}^2+\frac{1}{\bar r^2}A_\theta^2+k^2=0, \tag{11.4}$$

$$2A_{\bar r}B_{\bar r}+2\frac{A_\theta B_\theta}{\bar r^2}=0, \tag{11.5}$$

$$A_{\bar r\bar r}+\frac{1}{\bar r^2}A_{\theta\theta}+A_{\bar r}\left(2C_{\bar r}+\frac{1}{\bar r}\right)+2\frac{A_\theta C_\theta}{\bar r^2}=0. \tag{11.6}$$

If we now write $A=-irf(\theta)$, (11.4) gives

$$-f^2-f_\theta^2+k^2=0,$$

a nonlinear ordinary differential equation with two distinct types of solution. The simplest is $f=\pm k$, but there is also a family of solutions of the form $f=\pm k\cos(\theta+\phi_\pm)$, where $\phi_\pm$ is a free constant, which has $f=\pm k$ as its envelope. In (11.5) we take the simple solution $B=b_0$. In (11.6), when f is constant, we find that $C=-\frac{1}{2}\log r+\hat C_0(\theta)$. When $f=\pm k\cos(\theta+\phi_\pm)$, we notice that $A_{\theta\theta}/r^2$ cancels with A_r/r, which gives $A_rC_r+A_\theta C_\theta/r^2=0$ and we take the solution to be $C=\text{constant}=C_1^\pm$. We can now use the linearity of (11.1) to write

$$\begin{aligned}\Phi=&\exp\left\{\frac{-ik\bar r}{\delta}+b_0^+\log\delta-\frac{1}{2}\log\bar r+\hat C_0^+(\theta)+o(1)\right\}\\&+\exp\left\{\frac{+ik\bar r}{\delta}+b_0^-\log\delta-\frac{1}{2}\log\bar r+\hat C_0^-(\theta)+o(1)\right\}\\&+\exp\left\{\frac{-ik\bar r\cos(\theta+\phi_+)}{\delta}+b_1^+\log\delta+\hat C_1^++o(1)\right\}\\&+\exp\left\{\frac{+ik\bar r\cos(\theta+\phi_-)}{\delta}+b_1^-\log\delta+\hat C_1^-+o(1)\right\}.\end{aligned} \tag{11.7}$$

Returning to the original variable, r, if we choose $b_0^+=b_0^-=1/2$ and

$b_1^+ = b_1^- = 0$, we can remove the artificial parameter, δ, and arrive at a representation of the far field solution,

$$\Phi \sim C_1^+ \exp\{-ikr\cos(\theta + \phi_+)\} + C_1^- \exp\{ikr\cos(\theta + \phi_-)\}$$
$$+\frac{C_0^+(\theta)}{r^{\frac{1}{2}}}\exp(-ikr) + \frac{C_0^-(\theta)}{r^{\frac{1}{2}}}\exp(ikr) \quad \text{as } r \to \infty, \tag{11.8}$$

where $C_j^\pm = \exp(\hat{C}_j^\pm)$ for $j = 1, 2$. An application of the no flux condition, $\partial\phi/\partial y = 0$, on the barrier gives $dC_0^\pm/d\theta = 0$ at $\theta = 0, 2\pi$ and $\phi_- = \pi - \phi_+$, $C_1^- = C_1^+$. In region I, where there are incident waves, C_1^+ will be non-zero and the choice of $\phi_+ = -\alpha$ makes these satisfy the conditions at infinity. The terms involving $C_0^+(\theta)$ represent an incoming wave which is not the incident wave, so we need $C_0^+(\theta) = 0$. We now have estimates of the far field solution in each of the different regions.

In region I,

$$\Phi \sim \exp\{-ikr\cos(\theta - \alpha)\} + \exp\{-ikr\cos(\theta + \alpha)\}$$
$$+\frac{C_0^-(\theta)}{r^{\frac{1}{2}}}\exp(ikr) + o\left(\frac{1}{r^{\frac{1}{2}}}\right). \tag{11.9}$$

The first two terms are the incident and reflected plane waves. We have set the amplitude of these to one for convenience, something which we could have done at the outset, simply by scaling ϕ. The last term is the diffracted component which decays like $1/r^{1/2}$ as r, the distance from the edge, increases. In region II,

$$\Phi \sim \exp\{-ikr\cos(\theta - \alpha)\} + \frac{C_0^-(\theta)}{r^{\frac{1}{2}}}\exp(ikr) + o\left(\frac{1}{r^{\frac{1}{2}}}\right). \tag{11.10}$$

The first term is the incident wave and the second the diffracted component. In region III,

$$\Phi \sim \frac{C_0^-(\theta)}{r^{\frac{1}{2}}}\exp(ikr) + o\left(\frac{1}{r^{\frac{1}{2}}}\right). \tag{11.11}$$

In the shadow of the barrier we have only a diffracted component, as shown in figure 11.1.

We have now shown that the amplitude of the diffracted wave decays like $r^{-1/2}$ as $r \to \infty$. The function $C_0^-(\theta)$ is still undetermined. Rather more can be said about its structure by going to higher order in the WKB expansion. We will not do this as we will shortly demonstrate a method that gives the exact solution of this problem. All that we need from the far field solution in order to construct this exact solution is the fact that ϕ decays like $r^{-1/2}$.

The Near Field: $r \ll 1$

When $r \ll 1$ there is no need to distinguish between the different regions and we write $\hat{r} = r/\delta$, so $\hat{r} = O(1)$ when $r = O(\delta) \ll 1$, and (11.1) becomes Laplace's equation at leading order, with

$$\Phi_{\hat{r}\hat{r}} + \frac{1}{\hat{r}}\Phi_{\hat{r}} + \frac{1}{\hat{r}^2}\Phi_{\theta\theta} = 0. \tag{11.12}$$

The separable solution is

$$\Phi = \hat{r}^{\lambda}(A\cos\lambda\theta + B\sin\lambda\theta) \tag{11.13}$$

and the boundary condition $\Phi_\theta = 0$ on $\theta = 0, 2\pi$ gives $B = 0$, $2\lambda = n$, $n = \pm 1, \pm 2, \ldots$. As we require a solution that does not represent a source of mass at the tip of the barrier, we need $n = -1$, the least singular local solution. This gives

$$\Phi = A\frac{1}{\hat{r}^{1/2}}\cos\frac{1}{2}\theta. \tag{11.14}$$

In terms of the original variable, r,

$$\Phi \sim \frac{A\cos\frac{1}{2}\theta}{r^{\frac{1}{2}}} + O(1) \quad \text{as } r \to 0. \tag{11.15}$$

Note that we have absorbed the artificial parameter, δ, into the undetermined constant A. This shows that the solution can be weakly singular at the edge of the plate, $r = 0$. Since this singularity is integrable, it will not lead to any difficulty in taking the Fourier transform of Φ. That the singularity is unphysical is not a worry within the context of potential theory. The singularity can be resolved by including a viscous boundary layer, although we will not do this here.

11.1.2 Pre-transform Considerations

It is now convenient to subtract out the incident wave from the potential, and work with

$$\psi = \Phi - \exp\{-ikr\cos(\theta - \alpha)\}.$$

This satisfies

$$(\nabla^2 + k^2)\psi = 0, \tag{11.16}$$

and the boundary condition on the barrier is modified to

$$\psi_y(x, 0) = ik\sin\alpha\exp\{-ikx\cos\alpha\} \quad \text{at } y = 0,\ x \geq 0. \tag{11.17}$$

We can now try to solve (11.16) using Fourier integral transforms.

There are two problems that we need to be aware of before making any transformations. Firstly, we must ensure that in choosing the inversion contour we are in a domain of the transformed space where the transform actually exists. We must use the expansions for $r \gg 1$ and $r \ll 1$ to determine this domain *a priori*. Secondly, we must avoid having singularities of the transform actually on the inversion contour. This is most easily achieved by making ω have a small, positive imaginary part, so that, in (11.16), we can replace k by $\tilde{k} + i\epsilon$. We now have some freedom in choosing where we put the inversion contour, so that we can avoid any singularities and remain within the correct domain. Once we have constructed the Fourier transform of the solution, we can take the limit $\epsilon \to 0$ to recover our pure time harmonic solution.

Let's now use the solutions for small and large r and a complex k to decide where in the complex w-plane the Fourier transform of ψ,

$$\bar{\psi}(w, y) = \int_{-\infty}^{\infty} e^{iwx}\psi(x, y)\,dx,$$

exists. At a fixed, positive value of y, as $x \to \infty$ we have the far field solution (11.4) in region I where the reflected wave dominates, which gives

$$|e^{iwx}\psi(x, y)| \sim |e^{-w_2 x + i w_1 x} e^{-i(k+i\epsilon)(x\cos\alpha - y\sin\alpha)}| = O\left(e^{(-w_2+\epsilon\cos\alpha)x}\right), \quad (11.18)$$

with $w = w_1 + i w_2$. Provided $-w_2 + \epsilon\cos\alpha < 0$, the integrand is exponentially small as $x \to +\infty$. As $x \to -\infty$ with y fixed, we move from region I to region II, where there is only a diffracted component. Here we have

$$|e^{iwx}\psi(x, y)| \sim \left| e^{-w_2 x + i w_1 x} \frac{C_0^-(\pi)}{|x|^{\frac{1}{2}}} e^{i(k+i\epsilon)(-x)} \right| = O\left(\frac{e^{(-w_2+\epsilon)x}}{|x|^{\frac{1}{2}}} \right). \quad (11.19)$$

We now have an exponentially small integrand provided $-w_2 + \epsilon > 0$ and we can reasonably expect our Fourier transform to exist in the strip $\epsilon\cos\alpha < \text{Im}(w) < \epsilon$. We reach the same conclusion if $y < 0$ when we move from region III into region II. It is therefore convenient to define two overlapping half planes in the complex w-plane,

$$R^+ = \{w \mid \text{Im}(w) > \epsilon\cos\alpha\}, \quad R^- = \{w \mid \text{Im}(w) < \epsilon\}. \quad (11.20)$$

These planes are illustrated in figure 11.2, and overlap in the strip

$$R^+ \cap R^- = \{w \mid \epsilon\cos\alpha < \text{Im}(w) < \epsilon\}.$$

11.1.3 The Fourier Transform Solution

We now take Fourier transforms of (11.16) with respect to x, to obtain

$$\frac{\partial^2 \tilde{\psi}}{\partial y^2} - (w^2 - k^2)\tilde{\psi} = 0. \tag{11.21}$$

Since $\tilde{\psi}$ must be bounded as $|y| \to \infty$ in $R^+ \cap R^-$, the solution can be written as

$$\tilde{\psi} = \left\{ \begin{array}{ll} \beta_+(w)\exp\{-y(w^2 - k^2)^{\frac{1}{2}}\} & \text{for } y \geq 0, \\ \beta_-(w)\exp\{y(w^2 - k^2)^{\frac{1}{2}}\} & \text{for } y \leq 0, \end{array} \right\} \tag{11.22}$$

provided that $\text{Re}(w^2 - k^2)^{1/2} > 0$ in $R^+ \cap R^-$. The functions $\beta_\pm(w)$ are to be determined so as to satisfy the boundary conditions. This transform must be an analytic function of the complex variable w in the strip $R^+ \cap R^-$ so that the factor $(w^2 - k^2)^{\frac{1}{2}}$ will have branch cuts from the points $w = \pm(\tilde{k} + i\epsilon)$. The branch of the square root is chosen by fixing its value at one point. We choose $(w^2 - k^2)^{1/2}|_{w=0} = -ik$, which gives $\text{Re}(w^2 - k^2)^{1/2}|_{w=0} = \epsilon > 0$. The branch cuts and inversion contour are shown in figure 11.2. Turning to the boundary conditions on $y = 0$, which are different for x negative and x positive, we can immediately recognise that we have a difficulty determining $\beta_\pm(w)$. Let's summarise what we know.

The function ψ will be continuous across $y = 0$ for $x < 0$ but may be discontinuous across $y = 0$ for $x > 0$. The function ψ_y is continuous across $y = 0$ for all x, so its transform, $\tilde{\psi}_y$, must also be continuous across $y = 0$, so $\beta_-(w) = -\beta_+(w)$ and

$$\tilde{\psi}(w, y) = \text{sgn}(y)\beta_+(w)\exp\{-|y|(w^2 - k^2)^{\frac{1}{2}}\}. \tag{11.23}$$

The next step is to define the functions

$$\left. \begin{array}{rcl} A_-(w) & = & \displaystyle\frac{1}{2}\int_{-\infty}^{0} \{\psi(x, 0^+) - \psi(x, 0^-)\}e^{iwx}dx, \\ A_+(w) & = & \displaystyle\frac{1}{2}\int_{0}^{\infty} \{\psi(x, 0^+) - \psi(x, 0^-)\}e^{iwx}dx, \\ B_-(w) & = & \displaystyle\int_{-\infty}^{0} \psi_y(x, 0)e^{iwx}dx, \\ B_+(w) & = & \displaystyle\int_{0}^{\infty} \psi_y(x, 0)e^{iwx}dx. \end{array} \right\} \tag{11.24}$$

Since ψ is continuous at $y = 0$ for $x < 0$, $A_-(w) = 0$. The boundary

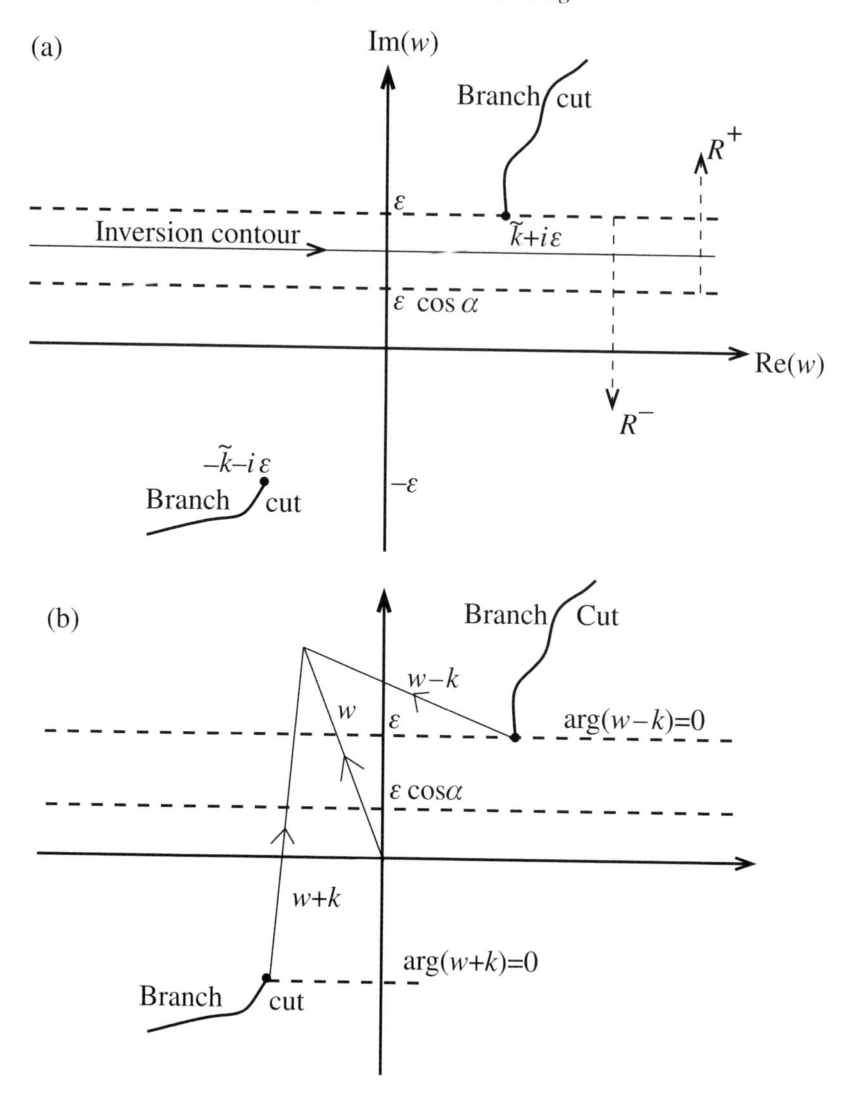

Fig. 11.2. (a) The inversion contour and strip of analyticity for the Fourier transform solution, (11.22). (b) In $R^+ \cap R^-$, $-\pi < \arg(w-k) < 0$ and $0 < \arg(w+k) < \pi$, so that $-\frac{\pi}{2} < \arg(w^2-k^2)^{1/2} < \frac{\pi}{2}$ and $\mathrm{Re}(w^2-k^2)^{1/2} > 0$.

condition (11.17) gives

$$B_+(w) = \int_0^\infty ik\sin\alpha \exp\{-ikx\cos\alpha + iwx\}dx = \frac{-k\sin\alpha}{w-k\cos\alpha} \qquad (11.25)$$

for $w \in R^+$. We do not know $A_+(w)$ or $B_-(w)$, the half range Fourier transforms of the jump in ψ across the barrier and in ψ_y on $y = 0$ for $x < 0$. However, (11.23) shows that

$$A_+(w) = \beta_+(w), \tag{11.26}$$

so once we have determined A_+, we will know $\tilde{\Psi}$, and

$$B_-(w) + B_+(w) = -(w^2 - k^2)^{\frac{1}{2}} A_+(w). \tag{11.27}$$

To solve this complex functional equation we will use what has become known as the **Wiener–Hopf technique**. This makes use of the fact that $\tilde{\psi}$ is analytic in the strip $R^+ \cap R^-$.

From the definitions of $A_\pm$ and $B_\pm$ as Fourier transforms and the estimates of $\psi(x, y)$ as $r \to \infty$, $A_+(w)$ is analytic in R^+ and $B_-(w)$ is analytic in R^-. If we note that

$$\sqrt{w^2 - k^2} = \sqrt{w - k}\sqrt{w + k} = \sqrt{w - (\tilde{k} + i\epsilon)}\sqrt{w + (\tilde{k} + i\epsilon)}, \tag{11.28}$$

we can make each individual square root analytic in one of these half planes by introducing the branch cuts illustrated in figure 11.2. Equation (11.27) can be rearranged into the form

$$\frac{1}{\sqrt{w - k}}\{B_+(w) + B_-(w)\} = -\sqrt{w + k}\, A_+(w). \tag{11.29}$$

The function $B_-(w)/\sqrt{w - k}$, which is unknown, is analytic in R^-, and the function $-\sqrt{w + k}\, A_+(w)$, also unknown, is analytic in R^+. The other function that appears in (11.29), $B_+(w)/\sqrt{w - k}$, is known, has a pole at $w = k\cos\alpha$ (see (11.25)), and can be conveniently written as

$$\begin{aligned}\frac{B_+(w)}{\sqrt{w - k}} &= \frac{B_+(w)}{\sqrt{k(\cos\alpha - 1)}} + B_+(w)\left\{\frac{1}{\sqrt{w - k}} - \frac{1}{\sqrt{k(\cos\alpha - 1)}}\right\} \\ &= C_+(w) + C_-(w).\end{aligned} \tag{11.30}$$

This splits the function into the sum of two functions that are each analytic in one or other of R^+ and R^-. Note that the pole in $C_+(w) = B_+(w)/\sqrt{k(\cos\alpha - 1)}$ is still there, but does not lie in the strict half plane, only on its boundary. In the definition of $C_-(w)$, the pole is cancelled out by the zero that we have introduced.

Using this definition we can rewrite (11.29) in the form

$$\frac{B_-(w)}{\sqrt{w - k}} + C_-(w) = -\sqrt{w + k}\, A_+(w) - C_+(w). \tag{11.31}$$

The left hand side of this is analytic in R^- and the right hand side is analytic in R^+. As there is an overlap between these domains in which the

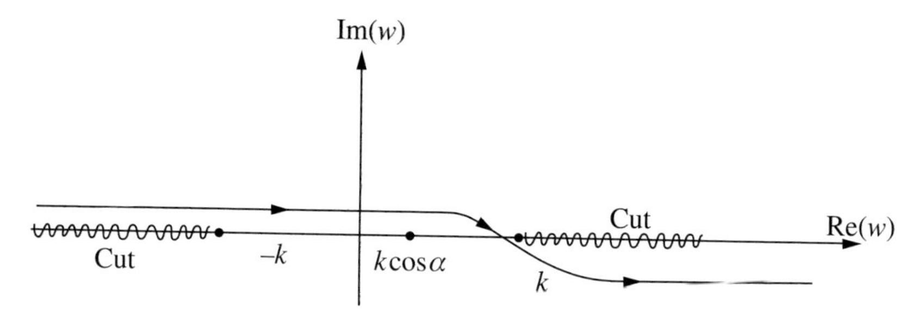

Fig. 11.3. The inversion contour $\mathscr{C}$.

two sides are equal, they must be analytic continuations of each other, and the function

$$D(w) = \left\{ \begin{array}{ll} \dfrac{B_-(w)}{\sqrt{w-k}} + C_-(w) & \text{for } w \in R^-, \\ -\sqrt{w+k}\, A_+(w) - C_+(w) & \text{for } w \in R^+, \end{array} \right\} \tag{11.32}$$

is analytic in the whole complex w-plane. From the asymptotic solution for $r \ll 1$, $D(w)$ must vanish as $|w| \to \infty$. By Liouville's theorem, $D(w)$ is therefore identically zero, and we have

$$C_-(w) = -\frac{B_-(w)}{\sqrt{w-k}}, \qquad C_+(w) = -\sqrt{w+k}\, A_+(w). \tag{11.33}$$

This leads to

$$A_+(w) = \frac{k \sin\alpha}{(w - k\cos\alpha)\sqrt{w+k}\sqrt{k(\cos\alpha - 1)}}, \tag{11.34}$$

and, using $1 - \cos\alpha = 2\sin^2 \frac{1}{2}\alpha$,

$$\tilde{\psi}(w, y) = \frac{-i\,\mathrm{sgn}(y)\sqrt{2k}\cos\frac{1}{2}\alpha \exp\{-|y|\sqrt{w^2 - k^2}\}}{(w - k\cos\alpha)\sqrt{w+k}}. \tag{11.35}$$

Our remaining task is to invert this Fourier transform. If we now let $\epsilon \to 0$ and deform our branch cuts onto the real axis, we must evaluate

$$\psi(x, y) = \frac{\mathrm{sgn}(y)\cos\frac{1}{2}\alpha}{i\pi}\sqrt{\frac{k}{2}}\int_{\mathscr{C}} \frac{e^{-iwx-|y|\sqrt{w^2-k^2}}}{(w - k\cos\alpha)\sqrt{w+k}}\,dw, \tag{11.36}$$

where the contour $\mathscr{C}$, shown in figure 11.3, lies above the pole at $w = k\cos\alpha$ before it crosses the real axis to pass beneath the branch cut at $w = k$. To proceed, it is convenient to convert this to an integral over the real line, since most straightforward approximation methods are based upon this type of integral.

We begin by defining the variable χ by $x = r\cos\chi$, $|y| = r\sin\chi$, and use χ and y as our independent variables. This definition of the angle χ can be compared with standard polar coordinates, with $\chi \equiv \theta$ for $y > 0$ and $\chi = -\theta$ for $y < 0$. The various regions now become region I: $0 < \chi < \pi - \alpha$, $y > 0$, region II: $\pi - \alpha < \chi < \pi$ and region III: $0 < \chi < \pi - \alpha$, $y < 0$. Now consider the contour H_1 defined by $w = -k\cos(\chi + it)$, $-\infty < t < \infty$. This is the branch of the hyperbola

$$\left(\frac{\text{Re}(w)}{k\cos\chi}\right)^2 - \left(\frac{\text{Im}(w)}{k\sin\chi}\right)^2 = 1,$$

with negative real part for $0 < \chi < \pi/2$ and positive real part for $\pi/2 < \chi < \pi$. When $\chi = \pi/2$, H_1 is the imaginary axis. We can deform our contour $\mathscr{C}$ onto H_1 by joining H_1 and $\mathscr{C}$ using the two circular arcs $\mathscr{C}_1$ and $\mathscr{C}_2$, shown in figure 11.4. If we then let the radii of these arcs tend to infinity, the integrals along them tend to zero, and the residue theorem allows us to write the integral along $\mathscr{C}$ in terms of the integral along H_1. If $k\cos\alpha > -k\cos\chi$, we must include the contribution from the simple pole at $w = k\cos\alpha$. When this is required, we find that the residue leads to a term that represents one of the plane waves. This transformation is effective because the exponential factor in the integrand becomes

$$\exp\left\{-iwx - |y|\sqrt{w^2 - k^2}\right\} = \exp(ikr\cosh t),$$

a form suitable for straightforward asymptotic approximation by the method of stationary phase. The integral along H_1 is transformed to

$$J = \frac{\text{sgn}(y)\cos\left(\frac{1}{2}\alpha\right)}{\pi}\int_{-\infty}^{\infty}\frac{e^{ikr\cosh t}\cos\frac{1}{2}(\chi + it)}{\cos\alpha + \cos(\chi + it)}dt, \tag{11.37}$$

which represents the potential due to the diffracted wave. We find that in the various regions, adding back in the incident wave,

— Region I: $\Phi = e^{-ik(x\cos\alpha + y\sin\alpha)} + e^{-ik(x\cos\alpha - y\sin\alpha)} + J$ – incident, reflected and diffracted components,
— Region II: $\Phi = e^{-ik(x\cos\alpha + y\sin\alpha)} + J$ – incident and diffracted components,
— Region III: $\Phi = J$ – the diffracted component.

We now have a formal solution of the problem. In principle, a numerical integration will give us the value of J throughout the domain. Of rather more interest is a simple approximation to J. For large r, we can use the method of stationary phase to infer that the major contribution to the integral comes from the region near $t = 0$. Provided

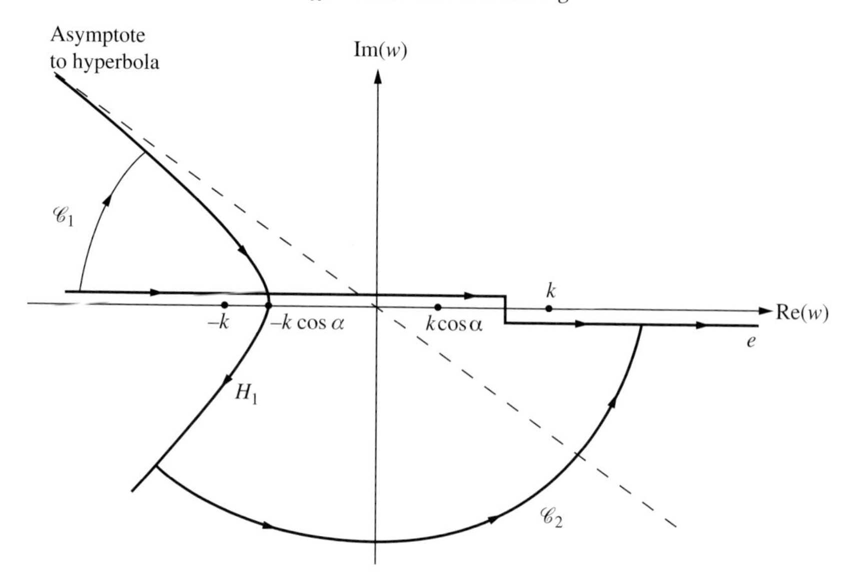

Fig. 11.4. The modified inversion contour, composed of circular arcs and a hyperbola.

that $\cos\alpha + \cos\chi \neq 0$, which occurs at the boundaries between the three regions, this gives the approximation

$$J \sim \operatorname{sgn}(y)\left\{\frac{\cos\left(\frac{1}{2}\alpha\right)\cos\left(\frac{1}{2}\chi\right)}{\cos\alpha+\cos\chi}\right\}\left\{\frac{2}{\pi kr}\right\}^{\frac{1}{2}} e^{i(kr+\frac{\pi}{4})} \text{ as } r\to\infty. \tag{11.38}$$

We can recognise the factor

$$\exp\left\{\frac{i\pi}{4}\right\}\operatorname{sgn}(y)\frac{\cos\left(\frac{1}{2}\alpha\right)\cos\left(\frac{1}{2}\chi\right)}{\cos\alpha+\cos\chi} = \exp\left\{\frac{i\pi}{4}\right\}\frac{\cos\left(\frac{1}{2}\alpha\right)\cos\left(\frac{1}{2}\theta\right)}{\cos\alpha+\cos\theta}$$

for $y > 0$ as the unknown function $C_0^-(\theta)$, which appeared in our estimate of the solution in the far field and is plotted in figure 11.5. There is a singularity in C_0^- at $\theta = \pi - \alpha$, where the method that we used to approximate the integral (11.37) is not valid. However, it is clear that the size of the diffracted field is greatest in the neighbourhood of the boundaries between the three regions. Although we will not try to refine our approximation method close to these boundaries, the diffracted field there leads to a smoothing of what would otherwise be a sharp distinction between regions, in particular the boundary between the shadow region III and the illuminated region II. For further details on diffraction by sharp edges, see Crighton (1981).

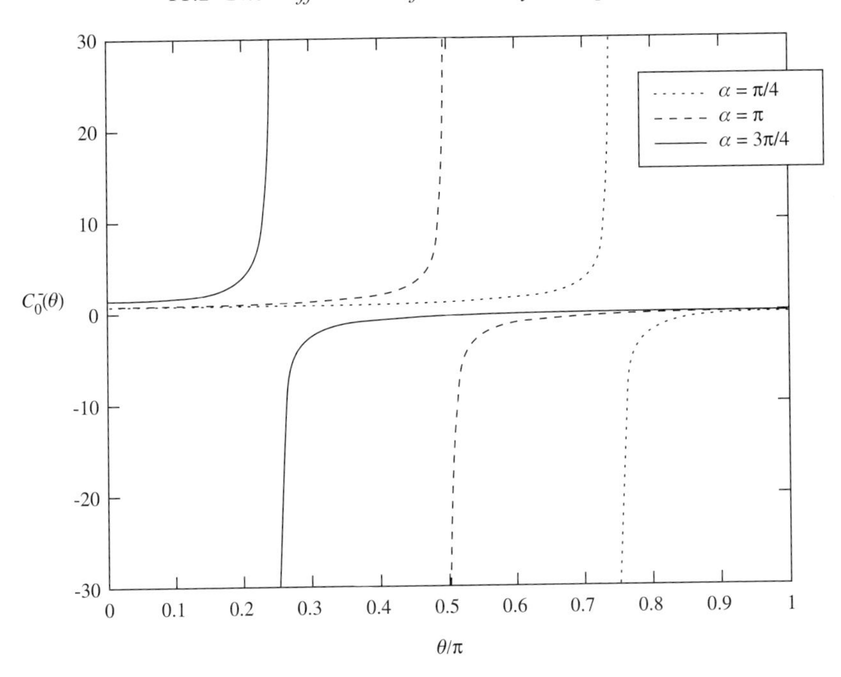

Fig. 11.5. The intensity of the diffracted field for $r \gg 1$.

11.2 The Diffraction of Waves by an Aperture

In this section we introduce more of the ideas involved in solving diffraction problems, and examine the differences between the diffraction of scalar and vector waves.

11.2.1 Scalar Diffraction: Acoustic Waves

Let the plane $z = 0$ be occupied by a thin, rigid wall, or screen, except for a bounded region, A, where there is a hole, or aperture. The acoustic potential, $\phi(x, y, z, t) = \phi(\mathbf{r}, t)$, must satisfy the three-dimensional wave equation, (3.15), and the no flux boundary condition, $\partial\phi/\partial z = 0$, at the rigid wall. Let's consider time harmonic waves, so that $\phi(\mathbf{r}, t) = \Phi(\mathbf{r})e^{i\omega t}$. (Note that the change in sign from the definition given in the previous section is for consistency with the results presented in chapter 3.) This means that Φ satisfies the Helmholtz equation,

$$\nabla^2\Phi + \frac{\omega^2}{c^2}\Phi = 0, \tag{11.39}$$

subject to

$$\frac{\partial \Phi}{\partial z} = 0 \quad \text{at } z = 0, \text{ for } (x, y) \notin A. \tag{11.40}$$

If we look back to our analysis of acoustic sources in a plane wall, we find that we have essentially already found an integral representation of the solution of (11.39) subject to (11.40). By writing (3.63) in terms of the acoustic potential, using the current coordinate system, we find that

$$\Phi(x, y, z) = -\frac{1}{2\pi} \int_A \frac{\partial \Phi}{\partial z}(x', y', 0) \frac{e^{-i\omega R/c}}{R} dx' dy', \tag{11.41}$$

where

$$R = \sqrt{(x - x')^2 + (y - y')^2 + z^2} = |\mathbf{r} - \mathbf{r}'|. \tag{11.42}$$

We obtained this expression by adding up elementary acoustic sources and their images in the wall. The more formal derivation of (11.41) is basically the same, but is stated in terms of the free space Green's function and its image in the wall. Now, given the normal derivative of the acoustic potential at the aperture, we can find the potential everywhere else. In section 3.7, we knew $\partial \phi / \partial n$ everywhere, and could find the solution directly. Here, we will have to make an approximation.

Let's assume that a plane sound wave is incident on the aperture from $z < 0$ with wavenumber, and hence direction, $\mathbf{k} = (k_x, k_y, k_z)$, with $|\mathbf{k}| = \omega / c$, and potential

$$\phi_{\text{inc}} = \alpha e^{i(\omega t - \mathbf{k} \cdot \mathbf{r})}.$$

We make the assumption that the normal derivative of the potential at the aperture is given purely by the incident plane wave, as if the screen were not present. This is called the **Kirchhoff approximation**, and gives

$$\frac{\partial \Phi}{\partial z}(x', y', 0) \approx -i\alpha k_z e^{-i(k_x x' + k_y y')}. \tag{11.43}$$

This appears at first sight to be a very crude approximation, but is surprisingly accurate in the forward direction, although less good for the backscattered wave. A useful mathematical way to think about it is as the first iteration of an iterative solution of the integral equation (11.41) with $z = 0$.

Equation (11.41) now becomes

$$\Phi(x, y, z) \approx \frac{i\alpha k_z}{2\pi} \int_A \frac{1}{R} \exp\left[-i\left(k_x x' + k_y y' + \frac{\omega R}{c}\right)\right] dx' dy'. \tag{11.44}$$

Just as we did in section 3.7, we concentrate on the far field solution, with $|\mathbf{r}| \gg |\mathbf{r}'|$ so that

$$R \sim r - \frac{xx' + yy'}{r}, \tag{11.45}$$

where $r^2 = x^2 + y^2 + z^2$. Solutions based on this approximation are referred to as **Fraunhofer diffraction**. If we were also to retain the next term in the expansion for R, the approximation is known as **Fresnel diffraction**, and is useful when r is large compared to the wavelength of the incident wave, but not compared to the dimensions of the aperture. We will not consider Fresnel diffraction here.

Substituting (11.45) into (11.44) gives, at leading order,

$$\Phi(x, y, z) = \frac{i\alpha k_z}{2\pi r} e^{-i\omega r/c} \int_A e^{i(nx'+my')} dx'dy', \tag{11.46}$$

where

$$\left.\begin{aligned} n &= \frac{\omega x}{rc} - k_x = \mathbf{e}_x \cdot \left(\frac{\omega \mathbf{r}}{rc} - \mathbf{k}\right) = \frac{\omega}{c}\mathbf{e}_x \cdot (\hat{\mathbf{r}} - \hat{\mathbf{k}}), \\ m &= \frac{\omega y}{rc} - k_y = \mathbf{e}_y \cdot \left(\frac{\omega \mathbf{r}}{rc} - \mathbf{k}\right) = \frac{\omega}{c}\mathbf{e}_y \cdot (\hat{\mathbf{r}} - \hat{\mathbf{k}}). \end{aligned}\right\} \tag{11.47}$$

Here, $\mathbf{e}_x$, $\mathbf{e}_y$ and $\mathbf{e}_z$ are unit vectors in the coordinate directions, and $\hat{\mathbf{k}} = \mathbf{k}/|\mathbf{k}|$ and $\hat{\mathbf{r}} = \mathbf{r}/r$ are the unit vectors in the directions of $\mathbf{k}$ and $\mathbf{r}$. Note that the integral in (11.46) is just the two-dimensional Fourier transform of a function that is unity over the aperture and zero elsewhere. When $k_x = k_y = 0$, and the wave is incident on the aperture normally, (11.46) is just proportional to the result that we obtained in section 3.7, with the aperture replaced by a uniformly vibrating piston.

Let's consider the case of a rectangular aperture, so that

$$A = \{(x', y') \mid -a < x < a, \ -b < y < b\},$$

and hence

$$\Phi(x, y, z) = \frac{2\alpha i k_z}{\pi r} e^{-i\omega r/c} \frac{\sin na}{n} \frac{\sin mb}{m}. \tag{11.48}$$

If $\omega a/c \ll 1$ and $\omega b/c \ll 1$, so that the aperture is acoustically compact, the far field just looks like an acoustic source, as we might expect. However, it is worth noting that the Kirchhoff approximation, that the normal derivative of the potential at the aperture can be taken to be that of the incident wave, is very poor when the aperture is acoustically compact. In fact, the accuracy is greatest when the aperture is much larger than a wavelength. Figure 11.6 shows the diffraction pattern when the wave is incident normal to the wall, with a square aperture, as given by (11.48) with $a = b$. The shading of the figure has been chosen to show up the side lobes of the far field sound, which actually have a much smaller amplitude than the main peak, which lies over the aperture. The figure also agrees well with pictures of the optical diffraction pattern when light is shone through a square aperture, which we shall discuss in the next section.

Finally, if the incident wave is not normally incident, the intensity of the diffracted wave is reduced by the factor $k_z = \mathbf{e}_z \cdot \mathbf{k}$ and, from (11.47), the position where the amplitude is greatest, $m = n = 0$, is now where $\hat{r}_x = \hat{k}_x$ and $\hat{r}_y = \hat{k}_y$. Moreover, since $\hat{\mathbf{r}}$ and $\hat{\mathbf{k}}$ are unit vectors, if two of their components agree, all three must agree, so that $\hat{\mathbf{r}} = \hat{\mathbf{k}}$ at the centre of the diffraction pattern. This simply means, unsurprisingly, that the intensity of the sound is greatest in the direction of the incident wave. In terms of the optical problem, if someone shines a torch through a hole, you need to line up the hole and the torch in order to see the brightest light.

11.2.2 Vector Diffraction: Electromagnetic Waves

We now consider the analogous problem to that of the previous subsection, but for electromagnetic waves. A thin, perfectly conducting plane wall lies at $z = 0$, except in the bounded region A, where there is an aperture. A plane electromagnetic wave is incident on the wall from $z < 0$. What are the diffracted fields on the other side of the screen? The electric and magnetic fields are vector fields, which are rather more difficult to deal with than scalar fields, such as the acoustic potential, particularly with regard to their polarisation. Let's see if we can generalise the results of the previous section.

We begin by writing the electric and magnetic fields in terms of incident, reflected and diffracted fields, with

$$\mathbf{E} = \mathbf{E}_{\mathrm{I}} + \mathbf{E}_{\mathrm{R}} + \mathbf{E}_{\mathrm{D}}, \quad \mathbf{B} = \mathbf{B}_{\mathrm{I}} + \mathbf{B}_{\mathrm{R}} + \mathbf{B}_{\mathrm{D}}. \tag{11.49}$$

Using the results of subsection 6.7.2, $\mathbf{E}_{\mathrm{R}}$ and $\mathbf{B}_{\mathrm{R}}$ are defined to be the reflected fields that would exist if there were no aperture in the wall, with

$$\mathbf{E}_{\mathrm{I}} + \mathbf{E}_{\mathrm{R}} = -2e^{-i(k_x x + k_y y)} \times \left(iE_{x0}\sin k_z z, iE_{y0}\sin k_z z, -E_{z0}\cos k_z z\right) H(-z), \tag{11.50}$$

where $\mathbf{k} = (k_x, k_y, k_z)$ is the wavenumber vector of the incident wave, whose electric field has amplitude $\mathbf{E}_0 = (E_{x0}, E_{y0}, E_{z0})$, with $\mathbf{k} \cdot \mathbf{E}_0 = 0$.

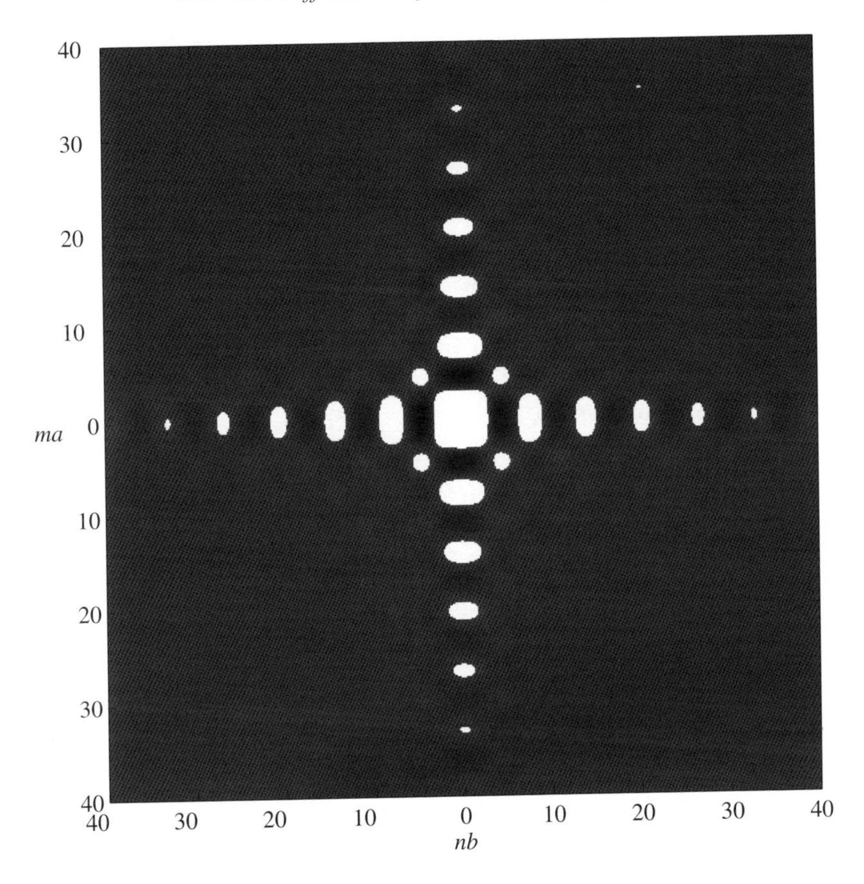

Fig. 11.6. The diffraction of a plane sound wave by a square aperture.

The amplitude, $\mathbf{E}_0$, can be complex, and the polarisation of the incoming wave elliptical. The magnetic fields can be found from the usual relations for plane waves as

$$\mathbf{B}_\mathrm{I} = \frac{1}{\omega}\mathbf{k} \times \mathbf{E}_\mathrm{I}, \quad \mathbf{B}_\mathrm{R} = \frac{1}{\omega}\mathbf{k}_\mathrm{R} \times \mathbf{E}_\mathrm{R},$$

where $\mathbf{k}_\mathrm{R} = (k_x, k_y, -k_z)$ is the wavenumber of the reflected wave. We can immediately see that the tangential components of $\mathbf{E}_\mathrm{I} + \mathbf{E}_\mathrm{R}$ vanish at $z = 0$, as required by the boundary conditions, and that the normal component is discontinuous at $z = 0$, which indicates that there is a distribution of charge at $z = 0^-$. A similar calculation for the magnetic field shows that the normal component of $\mathbf{B}_\mathrm{I} + \mathbf{B}_\mathrm{R}$ is zero at $z = 0$, as required, but that the tangential components are discontinuous, which indicates that there is a surface current at $z = 0^-$.

Now, even when we consider the total fields, **E** and **B**, any surface currents will be confined to the plane $z = 0$, although not of course at the aperture. Taking the point of view that we discussed briefly in subsection 6.7.2, that the surface charges and currents can be thought of as generating the electromagnetic fields, we can see that it is the difference between the surface densities when there is no aperture and those when there is an aperture that drives the diffracted fields, $\mathbf{E}_\mathrm{D}$ and $\mathbf{B}_\mathrm{D}$. We can make use of this to derive the vector equivalent of (11.41).

The forced wave equation, (6.94), for the vector potential associated with the diffracted field, $\mathbf{A}_\mathrm{D}$, is driven purely by the surface currents, now that we have subtracted out the incident and reflected fields. This means that its z-component, $A_{\mathrm{D}z}$, is zero, since there are no currents in the z-direction. The other two components of $\mathbf{A}_\mathrm{D}$ satisfy the Helmholtz equation for $z \neq 0$, so we can use (11.41) to write

$$\left.\begin{aligned} A_{\mathrm{D}x}(x,y,z) &= -\frac{1}{2\pi}\int_S \frac{\partial A_{\mathrm{D}x}}{\partial z}(x',y',0)\frac{e^{-i\omega R/c_0}}{R}dx'dy', \\ A_{\mathrm{D}y}(x,y,z) &= -\frac{1}{2\pi}\int_S \frac{\partial A_{\mathrm{D}y}}{\partial z}(x',y',0)\frac{e^{-i\omega R/c_0}}{R}dx'dy', \end{aligned}\right\} \qquad (11.51)$$

where S is the plane $z = 0$. However, since $A_{\mathrm{D}z} = 0$,

$$\mathbf{B}_\mathrm{D} = \nabla \times \mathbf{A}_\mathrm{D} = \left(-\frac{\partial A_{\mathrm{D}y}}{\partial z}, \frac{\partial A_{\mathrm{D}x}}{\partial z}, \frac{\partial A_{\mathrm{D}y}}{\partial x} - \frac{\partial A_{\mathrm{D}x}}{\partial y}\right),$$

so, from (11.51),

$$\mathbf{A}_\mathrm{D} = \frac{1}{2\pi}\int_S (\mathbf{n} \times \mathbf{B}_\mathrm{D})\frac{e^{-i\omega R/c_0}}{R}dx'dy',$$

and hence

$$\mathbf{B}_\mathrm{D} = \frac{1}{2\pi}\nabla \times \int_S (\mathbf{n} \times \mathbf{B}_\mathrm{D})\frac{e^{-i\omega R/c_0}}{R}dx'dy'. \qquad (11.52)$$

This is a vector equivalent of (11.41), which gives the diffracted magnetic field in terms of its tangential component on the screen. This is not, however, a convenient representation to use in this problem. We want to use the electric rather than the magnetic field, since its tangential component always vanishes on the screen.

How can we write an expression similar to (11.52) for the electric field? The trick is to notice that the source-free Maxwell equations, (6.29) to (6.32), are unchanged under the transformation $\mathbf{E} \mapsto -c_0\mathbf{B}$, $\mathbf{B} \mapsto \mathbf{E}/c_0$.

Therefore, since (11.52) relates the tangential components of $\mathbf{B}_\mathrm{D}$ on the screen to the total diffracted magnetic field away from the screen, the same equation holds for $\mathbf{E}_\mathrm{D}$, with

$$\mathbf{E}_\mathrm{D} = \frac{1}{2\pi}\nabla \times \int_S (\mathbf{n}\times\mathbf{E}_\mathrm{D})\,\frac{e^{-i\omega R/c_0}}{R}\,dx'dy'. \qquad (11.53)$$

The crucial difference is that the tangential components of $\mathbf{E}_\mathrm{I}+\mathbf{E}_\mathrm{R}$ vanish on the screen, as do the tangential components of the total field, $\mathbf{E}$. Since $\mathbf{E}_\mathrm{I}+\mathbf{E}_\mathrm{R} \equiv 0$ for $z \geq 0$, we have the exact expression for the electric field

$$\mathbf{E} = \frac{1}{2\pi}\nabla \times \int_A (\mathbf{n}\times\mathbf{E})\,\frac{e^{-i\omega R/c_0}}{R}\,dx'dy', \quad \text{for } z \geq 0. \qquad (11.54)$$

This is the key equation, and we can now make the Kirchhoff approximation, and assume that $\mathbf{E} \approx \mathbf{E}_\mathrm{I}$ on the aperture, so that (11.54) gives an approximation to $\mathbf{E}$ for $z \geq 0$.

By definition,

$$\mathbf{E}_\mathrm{I}(x', y', 0) = \mathbf{E}_{\mathrm{I}0}e^{-i\mathbf{k}\cdot\mathbf{r}'},$$

and hence from (11.54)

$$\mathbf{E}(x, y, z) \approx -\frac{1}{2\pi}(\mathbf{n}\times\mathbf{E}_0)\times\int_A \nabla\left(\frac{e^{-i\omega R/c_0}}{R}\right)e^{i\mathbf{k}\cdot\mathbf{r}'}\,dx'dy'. \qquad (11.55)$$

Now, in the far field, using the Fraunhofer approximation,

$$\nabla\left(\frac{e^{-i\omega R/c_0}}{R}\right) \sim -\frac{i\omega}{c_0 r^2}\mathbf{r}\exp\left\{-i\left(\frac{\omega r}{c_0}+\frac{xx'}{rc_0}+\frac{yy'}{rc_0}\right)\right\},$$

and hence, at leading order,

$$\mathbf{E}(x, y, z) = -\frac{i\omega}{2\pi c_0 r^2}e^{-i\omega r/c_0}\mathbf{r}\times(\mathbf{n}\times\mathbf{E}_0)\int_A e^{i(nx'+my')}\,dx'dy'. \qquad (11.56)$$

The integral in this expression is the same as that for scalar diffraction by an aperture, (11.46). For the rectangular aperture,

$$\mathbf{E}(x, y, z) = -\frac{2i\omega}{\pi c_0 r^2}e^{-i\omega r/c_0}\mathbf{r}\times(\mathbf{n}\times\mathbf{E}_0)\frac{\sin na}{n}\frac{\sin mb}{m}, \qquad (11.57)$$

with the diffraction pattern for normal incidence on a square aperture similar, but not completely identical, to that shown in figure 11.6. For general angles of incidence, we note that $\mathbf{r}\cdot\mathbf{E} = 0$. The magnetic field can be obtained from (6.31), using the Fraunhofer approximation to take only the leading order solution. We find that, at leading order, $\mathbf{B} = (\hat{\mathbf{r}}\times\mathbf{E})/c_0$, which is what we would obtain for a plane wave propagating in the $\mathbf{r}$-direction. We conclude that, in the far field, the diffracted wave is

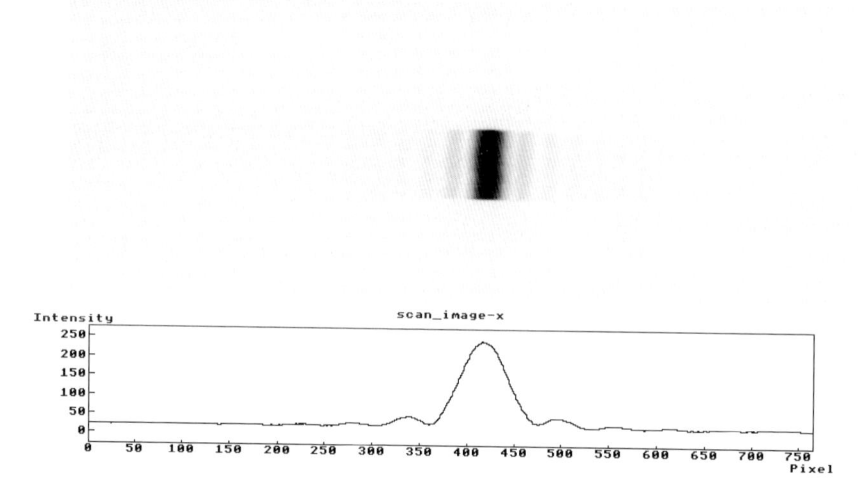

Fig. 11.7. The diffraction pattern obtained by shining light through a long, thin slit.

locally equivalent to a plane wave propagating radially, with plane of polarisation through $\mathbf{r}$ and $\mathbf{r} \times (\mathbf{n} \times \mathbf{E}_0)$, and hence dependent upon the point of observation. Furthermore, comparing (11.57) with (11.48), the angular variation differs by a factor of $\hat{\mathbf{r}} \times (\mathbf{n} \times \mathbf{E}_0)/\hat{k}_z$. This leads to a slightly different angular variation in the intensity of light from that of the intensity of sound in the equivalent acoustic problem. However, the distribution given by the trigonometric terms in (11.57) still gives a good approximation to the observed intensity. Figure 11.7 shows the experimentally measured diffraction pattern from a long, thin slit, which is in good agreement with the directivity pattern derived for the equivalent acoustic problem, shown in figure 3.10.

A more detailed discussion of diffraction is given in the book by Born and Wolf (1975). The appropriate sections of the books by Jackson (1975) and Elmore and Heald (1969) also contain good overviews of the subject. An alternative way of considering diffraction problems is provided by the **geometrical theory of diffraction** (Keller, 1958), which uses a WKB expansion and the idea of **rays** to construct the solution. The extension of these ideas to **complex rays** is a topic of considerable current interest (Chapman, Lawry, Ockendon and Tew, 1999). The physical equivalent of the Kirchhoff approximation may soon lead to a device for projecting three-dimensional images using nanotechnology to place tiny sources of

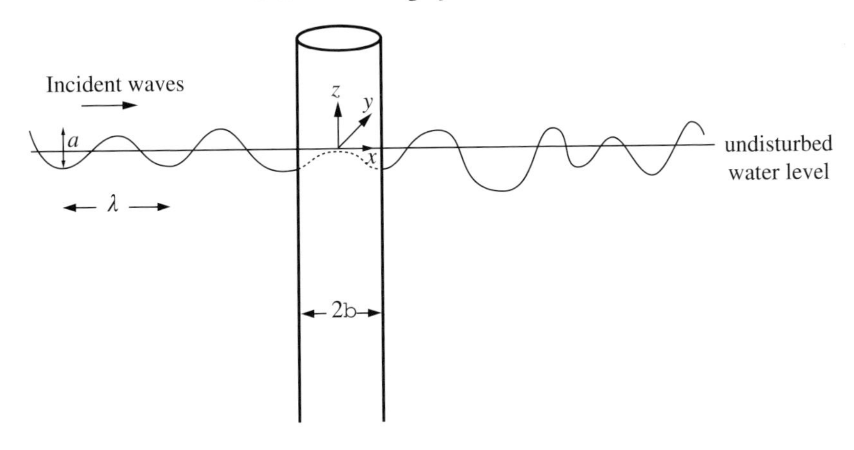

Fig. 11.8. The scattering of small amplitude water waves by a surface piercing cylinder.

light on a plane surface. The technique is called phased array optics, and is discussed by Wowk (1996).

11.3 The Scattering of Linear, Deep Water Waves by a Surface Piercing Cylinder

We now consider two-dimensional progressive gravity waves of small amplitude in deep water, incident upon a surface piercing cylinder of radius b. The cylinder extends vertically downwards and scatters the waves that are incident upon it, as shown in figure 11.8. As the incident wave decays exponentially with depth, the parts of the cylinder deep in the water do not have much effect on the scattered surface wave, and we idealize the cylinder as being infinitely long.

The boundary value problem that we must solve for the velocity potential, ϕ, is

$$\phi_{xx} + \phi_{yy} + \phi_{zz} = 0 \quad \text{in } z < 0,\ x^2 + y^2 > b^2, \tag{11.58}$$

with

$$\phi \to 0 \quad \text{as } z \to -\infty, \tag{11.59}$$

$$\phi_{tt} + g\phi_z = 0 \quad \text{on } z = 0, \quad \phi_r = 0 \quad \text{on } r = b, \tag{11.60}$$

$$\phi \sim \phi_{\rm I}, \quad \eta \sim \eta_{\rm I} \quad \text{as } x^2 + y^2 \to \infty. \tag{11.61}$$

In chapter 4, we showed that the incident potential and free surface displacement for a deep water, progressive gravity wave take the form

$$\phi_{\mathrm{I}} = \frac{ga}{\omega} e^{kz} \cos(kx - \omega t), \quad \eta_{\mathrm{I}} = -a \sin(kx - \omega t),$$

where $\omega = \sqrt{gk}$ is the deep water dispersion relation. It is convenient to do some preliminary manipulation of this result. We write

$$\phi_{\mathrm{I}} = \mathrm{Re}\left(\frac{ga}{\omega} e^{kz} e^{ikx - i\omega t}\right) = \mathrm{Re}\left(\Phi_{\mathrm{I}} e^{-i\omega t}\right) = \Phi_{\mathrm{I}} e^{-i\omega t},$$

with the understanding that we mean the real part, so that

$$\Phi_{\mathrm{inc}} = \frac{ga}{\omega} e^{kz} e^{ikx}.$$

If we introduce polar coordinates $x = r\cos\theta$, $y = r\sin\theta$ in the horizontal plane of the cylinder, we can write this in the form

$$\Phi_{\mathrm{inc}} = \frac{ga}{\omega} e^{kz} e^{ikr\cos\theta} = \frac{ga}{\omega} e^{kz} \sum_{n=0}^{\infty} \alpha_n i^n J_n(kr) \cos n\theta, \tag{11.62}$$

using a standard result from the theory of Bessel functions (Watson, 1922). Here $\alpha_0 = 1$, $\alpha_n = 2$ for $n \geq 1$.

Let's now return to the three-dimensional boundary value problem given by (11.58) to (11.61) and look for a solution in the form

$$\Phi = \Phi_{\mathrm{inc}} + \frac{ga}{\omega} e^{kz - i\omega t} \Phi(r, \theta).$$

This gives us a modified boundary value problem for Φ, which now satisfies the two-dimensional Helmholtz equation,

$$\left.\begin{aligned} &(\nabla^2 + k^2)\Phi = 0 \quad \text{for } r > b,\ 0 \leq \theta \leq 2\pi, \\ &\Phi \to 0 \quad \text{as } r \to \infty, \\ &\Phi_r = -\sum_{n=0}^{\infty} k\alpha_n i^n J_n'(kb) \cos n\theta \quad \text{on } r = b. \end{aligned}\right\} \tag{11.63}$$

Note that the free surface and deep water conditions are automatically satisfied for this choice of ϕ. If we look for a separable solution we find that $\Phi = \{AJ_n(kr) + BY_n(kr)\}\cos n\theta$. As both Bessel functions are bounded in the domain of solution, we must retain both constants A and B. If we now write this in terms of the Hankel functions, $H_n^{(1)}(kr) = J_n(kr) + iY_n(kr)$, $H_n^{(2)}(kr) = J_n(kr) - iY_n(kr)$, the subsequent analysis is easier. Both these functions are tabulated and their asymptotic properties are known (for example, Abramowitz and Stegun (1972)). In particular, they satisfy

$$\sqrt{r}\left(\frac{\partial}{\partial r} - ik\right) H_n^{(1)}(kr) \to 0, \quad \sqrt{r}\left(\frac{\partial}{\partial r} + ik\right) H_n^{(2)}(kr) \to 0 \quad \text{as } r \to \infty. \tag{11.64}$$

Remember that $ik = -(\omega/g)\partial/\partial t$, so we can see that $H_n^{(1)}(kr)$ represents an outgoing scattered wave, whilst $H_n^{(2)}(kr)$ represent an unphysical incoming wave. Accordingly, the solution can be written as

$$\Phi(r,\theta) = \sum_{n=0}^{\infty} C_n H_n^{(1)}(kr)\cos n\theta. \tag{11.65}$$

The boundary condition on $r = b$ is satisfied provided that

$$C_n = -\alpha_n i^n \frac{dJ_n}{dr}(kb) \Big/ \frac{dH_n^{(1)}}{dr}(kb), \tag{11.66}$$

and our solution can be written in the form

$$\phi = \frac{ga}{\omega} e^{kz-i\omega t} \sum_{n=0}^{\infty} \left\{\alpha_n i^n J_n(kr) + C_n H_n^{(1)}(kr)\right\} \cos n\theta. \tag{11.67}$$

From this we can compute other quantities of physical interest. The pressure in linear water waves is given by $p = -\rho\phi_t - \rho g z$. It is easy, but laborious, to show that the hydrodynamic part of this is

$$p_{\text{hydro}} = \frac{2\rho g a}{\pi k b} e^{kz-i\omega t} \sum_{n=0}^{\infty} \left\{ \alpha_n i^n \cos n\theta \Big/ \frac{dH_n^{(1)}}{dr}(kb) \right\}, \tag{11.68}$$

and quantities such as the horizontal component of the hydrodynamic force can be found by integration of this around the boundary of the cylinder. The result consists of just a single term (see exercise 11.5).

Figure 11.9 shows the amplitude of the scattered part of the free surface displacement for $kb = 0.1$, 1 and 10, for an incoming wave of unit amplitude. When $kb = 0.1$, the incident waves are fairly long compared with the radius of the cylinder, and there is little scattering. When $kb = 1$, the incident wavelength and the radius of the cylinder are comparable and there is a significant amount of scattering, fairly evenly distributed around the cylinder, but with the largest amplitude directly in the shadow of the cylinder. When $kb = 10$, the wavelength of the incident waves is fairly short compared with the radius of the cylinder, and there is a shadow region behind the cylinder, where the scattered wave dominates the flow, with a smaller disturbance elsewhere. This last case has much in common with the diffraction problems that we studied in the first two sections of this chapter, with the sides of the cylinder effectively presenting two edges to the incident wave. Indeed, behind the cylinder there are alternating radial bands of large and small amplitude disturbance, analogous to the light and dark regions formed in diffraction by an aperture.

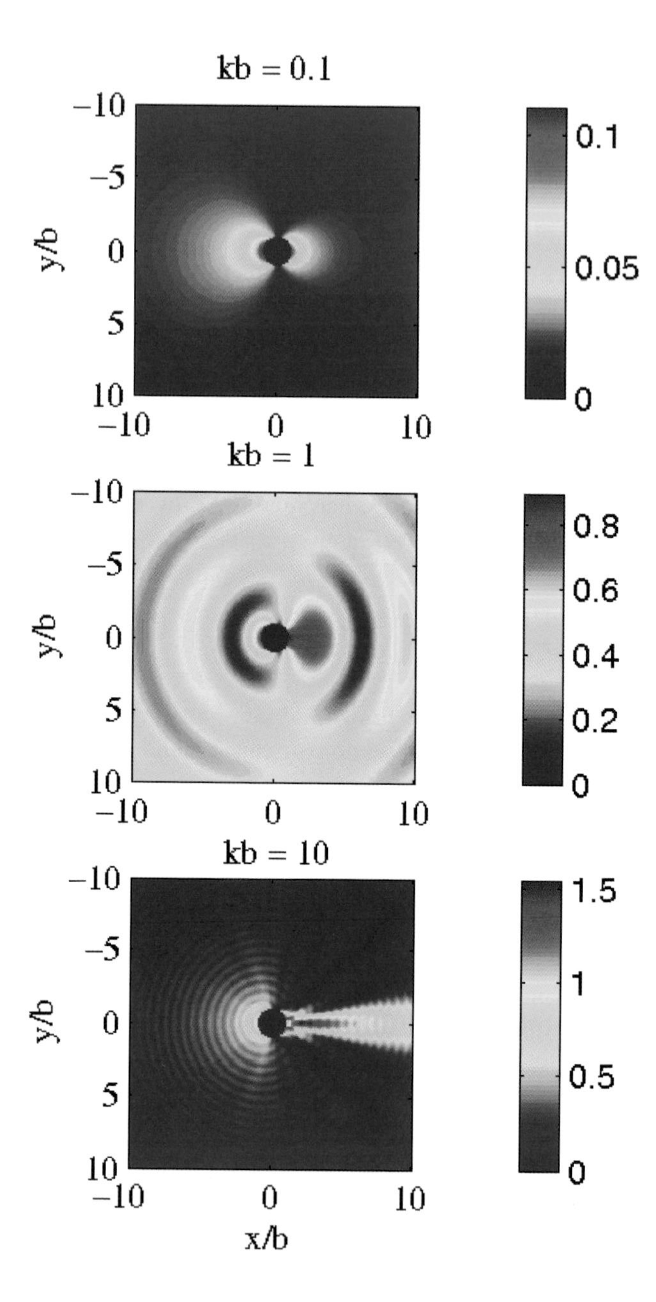

Fig. 11.9. The amplitude of the wave scattered by an upright cylinder subject to an incoming plane wave, with $kb = 0.1$, 1 and 10.

Although this result is useful, when waves from the sea are incident on the legs of an oil platform or the pillars of an estuary bridge, resonances can occur between the wave fields scattered from each leg. These are undesirable and designers of such structures choose spacings and sizes of the pillars or legs to avoid this. To perform this calculation, it is necessary to have results for the scattering of a wave by two or more cylinders. The method above does not give this solution for all spacings and approximate or numerical methods must be used (Evans and Porter, 1997).

Exercises

11.1 Identify the singularities of the function

$$f(w) = \frac{\sqrt{w-i}}{\alpha - iw},$$

where α is a real constant. Find an additive split of this function, $f(w) = f_+(w) + f_-(w)$, where f_+ is analytic in the half plane $\mathrm{Im}(w) > -\alpha$ and f_- is analytic in the half plane $\mathrm{Im}(w) < 1$. Repeat this for $f(w) = \sqrt{1+w^2}$.

11.2 A two-dimensional wave of wavenumber k is incident on a semi-infinite barrier at an angle α. If the impedance boundary conditions

$$\phi(x, 0^{\pm}) = \pm i\delta\phi_y(x, 0^{\pm}), \qquad x > 0,$$

hold, show that the derivative of the potential, $\Phi = (\phi - \phi_{\mathrm{I}})_y$, satisfies the Wiener–Hopf problem

$$\Phi_-(w, 0) = -\left(1 + i\delta\sqrt{w^2 - k^2}\right)\left\{\Phi_+(w, 0^+) - \Phi_-(w, 0^-)\right\}$$
$$+ \frac{2ik\delta\sqrt{w^2 - k^2}\sin\alpha}{(w - k\sin\alpha)}.$$

Investigate the solutions of this and comment on the limits $\delta \to 0$ and $\delta \to \infty$.

11.3 A two-dimensional plane wave with wavenumber k, which propagates left to right in the strip $-\infty < x < \infty$, $-b < y < b$, is incident on a barrier $y = 0$, $x > 0$. The wave field ϕ which consists of the incident wave and the waves diffracted by the strip satisfies the equations

$$(\nabla^2 + k^2)\phi = 0 \ \text{ for } -\infty < x < \infty,\ -b < y < b,$$
$$\phi_y(x, \pm b) = 0,$$
$$\phi(x, 0) = 0 \ \text{ for } 0 \le x < \infty.$$

Use Fourier transforms and the Wiener–Hopf technique to find an explicit expression for $\phi(x, y)$.

11.4 A three-dimensional plane wave with wavenumber k is incident on a barrier containing a circular aperture of radius a. Find the diffraction pattern at some distance from the hole in the Fraunhofer limit when (a) the wave is acoustic, (b) the wave is electromagnetic (see exercise 3.9 for some hints).

11.5 Small amplitude water waves are incident on a surface piercing cylinder of radius b which extends down to the sea bed at $z = -h$. Find an expression for the scattered wave field and show that the magnitude of the force per unit length can be written

$$|F_x| = \frac{4ga \cosh k(z+h)}{k \cosh kh \sqrt{[J_1'(kb)]^2 + [Y_1'(kb)]^2}}$$

where k and a are the wave number and amplitude of the incident wave.

12

Solitons and the Inverse Scattering Transform

The KdV equation, (8.66), and the nonlinear Schrödinger equation (NLS), which we will derive in subsection 12.2.1, are nonlinear partial differential equations. Although there is no general analytical method for solving such equations, they are members of a family of exceptional cases where the solution can be deduced from that of a related *linear* equation. Such partial differential equations are said to be **integrable**, other notable examples being the intermediate long wave equation, the Benjamin–Ono equation, the Kadomtsev–Petviashvili equation, the Klein–Gordon equation, the sine–Gordon equation, the Boussinesq equation, the N-wave interaction equations, the Toda lattice equations, the self-dual Yang–Mills equations, and various generalisations of the nonlinear Schrödinger equation (for a review see Ablowitz and Clarkson (1991)). These include partial differential equations in one and two spatial dimensions, integro-differential equations and differential–difference equations that arise in models of physical situations ranging from fluid and solid mechanics to particle physics. (The sine–Gordon equation arises as a simplified nonlinear field equation in particle physics, and its soliton solutions are often known as Skyrmions after Tony Skyrme, Professor of Mathematical Physics at the University of Birmingham from 1964 to 1987 (see, for example, Perring and Skyrme (1962)).)

We will not consider the deep underlying reasons for the integrability of these various systems, and restrict ourselves to showing how to solve the KdV equation and the NLS equation using the **inverse scattering transform**.

12.1 The Korteweg–de Vries Equation

Before we begin, it is helpful to rewrite (8.66) in the standard form used in the literature on this subject by defining

$$x = z, \; t = \frac{1}{6}\tau, \; u = -\frac{3}{2}H_0,$$

so that the KdV equation becomes

$$u_t - 6uu_x + u_{xxx} = 0. \tag{12.1}$$

Note that x and t are *not* the same as those used in section 8.3.1. In addition, rather perversely, waves of elevation with $H_0 > 0$ correspond to negative values of u, so we shall plot $-u(x,t)$ in the following sections.

12.1.1 The Scattering Problem

Consider the ordinary differential equation

$$\psi_{xx} + (\xi^2 - u(x))\psi = 0, \tag{12.2}$$

where ξ is a complex constant, the **eigenvalue**, and $u(x)$ is the **scattering potential**. The choice of the symbol u for the potential is no coincidence, but, for the moment, we simply require that u should be a bounded function of x, absolutely integrable on the real line, and decaying rapidly enough that

$$\int_{-\infty}^{\infty} (1 + |x|)|u(x)|dx$$

is bounded. The reasons for the exact form of this condition are technical and need not concern us here (Faddeev, 1958).

In general, since $u \to 0$ as $x \to \pm\infty$, we expect that

$$\psi \sim Ae^{i\xi x} + Be^{-i\xi x} \quad \text{as } x \to \pm\infty,$$

for some constants A and B. If we now define $\psi_-(x;\xi)$ to be the solution that satisfies

$$\psi_- \sim e^{-i\xi x} \quad \text{as } x \to -\infty, \tag{12.3}$$

we must have

$$\psi_- \sim a(\xi)e^{-i\xi x} + b(\xi)e^{i\xi x} \quad \text{as } x \to \infty, \tag{12.4}$$

for some complex functions $a(\xi)$ and $b(\xi)$. This is a **scattering** or **connection problem** and $a(\xi)$ and $b(\xi)$ are the **scattering data** or **connection coefficients**. The KdV equation was the first nonlinear partial differential equation to which this method was applied (Gardner, Greene, Kruskal and Miura, 1967), and (12.2) arises in the context of quantum mechanical scattering, hence the name of the method. In a scattering problem the scattering data characterise the effect of the potential u on an incident wave, with a giving the amplitude of the transmitted wave and b of the reflected wave. We will not think of the problem in these terms here, although we will refer to the complex functions $a(\xi)$ and $b(\xi)$ as the scattering data (see Drazin and Johnson (1989) for the more usual description in terms of a scattering problem).

Before looking at some examples, there are three properties of the scattering data that we will need later.

(i) Note that if $a(\xi) = 0$ at $\xi = \xi_0$ and $\text{Im}(\xi_0) > 0$, $\psi_- \to 0$ as $x \to \pm\infty$. Now, taking appropriate multiples of (12.2) and its complex conjugate, we find that

$$\psi^* \psi_{xx} - \psi\psi^*_{xx} + \left(\xi_0^2 - \xi_0^{*2}\right)|\psi|^2 = 0, \tag{12.5}$$

where a star denotes the complex conjugate. If we now integrate this over the real line, using the fact that $\psi_- \to 0$ as $x \to \pm\infty$, we obtain

$$\left(\xi_0^2 - \xi_0^{*2}\right)\int_{-\infty}^{\infty} |\psi_-|^2 dx = 0. \tag{12.6}$$

Since the integrand is non-negative and not identically equal to zero, we must have $\xi_0^2 = \xi_0^{*2}$, so that ξ_0^2 is real. Since we assumed that $\text{Im}(\xi_0) > 0$, ξ_0 must be purely imaginary. In other words,

any zeros of $a(\xi)$ in the upper half plane lie on the imaginary axis. (12.7)

These zeros play a vital role in the solution of the KdV equation. They are referred to as the **bound state eigenvalues** or the **discrete spectrum**, since for these values of ξ, $\psi \to 0$ as $x \to \pm\infty$.

(ii) Let's assume that $a(\xi)$ has a simple zero at $\xi = i\kappa$, with κ real and positive. Then

$$a(\xi) \sim id(\xi - i\kappa) \quad \text{as } \xi \to i\kappa, \tag{12.8}$$

for some complex constant d. When $\xi = i(\kappa + \epsilon)$, with ϵ real,

$$a \sim -d\epsilon \quad \text{as } \epsilon \to 0.$$

This means that

$$\psi_- \sim -d\epsilon e^{(\kappa+\epsilon)x} \quad \text{as } x \to \infty.$$

Since $u(x)$ and ξ^2 are real, and ψ_- is real as $x \to -\infty$, it must also be real as $x \to \infty$, and hence d is real. We conclude that

the residue at a pole of $1/a(\xi)$ with $\mathrm{Im}(\xi) > 0$ is purely imaginary. (12.9)

(iii) As $|\xi| \to \infty$, $\psi_{xx} + \xi^2\psi \sim 0$, and hence $\psi \sim Ae^{-i\xi x} + Be^{i\xi x}$ for all x. For (12.3) and (12.4) to be consistent with this, we must have

$$a(\xi) \to 1 \text{ and } b(\xi) \to 0 \text{ as } |\xi| \to \infty. \tag{12.10}$$

Here are a couple of examples of scattering by localised potentials.

Example 1: Scattering by a Delta Function. Just about the simplest example is $u(x) = -U_0\delta(x)$, with U_0 a constant, where the potential is concentrated at $x = 0$ as a delta function. We are therefore interested in the solution of

$$\psi_{xx} + (U_0\delta(x) + \xi^2)\psi = 0, \tag{12.11}$$

subject to

$$\psi \sim e^{-i\xi x} \quad \text{as } x \to -\infty. \tag{12.12}$$

For $x \neq 0$, $\delta(x) = 0$, and the solution of this ordinary differential equation is

$$\psi_- = \begin{cases} a(\xi)e^{-i\xi x} + b(\xi)e^{i\xi x} & \text{for } x > 0, \\ e^{-i\xi x} & \text{for } x < 0. \end{cases}$$

To determine what happens at $x = 0$, and hence determine the scattering data, we integrate over $-\epsilon \leq x \leq \epsilon$, and find that

$$[\psi_x]_{-\epsilon}^{\epsilon} = -U_0\psi(0) - \xi^2 \int_{-\epsilon}^{\epsilon} \psi dx.$$

If we now let ϵ tend to zero, we find that ψ_x is discontinuous at $x = 0$, with

$$[\psi_x]_{0^-}^{0^+} = -U_0\psi(0). \tag{12.13}$$

A further integration shows that ψ is continuous at $x = 0$. These two conditions determine a and b as

$$a(\xi) = 1 - \frac{iU_0}{2\xi}, \quad b(\xi) = \frac{iU_0}{2\xi}.$$

Note that, as expected from (12.7), when $U_0 > 0$ and $a(\xi)$ has a zero in the upper half plane, this zero lies on the imaginary axis at $\xi = \frac{1}{2}iU_0$, and the residue of $1/a(\xi)$ at this pole is purely imaginary (see (12.9)).

In addition, consistent with (12.10), it is clear that $a \to 1$ and $b \to 0$ as $|\xi| \to \infty$.

Example 2: Scattering by $u = -U_0 \text{sech}^2 x$. In this case the potential is of a more physically realistic form, which, given the functional form of the solitary wave solution (8.78), we might expect is of some importance when we come to consider the solution of the KdV equation. We are interested in the solutions of

$$\psi_{xx} + \left(U_0 \text{sech}^2 x + \xi^2\right) \psi = 0. \tag{12.14}$$

We can write this in a more familiar form by making the change of variable $T = \tanh x$, so that $dT/dx = \text{sech}^2 x$, and hence (12.14) becomes

$$\frac{d}{dT}\left\{(1-T^2)\frac{d\psi}{dT}\right\} + \left(U_0 + \frac{\xi^2}{1-T^2}\right)\psi = 0. \tag{12.15}$$

This is the **associated Legendre equation**. Note that $-\infty < x < \infty$ maps to $-1 < T < 1$. The solution ψ_- can be written in terms of the hypergeometric function. We will not go into the details here (see Morse and Feshbach (1953)), but note that

$$a(\xi) = \frac{\Gamma(\tilde{c})\Gamma(-i\xi)}{\Gamma(\tilde{a})\Gamma(\tilde{b})}, \quad b(\xi) = \frac{\Gamma(\tilde{c})\Gamma(i\xi)}{\pi}\cos\left(\pi\sqrt{U_0 + \frac{1}{4}}\right), \tag{12.16}$$

where

$$\tilde{a} = \frac{1}{2} - i\xi + \sqrt{U_0 + \frac{1}{4}}, \quad \tilde{b} = \frac{1}{2} - i\xi - \sqrt{U_0 + \frac{1}{4}}, \quad \tilde{c} = 1 - i\xi.$$

The two things we need to bear in mind about the gamma function, $\Gamma(z)$, are firstly that it has simple poles at $z = 0, -1, -2, \dots$, and secondly that it has no finite zeros. We can now see that $b(\xi)/a(\xi) = 0$ if and only if $\cos\left(\pi\sqrt{U_0 + \frac{1}{4}}\right) = 0$, and hence $U_0 = N(N+1)$ for N a positive integer. A scattering potential $u(x)$ for which $b(\xi)/a(\xi) = 0$ when ξ is real is called a **reflectionless potential**, and $b(\xi)/a(\xi)$ is known as the **reflection coefficient**. The series of potentials $u(x) = -N(N+1)\text{sech}^2 x$ for $N = 1, 2, 3 \dots$, are therefore all reflectionless. We can also use (12.16) to determine the bound state eigenvalues by locating the zeros of $a(k)$ on the positive imaginary axis. These lie where $\Gamma(\tilde{a})$ and $\Gamma(\tilde{b})$ have poles. The poles of $\Gamma(\tilde{a})$ lie in the lower half plane, whilst the poles of $\Gamma(\tilde{b})$ are given by

$$\xi = i\left(\sqrt{U_0 + \frac{1}{4}} - m - \frac{1}{2}\right) \quad \text{for } m = 0, 1, \dots.$$

There is therefore at least one bound state eigenvalue if and only if $\sqrt{U_0 + \frac{1}{4}} > \frac{1}{2}$, and hence $U_0 > 0$. In this case, the number of poles is given by

$$\left[\sqrt{U_0 + \frac{1}{4}} - \frac{1}{2}\right] + 1,$$

where the square brackets denote the integer part of the quantity enclosed. Note that when $U_0 = N(N+1)$, there are N bound state eigenvalues.

12.1.2 The Inverse Scattering Problem

The question now arises, given the scattering data, is it possible to reconstruct the potential, $u(x)$? This is called the **inverse scattering problem** and we will find that it has a unique solution for potentials that decay sufficiently rapidly as $x \to \pm\infty$. Moreover, for given scattering data, the potential satisfies a linear integral equation. In order to construct this integral equation, we begin by considering the function

$$\psi_+(x;\xi) = e^{i\xi x} + \int_x^{\infty} K(x,y)e^{i\xi y}\,dy, \tag{12.17}$$

with $K(x,y) = 0$ for $y < x$. What conditions must $K(x,y)$ satisfy for $\psi_+(x;\xi)$ to be a solution of the scattering problem? If we substitute (12.17) into (12.2), and note that the total derivative at $y = x$ is related to the partial derivatives by

$$\frac{dK}{dx}(x,x) = \frac{\partial K}{\partial x}(x,x) + \frac{\partial K}{\partial y}(x,x),$$

we find that

$$\frac{\partial^2 \psi_+}{\partial x^2} - (u(x) - \xi^2)\psi_+ = -e^{i\xi x}\left\{u(x) + 2\frac{dK}{dx}(x,x)\right\}$$
$$+ \int_x^{\infty}\left\{\frac{\partial^2 K}{\partial x^2}(x,y) - u(x)K(x,y)\right\}e^{i\xi y}\,dy$$
$$+ \left[\frac{\partial K}{\partial y}(x,x)e^{i\xi x} - i\xi K(x,x)e^{i\xi x} + \xi^2\int_x^{\infty} K(x,y)e^{i\xi y}\,dy\right].$$

If we now take the integral within the square brackets and integrate by parts twice, assuming that $K \to 0$ and $\partial K/\partial y \to 0$ as $y \to \infty$, we arrive at

$$\frac{\partial^2 \psi_+}{\partial x^2} - (u(x) - \xi^2)\psi_+ = -e^{i\xi x}\left\{u(x) + 2\frac{dK}{dx}(x,x)\right\}$$
$$+ \int_x^{\infty}\left\{\frac{\partial^2 K}{\partial x^2}(x,y) - \frac{\partial^2 K}{\partial y^2}(x,y) - u(x)K(x,y)\right\}e^{i\xi y}\,dy.$$

This means that ψ_+ will be a solution of (12.2) if $K(x,y)$ is a solution of the boundary value problem (usually called a **Goursat problem**)

$$\frac{\partial^2 K}{\partial x^2} - \frac{\partial^2 K}{\partial y^2} - u(x)K = 0 \quad \text{for } y > x, \tag{12.18}$$

subject to

$$\frac{dK}{dx} = -\frac{1}{2}u(x) \quad \text{on } y = x, \tag{12.19}$$

with

$$K \to 0, \quad \frac{\partial K}{\partial y} \to 0 \quad \text{as } y \to \infty. \tag{12.20}$$

The theory of hyperbolic partial differential equations (see, for example, the book by Kevorkian (1990)) shows that the solution of this problem exists and is unique. This is all that we need to know. For a given function, $u(x)$, the function $K(x,y)$ can, in principle, be determined. More importantly, if we *don't* know $u(x)$, but can somehow determine $K(x,x)$, in particular from a knowledge of the scattering data, we can calculate $u(x)$ from (12.19), since

$$u(x) = -2\frac{dK}{dx}(x,x). \tag{12.21}$$

By taking the complex conjugate of (12.2), we can see that if $\psi(x;\xi)$ is a solution, then so is $\psi^*(x;\xi^*)$. This means that, since $\psi_+ \sim e^{i\xi x}$ as $x \to \infty$, we must have

$$\begin{aligned}\psi_-(x;\xi) &= a(\xi)\psi_+^*(x;\xi^*) + b(\xi)\psi_+(x;\xi) = a(\xi)e^{-i\xi x} + b(\xi)e^{i\xi x} \\ &\quad + a(\xi)\int_{-\infty}^{\infty} K(x,y)e^{-i\xi y}\,dy + b(\xi)\int_x^{\infty} K(x,y)e^{i\xi y}\,dy,\end{aligned} \tag{12.22}$$

where we have used the fact that $K(x,y) = 0$ for $x < y$ to extend the range of one of the integrals down to minus infinity. We can now think of this integral as a Fourier transform. The Fourier inversion formula then shows that

$$\begin{aligned}K(x,y) = \frac{1}{2\pi}\int_{-\infty}^{\infty} &\left\{\frac{1}{a(\xi)}\psi_- - e^{-i\xi x}\right. \\ &\left. -\frac{b(\xi)}{a(\xi)}e^{i\xi x} - \frac{b(\xi)}{a(\xi)}\int_x^{\infty} K(x,z)e^{i\xi z}\,dz\right\} e^{i\xi y}\,d\xi.\end{aligned} \tag{12.23}$$

It is now convenient to define

$$\hat{B}(X) = \frac{1}{2\pi}\int_{-\infty}^{\infty} \frac{b(\xi)}{a(\xi)}e^{i\xi X}\,d\xi, \tag{12.24}$$

which is proportional to the Fourier transform of $b(\xi)/a(\xi)$ for ξ real. Using this definition in (12.23), we arrive at

$$K(x,y)+\hat{B}(x+y)+\int_x^{\infty} K(x,z)\hat{B}(y+z)dz = R(x,y), \tag{12.25}$$

where

$$R(x,y)=\frac{1}{2\pi}\int_{-\infty}^{\infty}\left\{\frac{1}{a(\xi)}\psi_-(x;\xi)e^{i\xi x}-1\right\}e^{i\xi(y-x)}d\xi. \tag{12.26}$$

Since we can calculate $\hat{B}(X)$ for given scattering data, $a(\xi)$ and $b(\xi)$, (12.25) is a linear integral equation for $K(x,y)$, whose solution allows us to determine the scattering potential from (12.21), *provided that we can determine* $R(x,y)$ *for* $y > x$ *in terms of the scattering data only*. This is what we will do next.

Since $b(\xi)/a(\xi) \to 0$ and $\psi_- \sim e^{-i\xi x}$ as $|\xi| \to \infty$, we can close the contour of integration in the upper half ξ-plane, so that $R(x,y)$ can be determined from the sum of the residues at the poles of $1/a(\xi)$ in the upper half plane. As we have seen, these all lie on the positive imaginary axis, at $\xi = i\kappa_n$ for $n = 1, 2, \dots, N$, so that

$$R(x,y)=i\sum_{n=1}^{N}\left(\text{Residue at } \xi=i\kappa_n \text{ of}\left\{\frac{1}{a(\xi)}\psi_-(x;\xi)e^{i\xi y}\right\}\right). \tag{12.27}$$

In order to calculate this, we note that since $a(i\kappa_n)=0$, (12.22) shows that

$$\psi_-(x;i\kappa_n)=b(i\kappa_n)\psi_+(x;i\kappa_n)=b(i\kappa_n)\left\{e^{-\kappa_n x}+\int_x^{\infty}K(x,y)e^{-\kappa_n y}dy\right\}.$$

We also know from (12.8), assuming that a has a simple zero at $\xi = i\kappa_n$, that

$$\frac{1}{a}\sim\frac{1}{id_n(\xi-i\kappa_n)} \quad \text{as } \xi\to i\kappa_n,$$

with d_n real. Although we will not prove it here, it can be shown (see the book by Drazin and Johnson (1989)) that $b(i\kappa_n)/d_n$ must be positive. We therefore write $b(i\kappa_n)/d_n = c_n^2$, and arrive at

$$R(x,y)=\sum_{n=1}^{N}c_n^2\left\{e^{-\kappa_n x}+\int_x^{\infty}K(x,y)e^{-\kappa_n z}\,dz\right\}e^{-\kappa_n y}. \tag{12.28}$$

The constants c_n are called the **coefficients of normalisation** or **normalisation constants**.

We can now use this in (12.25) and find that

$$K(x,y) + \hat{B}(x+y) + \int_x^\infty K(x,z)\hat{B}(y+z)dz$$
$$= -\sum_{n=1}^{N} c_n^2 \left\{ e^{-\kappa_n x} + \int_x^\infty K(x,z)e^{-\kappa_n z}dz \right\} e^{-\kappa_n y}. \quad (12.29)$$

Comparing this with (12.24), the definition of $\hat{B}$, we can finally see that by defining the function

$$B(X) = \frac{1}{2\pi}\int_{-\infty}^{\infty} \frac{b(\xi)}{a(\xi)} e^{i\xi X} d\xi + \sum_{n=1}^{N} c_n^2 e^{-\kappa_n X}, \quad (12.30)$$

we can write (12.25) as

$$K(x,y) + B(x+y) + \int_x^\infty K(x,z)B(y+z)dz = 0, \quad (12.31)$$

the **Gel'fand–Levitan–Marchenko (GLM) equation**. This is a linear integral equation that must be solved to find $K(x,y)$. The potential is then given by (12.21). Note that the equation only depends upon the N discrete eigenvalues, κ_n and their coefficients of normalisation, c_n, along with the reflection coefficient, $b(\xi)/a(\xi)$.

In general, (12.31) has no analytical solution in closed form. However, we shall see below that an analytical solution can be constructed if $B(z+y)$ is separable, in the sense that

$$B(y+z) = \sum_{j=1}^{N} Y_j(y)Z_j(z).$$

In particular if $b(\xi)/a(\xi)$ is zero when ξ is real, a reflectionless potential,

$$B(y+z) = \sum_{n=1}^{N} c_n^2 e^{-\kappa_n y} e^{-\kappa_n z}.$$

We conclude this subsection with two examples to illustrate the solution of inverse scattering problems relevant to the solution of the KdV equation.

Example 3: Inverse Scattering for a Reflectionless Potential with a Single Bound State. Consider a reflectionless potential, $b(\xi)/a(\xi) = 0$, with a single bound state eigenvalue, $\xi = i\kappa$, with normalisation constant c. In this case the GLM equation (12.31) is

$$K(x,y) + c^2 e^{-\kappa(x+y)} + \int_x^\infty K(x,z)c^2 e^{-\kappa(y+z)}dz = 0, \tag{12.32}$$

which we must solve for $K(x, y)$. We can do this by looking for a separable solution,

$$K(x,y) = L(x)e^{-\kappa y}.$$

This immediately allows us to factor the y-dependence out of (12.32) to leave

$$L(x) + c^2 e^{-\kappa x} + c^2 L(x) \int_x^\infty e^{-2\kappa z}dz = 0,$$

and hence

$$L(x) = -\frac{2\kappa c^2 e^{-\kappa x}}{2\kappa + c^2 e^{-2\kappa x}}.$$

This means that

$$u(x) = -2\frac{d}{dx}K(x,x) = -\frac{16\kappa^3 c^2}{\left(c^2 e^{-\kappa x} + 2\kappa e^{\kappa x}\right)^2}, \tag{12.33}$$

which we can rearrange into

$$u(x) = -2\kappa^2 \operatorname{sech}^2\{\kappa(x - x_0)\}, \tag{12.34}$$

where $x_0 = -\log(2\kappa/c^2)/2\kappa$, a constant that we can remove from (12.34) be redefining the origin to be at $x = x_0$. With $\kappa = 1$ and $c = \sqrt{2}$, this is simply the reflectionless potential with $N = 1$ that we studied in example 2. We can now see that this is one of a unique family of reflectionless potentials with a single bound state, given by (12.34). As κ increases, the minimum of the potential becomes more negative, whilst becoming localised in a narrower region about the origin.

Example 4: Inverse Scattering for a Reflectionless Potential with N Bound States. Consider a reflectionless potential, $b(\xi)/a(\xi) = 0$, with N bound states, $\xi = i\kappa_n$, and normalisation constants c_n, for $n = 1, 2, \dots, N$. In this case, the GLM equation is

$$K(x,y) + \sum_{n=1}^{N} c_n^2 e^{-\kappa_n(x+y)} + \int_x^\infty K(x,z)\sum_{n=1}^{N} c_n^2 e^{-\kappa_n(y+z)}dz.$$

We seek a separable solution of the form

$$K(x,y) = \sum_{m=1}^{N} L_m(x)e^{-\kappa_m y},$$

so that

$$\sum_{m=1}^{N} L_m(x)e^{-\kappa_m y} + \sum_{m=1}^{N} c_m^2 e^{-\kappa_m(x+y)} + \sum_{m=1}^{N} c_m^2 e^{-\kappa_m y} \sum_{n=1}^{N} L_n(x) \int_x^\infty e^{-(\kappa_n+\kappa_m)z}\,dz = 0.$$

Since this equation must hold for all x and y, we equate coefficients of $e^{-\kappa_m y}$ and, after evaluating the integral, arrive at a system of N equations in the N unknowns $L_m(x)$, given by

$$c_m^2 e^{-\kappa_m x} + L_m(x) + c_m^2 \sum_{n=1}^{N} L_n(x) \frac{e^{-(\kappa_n+\kappa_m)x}}{\kappa_n + \kappa_m} = 0 \text{ for } m = 1, 2, \dots, N.$$

The easiest way to deal with this system is to introduce a matrix formalism and use suffix notation, so that

$$C_m + M_{mn} L_n = 0,$$

with summation over the repeated index, $n = 1, 2, \dots, N$, and

$$M_{mn} = \delta_{mn} + \frac{c_m^2 e^{-(\kappa_m+\kappa_n)x}}{\kappa_m + \kappa_n}, \quad C_m = c_m^2 e^{-\kappa_m x}. \tag{12.35}$$

The solution is

$$L_n(x) = -M_{nm}^{-1} C_m,$$

and the quantity that we are interested in is

$$K(x, x) = \sum_{n=1}^{N} L_n(x) e^{-\kappa_n x}.$$

If we introduce the vector

$$h_n = e^{-\kappa_n x},$$

we can write

$$K(x, x) = h_n L_n = -h_n M_{nm}^{-1} C_m.$$

Now we note that the matrix M_{mn} has the helpful property

$$\frac{dM_{mn}}{dx} = -c_m^2 e^{-(\kappa_m+\kappa_n)x} = -C_m h_n,$$

and hence

$$K(x, x) = -M_{nm}^{-1} \frac{dM_{mn}}{dx}.$$

It is a standard result from linear algebra that

$$M_{nm}^{-1}\frac{dM_{mn}}{dx} = \frac{d}{dx}\left\{\log\left(\det M\right)\right\},$$

and hence

$$u(x) = -2\frac{d^2}{dx^2}\left\{\log\left(\det M\right)\right\}. \tag{12.36}$$

Note that Hirota (1973) used a formalism based upon this matrix notation to find N soliton solutions of other nonlinear partial differential equations (see the book by Ablowitz and Clarkson (1991) for a detailed discussion of Hirota's method).

When there are two bound states, it is feasible to evaluate (12.36) by hand. The potential is given by

$$\begin{aligned} u(x) = -\left(\det M\right)^{-2}\Bigg\{ & 4\kappa_1 c_1^2 e^{-2\kappa_1 x} + 4\kappa_2 c_2^2 e^{-2\kappa_2 x} \\ & + \frac{4c_1^2c_2^2(\kappa_1-\kappa_2)^2}{\kappa_1\kappa_2}e^{-2(\kappa_1+\kappa_2)x} + \frac{c_1^4c_2^2\kappa_2(\kappa_1-\kappa_2)^2}{\kappa_1^2(\kappa_1+\kappa_2)^2}e^{-(4\kappa_1+2\kappa_2)x} \\ & + \frac{c_1^2c_2^4\kappa_1(\kappa_1-\kappa_2)^2}{\kappa_2^2(\kappa_1+\kappa_2)^2}e^{-(2\kappa_1+4\kappa_2)x}\Bigg\}, \end{aligned} \tag{12.37}$$

where

$$\det M = 1 + \frac{c_1^2}{2\kappa_1}e^{-2\kappa_1 x} + \frac{c_2^2}{2\kappa_2}e^{-2\kappa_2 x} + \frac{c_1^2c_2^2}{4\kappa_1\kappa_2}\frac{(\kappa_1-\kappa_2)^2}{(\kappa_1+\kappa_2)^2}e^{-2(\kappa_1+\kappa_2)x}. \tag{12.38}$$

The potential when $\kappa_1 = 1.25$, $\kappa_2 = 1.75$, $c_1 = \sqrt{6}$ and $c_2 = 2\sqrt{3}$ is shown in figure 12.1. To determine the potential for $N > 2$, it is safer to use a computer algebra package (see example 7).

12.1.3 Scattering Data for KdV Potentials

The choice of the symbol u in (12.1) and (12.2) is no coincidence. Let's consider the eigensolutions of (12.2) when the potential is $u(x,t)$ and u satisfies the KdV equation, (12.1). At any time t, the potential $u(x,t)$ is associated with the scattering data $a(\xi,t)$ and $b(\xi,t)$ and solutions $\psi(x,t;\xi)$ of the scattering problem. As far as the scattering problem is concerned, the t dependence is simply a parameter. After finding the scattering data that corresponds to the initial condition for the KdV equation, $u = u(x,0)$, the crucial question is: how does this scattering data depend upon t if u satisfies the KdV equation? We will show that the time evolution of the scattering data is extremely simple. Finally, to

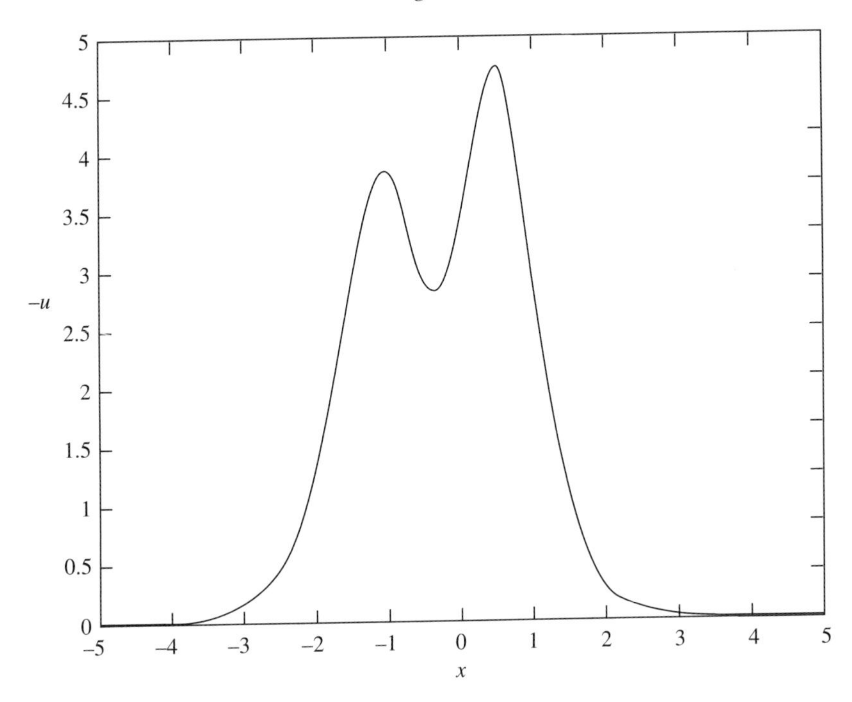

Fig. 12.1. The reflectionless scattering potential, $-u(x)$, when there are two bound states with $\kappa_1 = 1.25$, $\kappa_2 = 1.75$, $c_1 = \sqrt{6}$ and $c_2 = 2\sqrt{3}$.

find the solution of the KdV equation when $t > 0$ we must solve the inverse scattering problem for this time dependent scattering data, which may or may not result in a closed form analytical solution. This is the essence of the inverse scattering transform method. A useful analogy is with the solution of linear, second order partial differential equations using integral transforms, for example Fourier or Laplace. On taking the transform we obtain an ordinary differential equation that governs the evolution of the transformed variable and is straightforward to solve. We then invert the transform using the inversion formula, which we may or may not be able to reduce to a closed form analytical solution.

We can postulate any partial differential equation $\psi_t = F(u, \psi)$ for the time evolution of ψ, provided that it is consistent with (12.2) and the KdV equation. We will not go into how such an equation can be determined, but simply show that

$$\psi_t = 2(u + 2\xi^2)\psi_x - u_x\psi, \tag{12.39}$$

is the appropriate choice. For consistency we require that

$$\frac{\partial \psi_{xx}}{\partial t} = \frac{\partial^2 \psi_t}{\partial x^2},$$

where ψ_{xx} and ψ_t are given by (12.2) and (12.39). This leads to the requirement that

$$u_t\psi + (u - \xi^2)\psi_t = 2(u + 2\xi^2)\psi_{xxx} + 3u_x\psi_{xx} - u_{xxx}\psi.$$

If we now use (12.2) and (12.39) to write ψ_{xxx}, ψ_{xx} and ψ_t in terms of ψ and ψ_x, we find that

$$(u_t - 6uu_x + u_{xxx})\psi = 0.$$

Clearly, (12.39) is consistent with (12.2), provided that $u(x,t)$ satisfies the KdV equation.

Now it is straightforward to determine how $a(\xi,t)$ and $b(\xi,t)$ evolve with time. As $x \to \infty$, $u \to 0$, so (12.39) shows that

$$\psi_t \sim 4\xi^2\psi_x \quad \text{as } x \to \infty.$$

But we know that $\psi_- \sim ae^{-i\xi x} + be^{i\xi x}$ as $x \to \infty$, so that we must have

$$a_t = -4i\xi^3 a, \quad b_t = 4i\xi^3 b,$$

and hence

$$a(\xi,t) = a(\xi,0)e^{-4i\xi^3 t}, \quad b(\xi,t) = b(\xi,0)e^{4i\xi^3 t}. \tag{12.40}$$

In particular, the bound state eigenvalues (the zeros of $a(\xi,t)$) are independent of t, and from the definition of c_n we deduce that $c_n(t) = c_n(0)e^{4\kappa_n^3 t}$. To summarise, the scattering data evolves according to

$$\kappa_n(t) = \kappa_n(0), \; c_n(t) = c_n(0)e^{4\kappa_n^3 t}, \; b(\xi,t)/a(\xi,t) = b(\xi,0)e^{8i\xi^3 t}/a(\xi,0). \tag{12.41}$$

We conclude by going through some specific examples of the use of the inverse scattering transform to solve the KdV equation, based on the examples given in the previous subsection.

12.1.4 Examples: Solutions of the KdV Equation

Example 5: the Reflectionless Potential $u(x,0) = -2\kappa_1^2\mathrm{sech}^2\kappa_1(x-x_0)$*; the Single Soliton Solution.* We know from example 3 that $u(x,0) = -2\kappa_1^2\mathrm{sech}^2\kappa_1(x-x_0)$ is a reflectionless potential, $b(\xi,0)/a(\xi,0) = 0$, with a single bound state eigenvalue, $\kappa(0) = \kappa_1$, and normalisation constant $c(0) = \sqrt{2\kappa_1}e^{-\kappa_1 x_0}$ (see exercise 12.3). Equations (12.41) show that if the initial profile is a reflectionless potential, it remains reflectionless when $t \neq 0$, with $b(\xi,t)/a(\xi,t) = 0$. Also $u(x,t)$ has a single, constant bound state eigenvalue, $\kappa(t) = \kappa_1$. The only part of the scattering data that changes with time is the normalisation constant, with

$$c(t) = \sqrt{2\kappa_1}e^{-\kappa_1(x_0-4\kappa_1^2 t)}. \tag{12.42}$$

We can now find the development of this initial profile for $t \neq 0$ by solving the inverse scattering problem. Note that we can wind time backwards as well as forwards from $t = 0$ and obtain the solution for $-\infty < t < \infty$. Fortunately, we have already solved the inverse scattering problem for $u(x,0)$ in example 3. The time dependence of $c(t)$ does not affect the solution of the GLM equation, since t only appears as a parameter. All we have to do is substitute for $c(t)$ from (12.42) into (12.33) to give

$$u(x,t) = \frac{16\kappa_1^3 c(0)^2 e^{-2\kappa_1(x-4\kappa_1^2 t)}}{\left(c(0)^2 e^{-2\kappa_1(x-4\kappa_1^2 t)} + 2\kappa_1\right)^2}, \tag{12.43}$$

and hence

$$u(x,t) = -2\kappa_1^2 \text{sech}^2\left\{\kappa_1\left(x - x_0 - 4\kappa_1^2 t\right)\right\}. \tag{12.44}$$

This is just the initial profile (12.34) propagating to the right without change of form at speed $4\kappa_1^2$. In fact it is the solitary wave solution, (8.78), written in terms of u.

Example 6: Reflectionless potentials with Two Bound States; the Interaction of two Solitons. We have now seen how a reflectionless potential with a single bound state corresponds to the propagation of a solitary wave. Let's now consider the reflectionless potential given by (12.37). Such a potential remains reflectionless, $b(\xi,t)/a(\xi,t) = 0$, and retains its two bound state eigenvalues, κ_1 and κ_2, throughout the time development of the solution. If we look carefully, we can see that c_1 and c_2 only appear in (12.37) in the combinations $c_1 e^{-\kappa_1 x}$ and $c_2 e^{-\kappa_2 x}$. These two combinations evolve as

$$c_1 e^{-\kappa_1(x-4\kappa_1^2 t)} \text{ and } c_2 e^{-\kappa_2(x-4\kappa_2^2 t)}.$$

This means that all we have to do in (12.37) to obtain the solution when $t \neq 0$ is to make the substitutions

$$\kappa_1 x \mapsto \kappa_1(x - 4\kappa_1^2 t), \quad \kappa_2 x \mapsto \kappa_2(x - 4\kappa_2^2 t). \tag{12.45}$$

This is precisely what we did in the previous subsection to show that the initial profile simply propagated along as a solitary wave. The expression that we now obtain by using (12.45) in (12.37) is a function of the two travelling wave coordinates, $x-4\kappa_1^2 t$ and $x-4\kappa_2^2 t$. It is now instructive to consider the behaviour of the solution for $t \gg 1$. Unless $x-4\kappa_1^2 t = O(1)$ or $x-4\kappa_2^2 t = O(1)$, we have

$$u(x,t) \sim -\frac{16\kappa_1^3(\kappa_1+\kappa_2)^2}{c_1^2(\kappa_1-\kappa_2)^2} e^{-16\kappa_1^3 t} \quad \text{as } t \to \infty,$$

assuming that the bound state eigenvalues are ordered so that $\kappa_1 < \kappa_2$. The solution is therefore exponentially small away from an $O(1)$ neighbourhood of the two points $x = 4\kappa_1^2 t$ and $x = 4\kappa_2^2 t$.

When $x-4\kappa_2^2 t = O(1)$, we have $x-4\kappa_1^2 t \sim 4(\kappa_2^2-\kappa_1^2)t \gg 1$, and

$$u(x,t) \sim -2\kappa_2^2 \operatorname{sech}^2\left\{\kappa_2(x-4\kappa_2^2 t)+\delta\right\},$$

where δ is a constant (see exercise 12.5). This is just the single soliton solution corresponding to the bound state eigenvalue κ_2, translated by an $O(1)$ distance in the x-direction. Similarly, when $x-4\kappa_1^2 t = O(1)$ we obtain the solution for a single soliton with eigenvalue κ_1 at leading order. Analogous results hold for t large and negative. The solution corresponding to the reflectionless potential $u(x,0) = -6\operatorname{sech}^2 x$ is shown in figure 8.19. The solution for $\kappa_1 = 1.25$, $\kappa_2 = 1.75$, $c_1 = \sqrt{6}$ and $c_2 = 2\sqrt{3}$, with reflectionless potential $u(x,0)$, which is plotted in figure 12.1, is shown as a grey scale plot in figure 12.2. In each case, there are two solitary waves, widely spaced for t large and negative. The larger, faster wave catches the shorter, slower wave, there is a *nonlinear* interaction, and the waves separate as $t \to \infty$, whilst retaining their identities and emerging unchanged by the interaction. It is this particle-like behaviour that characterises these solitary waves as solitons. Figure 12.2 shows that there is a phase change due to the interaction, by which we mean that the faster wave emerges from the interaction displaced further to the right than it would have been in the absence of the slower wave, whilst the slower wave is displaced to the left. This phase change is a consequence of the nonlinearity of the interaction, and always occurs when solitons collide. When two *linear* waves interact, for example two counter-propagating solutions of the one-dimensional wave equation, $f(x-ct)+g(x+ct)$, although the waves emerge unchanged by the interaction, there is no phase change.

Example 7: General Reflectionless Potentials – the Interaction of N Solitons. The general reflectionless potential with N distinct bound state eigenvalues is given by (12.35) and (12.36). From the discussion in the

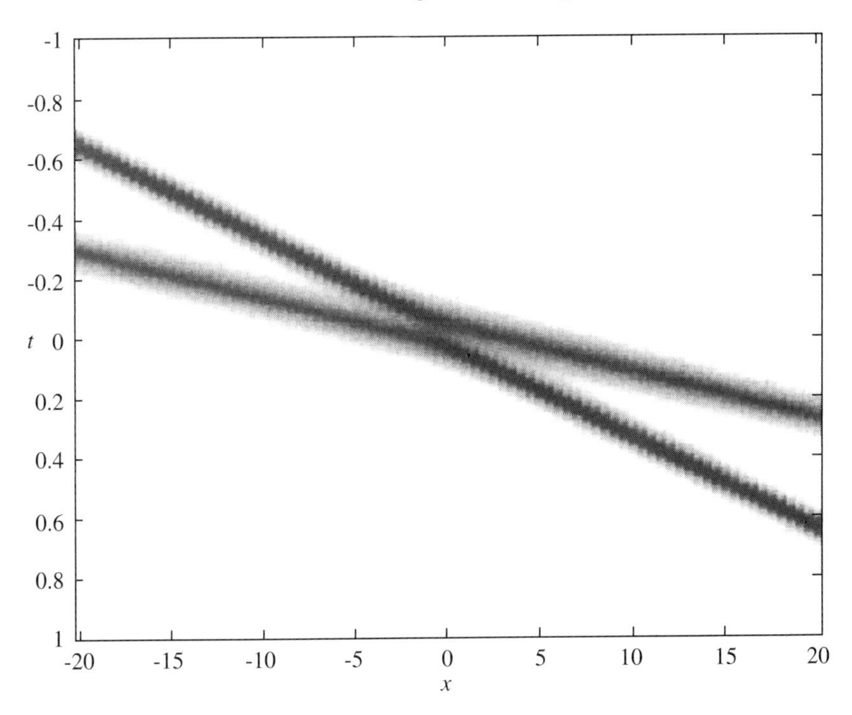

Fig. 12.2. The solution of the KdV equation corresponding to two bound states with $\kappa_1 = 1.25$, $\kappa_2 = 1.75$, $c_1(0) = \sqrt{6}$ and $c_2(0) = 2\sqrt{3}$. The larger $-u$, the darker the plot. The profile $u(x, 0)$ is shown in figure 12.1.

previous subsection, it is clear that to obtain $u(x, t)$ for $t \neq 0$ we simply need to make the substitution

$$\kappa_n x \mapsto \kappa_n(x - 4\kappa_n^2 t). \tag{12.46}$$

For t large and negative the solution consists of N solitons, which interact nonlinearly as t increases, eventually separating as $t \to \infty$. The reflectionless potential $u(x, 0) = -N(N + 1)\text{sech}^2 x$ is a special case where all of the solitons combine at $t = 0$ into a profile with a single local minimum, before emerging again, all with phase shifts. A more typical interaction between three solitons is plotted in figure 12.3 for the arbitrarily chosen values $\kappa_1 = 1.25$, $\kappa_2 = 1.75$, $\kappa_3 = 2.25$, $c_1 = \sqrt{6}$, $c_2 = 2\sqrt{3}$ and $c_3 = 1$. We have used (12.35) and (12.36) to determine $u(x, t)$, with the aid of a computer algebra package. Note that, since the KdV equation, (12.1), is unchanged by the transformation $x \mapsto -x$, $t \mapsto -t$, the solution must satisfy $u(x, t) = u(-x, -t)$. This symmetry is evident in figures 8.19, 12.2

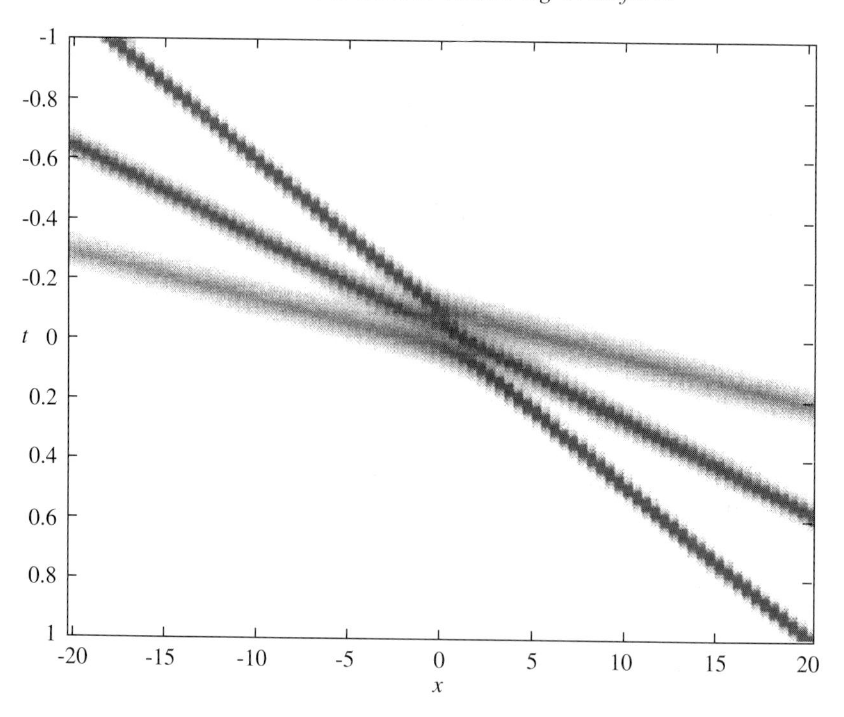

Fig. 12.3. The solution of the KdV equation corresponding to three bound states with $\kappa_1 = 1.25$, $\kappa_2 = 1.75$, $\kappa_3 = 2.25$, $c_1(0) = \sqrt{6}$, $c_2(0) = 2\sqrt{3}$ and $c_3(0) = 1$. The larger $-u$, the darker the plot.

and 12.3. The curious reader can consult the book by Shen (1993), where the first seven N soliton solutions are written out in all their glory. The $N = 7$ solution occupies nine printed pages!

Example 8: The Delta Function Potential, $u(x, 0) = -U_0\delta(x)$ – the Generation of Cnoidal Waves. We have seen in example 1 that, provided $U_0 > 0$, the initial scattering data for $u(x, 0) = -U_0\delta(x)$ is a single bound state eigenvalue, $\kappa = \frac{1}{2}U_0$, with normalisation constant $c(0) = \sqrt{\kappa}$ and reflection coefficient

$$\frac{b(\xi, 0)}{a(\xi, 0)} = -\frac{U_0}{U_0 + 2i\xi}.$$

The evolving scattering data retains its single bound state eigenvalue, whilst

$$c(t) = \sqrt{\kappa}e^{4\kappa^3 t}, \quad \frac{b(\xi, t)}{a(\xi, t)} = -\frac{U_0}{U_0 + 2i\xi}e^{8i\xi^3 t}. \tag{12.47}$$

We must now try to solve the inverse scattering problem to find the solution $u(x,t)$. The function $B(X,t)$, which is given by (12.30) and appears in the GLM equation, (12.31), is

$$B(X,t) = \kappa e^{8\kappa^3 t - \kappa X} - \frac{U_0}{2\pi}\int_{-\infty}^{\infty} \frac{e^{8i\xi^3 t + i\xi X}}{U_0 + 2i\xi}\,d\xi. \tag{12.48}$$

Now, since the GLM equation is linear, we can identify the first term in this expression, associated with the bound state eigenvalue, with a single soliton. Specifically, if we write

$$u(x,t) = -2\kappa_1^2 \text{sech}^2\left\{\kappa\left(x - x_0 - 4\kappa^2 t\right)\right\} + u_c(x,t), \tag{12.49}$$

then $u_c(x,t)$ is determined in the usual way from the GLM equation with

$$B(X,t) = B_c(X,t) = -\frac{U_0}{2\pi}\int_{-\infty}^{\infty} \frac{e^{8i\xi^3 t + i\xi X}}{U_0 + 2i\xi}\,d\xi. \tag{12.50}$$

Moreover, if $U_0 < 0$, the solution is given by $u = u_c(x,t)$ alone. We would now like to know how u_c behaves for $t \gg 1$. If we consider B_c evaluated at $X = vt$, then

$$B_c(vt,t) = -\frac{U_0}{2\pi}\int_{-\infty}^{\infty} \frac{e^{it(8\xi^3 + \xi v)}}{U_0 + 2i\xi}\,d\xi,$$

and the integral can be approximated using the method of stationary phase. We find that for $v > 0$, and hence $x > 0$, the integral is exponentially small as $t \to \infty$, since there are no real points of stationary phase, whilst for $v < 0$ the integral is of $O(t^{-1/2})$, with two real points of stationary phase at $\xi = \pm\sqrt{-v/24}$. This means that B_c is uniformly small, and we can neglect the integral involving the product of K_c and B_c in the GLM equation, and find that

$$u_c(x,t) \sim 2\frac{d}{dx}B_c(2x,t) = -\frac{2iU_0}{\pi}\int_{-\infty}^{\infty} \frac{e^{8i\xi^3 t + 2i\xi x}}{U_0 + 2i\xi}\,d\xi \text{ as } t \to \infty. \tag{12.51}$$

We can use the method of stationary phase to analyse this integral, and find that u_c is exponentially small for $x > 0$, and

$$u_c(x,t) \sim \frac{4U_0}{U_0^2 + 4\xi_0^2}\sqrt{\frac{\xi_0}{24\pi t}} \times \left\{U_0 \sin\left(\frac{4}{3}\xi_0 vt + \frac{\pi}{4}\right) - 2\xi_0 \cos\left(\frac{4}{3}\xi_0 vt + \frac{\pi}{4}\right)\right\} \text{ for } t \gg 1, \tag{12.52}$$

when $x = vt < 0$, with $\xi_0 = \sqrt{-v/12}$. We conclude that the component of the solution due to the reflection coefficient, $b(\xi,t)/a(\xi,t)$, represents a cnoidal wave, whose amplitude is of $O(t^{-1/2})$ for $t \gg 1$, when it decays away under the action of dispersion. When $U_0 > 0$ the full solution for $t \gg 1$ is dominated by the single soliton that propagates away at speed $4\kappa^2$. When $U_0 < 0$, no solitons are formed, and $u \to 0$ as $t \to \infty$ for all x. Somewhat better estimates of the amplitude of the cnoidal wave can be made when this analysis is carried out with greater rigour, as demonstrated in the book by Schuur (1986).

This analysis is dependent neither on the exact form of $b(\xi, 0)/a(\xi, 0)$ nor on the number of bound state eigenvalues. As $t \to \infty$, the cnoidal waves, which correspond to the contribution to $B(X, t)$ due to $b(k, t)$, are of $O(t^{-1/2})$ for $x < 0$ and exponentially small for $x > 0$. We can think of this as a decaying 'tail' of waves left behind by the initial disturbance. The contribution to $B(X, t)$ due to each bound state eigenvalue results in a soliton which propagates away in the positive x-direction at the appropriate speed.

This leads us to the only consistent way of defining a soliton, in general. A soliton is the contribution to the solution that results from a distinct, bound state eigenvalue of the associated scattering problem for the initial condition $u(x, 0)$. It is the invariance of both the number of eigenvalues and the size of each that gives solitons their ability to survive nonlinear interactions and propagate unchanged apart from a change of phase.

12.2 The Nonlinear Schrödinger Equation

The **nonlinear Schrödinger equation** (NLS) arises in many different contexts. It is the equation that governs the envelope of wavepackets in the presence of the competing effects of linear dispersion and nonlinear amplitude dependence of the material properties in a one-dimensional system. We begin by deriving NLS for the propagation of a plane electromagnetic wave in a nonlinear medium.

12.2.1 Derivation of the Nonlinear Schrödinger Equation for Plane Electromagnetic Waves

In chapter 6 we saw how Maxwell's equations are modified when the electric and magnetic fields exist in an insulating medium to give (6.41) to (6.45). In particular, we wrote $\mathbf{D} = \epsilon \mathbf{E}$, and assumed that the dielectric constant, ϵ, was indeed constant, or at most dependent on the frequency, but not the intensity, of an electromagnetic wave. This is a reasonable approximation if the electric field is not too large, as is the case for most naturally occurring visible light. Before 1960, linear optics was used successfully to study most systems of interest. The situation changed with the invention of the laser in 1960, since a laser can produce extremely intense visible light. In particular, the response of the almost transparent glass used in optical fibres is slightly, but for laser light significantly, nonlinear. In this case, we write $\mathbf{D} = \epsilon_0 \mathbf{E} + \mathbf{P}$, where $\mathbf{P}$ is the **polarisation**. For a material with an isotropic, centrosymmetric crystal structure, it can be shown that (see the books by Sauter (1996), and Newell and Moloney (1992))

$$\frac{1}{\epsilon_0}\mathbf{P} = \int_{-\infty}^{\infty} \chi_1(t-\tau)\mathbf{E}(\tau)d\tau$$
$$+ \int_{-\infty}^{\infty}\int_{-\infty}^{\infty}\int_{-\infty}^{\infty} \chi_3(t-\tau_1, t-\tau_2, t-\tau_3)\left(\mathbf{E}(\tau_1)\cdot\mathbf{E}(\tau_2)\right)\mathbf{E}(\tau_3)d\tau_1 d\tau_2 d\tau_3. \tag{12.53}$$

The first term is linear, and expresses the fact that $\mathbf{P}$ does not respond instantaneously to changes in $\mathbf{E}$. Causality requires that $\chi_1(T) = 0$ for $T < 0$ so that $\mathbf{P}$ only responds to changes in $\mathbf{E}$ that have already occurred. In addition, $\chi_1(T) \to 0$ as $T \to \infty$ over the time scale on which the electrons in the material respond to the effect of changes in the electric field. We shall see below that this leads to a frequency dependent refractive index, $n = n_0(\omega)$. In addition, χ_1, and hence n_0, is complex, with the imaginary part leading to decay, or attenuation, of a wave as it propagates through the material. This is an important effect when considering the propagation of wavepackets through an optical fibre hundreds of kilometres long, but we will not consider it here. The imaginary part of χ_1 is extremely small, since modern optical fibres are manufactured from very pure materials and are almost completely transparent.

The second term in (12.53) represents a cubic nonlinearity in the properties of the medium, and it can be shown that this leads to a nonlinear refractive index of the form

$$n = n_0(\omega) + n_2(\omega)|\mathbf{E}|^2. \tag{12.54}$$

This is known as the **Kerr effect** (see, for example, Hasegawa (1990)). Note that it is the requirement that the model should be unchanged by the reflection $\mathbf{E} \mapsto -\mathbf{E}$ that leads to a cubic, rather than quadratic, nonlinearity in the polarisation and hence to a quadratic nonlinearity in the refractive index. For $n_2 > 0$, this means that the larger the amplitude of the wave, the more slowly it propagates. This can lead to many interesting two and three-dimensional focusing effects, which we will not discuss here. For details, see the book by Newell and Moloney (1992).

We confine our attention to the one-dimensional propagation of electromagnetic waves in nonlinear media. Although NLS can be derived for the propagation of wavepackets in a cylindrical fibre, it is technically easier to consider the propagation of a plane wave, $\mathbf{E} = (0, E(z,t), 0)$. The governing equation is

$$\begin{aligned}\frac{\partial^2 E}{\partial z^2} - \frac{1}{c_0^2}\frac{\partial^2 E}{\partial t^2} = \frac{1}{c_0^2}\frac{\partial^2}{\partial t^2}\int_{-\infty}^{\infty}\chi_1(t-\tau)E(z,\tau)d\tau \\ +\frac{1}{c_0^2}\frac{\partial^2}{\partial t^2}\int_{-\infty}^{\infty}\int_{-\infty}^{\infty}\int_{-\infty}^{\infty}\chi_3(t-\tau_1, t-\tau_2, t-\tau_3) \\ \times E(z,\tau_1)E(z,\tau_2)E(z,\tau_3)d\tau_1 d\tau_2 d\tau_3,\end{aligned} \tag{12.55}$$

and we wish to solve it subject to the initial condition

$$E(z,0) = \delta A_0(z/L)e^{ikz}, \tag{12.56}$$

where A_0 may be complex. This form of initial condition represents a wavepacket with an underlying wave of wavenumber k, modulated by an initial **envelope function**, A_0, over a length scale L, as shown in figure 12.4. (Note that such a solution can be written in wavepacket form, (1.11), with the function $A(k)$ only non-zero for a range of wavenumbers of $O(L^{-1})$.) We will fix the small parameter, δ, later, by choosing the appropriate balance between terms in the governing equation. We also define $\epsilon = 1/kL \ll 1$, which expresses the fact that there is an underlying carrier wave whose amplitude is modulated over a length scale much longer than its wavelength.

We now define the dimensionless variables

$$\left.\begin{aligned}\bar{E} = n_2^{1/2}E,\ \bar{A}_0 = n_2^{1/2}A_0,\ \bar{z} = z/L,\ \bar{t} = c_0 t/L,\ \bar{\tau}_i = kc_0\tau_i, \\ \bar{\chi}_1(\bar{\tau}) = \chi_1(\tau)/kc_0,\ \bar{\chi}_3(\bar{\tau}_i) = \chi_3(\tau_i)/k^3c_0^3 n_2,\end{aligned}\right\} \tag{12.57}$$

with all barred variables of $O(1)$ for $\epsilon \ll 1$ and $\delta \ll 1$. In using $n_2^{-1/2}$ to scale the electric field, we have run slightly ahead of ourselves, since n_2 is, as we shall see, related to χ_3. However, if we accept that the only nonlinearity is in the refractive index, $n_0 + n_2|E|^2$, this is the only scaling available to us. Note that we have also assumed that the time scale associated with χ_1 and χ_3 is of $O(1/kc_0)$. This simply means that we assume that these functions vary by an $O(1)$ amount for an $O(1)$ change in the wavenumber, and hence that the underlying wave is dispersive at

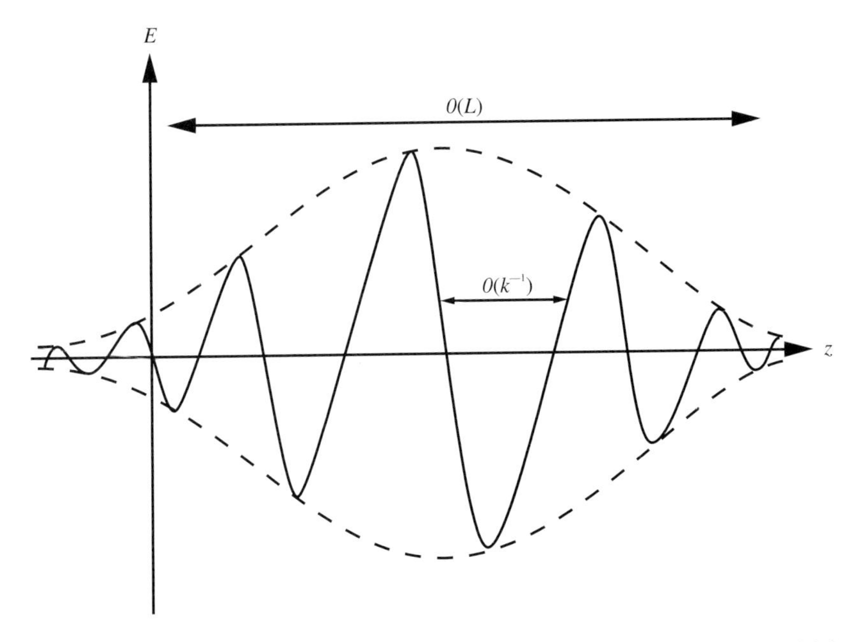

Fig. 12.4. A slowly modulated wavepacket solution. The dashed line is the initial envelope function, $A_0(z)$.

leading order. In terms of these variables

$$\begin{aligned}\frac{\partial^2 \bar{E}}{\partial \bar{z}^2} - \frac{\partial^2 \bar{E}}{\partial \bar{t}^2} = {} & \frac{\partial^2}{\partial \bar{t}^2}\int_{-\infty}^{\infty} \bar{\chi}_1(\bar{\tau})\bar{E}(\bar{z},\bar{t}-\bar{\tau})d\bar{\tau} \\ & + \frac{\partial^2}{\partial \bar{t}^2}\int_{-\infty}^{\infty}\int_{-\infty}^{\infty}\int_{-\infty}^{\infty} \bar{\chi}_3(\bar{\tau}_1,\bar{\tau}_2,\bar{\tau}_3) \\ & \times \bar{E}(\bar{z},\bar{t}-\bar{\tau}_1)\bar{E}(\bar{z},\bar{t}-\bar{\tau}_2)\bar{E}(\bar{z},\bar{t}-\bar{\tau}_3)d\bar{\tau}_1 d\bar{\tau}_2 d\bar{\tau}_3, \end{aligned} \tag{12.58}$$

subject to the initial condition

$$\bar{E}(\bar{z},0) = \delta \bar{A}_0(\bar{z})e^{i\bar{z}/\epsilon}. \tag{12.59}$$

We now seek a wavepacket solution of the form

$$\bar{E}(\bar{z},\bar{t}) = \delta \bar{A}(\bar{z},\bar{t})e^{i(\bar{z}-\bar{\omega}\bar{t})/\epsilon}, \tag{12.60}$$

where $\bar{\omega} = O(1)$ is the dimensionless frequency, which we must determine. Note that $(\bar{z}-\bar{\omega}\bar{t})/\epsilon = kz - kc_0\bar{\omega}t$, and hence that the frequency of the underlying wave is $\omega = kc_0\bar{\omega}$. We now substitute (12.60) into (12.58) and determine the first few terms in the asymptotic expansion of (12.58). There are two technical points to note here.

The first term on the right hand side of (12.58) becomes, on Taylor expanding the integrand,

$$\delta\frac{\partial^2}{\partial\bar{t}^2}\left[e^{i(\bar{z}-\bar{\omega}\bar{t})/\epsilon}\left\{\bar{A}\int_{-\infty}^{\infty}\bar{\chi}_1(\bar{\tau})e^{i\bar{\omega}\bar{\tau}}\,d\bar{\tau}-\epsilon\frac{\partial\bar{A}}{\partial\bar{t}}\int_{-\infty}^{\infty}\bar{\tau}\bar{\chi}_1(\bar{\tau})e^{i\bar{\omega}\bar{\tau}}\,d\bar{\tau}\right.\right.$$
$$\left.\left.+\frac{1}{2}\epsilon^2\frac{\partial^2\bar{A}}{\partial\bar{t}^2}\int_{-\infty}^{\infty}\bar{\tau}^2\bar{\chi}_1(\bar{\tau})e^{i\bar{\omega}\bar{\tau}}\,d\bar{\tau}+O(\epsilon^3)\right\}\right]. \tag{12.61}$$

We now define

$$\hat{\chi}_1(\bar{\omega})=\int_{-\infty}^{\infty}\bar{\chi}_1(\bar{\tau})e^{i\bar{\omega}\bar{\tau}}\,d\bar{\tau}, \tag{12.62}$$

the Fourier transform of $\bar{\chi}_1$, and note that

$$\hat{\chi}_1'=i\int_{-\infty}^{\infty}\bar{\tau}\bar{\chi}_1(\bar{\tau})e^{i\bar{\omega}\bar{\tau}}\,d\bar{\tau},$$

where a prime denotes $d/d\bar{\omega}$, and similarly for the second derivative. This allows us to write (12.61) in terms of $\hat{\chi}_1$ and its derivatives.

The second technical point concerns the nonlinear term on the right hand side of (12.58). As usual, we have been using the convention that we take the real part of the complex electric field, E, and so we must write

$$E=\frac{1}{2}\delta\left(\bar{A}e^{i(\bar{z}-\bar{\omega}\bar{t})/\epsilon}+\bar{A}^*e^{-i(\bar{z}-\bar{\omega}\bar{t})/\epsilon}\right),$$

where a star denotes the complex conjugate, before substituting into the nonlinear term. The integrand then consists of terms proportional only to $\exp(\pm i(\bar{z}-\bar{\omega}\bar{t})/\epsilon)$ or $\exp(\pm 3i(\bar{z}-\bar{\omega}\bar{t})/\epsilon)$. We only need to include the former, since the latter oscillate at a different frequency from the leading order terms, and will not lead to any secular terms when we use the method of multiple scales. This gives us the leading order approximation of the nonlinear term on the right hand side of (12.58) as

$$\frac{1}{4}\frac{\delta^3}{\epsilon^2}\bar{\omega}^2\bar{A}^2\bar{A}^*e^{i(\bar{z}-\bar{\omega}\bar{t})/\epsilon}\int_{-\infty}^{\infty}\int_{-\infty}^{\infty}\int_{-\infty}^{\infty}\bar{\chi}_3(\bar{\tau}_1,\bar{\tau}_2,\bar{\tau}_3)$$
$$\times\left\{e^{i\bar{\omega}(\bar{\tau}_1+\bar{\tau}_2-\bar{\tau}_3)}+e^{i\bar{\omega}(\bar{\tau}_1-\bar{\tau}_2+\bar{\tau}_3)}+e^{i\bar{\omega}(-\bar{\tau}_1+\bar{\tau}_2+\bar{\tau}_3)}\right\}d\bar{\tau}_1d\bar{\tau}_2d\bar{\tau}_3,$$

taking the real part only. Finally, the isotropy of the medium implies that

$$\bar{\chi}_3(\bar{\tau}_1,\bar{\tau}_2,\bar{\tau}_3)=\bar{\chi}_3(\bar{\tau}_2,\bar{\tau}_3,\bar{\tau}_1)=\bar{\chi}_3(\bar{\tau}_3,\bar{\tau}_1,\bar{\tau}_2),$$

and hence (noting that $\bar{A}\bar{A}^*=|\bar{A}|^2$) the leading order approximation to the nonlinear term is

$$\frac{3}{4}\frac{\delta^3}{\epsilon^2}\bar{\omega}^2|\bar{A}|^2\bar{A}e^{i(\bar{z}-\bar{\omega}\bar{t})/\epsilon}\hat{\chi}_3(\bar{\omega}),$$

where

$$\hat{\chi}_3(\bar{\omega}) = \int_{-\infty}^{\infty}\int_{-\infty}^{\infty}\int_{-\infty}^{\infty} \bar{\chi}_3(\bar{\tau}_1, \bar{\tau}_2, \bar{\tau}_3) e^{i\bar{\omega}(\bar{\tau}_1+\bar{\tau}_2-\bar{\tau}_3)} d\bar{\tau}_1 d\bar{\tau}_2 d\bar{\tau}_3.$$

Taking note of these two technical points, the first few terms in the expansion of (12.58) are

$$F(\bar{\omega})\bar{A} + i\epsilon\left\{2\frac{\partial \bar{A}}{\partial \bar{z}} + F'(\bar{\omega})\frac{\partial \bar{A}}{\partial \bar{t}}\right\} + \epsilon^2\left\{\frac{\partial^2 \bar{A}}{\partial \bar{z}^2} - \frac{1}{2}F''(\bar{\omega})\frac{\partial^2 \bar{A}}{\partial \bar{t}^2}\right\}$$
$$+\frac{3}{4}\delta^2\bar{\omega}^2\hat{\chi}_3(\bar{\omega})|\bar{A}|^2\bar{A} = O(\epsilon^3, \delta^2\epsilon), \tag{12.63}$$

where

$$F(\bar{\omega}) = \bar{\omega}^2\left(1 + \hat{\chi}_1(\bar{\omega})\right) - 1.$$

As we shall see below, we need to take $\delta = O(\epsilon)$ in order to obtain a balance between dispersion, which occurs through the $O(\epsilon^2)$ terms, and nonlinearity, through terms of $O(\delta^2)$. Without loss of generality, we therefore choose $\delta = \epsilon$.

At leading order, $F(\bar{\omega}) = 0$. Note that we do not need to expand the unknown frequency, $\bar{\omega}$, in powers of ϵ, since it only appears explicitly in the rapidly oscillating part of (12.60). Any higher order behaviour of $\bar{\omega}$ can be subsumed into the behaviour of $\bar{A}$. In terms of the physical variables,

$$k^2 = \frac{\omega^2}{c_0^2}\left(1 + \tilde{\chi}_1(\omega)\right), \tag{12.64}$$

where

$$\tilde{\chi}_1(\omega) = \int_{-\infty}^{\infty} \chi_1(\tau) e^{i\omega\tau} d\tau. \tag{12.65}$$

If we define $n_0(\omega) = \sqrt{1 + \tilde{\chi}_1(\omega)}$, we obtain the usual dispersion relation, $k = \omega n_0(\omega)/c_0$, and we can now see how the leading order refractive index, n_0, is related to χ_1.

Now that we know $\bar{\omega}$, (12.63) becomes

$$2\frac{\partial \bar{A}}{\partial \bar{z}} + F'(\bar{\omega})\frac{\partial \bar{A}}{\partial \bar{t}} - i\epsilon\left\{\frac{\partial^2 \bar{A}}{\partial \bar{z}^2} - \frac{1}{2}F''(\bar{\omega})\frac{\partial^2 \bar{A}}{\partial \bar{t}^2} + \frac{3}{4}\bar{\omega}^2\hat{\chi}_3(\bar{\omega})|\bar{A}|^2\bar{A}\right\} = O(\epsilon^2). \tag{12.66}$$

At leading order, the initial waveform, $\bar{A}_0(\bar{z})$, propagates without change of form with velocity $2/F'$. If we write this in terms of the physical variables, we find that

$$\frac{2}{F'(\bar{\omega})} = \frac{1}{c_0}\frac{d\omega}{dk} = \frac{c_g}{c_0} = \bar{c}_g,$$

where c_g is the group velocity, and $\bar{c}_g$ the dimensionless group velocity. Not surprisingly, the wavepacket propagates without change of form at the group velocity of the underlying wave.

The terms of $O(\epsilon)$ in (12.66) model the effect of dispersion and material nonlinearity on this propagating wavepacket. We expect that these effects will act over a time scale of $O(\epsilon^{-1})$, so we use the method of multiple scales. We begin by defining a slow time scale, $T = \epsilon\bar{t}$, and expand the envelope function as

$$\bar{A} = \bar{A}_1(\bar{z}, \bar{t}, T) + \epsilon\bar{A}_2(\bar{z}, \bar{t}, T) + O(\epsilon^2).$$

At leading order

$$\frac{\partial \bar{A}_1}{\partial \bar{z}} + \frac{1}{\bar{c}_g}\frac{\partial \bar{A}_1}{\partial \bar{t}} = 0,$$

and hence $A_1 = A_1(\bar{\eta}, T)$, where $\bar{\eta} = \bar{z} - \bar{c}_g\bar{t}$ is the travelling wave coordinate. Now, noting that

$$F''(\bar{\omega}) = \frac{d^2}{d\bar{\omega}^2}\left\{\bar{\omega}^2\left(1 + \hat{\chi}_1(\bar{\omega})\right)\right\} = c_0^2\frac{d^2}{d\omega^2}(k^2) = \frac{1}{\bar{c}_g^2}\left(1 - \frac{d\bar{c}_g}{d\bar{\omega}}\right),$$

we find that at $O(\epsilon)$

$$2\left(\frac{\partial \bar{A}_2}{\partial \bar{z}} + \frac{1}{\bar{c}_g}\frac{\partial \bar{A}_2}{\partial \bar{t}}\right) + \frac{2}{\bar{c}_g}\frac{\partial A_1}{\partial T} - i\left\{\frac{d\bar{c}_g}{d\bar{\omega}}\frac{\partial^2 \bar{A}_1}{\partial \bar{\eta}^2} + \frac{3}{4}\bar{\omega}^2\hat{\chi}_3(\bar{\omega})|\bar{A}_1|^2\bar{A}_1\right\} = 0.$$

To eliminate the secular terms, we need

$$\frac{\partial A_1}{\partial T} - \frac{1}{2}i\bar{c}_g\frac{d\bar{c}_g}{d\bar{\omega}}\frac{\partial^2 \bar{A}_1}{\partial \bar{\eta}^2} - \frac{3}{8}i\bar{c}_g\bar{\omega}^2\hat{\chi}_3(\bar{\omega})|\bar{A}_1|^2\bar{A}_1 = 0. \tag{12.67}$$

This is the nonlinear Schrödinger equation (NLS). In terms of the physical variables, for $E(z, t) = A(\eta, t)e^{i(kz-\omega t)}$ with $\eta = z - c_g t$,

$$\frac{\partial A}{\partial t} + \frac{1}{2}ic_g^3\frac{d^2k}{d\omega^2}\frac{\partial^2 A}{\partial \eta^2} - ic_g k\frac{n_2}{n_0}|A|^2 A = 0, \tag{12.68}$$

where $3\hat{\chi}_3(\bar{\omega}) = 8n_0(\bar{\omega})n_2(\bar{\omega})$. Although we will not go into the details here, this is the standard definition of n_2, and leads to the nonlinear dispersion relation, (12.54). Equation (12.68) is valid provided that $kL \gg 1$, $|E| = O(1/kLn_2^{1/2})$ and $d^2k/d\omega^2 = O(1/kc_0^2)$. These conditions ensure that the solution is of wavepacket form, the electric field is large enough for a weak Kerr effect to occur, and that this nonlinear effect can be balanced by weak linear dispersion. In addition, the electric field is not so strong that higher order terms appear in the nonlinear dispersion relation (12.54). The derivation of NLS for an optical fibre is rather more involved than that which we have given here. The only difference in the resulting equation is that the term that models the effect of linear dispersion, proportional to $\partial^2 A/\partial\eta^2$, includes the effect of both material dispersion and waveguide dispersion, which we discussed at the end of section 6.8.

We can now scale (12.68) to remove all of the constant coefficients, and obtain the two canonical forms of NLS, dependent on the relative sign of $d^2k/d\omega^2$ and n_2. We define

$$t = \tilde{t}\frac{1}{c_{\rm g}\text{sgn}(-d^2k/d\omega^2)}, \quad \eta = \tilde{\eta}\left(c_{\rm g}^2\left|\frac{d^2k}{d\omega^2}\right|\right)^{1/2}, \quad E = q\frac{1}{\left(k|n_2|n_0\right)^{1/2}},$$

so that

$$\frac{\partial q}{\partial \tilde{t}} - \frac{1}{2}i\frac{\partial^2 q}{\partial \tilde{\eta}^2} + i\,\text{sgn}\left(n_2\frac{d^2k}{d\omega^2}\right)|q|^2 q = 0. \qquad (12.69)$$

Finally, we write this as

$$\frac{\partial q}{\partial t} - \frac{1}{2}i\frac{\partial^2 q}{\partial x^2} \pm i|q|^2 q = 0, \qquad (12.70)$$

where t is not to be confused with the original time variable, and work with this form of the NLS equation from now on.

The other major application of NLS is to the analysis of nonlinear water wavepackets, and the associated Benjamin–Feir instability. For a discussion of this topic, see the book by Ablowitz and Segur (1981), and references therein.

We begin our analysis of NLS by considering the possible solitary wave solutions.

12.2.2 Solitary Wave Solutions of the Nonlinear Schrödinger Equation

The definition of a solitary wave solution of NLS is slightly different from that which we have applied earlier in the book, since the dependent variable, q, is complex. We seek a solution of the form

$$q(y,t) = \sqrt{\rho(y)}e^{i\sigma(y,t)}, \qquad (12.71)$$

where $y = x - vt$ is a travelling wave coordinate. In this way the amplitude of the wave envelope, $|q|^2 = \rho$, is a travelling wave, moving at the group velocity plus the velocity v. The phase of the wave is also a function of time, t, through the real function $\sigma(y,t)$. On substituting (12.71) into

(12.70), and equating real and imaginary parts, we find that

$$\frac{\partial}{\partial y}\left\{\rho\left(\sigma_y - v\right)\right\} = 0,$$

and hence

$$\sigma_y = \frac{f(t)}{\rho} + v, \tag{12.72}$$

and also

$$\sigma_t + \frac{1}{2}\sigma_y^2 - v\sigma_y = \frac{d}{d\rho}\left(\mp\frac{1}{2}\rho^2 + \frac{\rho_y^2}{8\rho}\right) \equiv g(y). \tag{12.73}$$

If we take $\partial^2/\partial t\partial y$ of (12.73) and eliminate σ_y using (12.72), we find that

$$\frac{f''}{2ff'} = -\frac{d}{dy}\left(\frac{1}{\rho}\right),$$

and hence that, unless $f(t)$ is a constant,

$$\frac{1}{\rho} = -\frac{f''}{2ff'}y + \text{constant.}$$

Since this means that ρ is singular for some finite value of y, we conclude that $f(t) = f_0$, a constant. Equation (12.72) then gives

$$\sigma = vy + f_1(t) + f_0\int^y \frac{ds}{\rho(s)}.$$

Substituting this into (12.73) shows that $f_1'(t)$ is a constant, and hence that $f_1 = k_0 t + k_1$, and

$$\sigma = vy + k_0 t + k_1 + f_0\int^y \frac{ds}{\rho(s)}. \tag{12.74}$$

Substituting (12.74) into (12.73) shows that

$$\frac{1}{2}\frac{d}{d\rho}\left(\mp\rho^2 + \frac{\rho_y^2}{4\rho}\right) = k_0 - \frac{1}{2}v^2 + \frac{f_0^2}{\rho^2},$$

and hence

$$\left(\frac{d\rho}{dy}\right)^2 = F(\rho) = \pm 4\rho^3 + 4(2k_0 - v^2)\rho^2 + 8k_2\rho - 8f_0^2. \tag{12.75}$$

The behaviour of the amplitude function, $\rho(y)$, is therefore qualitatively similar to that which we investigated for the KdV equation in subsection 8.3.2, since the square of the derivative of the amplitude is given by a cubic polynomial in the amplitude. As with the KdV equation, cnoidal wave solutions exist, but we will focus our attention on the possible solitary wave solutions. These are characterised by the existence of a positive, repeated root of $F(\rho)$. We shall see that the qualitative form of the solitary wave solutions is dependent upon the sign of the coefficient of ρ^3 in (12.75), and hence on the sign of the nonlinear term in (12.70).

(i) $n_2k'' < 0$, Normal Dispersion

When $n_2k'' < 0$, we shall see that localised solitary wave pulses can propagate without change of form. This is known as **normal dispersion**. In this case

$$F(\rho) = -4\rho^3 + 4(2k_0 - v^2)\rho^2 + 8k_2\rho - 8f_0^2.$$

Since $8f_0^2 \geq 0$, the product of the three roots of $F(\rho)$ cannot be strictly positive. The only possible repeated root is therefore $\rho = 0$, and hence we must take $f_0 = k_2 = 0$. This gives

$$\frac{d\rho}{dy} = 2\rho\sqrt{\rho_0 - \rho},$$

where $\rho_0 = 2k_0 - v^2$ is the other root. It is straightforward to integrate this equation, and we find that

$$\rho = \rho_0 \operatorname{sech}^2\left\{\sqrt{\rho_0}\,(y - y_0)\right\}, \tag{12.76}$$

$$\sigma = vy + \frac{1}{2}\left(\rho_0 + v^2\right)t + k_1, \tag{12.77}$$

and hence that there is a four parameter family of solitary wave solutions of the form

$$q = a \operatorname{sech}\left\{a(x - vt - x_0)\right\} \exp\left[i\left\{vx + \frac{1}{2}\left(a^2 - v^2\right)t - t_0\right\}\right]. \tag{12.78}$$

The parameters x_0 and t_0 are simply phase shifts in x and t. The remaining two parameters are the speed, v, and amplitude, a. Note that these two parameters are independent, in contrast to the solitary wave solutions of the KdV equation, given by (12.44), in which the speed is proportional to the amplitude. In addition, in a frame of reference moving with the solitary wave, the size of q has an explicit, periodic x and t dependence. In particular, the larger the velocity, v, the more rapid the spatial variation in x. This solitary wave is known as a **bright soliton**. The propagation of some typical bright solitons is illustrated in figure 12.5(a–c). It is clear from these examples that for sufficiently large values of v the envelope function, shown as a solid line, will start to have spatial oscillations of the same period as the underlying wave, and the approximations that led to NLS are not valid.

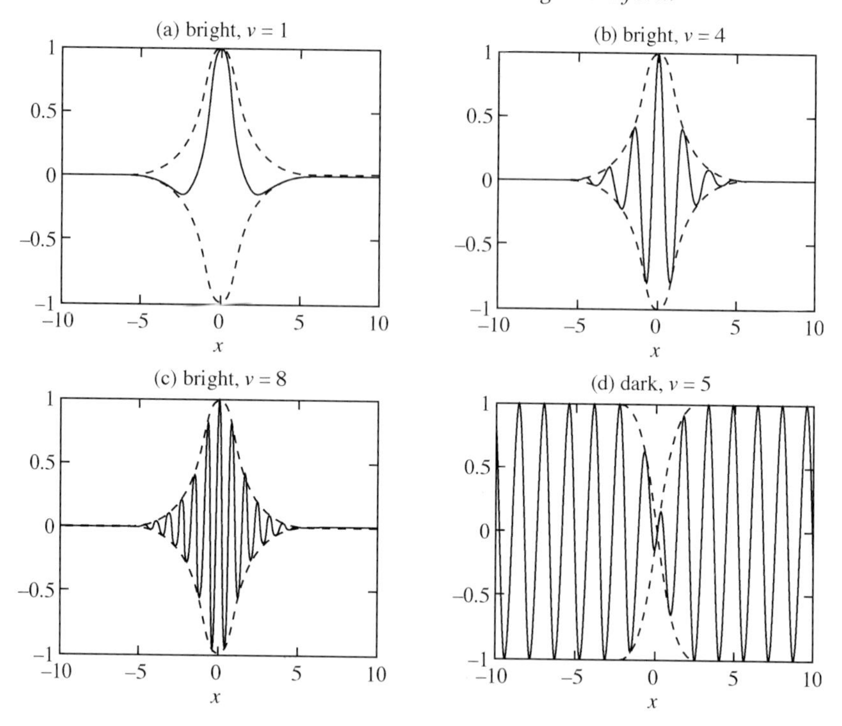

Fig. 12.5. Typical bright and dark soliton solutions of NLS. The solid line is the real part of q, and the broken line is $|q|$.

(ii) $n_2k'' > 0$*, Anomalous Dispersion*

When $n_2k'' > 0$, we shall see that localised solitary wave structures can only propagate as reductions in amplitude relative to a steady background wave train. This is known as **anomalous dispersion**. In this case

$$F(\rho) = 4\rho^3 + 4(2k_0 - v^2)\rho^2 + 8k_2\rho - 8f_0^2.$$

Now, when there is a repeated root of $F(\rho)$, and hence a solitary wave solution, we can write

$$\frac{d\rho}{dy} = 2(\rho - \rho_2)\sqrt{\rho - \rho_1},$$

with $\rho_2 > \rho_1 \geq 0$, and

$$f_0 = \rho_2\sqrt{\frac{1}{2}\rho_1}, \quad k_0 = \frac{1}{2}v^2 - \frac{1}{2}(\rho_1 + 2\rho_2), \quad k_2 = \frac{1}{2}\rho_2(2\rho_1 + \rho_2).$$

We can again evaluate ρ and hence σ, although the algebra is slightly more painful, and arrive at

$$\rho = \rho_2 \left[1 - K^2 \text{sech}^2 \left\{K\sqrt{\rho_2}(y - y_0)\right\}\right], \tag{12.79}$$

$$\sigma = \left(v + \sqrt{\rho_0(1-K^2)}\right) y + \left\{\frac{1}{2}v^2 - \frac{1}{2}\rho_2(3-K^2)\right\} t + k_1$$
$$+ \tan^{-1}\left[\frac{K}{\sqrt{1-K^2}} \tanh\left\{K\sqrt{\rho_2}(y-y_0)\right\}\right], \tag{12.80}$$

for $0 < K < 1$, where

$$K^2 = \frac{\rho_2 - \rho_1}{\rho_2}.$$

When $\rho_1 = 0$ and hence $K = 1$, the soliton takes the simpler form

$$\rho = \rho_2 \tanh^2\left\{\sqrt{\rho_2}(y - y_0)\right\}, \tag{12.81}$$

$$\sigma = vy + \left(\frac{1}{2}v^2 - \rho_2\right) t + k_2. \tag{12.82}$$

These solitary waves are rather different from those we have met before, since they asymptote to a periodic wave with amplitude ρ_2 as $y \to \pm\infty$. The solitary wave propagates as a reduction in the light intensity, and is known as a **dark soliton**. The propagation of the dark soliton with $K = 1$ is shown in figure 12.5(d). A soliton of this type, which alters the phase of a wave, is known as a **topological soliton**.

The solitary waves that we have constructed for both normal and anomalous dispersion are solitons in the sense that we defined above for the KdV equation. An inverse scattering transform is available to solve NLS, and this is the subject of the final subsection. We confine our attention to the case of normal dispersion, for which we have seen that there are bright soliton solutions, since these are the solitons that have a practical application in optical fibre technology. In the following, when we use the abbreviation NLS we refer to the equation

$$\frac{\partial q}{\partial t} - \frac{1}{2}i\frac{\partial^2 q}{\partial x^2} - i|q|^2 q = 0.$$

12.2.3 The Inverse Scattering Transform for the Nonlinear Schrödinger Equation

Although the inverse scattering transform for NLS is conceptually the same as that for the KdV equation, the details and qualitative results are somewhat different, as we shall see. The approach we use is a special case of the more general AKNS method (Ablowitz, Kaup, Newell and Segur, 1974), which generates inverse scattering transforms for a wide variety of equations.

The Scattering Problem

Consider the pair of equations

$$\psi_{1x} + i\xi\psi_1 = q\psi_2, \quad \psi_{2x} - i\xi\psi_2 = -q^*\psi_1, \tag{12.83}$$

for the moment regarding ψ_1 and ψ_2 as complex functions of x only. We assume that $q \to 0$ as $x \to \pm\infty$, with q a complex function of x. In these equations, ψ_1 and ψ_2 are the eigenfunctions and ξ is the eigenvalue. The AKNS method replaces $-q^*$ with another function, r, in the second equation, with different choices of r leading to inverse scattering transforms for different evolution equations for q and r. This leads to a matrix scattering problem. When $r = -q^*$ the equations have some symmetries that make the following approach less cumbersome than the matrix formulation. In particular, by taking the complex conjugate of (12.83) we find that

$$\text{if } (\psi_1(x;\xi), \psi_2(x;\xi)) \text{ is an eigensolution, so is } (\psi_2^*(x;\xi^*), -\psi_1^*(x;\xi^*)). \tag{12.84}$$

In general, since $q \to 0$, $(\psi_1, \psi_2) \sim (k_1 e^{-i\xi x}, k_2 e^{i\xi x})$ as $x \to \pm\infty$, for some complex constants k_1 and k_2. We define (ψ_{1+}, ψ_{2+}) to be the unique solution that satisfies

$$(\psi_{1+}(x;\xi), \psi_{2+}(x;\xi)) \sim (e^{-i\xi x}, 0) \quad \text{as } x \to \infty. \tag{12.85}$$

Using (12.84), another solution is $(\psi_{2+}^*(x;\xi^*), -\psi_{1+}^*(x;\xi^*))$, and from (12.85)

$$(\psi_{2+}^*(x;\xi^*), -\psi_{1+}^*(x;\xi^*)) \sim (0, -e^{i\xi x}) \quad \text{as } x \to \infty.$$

We conclude that the set

$$\{(\psi_{1+}(x;\xi), \psi_{2+}(x;\xi)),\ (\psi_{2+}^*(x;\xi^*), -\psi_{1+}^*(x;\xi^*))\}$$

$$\text{forms a basis for solutions of (12.83).} \tag{12.86}$$

We now define (ψ_{1-}, ψ_{2-}) to be the unique solution that satisfies

$$(\psi_{1-}(x;\xi), \psi_{2-}(x;\xi)) \sim (e^{-i\xi x}, 0) \quad \text{as } x \to -\infty. \tag{12.87}$$

By (12.86), we can express this as

$$\left.\begin{aligned}\psi_{1-}(x;\xi) &= a(\xi)\psi_{1+}(x;\xi) + b(\xi)\psi_{2+}^{*}(x;\xi^{*}),\\ \psi_{2-}(x;\xi) &= -b(\xi)\psi_{1+}^{*}(x;\xi^{*}) + a(\xi)\psi_{2+}(x;\xi).\end{aligned}\right\} \tag{12.88}$$

We also note that, since (12.83) shows that $\psi_1 \sim k_1 e^{-i\xi x}$ and $\psi_2 \sim k_2 e^{i\xi x}$ as $|\xi| \to \infty$, we must have

$$\left.\begin{gathered}(\psi_{1-},\ \psi_{2-}) \sim (e^{-i\xi x}, 0), \quad (\psi_{1+},\ \psi_{2+}) \sim (e^{-i\xi x}, 0),\\ a(\xi) \to 1, \quad b(\xi) \to 0, \quad \text{as } |\xi| \to \infty.\end{gathered}\right\} \tag{12.89}$$

Finally, we note an important property of the discrete spectrum. In general, if $q(x)$ is real, the bound state eigenvalues are purely imaginary (see exercise 12.10 and Satsuma and Yajima (1974)).

Example 1: Scattering by a Top Hat Potential. Let's consider what happens when

$$q(x) = \left\{\begin{array}{ll} 0 & \text{for } |x| > L,\\ q_0 & \text{for } |x| < L,\end{array}\right\} \tag{12.90}$$

with the constants L real and strictly positive and q_0 complex. We can find the scattering data $a(\xi)$ and $b(\xi)$ by constructing the solution (ψ_{1-}, ψ_{2-}). From (12.87),

$$\psi_{1-} = e^{-i\xi x}, \quad \psi_{2-} = 0 \quad \text{for } x < -L. \tag{12.91}$$

For $|x| < L$, where $q = q_0$, we seek exponential solutions and find that

$$\psi_{1-} = Ae^{ikx} + Be^{-ikx}, \quad \psi_{2-} = \frac{i}{q_0}\left\{A(k+\xi)e^{ikx} - B(k-\xi)e^{-ikx}\right\},$$

where

$$k = \sqrt{|q_0|^2 + \xi^2}. \tag{12.92}$$

We can determine the constants A and B by continuity of ψ_{1-} and ψ_{2-} at $x = -L$, and find that

$$\left.\begin{aligned}\psi_{1-} &= e^{i\xi L}\left\{\cos k(x+L) - \tfrac{i\xi}{k}\sin k(x+L)\right\},\\ \psi_{2-} &= -\tfrac{q_0^{*}}{k}e^{i\xi L}\sin k(x+L) \quad \text{for } |x| < L.\end{aligned}\right\} \tag{12.93}$$

Now, from (12.85) and (12.88),

$$\psi_{1-} = a(\xi)e^{-i\xi x}, \quad \psi_{2-} = -b(\xi)e^{i\xi x} \quad \text{for } x > L. \tag{12.94}$$

This finally allows us to determine a and b from continuity at $x = L$ as

$$a(\xi) = 2e^{i\xi L}\left(\cos 2kL - \frac{i\xi}{k}\sin 2kL\right), \quad b(\xi) = \frac{q_0^*}{k}\sin 2kL. \tag{12.95}$$

In order to determine the discrete spectrum, $\xi = \xi_n$ such that $a(\xi_n) = 0$, we must solve

$$\tan 2L\sqrt{|q_0|^2 + \xi^2} = \frac{\sqrt{|q_0|^2 + \xi^2}}{i\xi}. \tag{12.96}$$

We know that when q_0 is real the discrete spectrum is purely imaginary, and hence this equation has only imaginary roots. Since the value of these roots depends only on the absolute value of q_0, they must be purely imaginary for all complex q_0. We therefore define

$$\xi = i|q_0|\alpha, \quad \hat{L} = |q_0|L,$$

in terms of which (12.96) becomes

$$\tan 2\hat{L}\sqrt{1-\alpha^2} = -\frac{\sqrt{1-\alpha^2}}{\alpha}. \tag{12.97}$$

If $\alpha > 1$ we define $\gamma = \sqrt{\alpha^2 - 1}$, so that (12.97) becomes

$$\tanh 2\hat{L}\gamma = -\frac{\gamma}{\sqrt{1+\gamma^2}}.$$

Since the left hand side of this equation is monotone increasing with γ and the right hand side monotone decreasing, the unique solution is $\gamma = 0$, and hence $\xi = \xi_0 = i|q_0|$. This is not an acceptable root, since then $k = \sqrt{|q_0|^2 + \xi_0^2} = 0$, and the derivation of the solution given above breaks down.

If $\alpha < 1$ we define $\beta = \sqrt{1-\alpha^2}$, and hence

$$\tan 2\hat{L}\beta = -\frac{\beta}{\sqrt{1-\beta^2}}.$$

The right hand side of this equation is monotone decreasing with β between the singularities at $\beta = \pm 1$. The number of solutions, other than the trivial one at $\beta = 0$ which is again not acceptable, therefore depends upon the number of singularities of $\tan 2\hat{L}\beta$ that lie in the range $|\beta| < 1$. If $2\hat{L} < \pi/2$ there are no solutions, whilst for $(2N-1)\pi/2 < 2\hat{L} < (2N+1)\pi/2$, there are N solutions. In terms of the original variables there are

$$\left.\begin{array}{r}\text{no bound state eigenvalues for } 4L|q_0| < \pi, \\ N \text{ bound state eigenvalues for } (2N-1)\pi < 4L|q_0| < (2N+1)\pi.\end{array}\right\} \tag{12.98}$$

Each eigenvalue is purely imaginary, with $|\xi_n| < |q_0|$. We will use this information later when we come to consider the solution of NLS with this type of initial condition.

The Inverse Scattering Problem

Given the scattering data, $a(\xi)$ and $b(\xi)$, can we reconstruct the function $q(x)$? We proceed in the same manner as we did for the KdV equation, beginning by writing

$$\left.\begin{aligned} \psi_{1+}(x;\xi) &= e^{-i\xi x} + \int_x^{\infty} K_1(x,y)e^{-i\xi y}\,dy, \\ \psi_{2+}(x;\xi) &= \int_x^{\infty} K_2(x,y)e^{-i\xi y}\,dy, \end{aligned}\right\} \tag{12.99}$$

with $K_1(x,y) = K_2(x,y) = 0$ for $y < x$. By substituting these definitions into (12.83) and proceeding as we did for KdV (see exercise 12.8), we find that K_1 and K_2 satisfy the boundary value problem

$$\frac{\partial K_1}{\partial x} + \frac{\partial K_1}{\partial y} = qK_2, \quad \frac{\partial K_2}{\partial x} - \frac{\partial K_2}{\partial y} = -q^*K_1, \tag{12.100}$$

subject to

$$\left.\begin{aligned} K_1(x,y) \to 0, \quad K_2(x,y) \to 0, \quad \text{as } y \to \infty, \\ K_2(x,x) = \frac{1}{2}q^*(x). \end{aligned}\right\} \tag{12.101}$$

Although we will not do so here, it can be shown that the solution of (12.100) to (12.101) exists and is unique. Therefore, if we can determine $K_2(x,x)$ from our knowledge of the scattering data, we can reconstruct $q(x)$ from

$$q(x) = 2K_2^*(x,x). \tag{12.102}$$

If we now substitute (12.99) into (12.88), we find that

$$\left.\begin{aligned} \int_{-\infty}^{\infty} K_1(x,y)e^{-i\xi y}\,dy &= \frac{1}{a(\xi)}\psi_{1-} - e^{-i\xi x} - \frac{b(\xi)}{a(\xi)}\int_x^{\infty} K_2^*(x,y)e^{i\xi y}\,dy, \\ \int_{-\infty}^{\infty} K_2(x,y)e^{-i\xi y}\,dy &= \frac{1}{a(\xi)}\psi_{2-} + \frac{b(\xi)}{a(\xi)}e^{-i\xi x} + \frac{b(\xi)}{a(\xi)}\int_x^{\infty} K_1^*(x,y)e^{i\xi y}\,dy, \end{aligned}\right\} \tag{12.103}$$

where we have extended two of the integrals down to $-\infty$ using the fact that $K_1(x,y) = K_2(x,y) = 0$ for $y < x$. Since K_1, $K_2 \to 0$ as $y \to \pm\infty$, we can treat the left hand sides of these equations as Fourier transforms, and invert to give

$$\left.\begin{aligned} K_1(x,z) &= R_1(x,z) - \int_x^{\infty} K_2^*(x,y)\hat{f}(y+z)dy. \\ K_2(x,z) &= R_2(x,z) + \hat{f}(x+z) + \int_x^{\infty} K_1^*(x,y)\hat{f}(y+z)dy, \end{aligned}\right\} \tag{12.104}$$

where

$$\hat{f}(X) = \frac{1}{2\pi}\int_{-\infty}^{\infty} \frac{b(\xi)}{a(\xi)} e^{i\xi X}\, d\xi, \tag{12.105}$$

and

$$\left.\begin{aligned} R_1(x,z) &= \frac{1}{2\pi}\int_{-\infty}^{\infty} e^{i\xi(z-x)}\left\{\frac{1}{a(\xi)}\psi_{1-}(x;\xi)e^{i\xi x} - 1\right\} d\xi, \\ R_2(x,z) &= \frac{1}{2\pi}\int_{-\infty}^{\infty} e^{i\xi(z-x)}\left\{\frac{1}{a(\xi)}\psi_{2-}(x;\xi)e^{i\xi x}\right\} d\xi. \end{aligned}\right\} \tag{12.106}$$

Now (12.89) shows that we can evaluate R_1 and R_2 by closing the contour of integration in the upper half ξ-plane, and then using the residue theorem to show that

$$R_j(x,z) = i\sum_{n=1}^{N}\left(\text{Residue at } \xi_n \text{ of}\left\{\frac{1}{a(\xi)}\psi_{j-}(x;\xi)e^{i\xi z}\right\}\right), \quad \text{for } j = 1 \text{ or } 2, \tag{12.107}$$

where ξ_n are the N zeros of $a(\xi)$ in the upper half plane, which we refer to as the discrete spectrum. Since $a(\xi_n) = 0$, (12.88) shows that

$$\psi_{1-}(x;\xi_n) = b(\xi_n)\psi_{2+}^*(x;\xi_n^*), \quad \psi_{2-}(x;\xi_n) = -b(\xi_n)\psi_{1+}^*(x;\xi_n^*).$$

If we now assume that the zeros of $a(\xi)$ are simple, the definitions (12.99) show that

$$\left.\begin{aligned} R_1(x,z) &= i\sum_{n=1}^{N} c_n \int_x^{\infty} K_2^*(x,y)e^{i\xi_n y}dy, \\ R_2(x,z) &= -i\sum_{n=1}^{N} c_n\left\{e^{i\xi_n x} + \int_x^{\infty} K_1^*(x,y)e^{i\xi_n y}dy\right\}, \end{aligned}\right\} \tag{12.108}$$

where $c_n = b(\xi_n)/a'(\xi_n)$. Substituting this into (12.104), we find that

$$\left.\begin{aligned} K_1(x,z) + \int_x^{\infty} K_2^*(x,y)f(y+z)dy &= 0, \\ K_2(x,z) - f(x+z) - \int_x^{\infty} K_1^*(x,y)f(y+z)dy &= 0, \end{aligned}\right\} \tag{12.109}$$

where

$$f(X) = \frac{1}{2\pi}\int_{-\infty}^{\infty} \frac{b(\xi)}{a(\xi)} e^{i\xi X} d\xi - i\sum_{n=1}^{N} c_n e^{i\xi_n X}. \tag{12.110}$$

As we saw for the KdV equation, the function $b(\xi)/a(\xi)$ is the reflection coefficient, whose simple poles in the upper half plane represent the discrete spectrum. Finally, we can eliminate K_1 between equations (12.109) and, writing $K(x,y) = K_2^*(x,y)$, arrive at

$$K(x,z) - f^*(x+z) + \int_x^{\infty} G(x,y,z)K(x,y)dy = 0, \tag{12.111}$$

where

$$G(x,y,z) = \int_x^{\infty} f(y+Y)f^*(Y+z)dY, \tag{12.112}$$

and

$$q(x) = 2K(x,x). \tag{12.113}$$

Example 2: Inverse Scattering for a Reflectionless Potential with a Single Bound State. If $b(\xi)/a(\xi) = 0$ for ξ real and $a(\xi)$ has a single zero in the upper half plane at $\xi = \xi_1$ with $b(\xi_1)/a'(\xi_1) = c_1$, a simple calculation shows that

$$f(X) = -ic_1 e^{i\xi_1 X},$$

$$G(x,y,z) = -\frac{|c_1|^2}{i(\xi_1 - \xi_1^*)} e^{i\xi_1(y+x)} e^{-i\xi_1^*(x+z)},$$

and hence

$$K(x,z) - ic_1^* e^{-i\xi_1^*(x+z)} - \frac{|c_1|^2}{i(\xi_1 - \xi_1^*)}\int_x^{\infty} e^{i\xi_1(y+x)} e^{-i\xi_1^*(x+z)} K(x,y)dy = 0. \tag{12.114}$$

By seeking a solution in the form $K(x,z) = L(x)e^{-i\xi_1^* z}$, we find that

$$L(x) = \frac{ic_1^*(\xi_1 - \xi_1^*)^2 e^{-i\xi_1^* x}}{(\xi_1 - \xi_1^*)^2 - |c_1|^2 e^{2i(\xi_1 - \xi_1^*)x}}.$$

If we now write $\xi_1 = \alpha_1 + i\beta_1$ with α_1 and β_1 real, we can obtain

$$q(x) = 2L(x)e^{-i\xi_1^* x} = \frac{4i\beta_1^2 e^{-2(\beta_1 + i\alpha_1)x}}{4\beta_1^2 + |c_1|^2 e^{-4\beta_1 x}},$$

which we can rearrange to give

$$q(x) = ic_1^* e^{-2i\alpha_1 x}\frac{2\beta_1}{|c_1|}\operatorname{sech}\left\{2\beta_1 x - \log\left(\frac{|c_1|}{2\beta_1}\right)\right\}. \tag{12.115}$$

In particular, with $\alpha_1 = 0$, $c_1 = i$ and $\beta_1 = 1/2$, this shows that $q(x) = \operatorname{sech} x$ is a reflectionless potential with a single bound state eigenvalue. In fact the family of potentials, $q(x) = A \operatorname{sech} x$ is reflectionless for A an integer, and has A bound states. The analysis is similar to that discussed in KdV example 2 in section 12.1, but we will not go into the details here (Satsuma and Yajima, 1974).

Example 3: Inverse Scattering for a Reflectionless Potential with N bound states. For a reflectionless potential with N discrete eigenvalues in the upper half plane,

$$f(X) = -i \sum_{n=1}^{N} c_n e^{i\xi_n X},$$

$$G(x, y, z) = i \sum_{n=1}^{N} \sum_{m=1}^{N} \frac{c_n c_m^*}{(\xi_n - \xi_m^*)} e^{i\{\xi_n y - \xi_m^* z + (\xi_n - \xi_m^*)x\}},$$

and hence

$$K(x, z) - i \sum_{n=1}^{N} c_n^* e^{-i\xi_n^*(x+z)}$$

$$+ i \sum_{n=1}^{N} \sum_{m=1}^{N} \frac{c_n c_m^*}{(\xi_n - \xi_m^*)} e^{i\{-\xi_m^* z + (\xi_n - \xi_m^*)x\}} \int_x^\infty K(x, y) e^{i\xi_n y} dy = 0. \quad (12.116)$$

We can now seek a solution in the form

$$K(x, z) = \sum_{k=1}^{N} L_k(x) e^{-i\xi_k^*(x-z)}, \quad (12.117)$$

so that

$$q(x) = 2 \sum_{k=1}^{N} L_k(x). \quad (12.118)$$

Substituting (12.117) into (12.116), performing the integration and equating coefficients of $e^{-i\xi_k^* z}$ leads to

$$L_k - \sum_{n=1}^{N} \sum_{m=1}^{N} \frac{c_n c_k^*}{(\xi_n - \xi_k^*)(\xi_n - \xi_m^*)} L_m e^{2i(\xi_n - \xi_k^*)x} = i c_k^* e^{-2i\xi_k^* x}. \quad (12.119)$$

Solution of this system of equations, which can be written in matrix form, gives the potential via (12.118). As for the KdV equation, it is advisable to use a computer algebra package for $N > 2$. The reflectionless potential

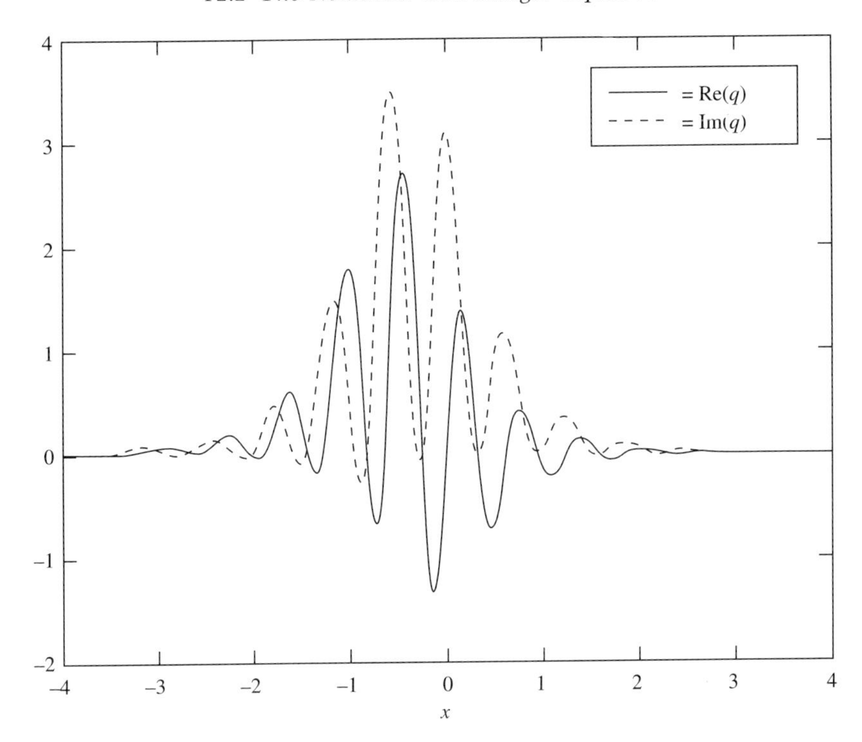

Fig. 12.6. The reflectionless potential with $\xi_1 = i$, $\xi_2 = 5 + i$, $c_1 = c_2 = 1$.

with two bound states and $\xi_1 = i$, $\xi_2 = 5 + i$, $c_1 = c_2 = 1$ is shown in figure 12.6.

Time Evolution of the Scattering Data

We now consider what happens to the eigenfunctions of (12.83) if $q = q(x,t)$, and hence $\psi_{1,2} = \psi_{1,2}(x,t;\xi)$, $a = a(\xi,t)$ and $b = b(\xi,t)$. Let's consider the time evolution given by

$$\left.\begin{aligned} \psi_{1t} &= \left(\frac{1}{2}i|q|^2 - i\xi^2\right)\psi_1 + \left(\frac{1}{2}iq_x + \xi q\right)\psi_2, \\ \psi_{2t} &= \left(\frac{1}{2}iq_x^* - \xi q^*\right)\psi_1 - \left(\frac{1}{2}i|q|^2 - i\xi^2\right)\psi_2. \end{aligned}\right\} \tag{12.120}$$

Again, for an explanation of how to choose the evolution equations appropriate to NLS, see the book by Drazin and Johnson (1989). For (12.120) to be compatible with (12.83), we need $\partial(\psi_{jt})/\partial x = \partial(\psi_{jx})/\partial t$.

On performing this cross-differentiation, we find that they are indeed compatible, provided that q satisfies NLS (see exercise 12.9).

Now consider the solution (ψ_{1-}, ψ_{2-}). From (12.85) and (12.88),

$$(\psi_{1-}, \psi_{2-}) \sim (a(\xi, t)e^{-i\xi x}, -b(\xi, t)e^{i\xi x}) \quad \text{as } x \to \infty.$$

If we substitute this asymptotic representation into (12.120) and use the fact that $q \to 0$ as $x \to \infty$, we find that

$$a_t = -i\xi^2 a, \quad b_t = i\xi^2 b,$$

and hence

$$a(\xi, t) = e^{-i\xi^2 t}a(\xi, 0), \quad b(\xi, t) = e^{i\xi^2 t}b(\xi, 0). \tag{12.121}$$

We conclude that the zeros of $a(x, t)$, and hence the discrete spectrum, are independent of t, and that

$$c_n(t) = e^{2i\xi_n^2 t}c_n(0). \tag{12.122}$$

Example 4: the Single Soliton Solution. If we substitute $c_1 = c_1(0)e^{2i\xi_1^2 t}$ into (12.115), we obtain

$$q(x, t) = ic_1^*(0)\exp\left[i\left\{-2\alpha_1 x + 2(\beta_1^2 - \alpha_1^2)t\right\}\right]\frac{2\beta_1}{|c_1(0)|}$$
$$\times \text{sech}\left\{2\beta_1(x + 2\alpha_1 t) - \log\left(\frac{|c_1(0)|}{2\beta_1}\right)\right\}. \tag{12.123}$$

By defining

$$v = -2\alpha_1, \quad a = 2\beta_1, \quad x_0 = a\log(|c_1(0)|/a), \quad t_0 = i\log(ic_1^*(0)/|c_1(0)|),$$

we now recover the single soliton solution given by (12.78). As for the KdV equation, a reflectionless potential with a single bound state eigensolution propagates as a soliton. Note that we can write

$$\xi_1 = \frac{1}{2}(-v + ia),$$

relating the real and imaginary parts of the eigenvalue to the amplitude and velocity of the soliton. In particular, if $q(x, 0)$ is real, ξ_1 is imaginary and v, the velocity of the soliton relative to the group velocity of the underlying wave, is zero.

Example 5: The N Soliton Solution If we substitute $c_m = c_m(0)e^{2i\xi_m^2 t}$ into

(12.119), we find that a reflectionless potential with N bound states evolves as

$$q(x,t) = 2\sum_{k=1}^{N} L_k(x,t), \tag{12.124}$$

with

$$L_k(x,t) - \sum_{n=1}^{N}\sum_{m=1}^{N} \frac{c_n(0)c_k^*(0)}{(\xi_n - \xi_k^*)(\xi_n - \xi_m^*)} L_m(x,t) e^{2i(\xi_n - \xi_k^*)\{x + (\xi_n + \xi_k^*)t\}}$$

$$= ic_k^*(0)e^{-2i\xi_k^*(x+\xi_k^* t)}. \tag{12.125}$$

The previous example would lead us to guess, correctly, that this solution represents the interaction of N solitons with velocities $-2\text{Re}(\xi_n)$ and amplitudes $2\text{Im}(\xi_n)$. Unfortunately, the analysis is rather more involved than for the equivalent solution of the KdV equation, and we simply describe the procedure for determining the leading order solution.

Let $y = x + (\xi_j + \xi_j^*)t$ be the travelling wave coordinate moving with the velocity of the jth soliton, in terms of which (12.125) can be written as

$$L_k e^{2i\xi_k^* y} e^{2i\xi_k^*(\xi_k^* - \xi_j^* - \xi_j)t}$$

$$-c_k^* \sum_{n=1}^{N} \frac{c_n e^{2i\xi_n y}}{\xi_n - \xi_k^*} e^{2i\xi_n(\xi_n - \xi_j - \xi_j^*)t} \sum_{m=1}^{N} \frac{L_m}{\xi_n - \xi_m^*} = ic_k^*. \tag{12.126}$$

If $|y| \gg 1$, we can see immediately that $L_k \ll 1$ and hence that $|q| \ll 1$. To determine the solution for $t \gg 1$ and $y = O(1)$, we note that if $\xi_n = \alpha_n + i\beta_n$,

$$|e^{2i\xi_n(\xi_n - \xi_j - \xi_j^*)t}| = |e^{-2i\xi_n^*(\xi_n^* - \xi_j^* - \xi_j)t}| = e^{-4\beta_n(\alpha_n - \alpha_j)t}.$$

The size of the exponential terms in (12.126) therefore depends on the relative size of the real parts of the eigenvalues. Let's number the eigenvalues so that they form a sequence with strictly increasing real parts, $\alpha_1 < \alpha_2 < \cdots < \alpha_N$. If two or more eigenvalues have the same real part there is more than one soliton moving at the same speed, the analysis is slightly different and we will not discuss it here. We now have

$$e^{2i\xi_n(\xi_n - \xi_j - \xi_j^*)t} \left\{ \begin{array}{ll} \ll 1 & \text{for } n > j, \\ = O(1) & \text{for } n = j, \\ \gg 1 & \text{for } n < j. \end{array} \right\} \tag{12.127}$$

Now that we know the sizes of the time dependent terms in (12.126) we

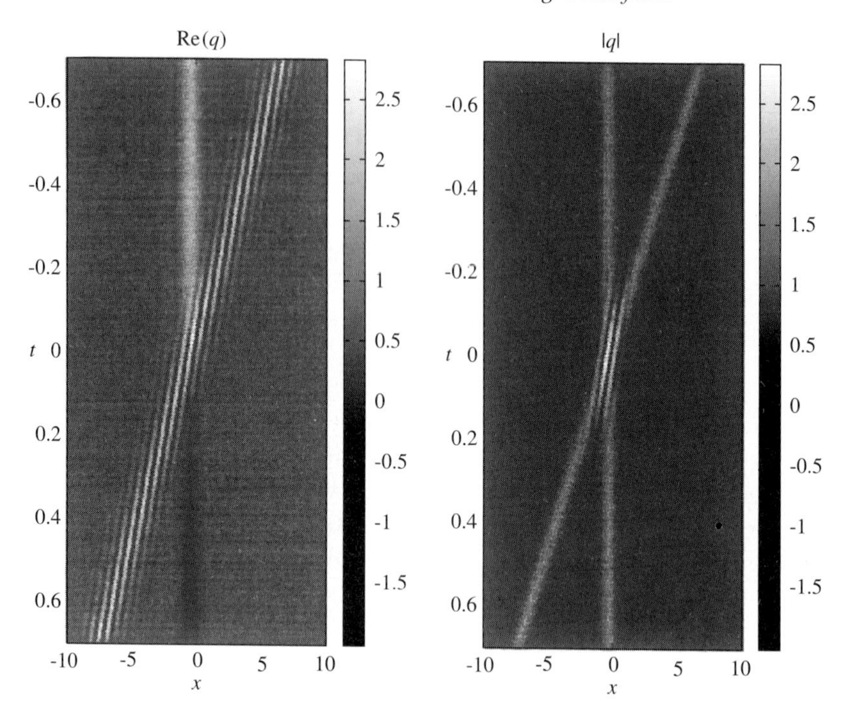

Fig. 12.7. The interaction of two bright solitons, with $\xi_1 = i$, $\xi_2 = 5 + i$, $c_1(0) = c_2(0) = 1$.

can construct its leading order asymptotic solution. For $j < k \leq N$, we need $L_k \ll 1$, whilst for $1 \leq k \leq j$ we need $L_k = O(1)$. Equation (12.126) shows that

$$\sum_{m=1}^{j} \frac{L_m}{\xi_n - \xi_m^*} = l_n e^{-2i\xi_n(\xi_n - \xi_j - \xi_j^*)t} \ll 1 \quad \text{for } 1 \leq n < j, \tag{12.128}$$

with l_n to be determined. At leading order, the right hand side of (12.128) is zero, and we have a system of linear equations that determines L_m for $1 \leq m < j$ as y and t independent multiples of L_j, and hence shows that $q(y,t)$ is proportional to L_j.

Now, substituting (12.128) into (12.126) for $1 \leq k < j$, we obtain a system of linear equations that determines l_n. On substituting for l_n into (12.126) with $k = j$, we obtain an equation for L_j. This is therefore a consistent algorithm for determining the leading order solution, and shows that for $y = O(1)$, $|q| = O(1)$. Moreover, it can be shown that the equation for L_j is of the same functional form as the equation that we

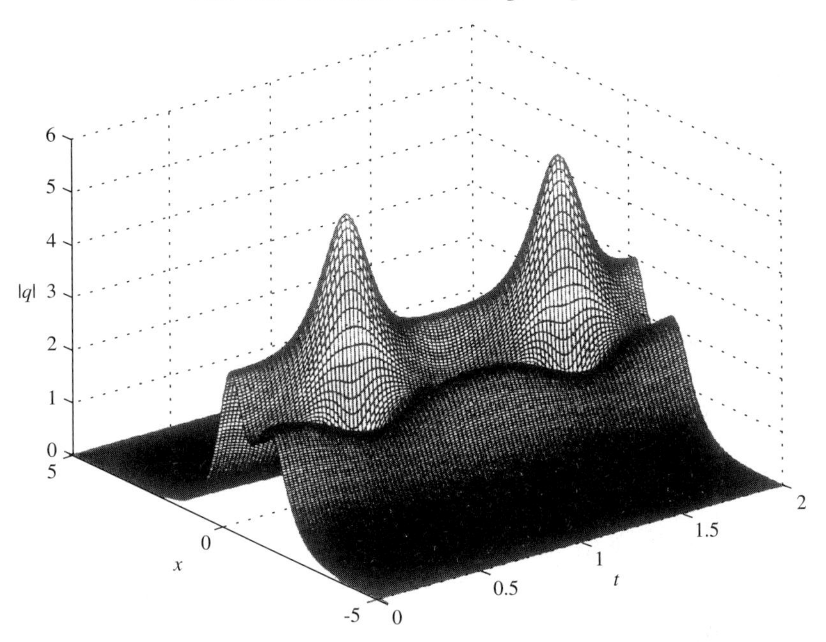

Fig. 12.8. The interaction of two bright solitons with $v = 0$. The scattering data when $t = 0$ are $\xi_1 = i$, $\xi_2 = 2i$, $c_1(0) = c_2(0) = 1$.

obtained for L when we considered the single soliton solution. Similar arguments apply for t large and negative. We conclude that for $y = O(1)$ and $|t| \gg 1$, the solution consists of N solitons of the form (12.78), with speeds $-2\text{Re}(\xi_n)$, amplitudes $2\text{Im}(\xi_n)$ and phase changes x_0 and t_0 determined by the details of the interaction between the solitons (see exercise 12.12). Figure 12.7 shows the interaction between two solitons when $\xi_1 = i$, $\xi_2 = 5 + i$ and $c_1(0) = c_2(0) = 1$. One soliton is stationary in this frame of reference, whilst the other has speed $v = 10$, and has a rapid change of phase along its length. The solution when $t = 0$ is shown in figure 12.6.

We have seen that if $q(x, 0)$ is real, the bound state eigenvalues are purely imaginary. Each therefore corresponds to a soliton with $v = 0$. The N solitons therefore do not separate, but remain where they are initially in the frame of reference moving with the group velocity. The resulting solution varies periodically with t. A typical example is shown in figure 12.8.

Most of the above examples involve reflectionless potentials, for which the solution consists of N solitons. By similar arguments to those that we used for the KdV equation, if $b(\xi)/a(\xi)$ is not identically zero for ξ real, a dispersive wave is also generated, but for $t \gg 1$, the solution is dominated by the solitons, just as it is for the KdV equation. For example, if $q(x,0)$ is the top hat potential (example 1 in this section), there is a dispersive wave component, since it is not a reflectionless potential, but for $t \gg 1$, the leading order solution consists of N solitons with zero velocity interacting in the neighbourhood of the origin, with N determined by the area under the initial potential, as given by (12.98).

We can now see why it is attractive to use solitons as the basic pulses transmitted along optical fibres. They owe their existence to a balance between dispersion and nonlinearity, so they do not suffer from the degradation due to dispersion alone that affects ordinary, linear pulses. In addition, solitons propagating at different velocities pass through each other without change of form, except for a small phase shift. Finally, if the properties of the pulse are of the correct orders of magnitude so that the nonlinear Schrödinger equation governs its propagation, it is straightforward to generate solitons.

Exercises

12.1 Show that if

$$u = w - \epsilon w_x - \epsilon^2 w^2,$$

and

$$w_t + 6(w - \epsilon^2 w^2) w_x + w_{xxx} = 0, \tag{12.129}$$

then u satisfies the KdV equation, (12.1). Show that

$$\int_{-\infty}^{\infty} w(x,t;\epsilon)dx \quad \text{is constant.}$$

By expanding w as

$$w(x,t;\epsilon) = \sum_{n=0}^{\infty} \epsilon^n w_n(x,t),$$

substituting into (12.129) and equating coefficient of ϵ, show that $C_1 = u$ and $C_2 = u_{xx} + u^2$ are conserved quantities in the sense that

$$\int_{-\infty}^{\infty} C_n(x,t)dx \quad \text{is constant.}$$

What do these represent physically? Deduce that the KdV equation has an infinite number of conserved quantities. Find the third conserved quantity. This property is intimately related to the integrability of the KdV equation, and is characteristic of integrable systems in general.

12.2 Determine the scattering data for (12.2) with the square potential

$$u(x) = U_0 H(1-x) H(x).$$

What is the qualitative difference between the scattering data for $U_0 > 0$ and $U_0 < 0$?

12.3 Show that the normalisation constant for the reflectionless potential $u(x) = -2\kappa^2 \mathrm{sech}^2 \kappa(x - x_0)$ is $c = \sqrt{2\kappa} e^{-\kappa x_0}$.

12.4 If the reflection coefficient is $b(k) = -U_0/(U_0 + 2ik)$ with $U_0 < 0$, solve the GLM equation and reconstruct the scattering potential, $u(x)$, in (12.2), without using KdV example 1 in section 12.1 to guide you.

12.5 Show that the two soliton solution discussed in KdV example 6 in section 12.1 takes the form

$$u(x,t) \sim -2\kappa_1^2 \mathrm{sech}^2 \left\{ \kappa_1 (x - 4\kappa_1^2 t) + \delta_1 - \Delta \right\}$$
$$-2\kappa_2^2 \mathrm{sech}^2 \left\{ \kappa_2 (x - 4\kappa_2^2 t) + \delta_2 + \Delta \right\}$$

as $t \to -\infty$, and

$$u(x,t) \sim -2\kappa_1^2 \mathrm{sech}^2 \left\{ \kappa_1 (x - 4\kappa_1^2 t) + \delta_1 + \Delta \right\}$$
$$-2\kappa_2^2 \mathrm{sech}^2 \left\{ \kappa_2 (x - 4\kappa_2^2 t) + \delta_2 - \Delta \right\}$$

as $t \to \infty$, and determine δ_1, δ_2 and Δ. What is the significance of the quantity Δ?

12.6 Consider the cnoidal wave solution of the KdV equation generated by a negative delta function initial condition, which we studied in example 8 in section 12.1. Show that when x is negative and of $O(t^{1/3})$, $u = O(t^{-1/3})$.

12.7 Use the method of multiple scales to show that the envelope of a wavepacket solution of the sine–Gordon equation

$$\frac{\partial^2 \phi}{\partial x^2} - \frac{1}{c^2} \frac{\partial^2 \phi}{\partial t^2} = a \sin \phi,$$

with a and c positive real constants, can satisfy the nonlinear Schrödinger equation for appropriate initial conditions and parameter values.

12.8 By substituting (12.99) into (12.83) and integrating the appropriate integrals by parts, show that K_1 and K_2 satisfy the boundary value problem given by (12.100) to (12.101).

12.9 By cross-differentiating (12.83) and (12.120), show that $q(x,t)$ must satisfy the nonlinear Schrödinger equation.

12.10 Use (12.83) and its complex conjugate to show that

$$\frac{\partial}{\partial x}\left\{\mathrm{Im}\left(\psi_1\psi_2^*\right)\right\} = \mathrm{Im}(q)\left(|\psi_1|^2 - |\psi_2|^2\right) - 2i\mathrm{Re}(\xi)\mathrm{Im}(\psi_1^*\psi_2).$$

Now use this equation to show that if $\xi = \xi_n$ is in the discrete spectrum and q is real, ξ_n is imaginary, provided that

$$\int_{-\infty}^{\infty} \mathrm{Im}(\psi_1^*\psi_2)dx \neq 0.$$

12.11 Determine the equation satisfied by the bound state eigenvalues when the initial conditions for NLS are

$$q(x,0) = \begin{cases} 1 & \text{for } |x - x_0| < 1, \\ e^{i\theta} & \text{for } |x + x_0| < 1, \\ 0 & \text{otherwise,} \end{cases}$$

for positive real constants $x_0 > 1$ and θ. Find the bound state eigenvalues at leading order when $x_0 \gg 1$. Describe the solution for $t \gg 1$ in this case.

12.12 Using the asymptotic procedure explained in the text, determine the leading order solution for the two NLS bright solitons ($N = 2$) when $|t| \gg 1$ for both t positive and t negative.

Appendix 1

Useful Mathematical Formulas and Physical Data

A1.1 Cartesian Coordinates

Differential operators in Cartesian, (x, y, z)-coordinates for a smooth scalar field $\phi(x, y, z)$ and vector field $\mathbf{u} = (u_x, u_y, u_z)$.

$$\nabla\phi = (\phi_x, \phi_y, \phi_z), \tag{A1.1}$$

$$\nabla \cdot \mathbf{u} = \frac{\partial u_x}{\partial x} + \frac{\partial u_y}{\partial y} + \frac{\partial u_z}{\partial z}, \tag{A1.2}$$

$$\nabla \times \mathbf{u} = \begin{vmatrix} \mathbf{e}_x & \mathbf{e}_y & \mathbf{e}_z \\ \frac{\partial}{\partial x} & \frac{\partial}{\partial y} & \frac{\partial}{\partial z} \\ u_x & u_y & u_z \end{vmatrix} \tag{A1.3}$$

$$\nabla^2\phi = \phi_{xx} + \phi_{yy} + \phi_{zz}. \tag{A1.4}$$

A1.2 Cylindrical Polar Coordinates

Differential operators in cylindrical polar coordinates, (r, θ, z) (see figure A1.1.

$$\nabla\phi = \left(\phi_r, \frac{1}{r}\phi_\theta, \phi_z\right), \tag{A1.5}$$

$$\nabla \cdot \mathbf{u} = \frac{1}{r}\frac{\partial}{\partial r}(ru_r) + \frac{1}{r}\frac{\partial u_\theta}{\partial \theta} + \frac{\partial u_z}{\partial z}, \tag{A1.6}$$

$$\nabla \times \mathbf{u} = \frac{1}{r}\begin{vmatrix} \mathbf{e}_r & r\mathbf{e}_\theta & \mathbf{e}_z \\ \frac{\partial}{\partial r} & \frac{\partial}{\partial \theta} & \frac{\partial}{\partial z} \\ u_r & ru_\theta & u_z \end{vmatrix} \tag{A1.7}$$

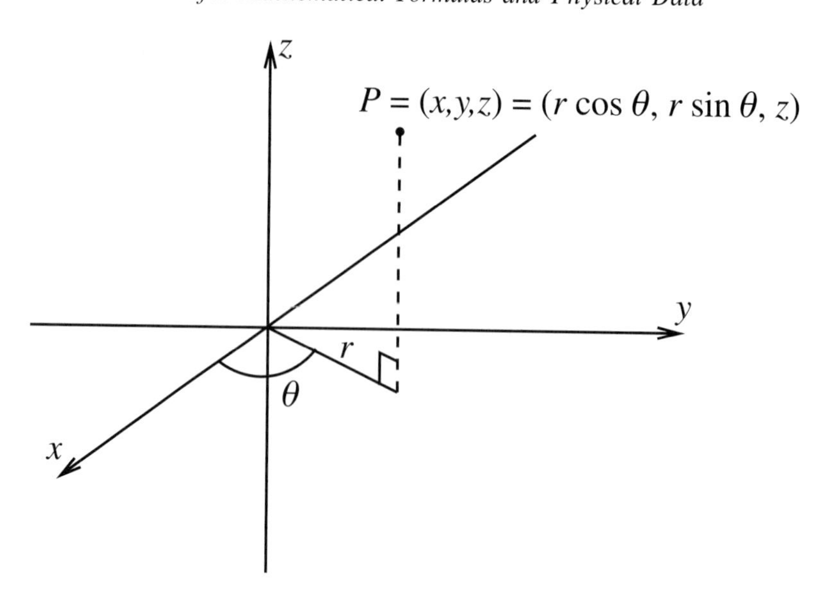

Fig. A1.1. Cylindrical polar coordinates.

$$\nabla^2\phi = \frac{1}{r}\frac{\partial}{\partial r}\left(r\frac{\partial\phi}{\partial r}\right) + \frac{1}{r^2}\frac{\partial^2\phi}{\partial\theta^2} + \frac{\partial^2\phi}{\partial z^2}. \qquad \text{(A1.8)}$$

A1.3 Spherical Polar Coordinates

Differential operators in spherical polar coordinates, (r, θ, χ) (see figure A1.2).

$$\nabla\phi = \left(\phi_r, \frac{1}{r}\phi_\theta, \frac{1}{r\sin\theta}\phi_\chi\right), \qquad \text{(A1.9)}$$

$$\nabla\cdot\mathbf{u} = \frac{1}{r^2}\frac{\partial}{\partial r}(r^2u_r) + \frac{1}{r\sin\theta}\frac{\partial}{\partial\theta}(\sin\theta u_\theta) + \frac{1}{r\sin\theta}\frac{\partial u_\chi}{\partial\chi}, \qquad \text{(A1.10)}$$

$$\nabla\times\mathbf{u} = \frac{1}{r^2\sin\theta}\begin{vmatrix} \mathbf{e}_r & r\mathbf{e}_\theta & r\sin\theta\mathbf{e}_\chi \\ \frac{\partial}{\partial r} & \frac{\partial}{\partial\theta} & \frac{\partial}{\partial\chi} \\ u_r & ru_\theta & r\sin\theta u_\chi \end{vmatrix} \qquad \text{(A1.11)}$$

$$\nabla^2\phi = \frac{1}{r^2}\frac{\partial}{\partial r}\left(r^2\frac{\partial\phi}{\partial r}\right) + \frac{1}{r^2\sin\theta}\frac{\partial}{\partial\theta}\left(\sin\theta\frac{\partial\phi}{\partial\theta}\right) + \frac{1}{r^2\sin\theta}\frac{\partial^2\phi}{\partial\chi^2}. \qquad \text{(A1.12)}$$

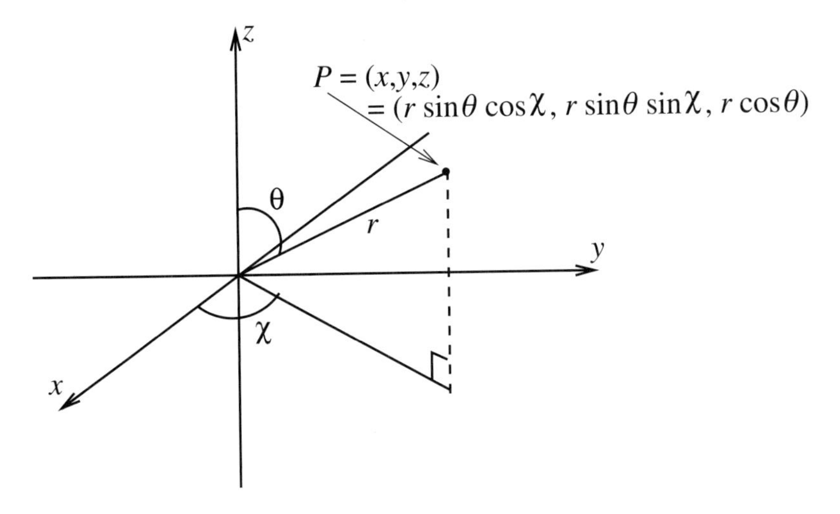

Fig. A1.2. Spherical polar coordinates.

A1.4 Some Vector Calculus Identities and Useful Results for Smooth Vector Fields

$$\nabla^2\phi = \nabla \cdot (\nabla\phi), \quad \nabla \times (\nabla\phi) = 0, \quad \nabla \cdot (\nabla \times \mathbf{u}) = 0, \tag{A1.13}$$

$$(\mathbf{u} \cdot \nabla)\mathbf{u} = \nabla\left(\frac{1}{2}|\mathbf{u}|^2\right) - \mathbf{u} \times (\nabla \times \mathbf{u}), \tag{A1.14}$$

$$\nabla^2\mathbf{u} = \nabla(\nabla \cdot \mathbf{u}) - \nabla \times (\nabla \times \mathbf{u}). \tag{A1.15}$$

The **divergence theorem** states that if a region V is bounded by a simple, closed surface, S, with unit outward normal $\mathbf{n}$, then, for a smooth vector field $\mathbf{u}$,

$$\int_S \mathbf{u} \cdot \mathbf{n}\, dS = \int_V \nabla \cdot \mathbf{u}\, dV. \tag{A1.16}$$

A useful variant of this is

$$\int_S \phi\mathbf{n}\, dS = \int_V \nabla\phi\, dV, \tag{A1.17}$$

for any smooth scalar field, ϕ.

Stokes' theorem states that if C is a simple, closed curve spanned by a surface S with unit normal $\mathbf{n}$, then

$$\int_C \mathbf{F} \cdot d\mathbf{r} = \int_S (\nabla \times \mathbf{F}) \cdot \mathbf{n}\, dS. \tag{A1.18}$$

This is most commonly used in two dimensions as **Green's theorem** in the plane. Taking $\mathbf{F} = (u_x, u_y, 0)$, we have

$$\int_C u_x dx + u_y dy = \int_S \left(\frac{\partial u_y}{\partial x} - \frac{\partial u_x}{\partial y} \right) dx\, dy. \tag{A1.19}$$

A1.5 Physical constants

Table A1.1. *Useful Physical Constants at Atmospheric Temperature and Pressure.*

Normal atmospheric temperature	$15\,^\circ\mathrm{C} \approx 288$ K
Normal atmospheric pressure	1 bar $= 10^5$ N m^{-2}
Gravitational acceleration (g)	9.81 m s^{-1}
Density of air (ρ_{air})	1.2 kg m^{-3}
Density of water (ρ_{water})	10^3 kg m^{-3}
Viscosity of air (μ_{air})	1.78×10^{-5} kg m^{-1} s^{-1}
Viscosity of water (μ_{water})	1.14×10^{-3} kg m^{-1} s^{-1}
Surface tension of water (σ_{water})	0.07 N m^{-1}
Ratio of specific heats of air (γ)	1.4
Universal gas constant (R)	8.3 J mol^{-1} K^{-1}
Density of mild steel	7000 kg m^{-3}
Young's modulus of mild steel (E)	2×10^{11} N m^{-1}
Poisson's ratio of mild steel (ν)	0.26
Lamé constants of mild steel	$\lambda \approx 8.6 \times 10^{10}$ N m^{-1}, $\mu \approx 7.9 \times 10^{10}$ N m^{-2}
Electrical permittivity of a vacuum (ϵ_0)	8.854×10^{-12} F^{-1}
Magnetic permeability of a vacuum (μ_0)	$4\pi \times 10^{-7}$ H^{-1}
Speed of light in a vacuum (c_0)	3.0×10^8 m s^{-1}
Typical solution phase reaction rate ($\dot{r}$)	10^{-8} mol m^{-3} s^{-1}
Typical diffusivity of ions in solution (D)	10^{-9} m^2 s^{-1}

Bibliography

Ablowitz, M.J. & Clarkson, P.A. (1991). *Solitons, Nonlinear Evolution Equations and Inverse Scattering*, Cambridge University Press.

Ablowitz, M.J., Kaup, D.J., Newell, A.C. & Segur, H. (1974). 'The inverse scattering transform – Fourier analysis for nonlinear problems', *Stud. Appl. Math.* **53**, 249–315.

Ablowitz, M.J. & Segur, H. (1981). *Solitons and the Inverse Scattering Transform*, SIAM.

Abramowitz, M. & Stegun, I.A. (1972). *Handbook of Mathematical Functions*, Dover Publications.

Achenbach, J.D. (1973). *Wave Propagation in Elastic Solids*, North-Holland.

Auld, B.A. (1973). *Acoustic Fields and Waves in Solids*, Wiley.

Barenghi, C.F., Mellor, C.J., Muirhead, C.M. & Vinen, W.F. (1986). 'Experiments on ions trapped below the surface of superfluid He^4', *J. Phys. C* **19**, 1135–1144.

Benjamin, T.B. & Feir, J.F. (1967). 'The disintegration of wave trains on deep water', *J. Fluid Mech.* **27**, 417–430.

Benjamin, T.B. & Lighthill, M.J. (1954). 'On cnoidal waves and bores', *Proc. R. Soc. Lond. A* **224**, 448–466.

Benjamin, T.B. & Ursell, F. (1954). 'The stability of the plane free surface of a liquid in vertical periodic motion', *Proc. R. Soc. Lond. A* **225**, 505–517.

Billingham, J. & King, A.C. (1995). 'The interaction of a moving fluid/fluid interface with a flat plate', *J. Fluid Mech.* **296**, 325–351.

Billingham, J. & Needham, D.J. (1992). 'The development of travelling waves in quadratic and cubic autocatalysis with unequal diffusion rates. III. Large time development in quadratic autocatalysis', *Q. Appl. Math.* **50**, 347–372.

Billingham, J. & Needham, D.J. (1993). 'Mathematical modelling of chemical clock reactions. II. A class of autocatalytic clock reaction schemes', *J. Eng. Math.* **27**, 113–145.

Bloor, M.I.G. (1970). 'The effect of viscosity on long waves in shallow water', *Phys. Fluids* **13**, 1435–1439.

Bloor, M.I.G. (1978). 'Large amplitude surface waves', *J. Fluid Mech.* **85**, 167.

Bluman, G.W. & Cole, J.D. (1974). *Similarity Methods for Differential Equations*, Springer-Verlag.

Born, M. & Wolf, E. (1975). *Principles of Optics* (5th Edition), Pergamon.

Brailovsky, I. & Sivashinsky, G.I. (1997). 'On deflagration-to-detonation transition', *Comb. Sci. Tech.* **130**, 201–231.

Burgers, J.M. (1948). 'A mathematical model illustrating the theory of turbulence', *Adv. Appl. Mech.* **1**, 171–199.
Cagniard, L. (1962). *Reflection and Refraction of Progressive Seismic Waves*, McGraw-Hill.
Carpenter, G. (1977). 'A geometric approach to singular perturbation problems with application to the nerve impulse equations', *J. Diff. Eqns* **23**, 335–367.
Chapman, S.J., Lawry, J.M.H., Ockendon, J.R. & Tew, R.H. (1999). 'On the theory of complex rays', *SIAM Rev.* **41**, 417–511.
Chaudry, A.N., Coveney, P.V. & Billingham, J. (1994). 'Exploring complexity in some simple nonlinear chemical kinetic schemes', *J. Chem. Phys.* (1994), **100**, 1921–1935.
Chree, C. (1889). *Trans. Camb. Phil. Soc.* **14**, 250.
Clemmow, P.C. (1973). *An Introduction to Electromagnetic Theory*, Cambridge University Press.
Cokelet, E.D. (1977). 'Steep gravity waves in water of arbitrary depth', *Phil. Trans. R. Soc. Lond. A* **286**, 183–201.
Crapper, G.D. (1957). 'An exact solution for progressive capillary waves of arbitrary amplitude', *J. Fluid Mech.* **2**, 532–540.
Crighton, D.G. (1981). 'Acoustics as a branch of fluid mechanics', *J. Fluid Mech.* **106**, 261–299.
Crighton, D.G. & Scott, J.F. (1979). 'Model equations in nonlinear acoustics' *Phil. Trans. R. Soc. Lond. A* **292**, 102–133.
Crosswell, K. (1995). *Alchemy of the Heavens*, Doubleday.
Dowling, A.P. & Ffowcs Williams, J.E. (1983). *Sound and Sources of Sound*, Ellis Horwood.
Drazin, P.G. & Johnson, R.S. (1989). *Solitons: an introduction*, Cambridge University Press.
Elmore, W.C. & Heald, M.A. (1969). *Physics of Waves*, McGraw-Hill.
Erdélyi, A. (1954). *Tables of Integral Transforms*, McGraw-Hill.
Evans, D.V. & Kuznetsov, N. (1997). 'Trapped modes', in *Gravity Waves in Water of Finite Depth*, Hunt J.N. (ed.), Computational Mechanics Publications.
Evans, D.V. & Porter, R. (1997). 'Near-trapping of waves by circular arrays of vertical cylinders', *Appl. Ocean Res.* **19**, 83–99.
Evans, R.A. & Bloor, M.I.G. (1977). 'The starting mechanism of wave-induced flow through a sharp-edged orifice', *J. Fluid Mech.* **82**, 115–128.
Faddeev, L.D. (1958). 'On the relation between the S-matrix and potential for the one-dimensional Schrödinger operator', *Dokl. Akad. Nauk SSSR*, **121**, 63–66.
Fitzhugh, R. (1961). 'Impulses and physiological states in theoretical models of the nerve membrane', *Biophys. J.* **1**, 445–466.
Fletcher, N.H. (1976). *Physics and Music*, Heinemann.
Fletcher, N.H. & Rossing, T.D. (1991). *The Physics of Musical Instruments*, Springer-Verlag.
Gardner, C.S., Greene, J.M., Kruskal, M.D. & Miura, R.A. (1967). 'Method for solving the Korteweg–de Vries equation', *Phys. Rev. Lett.* **19**, 1095–1097.
González, F.I. (1999). 'Tsunami!', *Scientific American* **280**, 44–55.
Hasegawa, A.H. (1990). *Optical Solitons in Fibers* (2nd Edition), Springer-Verlag.
Hastings, S.P. (1975). 'Some mathematical problems from neurobiology', *Am. Math. Monthly* **14**, 881–895.
Hinch, E.J. (1991). *Perturbation Methods*, Cambridge University Press.

Hirota, R. (1973). 'Exact N-soliton solution of the wave equation of long waves in shallow water and in nonlinear lattices', *J. Math. Phys.* **14**, 810–814.

Hodgkin, A.L. & Huxley, A.F. (1952). 'A quantitative description of the membrane current and its application to conduction and excitation in the nerve', *J. Physiology* **117**, 500–544.

Hudson, J.A. (1980). *The Excitation and Propagation of Elastic Waves*, Cambridge University Press.

Jackson, J.D. (1975). *Classical Electrodynamics* (2nd Edition), Wiley.

Keener, J. & Sneyd, J. (1998). *Mathematical Physiology*, Springer-Verlag.

Keller, J.B. (1958). 'A geometrical theory of diffraction', in *Proc. Sympos. Appl. Math.* **8**, 27–32.

Keller, J.B. (1977). 'Elastic Waveguides' in *Modern Problems in Elastic Wave Propagation*, ed. Miklowitz, J. & Achenbach, J.D. Wiley.

Kerner, B.S. (1999). 'The physics of traffic', *Physics World*, August 1999, 25–30.

Kevorkian, J. (1990). *Partial Differential Equations: Analytical Solution Techniques*, Wadsworth and Brooks.

Kevorkian, J & Cole, J.D. (1981). *Perturbation Methods in Applied Mathematics*, Springer-Verlag.

King, A.C. & Moni, J.N. (1995). 'Guided and unguided interfacial solitary waves', *Q. J. Mech. Appl. Math.* **48**, 21.

Kolodner, I.I. & Pederson, R.N. (1966). 'Pointwise bounds on solutions of semilinear parabolic equations', *J. Diff. Eqns* **2**, 353–364.

Lamb, G.L. (1980). *Elements of Soliton Theory*, Wiley-Interscience.

Lamb, H. (1904). 'On the propagation of tremors over the surface of an elastic solid', *Phil. Trans. R. Soc. Lond. A* **203**, 1.

Lamb, H. (1932). *Hydrodynamics* (6th Edition), Cambridge University Press.

Larson, D.A. (1978). 'Transient bounds and time asymptotic behaviour of solutions of nonlinear equations of Fisher type', *SIAM J. Appl. Math.* **34**, 93–103.

LeVeque, R.J. (1990). *Numerical Methods for Conservation Laws*, Birkhäuser Verlag.

Lighthill, M.J. (1978). *Waves in Fluids*, Cambridge University Press.

Longuet-Higgins, M.S. (1973). 'On the form of highest progressive and standing waves in deep water', *Proc. R. Soc. Lond. A* **331**, 445–456.

Mandl, F. (1971). *Statistical Physics*, Wiley.

Miklowitz, J. (1977). *The Theory of Elastic Waves and Waveguides*, North-Holland.

Morse, P.M. & Feshbach, H. (1953). *Methods of Mathematical Physics*, McGraw-Hill.

Murray, J.D. (1993). *Mathematical Biology* (2nd Edition), Springer.

Nagumo, J., Arimoto, S. & Yoshizawa (1962). 'An active pulse transmission line simulating nerve axons', *Proc. I. R. E.* **50**, 2061–2071.

Nayfeh, A.H. (1993). *Introduction to Perturbation Techniques*, Wiley.

Newell, A.C. & Moloney, J.V. (1992). *Nonlinear Optics*, Addison-Wesley.

Pask, C. (1992). 'Introduction to fiber optics and nonlinear effects in optical fibers', in *Guided Wave Nonlinear Optics*, Ostrowsky, D.B. & Reinisch, R. (eds.), Kluwer.

Peregrine, D.H. (1967). 'Long waves on a beach' *J. Fluid Mech.* **27**, 815–827.

Perring, J.K & Skyrme, T.H.R. (1962). 'A model unified field equation', *Nucl. Phys.* **31**, 550–555.

Pochhammer, L. (1876). *J. reine angew. Math.* **81**, 324.

Rinzel, J. & Terman, D. (1982). 'Propagation phenomena in a bistable reaction–diffusion system' *SIAM J. Appl. Math.* **42**, 1111–1137.
Sander, K.F. & Reed, G.A.L. (1986). *Transmission and Propagation of Electromagnetic Waves* (2nd Edition), Cambridge University Press.
Satsuma, J. & Yajima, N. (1974). 'Initial value problems of one-dimensional self-modulation of nonlinear waves in dispersive media', *Suppl. Prog. Theor. Phys.* **55**, 284–306.
Sauter, E.G. (1996). *Nonlinear Optics*, Wiley.
Schuur, P.C. (1986). *Asymptotic Analysis of Soliton Problems*, Springer-Verlag.
Sharpe, G.J. & Falle, S.A.E.G. (1999) 'One-dimensional numerical simulations of idealized detonations', *Proc. R. Soc. Lond. A* **455**, 1203–1214.
Shen, S.S. (1993). *A Course on Nonlinear Waves*, Kluwer.
Stokes, G.G. (1847). 'On the theory of oscillatory waves', *Camb. Trans.* **8**.
Taylor, G.I. (1963). *The Scientific Papers of Sir Geoffrey Taylor*, Batchelor G.K. (ed.), Cambridge University Press.
Turner, R.E.L. & Vanden-Broeck, J.M. (1988). 'Broadening of interfacial solitary waves', *Phys. Fluids* **31**, 2486–2490.
Watson, G.N. (1922). *A Treatise on the Theory of Bessel Functions*, Cambridge University Press.
Weiss, T.F. (1996). *Cellular Biophysics*, Vol. 2, M.I.T. Press.
Whitham, G.B. (1974). *Linear and Nonlinear Waves*, Wiley.
Wiggins, D.J.R., Sharpe, G.J. & Falle, S.A.E.G. (1998). 'Double detonations at the core–envelope boundary in Type Ia supernovae', *Month. Not. R. Astr. Soc.* **301**, 405–413.
Wowk, B. (1996). 'Phased array optics', in *Nanotechnology*, MIT Press. ed. B.C. Crandall.

Index